AA001077

2007 Device Research Conference

South Bend, IN
18-20 June 2007

IEEE Catalog Number: CFP07DRC-PRT
ISBN 10: 1-4244-1101-7
ISBN 13: 978-1-4244-1101-6

Copyright © 2007 by The Institute of Electrical and Electronics Engineers, Inc.
All Rights Reserved

Copyright and Reprint Permissions: Abstracting is permitted with credit to the source. Libraries are permitted to photocopy beyond the limit of U.S. copyright law for private use of patrons those articles in this volume that carry a code at the bottom of the first page, provided the per-copy fee indicated in the code is paid through Copyright Clearance Center, 222 Rosewood Drive, Danvers, MA 01923.

For other copying, reprint or republications permission, write to IEEE Copyrights Manager, IEEE Operations Center, 445 Hoes Lane, Piscataway, New Jersey USA 08854. All rights reserved.

IEEE Catalog Number:	CFP07DRC-PRT
ISBN 10:	1-4244-1101-7
ISBN 13:	978-1-4244-1101-6
ISSN:	1548-3770

Additional Copies of This Publication Are Available from:

IEEE Service Center
445 Hoes Lane
Piscataway, NJ 08854

Phone:	(800) 678-IEEE
	(732) 981-1393
Fax:	(732) 981-9667
E-mail:	customer-service@ieee.org

June 18– 20, 2007
The University of Notre Dame
South Bend, Indiana

DEVICE
RESEARCH
CONFERENCE

OFFICERS

Bobby Brar .... General Program Chair
Joerg Appenzeller .... Technical Program Chair
Steve Koester .... Technical Program Vice Chair
Theresa Mayer .... Past Chair
Patrick Fay .... Local Arrangements Chair
Tom Jackson .... Treasurer

TECHNICAL PROGRAM COMMITTEE

Prabhat Agarwal, NXP Semiconductors
Paul Baude, 3M Company
David Chow, HRL Laboratories
Patrick Fay, University of Notre Dame
Michael Flatte, University of Iowa
David Gundlach, National Institute of Standards and Technology
Jing Guo, University of Florida
Judy Hoyt, Massachusetts Institute of Technology
Pranav Kalavade, Intel
Mike Larson, Agility Communication, Inc.
Sang-Hyun Oh, University of Minnesota
Taiichi Otsuji, Tohoku University
Nitin Samarth, Penn State University
Seigo Tarucha, Tokyo University
Miguel Urteaga, Teledyne
Lars-Erik Wernersson, University of Lund
Janos Voros, Technical University Zuerich
Mike Wojtowicz, Northrup Grumman
John Zolper, DARPA/MTO

Sponsored by the IEEE Electron Devices Society

iii

Schedule of Events

SUNDAY PM, JUNE 17TH, 2007

Registration ....5:00 – 8:00 PM
Location .... McKenna Hall Information Desk

Welcoming Reception ....6:00 PM – 8:00 PM
Location .... McKenna Hall Atrium

MONDAY AM, JUNE 18TH, 2007

Registration..7:30 AM – 5:00 PM
Location... ..... McKenna Hall Information Desk

Plenary Session...8:20 AM
Location.. DeBartolo Hall Room 102

MONDAY PM, JUNE 18TH, 2007

Session II.A. Thin Film Transistors..1:30 PM
Location.. DeBartolo Hall Room 102

Session II.B. Wide Bandgap Devices..1:30 PM
Location.. DeBartolo Hall Room 141

Session II.C. Silicon CMOS..1:30 PM
Location.. DeBartolo Hall Room 155

Session III. Poster Session...7:00 – 9:00 PM
Location.. McKenna Hall, Main Floor

TUESDAY AM, JUNE 19TH, 2007

Registration..7:30 AM – 5:00 PM
Location.. .... McKenna Hall Information Desk

Session IV.A. High-Speed and Terahertz Devices.................................8:20 AM
Location.. DeBartolo Hall Room 102

Session IV.B. Semiconducting Nanowire Devices.................................8:20 AM
Location.. DeBartolo Hall Room 141

TUESDAY PM, JUNE 19TH, 2007

Session V.A. Photonic Devices...1:30 PM
Location.. DeBartolo Hall Room 102

Session V.B. III-V CMOS...1:30 PM
Location.. DeBartolo Hall Room 141

Session V.C. Memory Devices...1:30 PM
Location.. DeBartolo Hall Room 155

Conference Picnic ....6:00-8:00 PM
Location... Irish Courtyard (Enter through Morris Inn lobby)

Schedule of Events

TUESDAY PM, JUNE 19TH, 2007

Rump Sessions ....8:00-10:00 PM
I. The Future of Nanowire Devices: Top-down or bottom-up?...................... McKenna Hall,
Rooms 100-104
II. Why do we need nano for biosensors?.. .... McKenna Hall, Rooms 112-114
III. The THz Bridge... McKenna Hall, Auditorium

WEDNESDAY AM, JUNE 20TH, 2007

Registration..7:30 AM – 1:00 PM
Location .... McKenna Hall Information Desk

Joint EMC/DRC Plenary Session...8:20 AM
Location... Leighton Hall - DeBartolo Performing Arts Center

Session VI.A. Nanobiotechnology .... 10:00 AM
Location.. McKenna Auditorium

WEDNESDAY PM, JUNE 20TH, 2007

Session VII.A. Magnetoelectronics Devices and Optical Resonators ....1:30 PM
Location... DeBartolo Hall Room 141

Session VII.B. Carbon Nanotube and Graphene Devices.......................................1:30 PM
Location... DeBartolo Hall Room 136

Table of Contents

Session I. PLENARY SESSION — 1-10

I.-1 CMOS Integrated Nano-Photonics is now a Commercial Technology
E. Yablonovitch, Electrical Engineering Department, University of California, Los Angeles, Los Angeles, California, USA

I.-2 Mobility-Enhanced MOS Device Technologies in Nano-CMOS era
S.Takagi, MIRAI-AIST, Kawasaki, JAPAN and Department of Electronic Engineering, School of Engineering, The University of Tokyo, Tokyo, JAPAN

I.-3 Hybrid CMOS/Nanodevice Circuits: Architectures, Applications, and Device Needs
K. K. Likharev, Stony Brook University, Stony Brook, New York, USA

Session II.A. Thin Film Transistors — 11-24

II.A-1 Uniform ZnO Thin-Film Transistors by an Ambient Process
S. F. Nelson, D. Levy, D. Freeman, P. Cowdery-Corvan, L. Tutt, M. Burberry, L. Irving, Eastman Kodak Company, Rochester, New York, USA

II.A-2 Flexible Microwave Single-Crystal Si TFTs with fmax of 5.5 GHz
H. Pang[1], H.-C. Yuan[1], M. G. Lagally[2], G. K. Celler[3] and Z. Ma[1], [1]Department of Electrical and Computer Engineering, [2]Department of Materials Science and Engineering, University of Wisconsin-Madison, Madison, Wisconsin USA, and [3]Soitec USA, Peabody, Massachusetts, USA

II.A-3 A 'Bottom-up' Redefinition for Mobility and the Effect of Poor Tube-Tube Contact on the Performance of CNT Nanobundle Thin Film Transistors
N. Pimparkar and M. A. Alam , School of Electrical and Computer Engineering, Purdue University, West Lafayette, Indiana, USA

II.A-4 Fast ZnO Thin-Film Transistor Circuits
J. Sun[1], D. Mourey[1], D. Zhao[1], S. Park[1], S. F. Nelson[2], David H. Levy[2], D. Freeman[2], Peter Cowdery-Corvan[2], Lee Tutt[2], and T. N. Jackson[1], [1]Center for Thin Film Devices and Materials Research Institute, Department of Electrical Engineering, Penn State University, University Park, Pennsylvania, USA and [2]Eastman Kodak Company, Rochester, New York, USA

II.A-5 Electrical and chemical stability of polymer transistors
R. A. Street and M. Chabinyc, Palo Alto Research Center, Palo Alto, California, USA

II.A-6 Solution processed OTFT circuits on plastic substrates
S. K. Park[1], S. Subramanian[2], J. Anthony[2], and T. N. Jackson[1], [1]Center for Thin Film Devices and Materials Research Institute, Department of Electrical Engineering, Penn State University, University Park, Pennsylvania, USA and [2]Department of Chemistry, University of Kentucky, Lexington, Kentucky, USA

Session II.B. Wide Bandgap Devices — 25-44

II.B-1 GaN Power Devices for Microwave/Switching Applications
D. Ueda, Semiconductor Device Research Center, Matsushita Electric Industrial Co., Ltd., Takatsuki, Osaka, JAPAN

II.B-2 Novel 1 kV, normally-off, vertically integrated, dual-gate VJFET power switch with a low 4.6 mOcm2 on-state resistance
B. Nechay[1], E. Stewart[1], V. Veliadis[1], T. McNutt[1], H. Hearne[1], G. DeSalvo[1], C. Clarke[1], and S. Scozzie[2], [1]Northrop Grumman Science and Technology Center, Linthicum, Maryland, USA and [2]U.S. Army Research Laboratory, Adelphi, Maryland, USA

II.B-3   **AlGaN/GaN HEMTs on Diamond Substrate**
D. C. Dumka and P. Saunier, TriQuint Semiconductor, Inc., Richardson, Texas, USA

II.B-4   **10 W/mm and High PAE Field-plated AlGaN/GaN HEMTS at Ka-band with n+GaN Source Contact Ledge**
J. S. Moon, P. Hashimoto, D. Wong, M. Hu, M. Antcliffe, C. McGuire, M. Micovic, P. Willadsen, and D. Chow, HRL Laboratories LLC, Malibu, California, USA

II.B-5   **Progress in GaN Performances and Reliability**
P. Saunier[1], C. Lee[1], A. Balistreri[1], D. Dumka[1], J. Jimenez[1], H. Q. Tserng[1], M.Y. Kao[1], P.C. Chao[2], K. Chu[2], A. Souzis[3], I. Eliashevich[4], S. Guo[4], J. del Alamo[5], J. Joh[5], and M. Shur[6], [1]TriQuint Semiconductor Texas, Richardson, Texas, USA, [2]BAE Systems, Nashua, New Hampshire, [3]II-VI, Incorporated, Pine Brook, New Jersey, USA, [4]IQE RF, Somerset, New Jersey, USA, [5]MIT, Cambridge, Massachusetts, USA, and [6]RPI, Troy, New York, USA

II.B-6   **AlGaN/ GaN HEMTs with Large Angle Implanted Nonalloyed Ohmic Contacts**
F. Recht[1], L. McCarthy[1], L. Shen[1], C. Poblenz[2], A. Corrion[2], J. S. Speck[2], and U. K. Mishra[1], [1]Department of Electrical and Computer Engineering and [2]Department of Materials, University of California, Santa Barbara, California, U.S.A

II.B-7   **Self-Aligned AlGaN/GaN High Electron Mobility Transistors**
V. Kumar, D-H. Kim, A. Basu, and I. Adesida, Micro and Nanotechnology Laboratory and Department of Electrical and Computer Engineering, University of Illinois at Urbana-Champaign, Urbana, Illinois, USA

II.B-8   **Analysis of lateral surface leakage in the vicinity of Schottky gates in AlGaN/GaN HEMTS**
J. Kotani, M. Tajima and T. Hashizume, Research Center for Integrated Quantum Electronics (RCIQE) and Graduate School of Information Science and Technology, Hokkaido University, Sapporo, Hokkaido, JAPAN

II.B-9   **Drain-to-Gate Field Engineering for Improved Frequency Response of GaN-based HEMTs**
N. Pala[1,2], Z. Yang[2], A. Koudymov[2], X. Hu[1], J. Deng1, R. Gaska[1], G. Simin[3] and M. S. Shur[2], [1]Sensor Electronic Technology, Columbia, South Carolina, USA, [2]ECSE Department, Rensselaer Polytechnic Institute, Troy, New York, USA, and [3]ECE Department, University of South Carolina, Columbia, South Carolina, USA

Session II.C.   Silicon CMOS   45-60

II.C-1   **Silicon Nanowire Field Effect Devices By Top-Down CMOS Technology**
N. Balasubramanian, N. Singh, S. C. Rustagi, Kavitha, A. Agarwal, G. Zhiqiang, G. Q. Lo, and D. L. Kwong, Institute of Microelectronics, Science Park II, SINGAPORE

II.C-2   **SiGe cantilever channel gate-all-around (GAA) fully depleted (FD) PMOSFET with high-? and metal gate**
S.-H. Lee[1], S. Dey[1], S V. Joshi[1], P. Majhi[2] and S. K. Banerjee[1], [1]Microelectronics Research Center, Electrical and Computer Engineering, University of Texas at Austin, Austin, Texas, USA and [2]Intel Assignee at Sematech, 2, Austin, Texas, USA

II.C-3   **Hole Transport in Nanoscale p-type MOSFET SOI Devices with High Strain**
H. M. Nayfeh[1], S. Jeng[1], S. Narasimha[1], S. Butt[1], R. Pal[2], A. Waite[2], K. Tabakman[1], J. B. Johnson[1], J. Liu[1], J. Holt[1], T. Adam[1], A. Madan[1], A. Domenicucci[1], [1]IBM Semiconductor Research and Development Center (SRDC), Hopewell Junction, New York, USA and [2]IBM Advanced Micro Devices (AMD), Hopewell Junction, New York, USA

II.C-4   **3X hole mobility enhancement in epitaxially grown SiGe PMOSFETs on (110) Si substrates with high k / metal gate for hybrid orientation technology**
S. Joshi[1], S. Dey[1], S.-h. Lee[1], C. Krug[2], H. Joo Na[2], P. Sivasubramani[2], P. D. Kirsch[2,3], P. Majhi[2,4], W. Wang[1], A. Campion[1] and S. K. Banerjee[1], [1]University of Texas at Austin, Austin, Texas, USA, [2]SEMATECH, Austin, Texas, USA, [3]IBM, and [4]Intel

II.C-5 **High – Mobility, Low Parasitic Resistance Si/Ge/Si Heterostructure Channel Schottky Source/Drain PMOSFETs**
A. Pethe and K. Saraswat, Department of Electrical Engineering, Stanford University, Stanford, California, USA

II.C-6 **Band to Band Tunneling Study in High Mobility Materials : III-V, Si, Ge and strained SiGe**
D. Kim[1], T. Krishnamohan[1], L. Smith[2], H.-S. Philip Wong[1], K. C. Saraswat[1], [1]Department of Electrical Engineering, Stanford University, Stanford, California, USA and [2]Synopsys Corporation, Mountain View, California, USA

II.C-7 **Process Integration and Electrical Properties of Bilayer Metal Gate/High-k MOSFETs**
C.-H. Lu[1], G. M. T. Wong[1], M. Deal[2], B. M. Clemens[1], and Y. Nishi[1,2], [1]Department of Materials Science and Engineering, Stanford University, Stanford, California, USA and [2]Department of Electrical Engineering, Stanford University, Stanford, California, USA

Session III. Poster Session   61-144

III-1 **Effects of Source Access Resistance on Gate lag in AlGaN/GaN HEMTs and Current Slump Behavior**
K. Horio, A. Nakajima and K. Itagaki, Faculty of Systems Engineering, Shibaura Institute of Technology, Saitama, JAPAN

III-2 **Low-Voltage Organic Thin-Film Transistors with Improved Stability and Large Transconductance**
H. Klauk[1], U. Zschieschang[1], R. Thomas Weitz[1], H. Meng[2], F. Sun[2], D. E. Keys[2], and C. R. Fincher[2], [1]Max Planck Institute for Solid State Research, Stuttgart, GERMANY and [2]Central Research and Development, Experimental Station, E. I. DuPont Company, Wilmington, Delaware, USA

III-3 **4H-SiC RF BJTs with Long Pulse L-band Operation**
F. Zhao[1], T. Shi[2], M. Mallinger[2], and B. Van Zeghbroeck[3], [1]Power Products Group, Microsemi Corporation, Bend, Oregon, USA, [2]Power Products Group, Microsemi Corporation, Santa Clara, California, USA, and [3]Department of Electrical and computer Engineering, University of Colorado, Boulder, Colorado, USA

III-4 **Room-temperature lasing of type-II "W" GaSb/GaAs quantum dots embedded in InGaAs quantum well**
J. Tatebayashi, A. Khoshakhlagh, G. Balakrishnan, S. H. Huang, M. Mehta, L. R. Dawson, and D. L. Huffaker, Center for High Technology Materials, University of New Mexico, Albuquerque, New Mexico, USA

III-5 **SWCNT-SET fabricated by dispersion method with CMC solvent**
T. Mori[1], K. Omura[1,2], S. Sato[1,3], M. Suzuki[1,4], K. Uchida[2], H. Yajima[2], and K. Ishibashi[1,4], [1]Advanced Device Laboratory, RIKEN, Saitama, JAPAN, [2]Department of Applied Chemistry, Tokyo University of Science, Tokyo, JAPAN, [3]Department of Physics, Tokyo University of Science, Tokyo, JAPAN, and [4]CREST, Japan Science and Technology (JST), Saitama, JAPAN

III-6 **Microwave Noise Characterization of Enhancement-Mode AlGaN/GaN/InGaN/GaN Double-Heterojunction HEMTs**
J. Liu, D. Song, Z. Cheng, W. C.–W. Tang, K. M. Lau and K. J. Chen, Department of Electronic and Computer Engineering, Hong Kong University of Science and Technology Clear Water Bay, Kowloon, HONG KONG

III-7 **Electro-Thermally Coupled Power Optimization for Future Transistors**
A. K. Chao, P. Kapur, E. Morifuji, K. C. Saraswat, and Y. Nishi, Department of Electrical Engineering, Stanford University, Stanford, California, USA

III-8 **Dispersion Design of a Left-Handed Microstrip Line with Planar Double-Stub and Split-Ring Structures for Leaky Wave Radiation toward Functional RF Wireless Interconnect**
M. Suhara[1,2], A. Shimizu[1], and T. Okumura[1,2], [1]Electrical Engineering, Graduate School of Engineering and [2]Electrical and Electronic Engineering, Graduate School of Science and Engineering, Tokyo Metropolitan University, Tokyo, JAPAN

viii

III-9 **High Power Vertical-structure GaN-based LEDs with Improved Current Spreading and Blocking Designs**
T.-M. Chen[1,2], S.-J. Wang[1], K.-M. Uang[2], S.-L. Chen[1], C.-C. Tsai[1], H.-Y. Kou[1], W.-C. Lee[1], and H. Kuan[3], [1]Institute of Microelectronics, Dept. of Electrical Engineering, National Cheng Kung University, Tainan, TAIWAN, [2]Dept. of Electrical Engineering, Wu-Feng Institute of Technology, Chia-yi, TAIWAN, and [3]Optoelectronics Center of Far East University, Tainan, TAIWAN

III-10 **Feasibility Study of Composite Dielectric Tunnel Barriers for Flash Memory**
S. Verma[1], E. Pop[2], P. Kapur[1], P. Majhi[2], K. Parat[2], and K. C.Saraswat[1], [1]Center for Integrated Systems, Stanford University, Stanford, California, USA and [2]Intel Corporation,

III-11 **Confined Optical Phonon Scattering in p-Silicon Nanowires**
M. Nawaz[1], and J.-P. Leburton[2], [1]Department of Electrical and Computer Engineering, University of Illinois at Urbana-Champaign, Urbana, Illinois, USA and [2]Beckman Institute, Department of Electrical and Computer Engineering, University of Illinois at Urbana-Champaign, Urbana, Illinois, USA

III-12 **Mid-Wavelength Infrared (MWIR) Avalanche Photodiode (APD) using InAs-GaSb type-II Strain layer Superlattice (SLS)**
S. Mallick[1], K. Banerjee[1], S. Ghosh[1,] S. Krishna[2], and J. B. Rodriguez[2], [1]Lab for Photonics and Spintronics, Electrical and Computer Engineering Department, University of Illinois at Chicago, Chicago, Illinois, USA and [2]Center for High Technology Materials (ECE Dept), University of New Mexico, Albuquerque, New Mexico, USA

III-13 **Body Thickness Optimization and Sensitivity Analysis for High Performance FinFETs**
D. Lekshmanan, A. Bansal and K. Roy, School of ECE, Purdue University, West Lafayette, Indiana, USA

III-14 **Flash Memory Fabricated with Protein-Mediated PbSe Nanocrystal Assembly as Floating Gate**
S. Tang[1], C. Hun Lee[1], X. Gao[2] and S. K. Banerjee[1], [1]Microelectronics Research Center, The University of Texas at Austin, Austin, Texas, USA and [2]Texas Materials Institute, The University of Texas at Austin, Austin, Texas, USA

III-15 **Novel Amorphous-Si AMOLED Pixels with OLED-independent Turn-on Voltage and Driving Current**
B. Hekmatshoar, A. Z. Kattamis, K..e Cherenack, S. Wagner and J. C. Sturm, Princeton Institute for the Science and Technology of Materials (PRISM), Department of Electrical Engineering, Princeton University, Princeton, New Jersey, USA

III-16 **AlGaN/GaN Bidirectional Power Switch**
N. Tipirneni, B. Wang, A. Monti and G. Simin. University of South Carolina, Dept. of Electrical Engineering, Columbia, South Carolina, USA

III-17 **n- and p-channel TaN/HfO2 MOSFETs on GaAs substrate using a germanium interfacial passivation layer**
H.-S. Kim, I. Ok, F. Zhu, M. Zhang, S. Park, J. Yum, H. Zhao, and J. C. Lee, Microelectronics Research Center, Department of Electrical and Computer Engineering, The University of Texas at Austin, Austin, Texas, USA

III-18 **Hexagonal Prism Blue Laser Diode with Low Threshold Power using Whispering Gallery Mode (WGM) Resonances**
S. Kim[1] and T. D. Sands[2], [1]School of Electrical and Computer Engineering, West Lafayette, Indiana, USA and [2]Birck Nanotechnology Center, West Lafayette, Indiana, USA

III-19 **Analytical Modeling of the Suspended-Gate FET and Design Insights for Digital Logic**
K. Akarvardar[1], C. Eggimann[2], D.Tsamados[2], Y. Chauhan[2], G. C. Wan[1], A. M. Ionescu[2],and H. S. P. Wong[1], [1]Center for Integrated Systems, Stanford University, Stanford, California, USA and [2]Swiss Federal Institute of Technology, Lausanne, SWITZERLAND

III-20 **Dynamic Two-Port Parameters of Ballistic Carbon Nanotube FETs: A Quantum Simulation Study**
Y. Ouyang and J. Guo, Electrical and Computer Engineering, University of Florida, Gainesville, Florida, USA

III-21 **Schottky Drain AlGaN/GaN HEMTs for mm-wave Applications**
X. Zhao, J.W. Chung, H. Tang, T. Palacios, Department of EECS and Microsystems Technology Laboratories, Massachusetts Institute of Technology, Cambridge, Massachusetts, USA

III-22 **Barrier layer downscaling of InAlN/GaN HEMTs**
F. Medjdoub[1], J.-F. Carlin[2], M. Gonschorek[2], E. Feltin[2], M.A. Py[2], M. Knez[3], D. Troadec[4], C. Gaquière[4], A. Chuvilin[5], U. Kaiser[5], N. Grandjean[2], and E. Kohn[1], [1]University of Ulm (EBS), Ulm, GERMANY, [2]EPFL, Lausanne, SWITZERLAND, [3]Max Planck Institute of Microstructure Physics, Weinberg, GERMANY, [4]IEMN, Villeneuve d'ascq, FRANCE, and [5]University of Ulm (ME), Ulm, GERMANY

III-23 **Estimation of Trap Density in AlGaN/GaN HEMTs from Subthreshold Slope Study**
J. W. Chung, X. Zhao, and T. Palacios Department of Electrical Engineering and Computer Science, Microsystems Technology Laboratories, Massachusetts Institute of Technology, Cambridge, Massachusetts, USA

III-24 **High Performance ZnO Nanowire FET with ITO Contacts**
M. A. Hollister, J. D. Le, G. Xiao, X. Lu, and R. A. Kiehl, Department of Electrical & Computer Engineering, The University of Minnesota, Minneapolis, Minnesota, USA

III-25 **High Efficiency Oxide-Confined High-Index-Contrast Broad-Area Lasers with Reduced Threshold Current Density and Improved Near-Field Profile**
Di Liang and Douglas C. Hall, Department of Electrical Engineering, University of Notre Dame, Notre Dame, Indiana, USA

III-26 **Inversion-type enhancement-mode InP MOSFETs with ALD Al2O3, HfO2 and HfAlO nanolaminates as high-k gate dielectrics**
Y.Q. Wu[1], Y. Xuan[1], P.D. Ye[1], Z. Cheng[2], A. Lochtefeld[2], [1]School of Electrical and Computer Engineering, Purdue University, West Lafayette, Indiana, USA and [2]AmberWave Systems Corp., Salem, New Hampshire, USA

III-27 **Barrier Lowering and Widening of Schottky Barrier MOSFETs by Self-Aligned Multiple Workfunction Gate**
S.-P. Yeh[1] and C.-Hsing Shih[2], [1]Institute of Electronics Engineering, National Tsing Hua University, Hsinchu, TAIWAN and [2]Department of Electrical Engineering, Yuan Ze University, Taoyuan, TAIWAN

III-28 **Reliability of 4H-SiC DMOSFETs Evaluated by Bias Stressing**
T. Okayama[1], S. D. Arthur[2], J. L. Garrett[2], and M.V. Rao[1], [1]Department of Electrical and Computer Engineering, George Mason University, Fairfax, Virginia, USA and [2]Semiconductor Technology Division, GE Global Research Center, Niskayuna, New York, USA

III-29 **A 53% High Efficiency GaAs Vertically Integrated Multi-junction Laser Power Converter**
D. Krut, R. Sudharsanan, T. Isshiki, R. King and N. H. Karam, Spectrolab Inc., a Boeing Company, Sylmar, California, USA

III-30 **Native-Oxide-Confined High Index Contrast InAs Quantum-Dot Laser Diodes**
J. Wang[1], D. C. Hall[1], V. Tokranov[2] and S. Oktyabrsky[2], [1]Department of Electrical Engineering, University of Notre Dame, Notre Dame, Indiana, USA and [2]School of NanoSciences and NanoEngineering, University at Albany-SUNY, Albany, New York, USA

III-31 **Surface Treatment for Leakage Reduction in AlGaN/GaN HEMTs**
R. Chu[1], L. Shen[1], N. Fichtenbaum[1], S. Keller[1], A. Corrion[2], C. Poblenz[2], J. Speck[2], and U. Mishra[1], [1]ECE Department, University of California, Santa Barbara, California, USA and [2]Materials Department, University of California, Santa Barbara, California, USA

III-32 **AlGaN/GaN HEMT with High PAE and Breakdown Voltage Grown by Ammonia MBE**
Y. Pei[1], C. Suh[1], R. Chu[1], F. Recht[1], L. Shen[1], A. Corrion[2], C. Poblenz[2], J. Speck[2] and U.K. Mishra[1], [1]Electrical and Computer Engineering Department, University of California, Santa Barbara, California, USA, and [2]Materials Department, University of California, Santa Barbara, California, USA

III-33   **Analytical Model of Apparent Threshold Voltage Lowering Induced by Contact Resistance in Amorphous Silicon Thin Film Transistors**
B. Hekmatshoar, K. Long, S. Wagner and J. C. Sturm, Princeton Institute for the Science and Technology of Materials (PRISM), Department of Electrical Engineering, Princeton University, Princeton, New Jersey, USA

III-34   **On-Chip Clocking Scheme for Nanomagnet QCA**
M. T. Alam[1], M. Niemier[2], W. Porod[1], S. Hu[2], M. Putney[2], J. DeAngelis[2], and G. H. Bernstein[1], [1]Center for Nano Science and Technology, Dept. of Electrical Engineering, University of Notre Dame, Notre Dame, Indiana, USA and [2]Dept. of Comp. Sci. and Eng., University of Notre Dame, Notre Dame, Indiana, USA

III-35   **Electrical Characterization of Vertical InAs Nanowires on Si**
C. Rehnstedt, T. Mårtensson, C. Thelander, L. Samuelson and L.-E. Wernersson, Solid State Physics / Nanometer Consortium, Lund University, Lund, SWEDEN

III-36   **Band-gap engineering of enhanced spin-orbit interactions in InAs/AlGaAs heterostructures for Datta-Das spin transistor**
T. Matsuda, M. Ohno and K. Yoh, Research Center for Integrated Quantum Electronics, Hokkaido University, Sapporo, Hokkaido, JAPAN

III-37   **Oxide-Induced Noise in Carbon Nanotube Devices**
Y.-M. Lin and P. Avouris, IBM T. J. Watson Research Center, Yorktown Heights, New York, USA

III-38   **In-Situ Inelastic Electron Tunneling Spectroscopy of Bistable Molecular Junction Devices**
H. Yoon[1], L. Cai[1], M. Maitani[2], D. L. Allara[3], and T. S. Mayer[1], [1]Department of Electrical Engineering, [2]Department of Materials Science and Engineering, and [3]Department of Chemistry, The Pennsylvania State University, University Park, Pennsylvania, USA

III-39   **Electric-Field Dependence of Junction Temperature in GaN HEMTs**
V. Mehrotra, K. Boutros, and B. Brar, Teledyne Scientific Company, Thousand Oaks, California, USA

Session IV.A.   High Speed and Terahertz Devices   145-160

IV.A-1   **Advanced InP and GaAs HEMT MMIC technologies for MMW commercial products**
M. Barsky, M. Biedenbender, X. Mei, P.-H. Liu, and R. Lai, Northrop Grumman Space Technology, Redondo Beach, California, USA

IV.A-2   **Ultra-Low Resistance Ohmic Contacts to InGaAs/InP**
U. Singisetti[1], A. M. Crook[1], E. Lind[1], J. D. Zimmerman[1], M. A. Wistey[1], A. C. Gossard[1], M. J. W. Rodwell[1], and S. R. Bank[2], [1]ECE and Materials Departments, University of California, Santa Barbara, California, USA and [2]ECE Department, University of Texas at Austin, Austin, Texas, USA

IV.A-3   **Sb-heterostructure Millimeter-Wave Detectors with Improved Noise Performance**
N. Su[1], Z. Zhang[1], H. P. Moyer[2], R. D. Rajavel[2] J. N. Schulman[2] and P. Fay[1], [1]Department of Electrical Engineering, University of Notre Dame, Notre Dame, Indiana, USA and [2]HRL Laboratories LLC, Malibu, California, USA

IV.A-4   **Delta-Doped Si/SiGe Zero-Bias Backward Diodes for Micro-Wave Detection**
S.-Y. Park[1], R. Yu[2], S.-Y. Chung[1], P. R. Berger[1,2], P. E. Thompson[3], and P. Fay[4] [1]The Ohio State University, Department of Electrical and Computer Engineering, Columbus, Ohio, USA , [2]The Ohio State University, Department of Physics, Columbus, Ohio, USA, [3]Naval Research Laboratory, Washington, District of Columbia, USA, and [4]The University of Notre Dame, Department of Electrical Engineering, Notre Dame, Indiana, USA

IV.A-5   **Room-Temperature Terahertz Oscillators Using Resonant Tunneling Diodes**
M. Asada, Tokyo Institute of Technology, CREST-Japan Science and Technology Agency 2-12-1-S9-3 Ookayama, Meguro-Ku, Tokyo, JAPAN

IV.A-6   **Novel Plasmon-Resonant Terahertz-Wave EmitterUsing a Double-Decked HEMT Structure**
T. Suemitsu[1], Y. M. Meziani[1], Y. Hosono[1], M. Hanabe[1], T. Otsuj[1], and E. Sano[2], [1]Research Institute of Electrical Communication, Tohoku University, Aoba, Sendai, JAPAN and [2]Research Center for Integrated Quantum Electronics, Hokkaido University, Sapporo, JAPAN

IV.A-7   **THz front-side illuminated quantum well photodetector**
M. Patrashin and I. Hosako, National Institute of Information and Communications Technology, Tokyo, JAPAN

Session IV.B. Semiconducting Nanowire Devices   161-180

IV.B-1   **Towards vertical III-V nanowire devices on silicon**
E. Bakkers[1], M. Borgström[1], W. van den Einden[1], M. van Weert[1], E. Minot[1], F. Kelkensberg[1], M. van Kouwen[1], J. van Dam[1], L. Kouwenhoven[1], V. Zwiller[1], A. Helman[2], O. Wunnicke[2], and M. Verheijen[2], [1]Philips Research Laboratories, Eindhoven, THE NETHERLANDS and [2]Kavli Institute of Nanoscience, Delft, THE NETHERLANDS

IV.B-2   **Control of Threshold Voltage in 80 nm Gate Length InAs Vertical Nanowire WIGFETs**
T. Löwgren[1], J. Ohlsson[1], L. Samuelson[2], and L.-E. Wernersson[2], [1]QuMat Technologies AB, Lund, SWEDEN and [2],Solid State Physics / Nanometer Consortium, Lund University, Lund, SWEDEN

IV.B-3   **Gallium Nitride Nanowire Devices – Fabrication, Characterization, and Simulation**
A. Motayed, A. V. Davydov, M. He, and S. N. Mohammad, National Institute of Standards and Technology, Material Science and Engineering Laboratory, Gaithersburg, Maryland, USA

IV.B-4   **High performance In2O3 nanowire transistors using organic gate nanodielectrics**
S. Ju[1], G. Lu[3], P.-C. Chen[2], A. Facchetti[3], C. Zhou[2], T. J. Marks[3], and D. B. Janes[1], [1]School of Electrical and Computer Engineering and Birck Nanotechnology Center, Purdue University, West Lafayette, Indiana, USA, [2]Dept. of Electrical Engineering, University of Southern California, Los Angeles, California, USA, and [3]Dept. of Chemistry and the Materials Research Center, Northwestern University, Evanston, Illinois, USA

IV.B-5   **Impact ionization FETs based on silicon nanowires**
M. T. Björk, O. Hayden, J. Knoch, H. Riel, H. Schmid, and W. Riess, IBM Research GmbH, Zurich Research Laboratory, Rueschlikon, Switzerland

IV.B-6   **A Low Power, Highly Scalable, Vertical Double Gate MOSFET Using Novel Processes**
H. Cho[1], P. Kapur[1], P. Kalavade[2], and K. C. Saraswat[1], [1]Department of Electrical Engineering, Stanford University, CIS, Stanford, California, USA and [2]Intel Corporation, Santa Clara, California, USA

IV.B-7   **An n-FET with a Si nanowire channel and doped epitaxially-thickened source and drain regions**
G. M. Cohen, P. M. Solomon, and S. E. Laux, J. O. Chu, M. J. Rooks, and W. Haensch, IBM T. J. Watson Research Center, Yorktown Heights, New York, USA

IV.B-8   **Reduction of Acoustic Phonon Limited Electron Mobility due to Phonon Confinement in Silicon Nanowire MOSFETs**
J. Hattoria[1,3], S. Uno[1,3], N. Mori[2], and K. Nakazato[1,3], [1]Department of Electrical Engineering and Computer Science, Nagoya University, Nagoya, JAPAN, [2]Department of Electronic Engineering, Osaka University, Osaka, Japan, and [3]SORST JST

IV.B-9   **THz probe for nanowire FETs: simulation of few-electron fingerprints**
K. M. Indlekofer[1], R. Németh[1], and J. Knoch[2], [1]Institute for Bio and Nanosystems, CNI, Research Center, Jülich, GERMANY and [2]IBM Research GmbH, Zurich Research Laboratory, Rueschlikon, SWITZERLAND

xii

Session V.A Photonic Devices   181-198

V.A-1 **The first commercial large-scale InP photonic integrated circuits: current status and performance**
S. Hurtt, A. G. Dentai, J. L. Pleumeekers, A. Mathur, R. Muthiah, C. Joyner, R. P. Schneider, R. Nagarajan, F. A. Kish, and D. F. Welch, Infinera Corporation, Sunnyvale, California, USA

V.A-2 **A hybrid silicon evanescent photodetector**
H. Park[1], A. W. Fang[1], R. Jones[2], O. Cohen[3], O. Raday[3], M. N. Sysak[1], M. J. Paniccia[2], and J. E. Bowers[1], [1]University of California Santa Barbara, ECE Department, Santa Barbara, California, USA, [2]Intel Corporation, Santa Clara, California, USA and [3]Intel Corporation, Jerusalem, ISRAEL

V.A-3 **High-Frequency Performance of a High-Power Traveling Wave Photodetector with Parallel Optical Feed**
A. Beling[1], H.-G. Bach[2], G. G. Mekonnen[2], R. Kunkel[2], D. Schmidt[2] and J. C. Campbell[1], [1]Department of Electrical and Computer Engineering, University of Virginia, Charlottesville, Virginia, USA, and [2]Fraunhofer Institute for Telecommunications, Heinrich-Hertz-Institut, Berlin, GERMANY

V.A-4 **Single Ultra Violet Photon Detection with 4H-SiC Avalanche Photodiodes**
X. Bai, D. Mcintosh, H. Liu, J. Campbell, Electrical and Computer Engineering, University of Virginia, Charlottesville, Virginia, USA

V.A-5 **The challenges and opportunities for dilute nitride antimonides in photonic devices**
Jim Harris , Stanford University Jim Harris , Stanford University, Stanford, California, USA

V.A-6 **1.65 µm buffer-free GaSb/AlGaSb quantum-well diode lasers grown on a GaAs substrate operating at room temperature**
M. Mehta, G. Balakrishnan, A. Jallapali, M. N. Kutty, L. R. Dawson, and D. L. Huffaker, Center for High Technology Materials, University of New Mexico, Albuquerque, New Mexico, USA

V.A-7 **High Temperature CW Operation of Interband Cascade Lasers at ? ˜ 4.0 µm**
C. S. Kim, M. Kim, W. W. Bewley, C. L. Canedy, D. C. Larrabee, J. A. Nolde, J. R. Lindle, I. Vurgaftman, and J. R. Meyer, Naval Research Laboratory, Washington District of Columbia, USA

V.A-8 **Strain-Compensated AlAs-InGaAs Quantum-Cascade Lasers with Emission Wavelength 3–5 µm**
M.P. Semtsiv[1], S. Dressler[1], M. Wienold[1], I. Bayrakli[1], M. Ziegler[2], K. Kennedy[3], R. Hogg[3], and W.T. Masselink[1], [1]Humboldt University, Berlin, Germany, [2]Max-Born-Institut, Berlin, Germany, and [3]University of Sheffield, Sheffield, UK

Session V.B III-V CMOS   199-212

V.B-1 **InGaAs CMOS: a "Beyond-the-Roadmap" Logic Technology?**
J. A. del Alamo and D. H. Kim, Massachusetts Institute of Technology, Cambridge, Massachusetts, USA

V.B-2 **InGaAs and GaAs/InGaAs Channel Enhancement Mode n-MOSFETs With HfO2 Gate Oxide and a-Si Interface Passivation Layer**
S. Oktyabrsky[1,] S. Koveshnikov[2], V. Tokranov[1], M.l Yakimov[1], R. Kambhampati[1], H. Bakhru[1], F. Zhu[2], J. Lee[3], and W. Tsaib[2], [1]College of Nanoscale Science and Engineering, University at Albany-SUNY, New York, USA, [2]Intel Corporation, Santa Clara, California, USA, and [3]The University of Texas at Austin, Department of Electrical and Computer Engineering, Austin, Texas, USA

V.B-3 **Enhancement Mode n-MOSFET with High-? Dielectric On GaAs Substrate**
K. Rajagopalan[2], P. Zurcher[2], J. Abrokwah[2], R. Droopad[2], D. A. J. Moran[1], R. J. W. Hill[1], X. Li[1], H. Zhou[1], D. McIntyre[1], S. Thoms[1], I.G. Thayne[1] and M. Passlack[2], [1]Department of Electronics & Electrical Engineering, University of Glasgow. Glasgow, UK and [2]Freescale Semiconductor Inc., Tempe, Arizona, USA

xiii

V.B-4   **High-performance submicron inversion-type enhancement-mode InGaAs MOSFET with maximum drain current of 360 mA/mm and transconductance of 130 mS/mm**
Y. Xuan, Y. Q. Wu, H. C. Lin, T. Shen and P. D. Ye, School of Electrical and Computer Engineering, Purdue University, West Lafayette, Indiana, USA

V.B-5   **Enhancement-mode In0.70Ga0.30As-channel MOSFETs with ALD Al2O3**
Y. Sun, E. W. Kiewra, J. P. De Souza, S. J. Koester, K. E. Fogel, D. K. Sadana, IBM Thomas J. Watson Research Center, Yorktown Heights, New York, USA

V.B-6   **Performance of Sub-micron Gate Length InAlP Native Oxide GaAs-channel MOSFETs**
J. Zhang, T. H. Kosel, D. C. Hall, and P. Fay, Dept. of Electrical Engineering, University of Notre Dame, Notre Dame, Indiana, USA

Session V.C   Memory Devices   213-228

V.C-1   **Hybrid ALD-SiN/Si-nanocrystals/ALD-SiN FinFET device with large P/E window for MLC NAND Flash memory application**
J.-D. Choe[1], S.-H. Lee[2], Y. J. Ahn[2], D. Jang[2], Y.-B. Yoon[2], J. J. Lee[2], I. Chung[1], K. Park[2] and D. Park[2], [1]School of Information and Communication Engineering, Sungkyunkwan University, Kyungki-Do, KOREA and [2]Technology Development Team 2, Samsung Electronics Co., Yongin-City, Kyungki-Do, KOREA

V.C-2   **Ge/Si hetero-nanocrystalnonvolatile floating gate memory**
B. Li, Y. Zhu and J. Liu, Quantum Structures Laboratory, Department of Electrical Engineering, University of California, Riverside, California, USA

V.C-3   **Memory Effects in Metal-Oxide-Semiconductor Capacitors Incorporating Dispensed Highly Mono-disperse One-Nanometer Silicon Nanoparticles**
O. M. Nayfeh[1], D. A. Antoniadis[1], K.Mantey[2] and M. H. Nayfeh[2,] [1]Microsystems Technology Laboratories, Massachusetts Institute of Technology, Cambridge, Massachusetts and [2]Department of Physics, University of Illinois at Urbana-Champaign, Urbana, Illinois, USA

V.C-4   **Modeling of Multi-layer Nanocrystal Memory**
T.-H. Hou, C. Lee, and E. C. Kan, School of Electrical and Computer Engineering, Cornell University, Ithaca, New York, USA

V.C-6   **Phase-change Memory**
C. Lam, IBM Qimonda Macronix PCRAM Joint Project, IBM T.J. Watson Research Center, Yorktown Heights, New York, USA

V.C-7   **Novel Cross Point Switch based on Zn1-xCdxS memory devices for FPGA**
K. Abe[1], Z. Wang[2], S. Fujita[1], T. H. Lee[2] and Y. Nishi[2], [1]Frontier Research Laboratory, Corporate R&D Center, Toshiba Corporation, JAPAN, and [2]Center for Integrated Systems, Stanford University, Stanford, California, USA

Rump Sessions   229-230

R.1   **The Future of Nanowire Devices: Top-down or bottom-up?**
Session Organizers: Prabhat Agarwal, NXP, Steven Koester, IBM, and Eric Pop, Univ. Illinois/Urbana Champaign

R.2   **Why do we need nano for biosensors?**
Session Organizers: Sang-Hyun Oh, University of Minnesota and Janos Voros, ETH Zurich

R.3   **The THz Bridge**
Session Organizer: Mike Wojtowicz, Northrup Grumman

Joint DRC/EMC Plenary Session 231-232

Low Cost "Plastic" Solar Cells: A Dream or Reality???
Alan J. Heeger, University of California, Santa Barbara

Session VI.A Nanobiotechnology 233-242

VI.A-1 **The brain and the computer**
K. Boahen, Stanford University, Bioengineering Dept., Stanford, California, USA

VI.A-2 **Joining Microelectronics and Microionics: Nerve Cells and Brain Tissue on Semiconductor Chips**
P. Fromherz, Department of Membrane and Neurophysics, Max Planck Institute for Biochemistry, Munich, GERMANY

VI.A-3 **Biolithography: DNA-assisted Manufacturing of Nanodevices for Optical and Electronic Biosensing**
J. Vörös, Laboratory of Biosensors and Bioelectronics, Institute for Biomedical Engineering, Zurich, SWITZERLAND

Session VII.A Magnetoelectronics Devices & Optical Resonators 243-256

VII.A-1 **Recent Advances in MRAM Technology**
J. M. Slaughter, Freescale Semiconductor, Inc., Chandler, Arizona, USA

VII.A-2 **Magnetic Sensitivity in Mesoscopic EMR Devices in I-V-I-V Configuration**
T. Boone, L. Folks, J. A. Katine, E. Marinero, N. Smith and B. A. Gurney, Hitachi Global Storage Technologies , San Jose Research Center, San Jose, California, USA

VII.A-3 **Simulation of Spin Torque Devices with Inelastic Spin flip Scattering**
S. Salahuddin and S. Datta, School of Electrical and Computer Engineering and NSF Network for Computational Nanotechnology, Purdue University, West Lafayette, Indiana, USA

VII.A-5 **Characterization and Application of Large Magnetoresistance in Organic Semiconductors**
M. Wohlgenannt, G. Veeraraghavan, Y. Sheng, O. Mermer, and T. D. Nguyen, Department of Physics and Astronomy, The University of Iowa, Iowa City, Iowa, USA

VII.A-6 **Metamaterials - Negative Indices with Positive Benefits?**
M. C. K. Wiltshire, Imaging Sciences Department, Imperial College London, London, UK

VII.A-7 **High Performance Polycrystalline Diamond Micro Resonators**
N. Sepúlveda[1], J. Lu[1], D. M. Aslam[1], and J. P. Sullivan[2] , [1]Electrical and Computer Engineering, Michigan State University, E. Lansing, Michigan, USA and [2]Sandia National Laboratories, Albuquerque, New Mexico, USA

Session VII.B Carbon Nanotube & Graphene Devices 257-274

VII.B-1 **Aligned Arrays Single Walled Carbon Nanotubes for Thin Film Electronics**
J. A. Rogers, University of Illinois at Urbana/Champaign, Urbana, Illinois, USA

VII.B-2 **Quantum Capacitance Measurement for SWNT FET with Thin ALD High-k Dielectric**
Y. Lu[1], H. Dai[1] and Y. Nishi[2], [1]Department of Chemistry and Laboratory for Advanced Materials, Stanford University, Stanford, California, USA and [2]Electrical Engineering, Stanford University, Stanford, California, USA

VII.B-3   **The study of low frequency noise of single-walled carbon nanotube transistors**
S. Kim, D. Chang, Y. Xuan, P. Ye and S. Mohammadi, School of Electrical and Computer Engineering and Birck Nanotechnology Center, Purdue University, West Lafayette, Indiana, USA

VII.B-4   **Semiconducting Graphene Ribbon Transistor**
Z. Chen, P. Avouris, IBM T. J. Watson Research Center, Yorktown Heights, New York, USA

VII.B-5   **Performance Limits of Nanocomposite Transistors & Nanobio Sensors: A Bottom-up Perspective**
M. A. Alam, N. Pimparkar, P. Nair, S. Kumar, and J. Murthy, School of Electrical and Computer Engineering, Purdue University, West Lafayette, Indiana, USA

VII.B-6   **Impact of Process Variation on Nanowire and Nanotube Device Performance**
B. C. Paul[1,2], S. Fujita[2], M. Okajima[2], T. Lee[1], H.S.P. Wong[1], and Y. Nishi[1], [1]Center for Integrated Systems, Stanford University, Stanford, California, USA and [2]Toshiba America Research Inc., San Jose, California, USA

VII.B-7   **Scaling Behaviors of Graphene Nanoribbon FETs**
Y. Yoon, Y. Ouyang, and J. Guo, Electrical and Computer Engineering, University of Florida, Gainesville, Florida, USA

VII.B-7   **Role of Electrical and Thermal Contact Resistance in the High-Bias Joule Breakdown of Single-Wall Carbon Nanotube Devices**
E. Pop, Dept. of Electrical and Computer Engineering and Micro and Nanotechnology Lab, University of Illinois, Urbana-Champaign, Urbana, Illinois, USA

xvi

(DeBartolo Hall Room 102)

Plenary Session

Monday AM, June 18th, 2007

Session Organizer: Bobby Brar, Teledyne Scientific & Imaging
Session Chair: Joerg Appenzeller, IBM

8:20 AM Welcoming Remarks
Presentations: IEEE Fellows and Best Student Paper Awards

8:50 AM I.-1 Plenary Paper
CMOS Integrated Nano-Photonics is now a Commercial Technology
E. Yablonovitch, Electrical Engineering Department, University of California, Los Angeles, Los Angeles, California, USA

9:50 AM Break

10:10 AM I.-2 Plenary Paper
Mobility-Enhanced MOS Device Technologies in Nano-CMOS era
S.Takagi, MIRAI-AIST, Kawasaki, JAPAN and Department of Electronic Engineering, School of Engineering, The University of Tokyo, Tokyo, JAPAN

11:10 AM I.-3 Plenary Paper
Hybrid CMOS/Nanodevice Circuits: Architectures, Applications, and Device Needs
K. K. Likharev, Stony Brook University, Stony Brook, New York, USA

2

CMOS Integrated Nano-Photonics is now a Commercial Technology

E. Yablonovitch
Electrical Engineering Department
University of California, Los Angeles
Los Angeles, CA, 90095-1594

It has become apparent, that Silicon technology can provide many of the requirements for nano-photonic integration, including many of the common opto-electronic components. Intel and Luxtera have both recently announced 10Gb/sec optical modulators, integrated into Silicon. Actually, ALL the other customarily required opto-electronic components are available NOW in Silicon, as well. Continuous wave optical power can be provided from off-chip, just as dc power is currently provided from off-chip. The first commercial applications are emerging, are optical 10Gb/s Ethernet, and Infiniband, a standard for communicating among multi-processors in an array, and in general, optical substitutes for electrical cables.

Mobility-Enhanced MOS Device Technologies in Nano-CMOS era

Shinichi. Takagi

MIRAI-AIST, 1, Komukai Toshiba-cho, Saiwai-ku, Kawasaki, 212-8582, Japan
Department of Electronic Engineering, School of Engineering, The University of Tokyo

Tel: +81-44-549-8217, Fax: +81-44-549-8216, E-mail: s-takagi@mirai.aist.go.jp

It has been well recognized that, under sub-100 nm regime, conventional device scaling concept has confronted with several physical and essential limitations. Therefore, any new device engineering to realize advanced CMOS by overcoming these difficulties is strongly needed. A group of theses new device technologies called the technology boosters can be classified mainly into three categories, as schematically shown in Fig. 1, gate stack engineering, source engineering and channel engineering. Particularly, the channel engineering includes carrier-transport-enhanced channels aiming at high current drive and multi-gate channels aiming at high immunity for short channel effects. Among them, the carrier-transport-enhanced channels, typically seen in strained-Si channels, are recently becoming more important.

The improvement of carrier transport properties can be obtained through a variety of ways including the optimal choices of surface orientations, channel directions, strain configurations and channel materials, which are summarized in Table 1. While the mobility is still an important parameter to describe the current drive under sub-100 nm regime, the reduction in the effective mass can be more essential in increasing the on-current under ballistic or near-ballistic transport regime, where the on-current can be described by the product of surface carrier concentration, N_s, at the bottle neck point and the injection velocity. Fig. 2 shows the calculated injection velocity under full ballistic transport as a function of N_s [1, 2]. The lower effective mass can lead to the increase in both thermal velocity in low N_s region and Fermi velocity in high N_s region. It can be interpreted in this sense that the effectiveness of high mobility channels in ultra-short channel MOSFETs is attributable mainly to the low effective mass. On the other hand, too low effective masses of channel materials degrade the gate capacitance, because of the reduction in inversion-layer capacitance, C_{inv}, associated with lower density-of-states [3]. As a result, the optimization of the effective mass is necessary to maximize the on-current under a give gate oxide thickness. This situation is typically seen in III-V semiconductor channel MOSFETs, as discussed in the last part of this paper.

When applying those technologies to future technology nodes, we also need to take into account the following issues; (1) successive increase in carrier transport properties with a progression in technology nodes, (2) individual optimization of nMOS and pMOS channel structures and (3) compatibility with multi-gate structures. This paper reviews our recent results on these carrier-transport-enhanced CMOS structures on the Si platform for future high performance and low power LSIs.

One of the most effective ways in enhancing hole transport properties is to use compressive SiGe or Ge channels. Actually, we have demonstrated that almost pure GOI MOSFETs with compressive strain (SGOI with Ge content of 93 % and compressive strain of 1.5 %), fabricated by local Ge condensation, exhibit the hole mobility enhancement of as high as 10 [4]. According to the comparison with the theoretical calculations, factors of 5x and 2x in this 10x enhancement are ascribed to the high Ge content and remaining compressive strain, respectively. Fig. 3 shows the comparison of the hole mobility obtained in the SGOI MOSFETs with electron mobility in Si and strained Si nMOSFETs. The hole mobility in the SGOI of 93% Ge content is comparable to or even higher than the electron mobility in strained-Si nMOSFETs. As a result, the combination of tensile strain Si nMOS and compressive strain SiGe (Ge) pMOS can be idealistic as CMOS with bi-axial strain. We have succeeded in this integration of strained-SOI n-MOSFETs and strained-SGOI p-MOSFETs with Ge content of 66 % on the same SGOI substrates by using local Ge condensation and selective epitaxy of strained-Si layers [5].

Another technique to boost pMOS performance is the introduction of uni-axial compressive stress. We have proposed and demonstrated a novel uni-axial strain technique enabling to introduce uniform and high stress through the combination of bi-axial strain structures and lateral strain relaxation [6, 7]. Here, fully-strained SGOI substrates [8] are used as starting materials and the active area (AA) of pMOSFETs is patterned into the shape with narrow and sufficiently long AA. Since the elastic strain relaxation occurs from the edge of islands [9], uni-axial relaxation of only strain transverse to current flow direction in the channel can be realized. Actually, the uni-axial strain configuration has been confirmed by the nano-electron diffraction method. The I_{on} enhancement of 80 % has been obtained in the planar uni-axial SGOI p-MOSFETs with 40 nm L_g (Fig. 4). This technique is also applicable to multi-channel FETs such as FinFETs and Tri-gate MOSFETs [10], as schematically shown in Fig. 5. It should be noted that this device structure can simultaneously combine three hole mobility booster techniques, (110) surfaces, uni-axial

978-1-4244-1101-6/07/$25.00 ©2007 IEEE   5

compressive strain and Ge (SiGe) channels. It has been shown in Fig. 6 that the G_m increase of the SGOI FinFETs amounts to 200 % and 60 % against planer Si MOSFETs and SOI FinFETs, respectively, owing to the additive contribution of each hole mobility booster. Also, higher performance is expected in SGOI MOSFETs with higher Ge content or pure Ge MOSFETs.

The present uni-axial strain technique is also applicable to n-channel FinFETs by using SSOI as a starting substrate [11, 12]. In addition, we have proposed and demonstrated [13] that new subband structure engineering for electrons on (110) surfaces utilizing uni-axial tensile strain along <110>, shown in Fig. 7, can effectively enhance the electron mobility along <110> on (110), which is known to be much lower than that on (100) and, thus, one of the problems in n-channel FinFETs along <110>. According to this engineering, the application of the uni-axial tensile strain along <110> can lead to the transfer of electrons from the 4-fold valleys with higher effective mass into the 2-fold valleys with lower effective mass and the resulting increase in the mobility [13, 14]. In addition, this uni-axial strain can also contribute to the increase in electron mobility on (100), because of the reduction in the effective mass as well as the increase in the occupancy of 2-fold valley electrons [15]. Actually, we have experimentally observed that uni-axial tensile strain SSOI FinFETs along <110> axis can provide 2.2 time higher G_m than conventional FinFETs (Fig. 8) and that the SSOI FinFETs along <110> have higher G_m than those along <100>, indicating the effectiveness of the present subband engineering. As a result, the optimum CMOS with uni-axial strain can be realized by combining tensile strain SSOI n-MOSFETs and compressive strain SGOI p-MOSFETs along the same <110> direction.

Furthermore, attentions have recently been paid to III-V channels as a candidate of high performance n-MOSFETs beyond strained-Si devices. A possible ultimate CMOS structure on Si platform using III-V MOSFETs is shown in Fig. 9 [8]. However, it should be noted here that the low effective mass of the III-V MIS channels reduces C_{inv} and resulting increase in EOT [3], while it can provide higher velocity. Fig. 10 shows the calculated I_{on}-V_g characteristics under full ballistic transport with gate oxide physical thickness (T_{ox}) of 3 and 0.5 nm [9]. It is confirmed that the III-V channels become less effective in the I_{on} increase with decreasing T_{ox}. Thus, higher performance can be expected in III-V MOSFETs with thicker T_{ox}. While further optimization of III-V channel structures is still needed, these results suggest that there exists an optimized channel material, depending on T_{ox}.

In summary, CMOS family utilizing SiGe/Ge channels, uni-axial strain and III-V channels can be key devices for high performance and low power advanced LSIs in the future.

Acknowledgement: This work was partly supported by the New Energy and Industrial Technology Development Organization (NEDO) and a Grant-in-Aid for Scientific Research from the Ministry of Education, Culture, Sports, Science and Technology.

References: [1] K. Natori, J. Appl. Phys., vol. 76 (1994) 4879 [2] S. Takagi, T. Mizuno, T. Tezuka, N. Sugiyama, S. Nakaharai, T. Numata, J. Koga and K. Uchida, Solid State Electron., vol. 49 (2005) 684 [3] M. Fishcetti, IEEE Trans. Electron Devices, vol. 38 (1991) 634 [4] T. Tezuka, S. Nakaharai, Y. Moriyama, N. Hirashita, E. Toyoda, N. Sugiyama, T. Mizuno and S. Takagi, Proc.VLSI Symp. (2005) 80 [5] T. Tezuka, S. Nakaharai, Y. Moriyama, N. Hirashita, E. Toyoda, T. Numata, T. Irisawa, K. Usuda, N. Sugiyama, T. Mizuno and S. Takagi, Semicond. Sci. Technol., vol. 22 (2007) S93 [6] T. Irisawa, T. Numata, T. Tezuka, K. Usuda, N. Hirashita, N. Sugiyama, E. Toyoda and S. Takagi, Proc. VLSI Symp. (2005) 178 [7] T. Irisawa, T. Numata, T. Tezuka, K. Usuda, N. Hirashita, N. Sugiyama, E. Toyoda and S. Takagi, IEEE Trans. Electron Devices, vol. 53, (2006) 2809 [8] T. Tezuka, N. Sugiyama T. Mizuno and S. Takagi, Tech. Dig. IEDM (2001) 946 [9] T. Tezuka, N. Sugiyama and S. Takagi, J. Appl. Phys., vol. 94 (2003) 7553 [10] T. Irisawa, T. Numata, T. Tezuka, N. Sugiyama, and S. Takagi, Tech. Dig. IEDM (2006) 457 [11] N. Collaert, R. Rooyackers, F. Clemente, P. Zimmerman, I. Cayrefourcq, B. Ghyselen, K. T. San, N. Eyckens, M. Jurczak, S. Biesemans, Proc. VLSI Symp. (2006) 80 [12] A.V-Y. Thean, D. Zhang, V. Vartanian, V. Adams, J. Conner, M. Canonico, H. Desjardin, P. Grudowski, B. Gu, Z.-H. Shi, S. Murphy, G. Spencer, S. Filipiak, D. Goedeke, X-D. Wang, B. Goolsby, V. Dhandapani, L. Prabhu, S. Backer, L-B La, D. Burnett, T. White, B.-Y. Nguyen, B. E. White, S. Venkatesan, J. Mogab, I. Cayrefourcq and C. Mazure, Proc. VLSI Symp. (2006) 130 [13] T. Irisawa, T. Numata, T. Tezuka, K. Usuda, S. Nakaharai, N. Hirashita, N. Sugiyama, E. Toyoda and S. Takagi, Tech. Dig. IEDM (2005) 727 [14] K. Uchida, T. Krishnamohan, K. C. Saraswat, and Y. Nishi, Tech. Dig. IEDM (2005) 135 [15] K. Uchida, A. Kinoshita, and M. Saitoh, Tech. Dig. IEDM Tech. Dig. (2006) 135 [16] S. Takagi, Nikkei Micro Device 22 (2005) 54 [17] S. Takagi and S. Sugahara, Ext. Abs. SSDM (2006) 1056

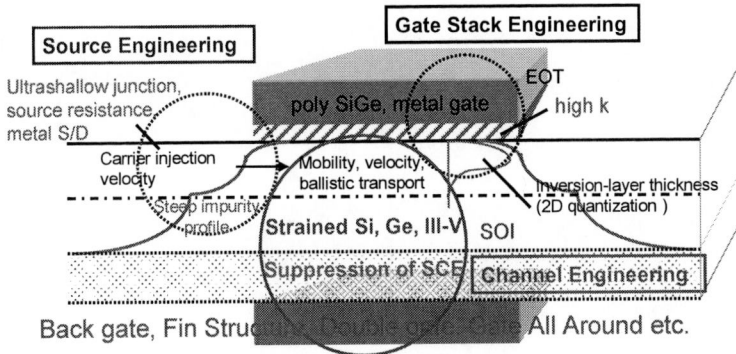

Fig. 1. Schematic diagram of three types of device engineering

	nMOSFET	pMOSFET
Channel Direction	-	<100> on (100) surface <110> on (110) surface
Surface Orientation	(100)	(110)
Strain in Si/Ge	bi- or uni-axial tensile	bi-axial tensile uni-axial compressive
Materials	III-V	SiGe/Ge

Table 1. Ways to enhance carrier transport properties in MOS channels

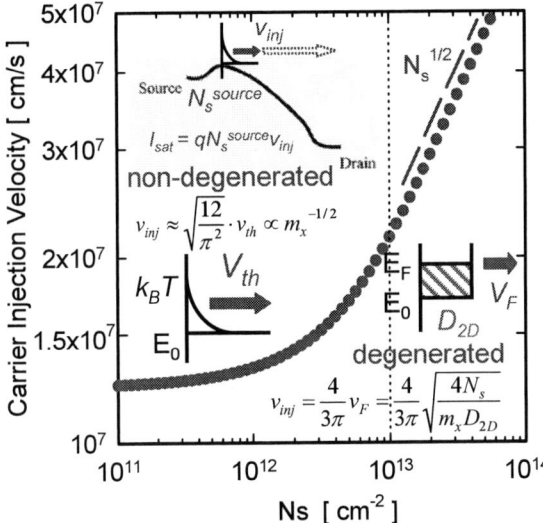

Fig. 2 Calculated injection velocity under full ballistic transport as a function of N_s for the (100) Si lowest subband

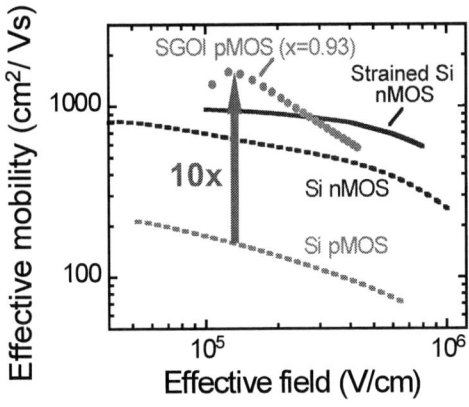

Fig. 3 Comparison of hole mobility in 93% Ge SGOI p-MOSFETs with electron mobility in strained Si n-MOSFETs and the universal electron and hole mobilities

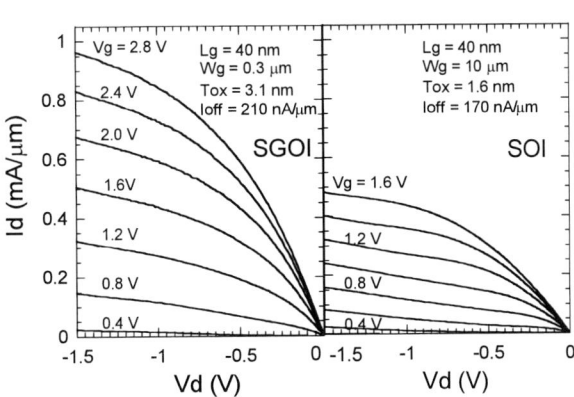

Fig. 4 I_d-V_d characteristics of 40 nm L_g uniaxially strained SGOI and control SOI MOSFET.

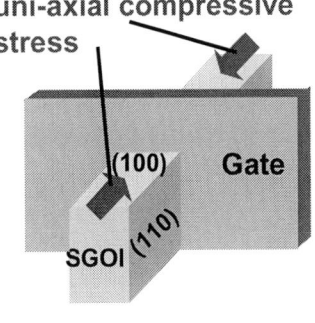

Fig. 5 Proposed SGOI pMOSFETs with uniaxial compressive strain along <110> axis

7

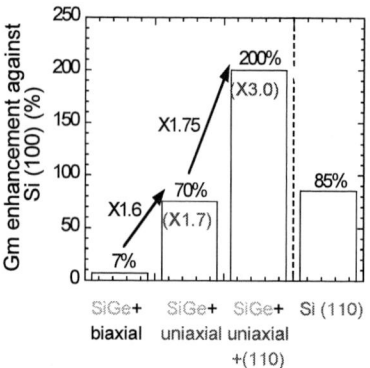

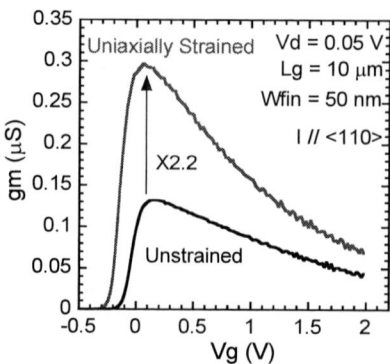

Fig. 6 G_m enhancement against Si (100) resulting from each performance enhancement booster; SiGe channels, uniaxial strain and (110) surfaces.

Fig. 8 Comparison of g_m between SSOI and control SOI Tri-Gate MOSFETs with W_{fin} of 50 nm. Large gm enhancement confirms that the uniaxial strain is preserved down to W_{fin} of 50 nm.

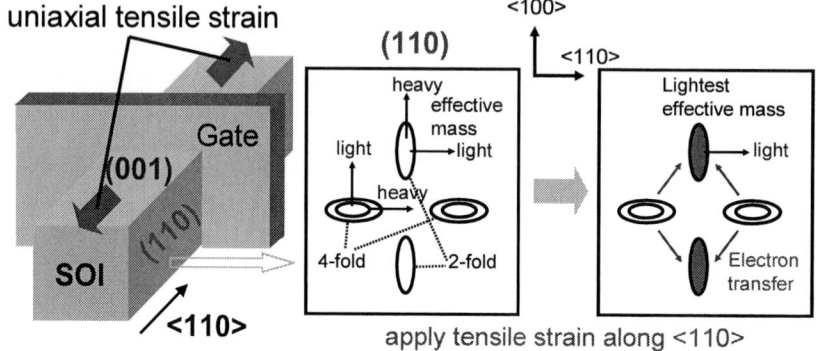

Fig. 7 Schematic illustration of the fabricated uni-axially strained SSOI Tri-gate MOSFETs with uni-axial tensile strain along <110> axis and schematic diagrams of equi-energy surfaces of 2D electrons on (110) surfaces with and without tensile strain along <110> axis, which transfers electrons in 4-dold valleys into 2-fold ones

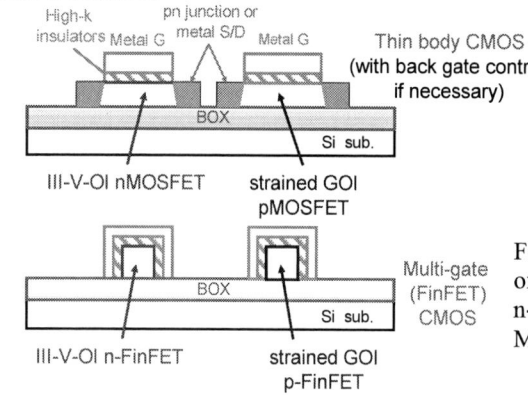

Fig. 9 Ultimate CMOS structure composed of the combination of III-V semiconductor n-channel MOSFETs and Ge p-channel MOSFETs on insulators

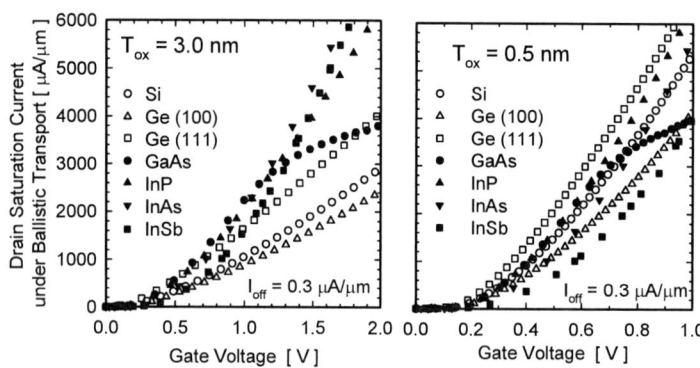

Fig. 10 Calculated I_{on} under full ballistic transport versus V_g with physical gate oxide thickness of 3.0 and 0.5 nm

8

Hybrid CMOS/Nanodevice Circuits: Architectures, Applications, and Device Needs

Konstantin K. Likharev

Stony Brook University, Stony Brook, NY 11794-3800, U. S. A.

Phone +1-631-632-8159, e-mail klikharev@notes.cc.sunysb.edu

I will review the recent work on devices, circuits and architectures for possible hybrid semiconductor /nanodevice integrated circuits [1-4], especially those based on nanowire crossbars, with similar, simple, two-terminal devices formed at each crosspoint (Fig. 1). Such add-on may be especially efficient if it is connected to the underlying CMOS circuit with an area-distributed, pin-based interface [1, 4-8] – Fig. 2a. A rotation of the nanowire crossbar by a certain, well-defined angle with respect to the interface pin grid (Fig. 2b) allows the CMOS subsystem to address each and every of the crosspoint devices, even with no nanoscale alignment between the CMOS and crossbar subsystems [5]. This feature liberates the high-resolution patterning technologies (such as nanoimprint or EUV interference lithography) from the burden of overlay requirements and may enable their fast scaling beyond the 10-nm frontier.

The recent detailed studies [6-12] have shown that if the crosspoint devices feature the functionality of programmable diodes, the hybrid circuits may enable (at least) the following applications:

(i) terabit-scale memories with access time below 100 ns and defect tolerance up to 10% [9],

(ii) FPGA-like reconfigurable logic circuits with the area-by-delay product at least two orders of magnitude lower than that of CMOS FPGAs fabricated with similar design rules and power per unit area [6, 7, 10], and

(iii) mixed-signal neuromorphic networks ("CrossNets") which may provide unparalleled performance for some important information processing tasks including online image recognition [11], and in future may become the first hardware basis for challenging the human cerebral cortex in both density and speed, at manageable power [12].

Recently, the hybrid circuit concept has received a strong boost from the announcement of reproducible fabrication of the necessary crosspoint devices (programmable diodes) using copper oxide [13]. However, these and other demonstrated devices with similar functionality can hardly be scaled below 10 nm. In my talk I will discuss prospects and problems of creating reproducible programmable diodes beyond that frontier.

The work has been supported in part by AFOSR, DTO, MARCO via FENA Center, and NSF.

[1] K. K. Likharev and D. B. Strukov, "CMOL: Devices, Circuits, and Architectures", in *Introducing Molecular Electronics,* eds. G. Cuniberti *et al.*, Springer, Berlin, 2005, pp. 447-477.
[2] A. DeHon and K. K. Likharev, "Hybrid CMOS/Nanoelectronic Digital Circuits: Devices, Architectures, and Design Automation", in: *Proc. of ICCAD-2005* , pp. 375-382.
[3] P. J. Kuekes, G. S. Snider, and R. S. Williams, "Crossbar Nanocomputers", *Sci. American.*, vol. 293, pp. 72-76, Nov. 2005.
[4] K. K. Likharev, "CMOL: Second Life for Silicon", *Microelectronics J.*, vol. 38, March 2007.
[5] K. K. Likharev, "Electronics below 10 nm", in: *Nano and Giga Challenges for Microelectronics*, eds. J. Greer *et al.* , Springer, Berlin, 2003, pp. 27-68.
[6] D. B. Strukov and K. K. Likharev, "A Reconfigurable Architecture for Hybrid CMOS/Nanodevice Circuits", in: *Proc. FPGA'06*, pp. 131-140.
[7] G. Snider, and R. S. Williams, *Nanotechnology*, "Nano/CMOS Architectures Using a Field-Programmable Nanowire Interconnect", vol. 18, art. 035204, Jan. 2007.
[8] D. Tu, M. Liu, W. Wang, and S. Haruehanroengra, "3D CMOL: A 3D FPGA Using CMOS/Nanomaterial Hybrid Digital Circuits", preprint, Feb. 2007.
[9] D. B. Strukov and K. K. Likharev, "Defect-Tolerant Architectures for Nanoelectronic Crossbar Memories", *J. of Nanoscience and Nanotechnology*, vol. 7, pp. 151-167 , Jan. 2007.
[10] D. B. Strukov and K. K. Likharev, "CMOL FPGA: A Reconfigurable Architecture for Hybrid Digital Circuits with Two-terminal Nanodevices", *Nanotechnology*, vol. 16, pp. 888-900, June 2005.
[11] J. H. Lee and K. K. Likharev, "CMOL CrossNets as Pattern Classifiers", *Lecture Notes in Computer Science*, vol. 3512, pp. 446-454, 2005.
[12] Ö. Türel, J. H. Lee, X. Ma, and K. K. Likharev, "Neuromorphic Architectures for Nanoelectronic Circuits", *Int. J. of Circ. Theor. Appl.*, vol. 32, pp. 277-302, Sep./Oct. 2004.
[13] A. Chen *et al.*, "Non-Volatile Resistive Switching for Advanced Memory Applications", in *IEDM'05 Tech. Digest*, Report 31.3.

978-1-4244-1101-6/07/$25.00 ©2007 IEEE

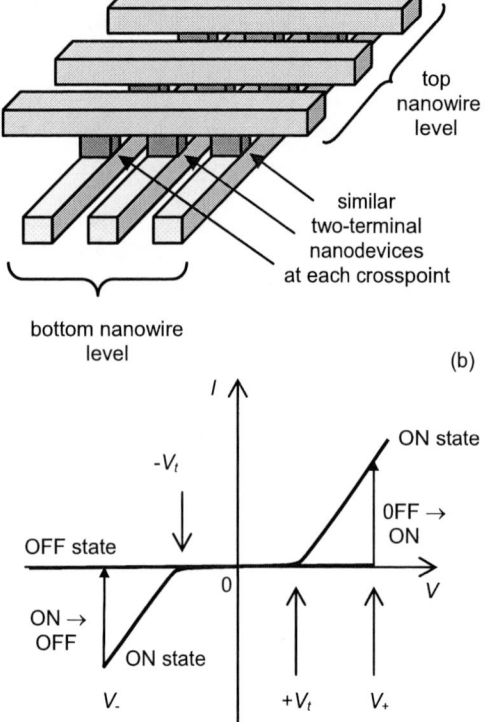

Fig. 1. Nanowire/nanodevice add-on: (a) crossbar structure and (b) I-V curve of a crosspoint device (programmable diode) – schematically.

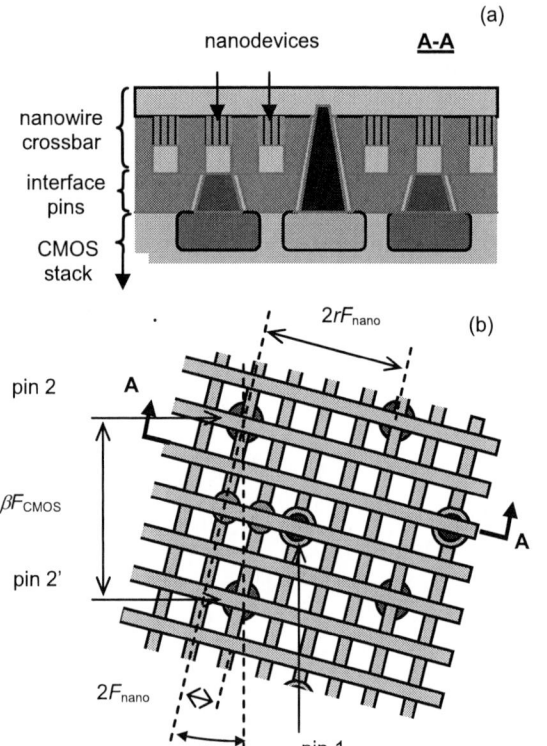

Fig. 2. Area-distributed "CMOL" interface between the CMOS and nano subsystems: (a) side and (b) top view.

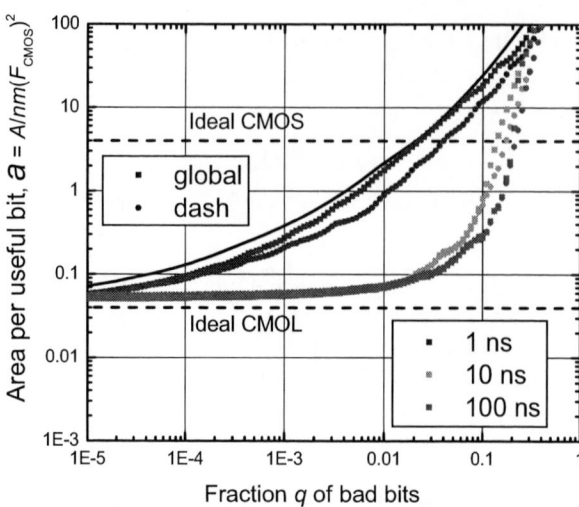

Fig. 3. Calculated density of CMOL memories as a function of bad bit density for an old (black lines) and a new architecture [9].

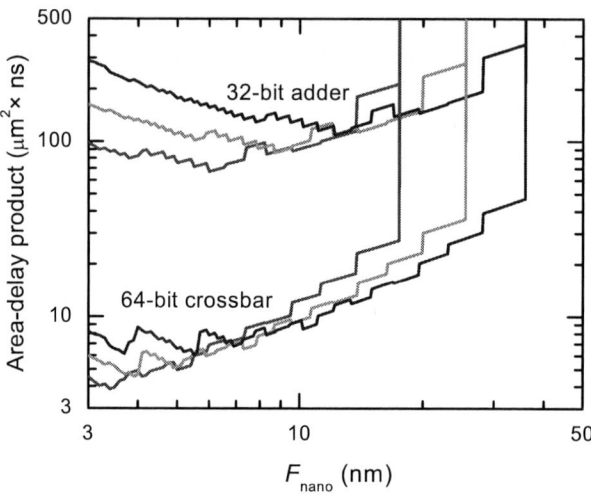

Fig. 4. Calculated area-by-delay product of two representative digital logic circuits implemented in CMOL technology, as a function of design rules [10].

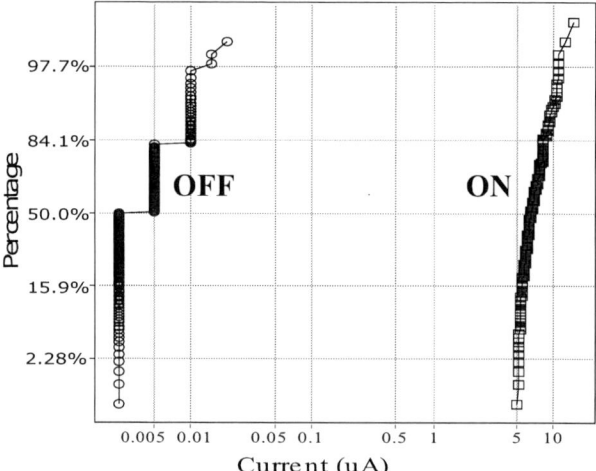

Fig. 5. Histogram of ON and OFF currents of CuO_x crosspoint devices [13].

Session II.A (DeBartolo Hall Room 102)

Thin Film Transistors

Monday PM, June 18th, 2007

Session Organizer: David Gundlach, National Institute of Standards and Technology
Session Chairs: H. Klauk, Max Planck Institute for Solid State Research and H. Pang, University of Wisconsin-Madison

1:30 PM II.A-1 Invited Paper
Uniform ZnO Thin-Film Transistors by an Ambient Process
S. F. Nelson, D. Levy, D. Freeman, P. Cowdery-Corvan, L. Tutt, M. Burberry, L. Irving, Eastman Kodak Company, Rochester, New York, USA

2:10 PM II.A-2 Student Paper
Flexible Microwave Single-Crystal Si TFTs with fmax of 5.5 GHz
H. Pang[1], H.-C. Yuan[1], M. G. Lagally[2], G. K. Celler[3] and Z. Ma[1], [1]Department of Electrical and Computer Engineering, [2]Department of Materials Science and Engineering, University of Wisconsin-Madison, Madison, Wisconsin USA, and [3]Soitec USA, Peabody, Massachusetts, USA

2:30 PM II.A -3 Student Paper
A 'Bottom-up' Redefinition for Mobility and the Effect of Poor Tube-Tube Contact on the Performance of CNT Nanobundle Thin Film Transistors
N. Pimparkar and M. A. Alam , School of Electrical and Computer Engineering, Purdue University, West Lafayette, Indiana, USA

2:50 PM II.A -4 Student Paper
Fast ZnO Thin-Film Transistor Circuits
J. Sun[1], D. Mourey[1], D. Zhao[1], S. Park[1], S. F. Nelson[2], David H. Levy[2], D. Freeman[2], Peter Cowdery-Corvan[2], Lee Tutt[2], and T. N. Jackson[1], [1]Center for Thin Film Devices and Materials Research Institute, Department of Electrical Engineering, Penn State University, University Park, Pennsylvania, USA and [2]Eastman Kodak Company, Rochester, New York, USA

3:10 PM Break

3:30 PM II.A-5 Invited Paper
Electrical and chemical stability of polymer transistors
R. A. Street and M. Chabinyc, Palo Alto Research Center, Palo Alto, California, USA

4:10 PM II.A-6 Student Paper
Solution processed OTFT circuits on plastic substrates
S. K. Park[1], S. Subramanian[2], J. Anthony[2], and T. N. Jackson[1], [1]Center for Thin Film Devices and Materials Research Institute, Department of Electrical Engineering, Penn State University, University Park, Pennsylvania, USA and [2]Department of Chemistry, University of Kentucky, Lexington, Kentucky, USA

12

Uniform ZnO Thin-Film Transistors by an Ambient Process

Shelby F. Nelson, David Levy, Diane Freeman, Peter Cowdery-Corvan, Lee Tutt,
Mitchell Burberry, Lyn Irving

Eastman Kodak Company, 1999 Lake Avenue, Rochester NY 14650 USA
Phone: 585-477-8417. E-mail: Shelby.nelson@kodak.com

Backplane technologies for liquid crystal displays are well established and successful, and most current research in that area is conducted towards reducing costs. Backplanes for OLEDs, on the other hand, have yet to reach the overwhelming success in performance and yield enjoyed by the LCD industry. Of the prominent competing semiconductors for OLED backplanes, amorphous silicon has the drawback of low mobility (~ 1 cm^2/Vs) and threshold voltage variation upon bias stress [1], while low-temperature polysilicon struggles with achieving adequate threshold uniformity across a display backplane. One approach to solving these issues is to consider material sets other than silicon. We have fabricated zinc oxide thin-film transistors using a novel ambient deposition process, with maximum temperature of 200°C. The TFTs deposited this way show sufficiently good properties to make them potentially applicable to OLED display backplanes.

The deposition system uses a proprietary process that operates in an open environment, *i.e.* requiring no containment chamber. This allows for a small footprint and an easy translation to a continuous roll-to-roll process. Typical deposition rates under the head are near 200 Å/min. The TFTs presented here are grown on glass substrates coated uniformly with a gate conductor. The gate conductors are usually ITO (commercially available), but can be metals such as chromium or molybdenum. Both the gate dielectric and the semiconductor layer were coated using Kodak's apparatus. The gate dielectric was formed with Trimethylaluminum as an aluminum source and water as an oxidizer to produce aluminum oxide. The semiconductor was produced using diethylzinc as a zinc source and water as an oxidizer to produce zinc oxide. Aluminum source-drain contacts, patterned by shadow mask or lithographically, are thermally evaporated, with a width of 600 µm and a length ranging from 50 to 150 µm. The semiconductor was isolated in each TFT by a simple photolithographic process. Normally no passivation layer is applied.

With 100-nm-thick alumina gate dielectric, devices typically have saturation mobility above 10 cm^2/Vs, and threshold voltage of between 6 and 7 V, subthreshold swing of 0.5 V/decade, and on/off ratio of 10^8. The gate leakage is less than 100 pA at 30 V, and because the leakage is over the entire 0.4 mm^2 isolation patch of ZnO, it translates to a gate leakage current density of 2.5×10^{-8} A/cm^2. The devices show good linear characteristics, with the linear mobility extracted at drain voltage of 0.1 V matching the saturation mobility. This suggests that there is negligible contact resistance.

The uniformity of TFTs over the deposited area of 3 cm $\times$ 5 cm is tested on a large array of TFTs. For a typical example, the mobility is 15 ± 2 cm^2/Vs, the threshold voltage is 6.55 ± 0.24 V, and the hysteresis, measured in the steep portion of the transfer curve, is 0.1 ± 0.1 V. When carefully processed, the threshold voltage variation over the growth area can be less than 0.1 V.

We have also tested the stability of these devices. Short tests of "shelf life," in which the sample is tested, exposed to room air and light, and retested after six months, showed the repeated measurements were well within the error bars of the initial measurements. A harsher test, of course, is a bias stress test. Devices were tested periodically while being stressed with gate bias of 20 V and drain bias of 10 V. Over an 18 h test, shifts as low as 1 V in threshold have been measured. This robustness to bias stress is critical for uncomplicated display pixel designs. However, not all our samples show equally good behavior, and the variables contributing to this are being explored.

In summary, we have shown promising early data on the electrical performance of ambient-deposited ZnO. The novel deposition process uses a maximum temperature of 200°C, and is capable of producing TFTs with mobility, threshold uniformity, and bias stability values compatible with OLED backplanes.

[1] M. J. Powell, *IEEE Trans. Electron Dev.*, "The physics of amorphous-silicon thin-film transistors," vol. 36(12), p. 2753 (1989).

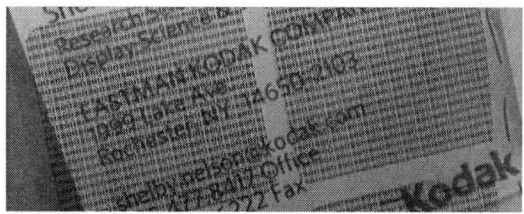

Fig. 1. Photograph of transparent ITO/Al$_2$O$_3$/ZnO sample on top of a business card. All that is visible is the aluminum of the contacts.

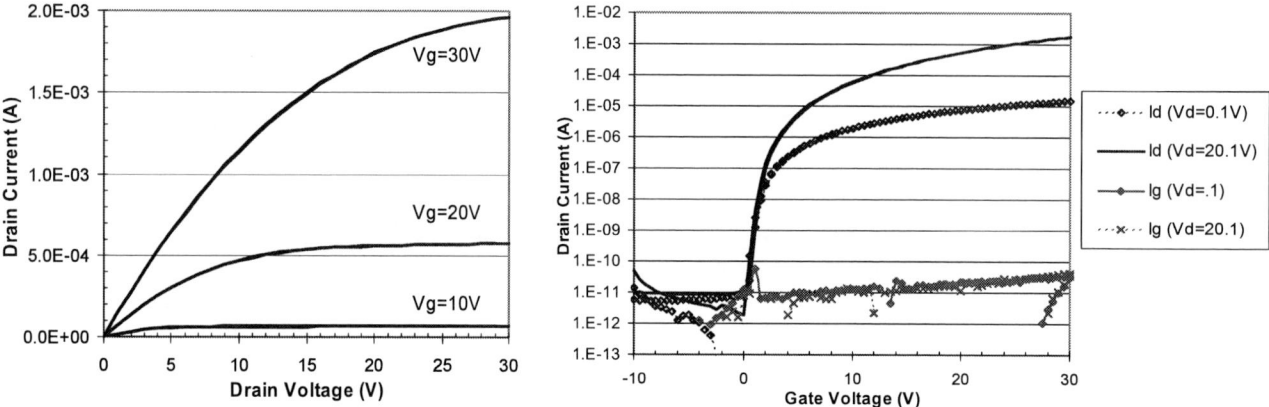

Fig. 2. Typical device and transfer characteristics for TFT with W/L = 600/100. The saturation mobility is 15.0 cm^2/Vs and the threshold voltage is 6.2 V.

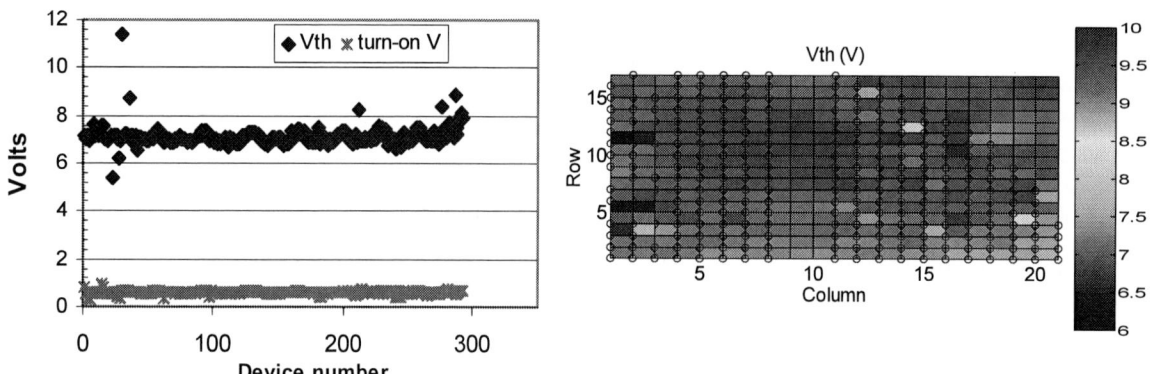

Fig. 3. Example of sample uniformity. In this sample, L = 50 μm, W = 600 μm. Over more than 300 devices, and including the obvious "outliers," Vth = 7.12 ± 0.43 V. The turn-on voltage is 0.6 V.

Fig. 4. Bias stability for 20 V gate bias, 10 V on drain, for 15 h. Three curves are shown; one before stress, one at 18,000 s, and the third at 54,000 s (15 h.) The shift in threshold voltage is 1.0 V.

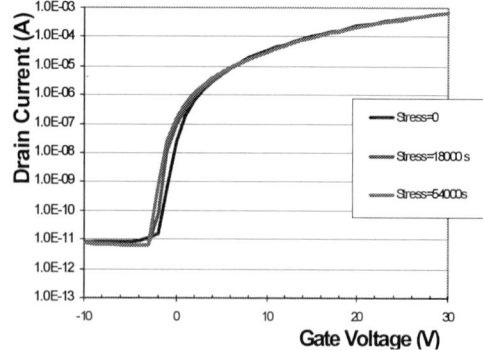

14

Flexible Microwave Single-Crystal Si TFTs with f_{max} of 5.5 GHz

H. Pang[a], H.-C. Yuan[a], Max G. Lagally[b], G. K. Celler[c] and Z. Ma[a,*]

[a] Department of Electrical and Computer Engineering, [b] Department of Materials Science and
Engineering, University of Wisconsin-Madison, Madison, WI 53706, USA,
[c] Soitec USA, 2 Centennial Dr., Peabody, MA 01960, USA
[*] Phone/Fax: (608) 261-1095 mazq@engr.wisc.edu

Microwave thin-film transistors on flexible plastic substrate using single-crystal Si as active channel material are reported in this paper. By employing a new TFT design, a maximum oscillation frequency (f_{max}) of 5.5 GHz, the highest reported speed for any devices on flexible substrate, has been achieved with an associated cut-off frequency (f_T) of 600 MHz. A maximum value of transconductance g_m of 22.28 μS with a gate length of about 1.5μm is measured. The degradation of g_m in large gate-length device due to poor Si/SiO interface is described.

Flexible, large-area electronics—macroelectronics—using amorphous silicon (α-Si), low-temperature poly-crystal silicon, or various organic and inorganic nano-crystalline semiconductor materials is beginning to show tremendous promise [1]. While flexible displays and photovoltaic industries are developing rapidly, high performance flexible electronics using a broad class of materials are demonstrated [2]. Recently, single-crystal Si TFTs made on plastic substrate have been reported with f_T and f_{max} of 1.9 and 3.1 GHz, respectively [3], providing great prospects for integrating RF systems into the macroelectronics. In this paper, we report TFTs with substantially reduced source access resistance for improved device speed (f_{max}).

The fabrication process that we developed to achieve flexible high-speed RF TFTs consists of both "hot process" applicable for Si substrate and "cold process" suitable for plastic substrate. A detailed process description can be found in a previous publication [3]. In this work, a 270-nm Si template layer on (001) SOI substrate was patterned into 30μm wide strips, followed by phosphorous ion implantation to create low-resistance source and drain regions. We employed a new layout design that provides overlap between gate and the source/drain regions in order to reduce the source-to-gate series parasitic resistances. The top-gate structure consists of 120 nm amorphous SiO, 30 nm Ti and 300 nm Au. Fig. 1 shows the optic-microscopic image of the transferred Si strips on plastic substrate. Figs. 2(a) and 2(b) show an optic-microscopic image of a finished TFT and an array of TFTs on a bendable plastic substrate, respectively.

The DC and small-signal properties of TFTs with same gate width (W_G) but different gate lengths (L_G) were measured. Figs. 3(a) and 3(b) show the transfer characteristics and I-V characteristics of a device with W/L as 100 μm/1.5 μm, respectively. The maximum g_m reaches 22.28 μS. The measured current gain (H_{21}) and available power gain (G_{max}) were plotted in Fig. 4 with extracted f_T and f_{max} of 600 MHz and 5.5 GHz, respectively, biased at V_G=1.4V and V_D=5V. The reduced f_T values (from previous 1.9 GHz [3]) are due to the increased gate-to-source capacitance (gate-to-source overlap to reduce source access resistance). Fig. 5 shows the dependence of f_T and f_{max} on drain current I_D, the peak values of which are both at I_D= 0.85mA, under V_G=1.4V and V_D=5V. The higher f_{max} is obtained owing to the reduced series resistances, with the consequence of increased parasitic source/drain-gate capacitances that reduces f_T. Such TFTs with low f_T and high f_{max} are more suitable for oscillators and power amplifier applications than those with high f_T and low f_{max}. Table 1 summarizes the flexible RF TFTs of different dimensions, where g_m, f_T and f_{max} increase as devices scale down. It is also found that the g_m decreases greatly as L_G is increased, which may be due to the poor Si/SiO surface property. As can be seen from the log-scale I_D curves in Fig. 6, the subthreshold swing (S) factor is observed to degrade from 1.17V/dec for 1.5μm device to 1.58V/dec for 3μm one. It indicates that even higher interface states in the larger gate-length devices limit the channel mobility hence the g_m.

This work was supported by NSF and AFOSR. The authors thank Paul Baude at 3M Company for the helpful discussions and suggestions.

[1] R. H. Reuss, et al., *IEEE Proceedings*, Vol. 93, No. 7, July 2005.
[2] J. Ahn, J. A. Rogers, et al., *Science*, Vol. 314, Dec. 2006.
[3] H. Yuan and Z. Ma, *Appl. Phys. Lett.*, 89, 212105, 2006.

978-1-4244-1101-6/07/$25.00 ©2007 IEEE

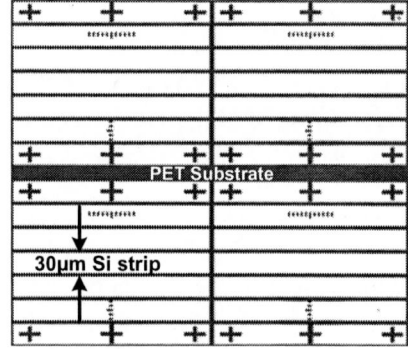

Fig. 1. Si strips transferred onto PET substrate.

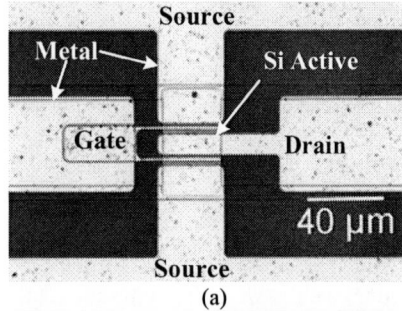

(a)

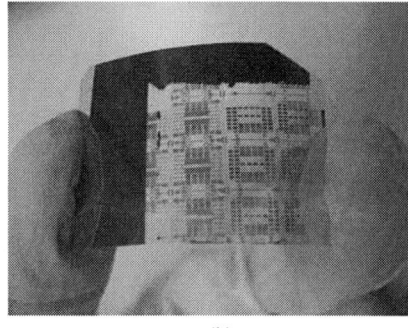

(b)

Fig. 2. (a) Optical microscopic image and (b) TFT array on bendable plastic substrate.

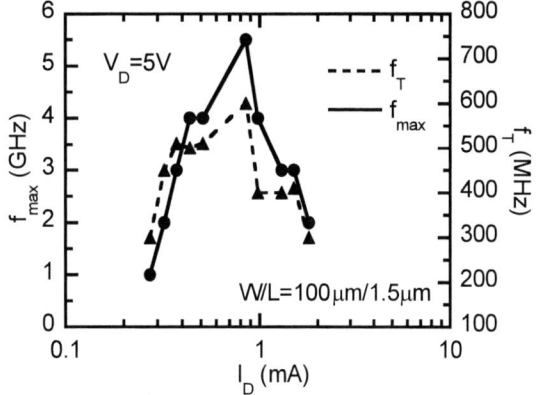

Fig. 5. f_T and f_{max} as a function of I_D.

Table 1. g_m, f_T and f_{max} of RF TFTs with different L_G.

L_G (μm)	g_m (μS)	f_{max} (GHz)	f_T (MHz)
1.5	22.28	5.5	600
2	15.67	1	350
3	7.48	0.7	<45

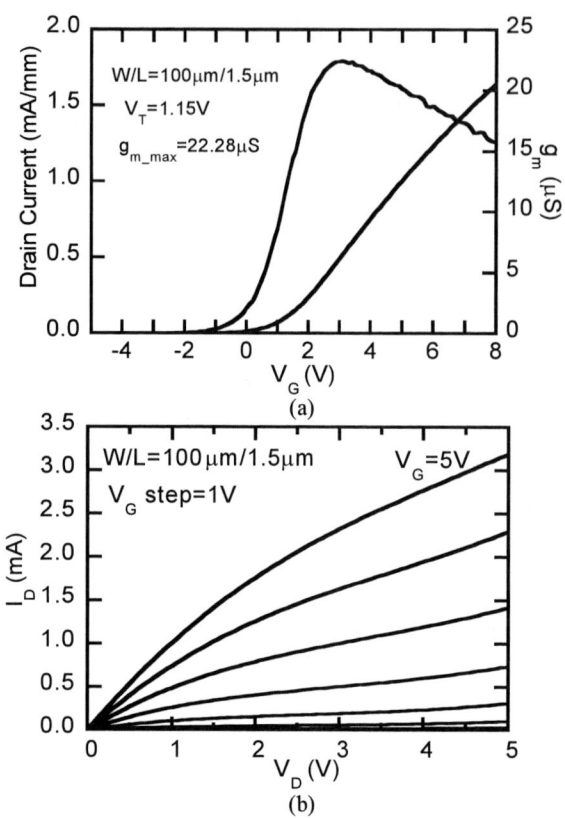

(a)

(b)

Fig. 3. (a) Transfer and (b) I-V characteristics of a RF TFT on plastic with W/L=100μm /1.5μm.

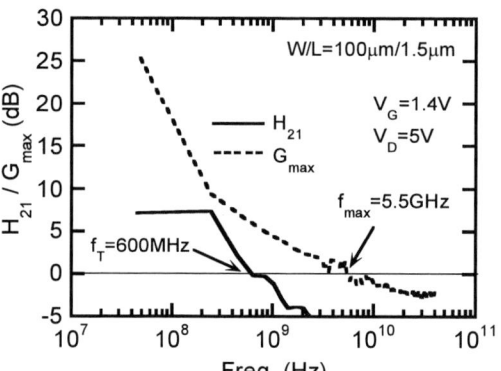

Fig. 4. Small-signal characteristics of a RF TFT on plastic.

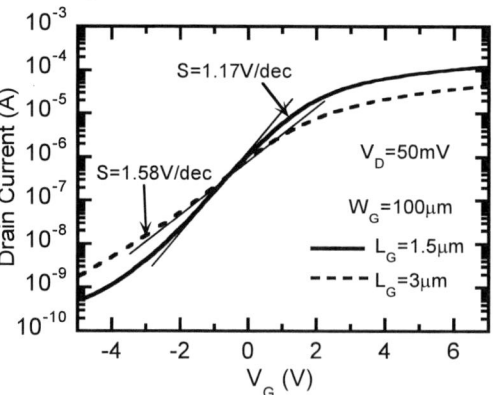

Fig. 6. Subthreshold I_D as a function of V_G.

A 'Bottom-up' Redefinition for Mobility and the Effect of Poor Tube-Tube Contact on the Performance of CNT Nanobundle Thin Film Transistors

Ninad Pimparkar* and Muhammad Ashraful Alam

School of Electrical and Computer Engineering, Purdue University, West Lafayette, IN 47907-1285, USA.

*Phone: (765) 409 3109   Fax: (765) 494 2706   Email: ninad@purdue.edu

Background: There have been many recent reports on Nanobundle Thin Film Transistors (NB-TFTs) based on percolating network of randomly-oriented Silicon nanowires (NW) and Carbon nanotubes (NT) or sticks in general (Fig. 1) with hopes of approaching mobility (μ) of *single* CNT/NW transistors (μ_I), without being limited by placement issues and low on current, I_{ON}. High-μ and highly homogenous NB-TFTs have potential to replace currently-dominant materials like amorphous (a-)Si or poly (p-) Si in applications in macroelectronics such as displays, e-paper, bio-chemical sensors, conformal radar, solar cells and others [1-4]. Puzzling, however, is the fact that the reported values of μ of NB-TFT (μ_{NB}) – calculated by traditional 'top-down' effective media approach (EMA) -- is not only far poorer than single CNT transistors, but also appears to be a random function of experimental conditions[2, 3, 7, 8]. In this paper, we show that (a) the randomness of μ_{NB} is not intrinsic, but rather signals the breakdown of 'top-down' definition μ_{NB} and a percolation-theory based 'bottom-up' definition of μ_{NB} can consistently interpret the results, and (b) the difference between μ_I and μ_{NB} can be attributed to geometrical parameters of transistor (!) such as tube density (D), tube length (L_S), channel length (L_C), etc. and tube-tube contact (C_{ij}). Our results not only provide specific guidance to achieve geometry-specific theoretical limits of μ_{NB}, but also suggest simple characterization of technology-critical C_{ij} from a few simple measurements.

Stick Percolation Model: We constructed a sophisticated first-principle numerical stick-percolation model for the above NB-TFTs by generalizing the random-network theory. The model [1] randomly populates a 2D grid by sticks of fixed length (L_S) and random orientation (θ) (Fig. 1) and determines I_{ON} through the network by solving the percolating electron transport through individual sticks. In contrast to classical percolation, the NB-TFT is a heterogeneous network: 1/3 of the CNTs are metallic and remaining 2/3 are semiconducting. Since, L_C and L_S are much larger than the phonon mean free path, linear-response transport (small V_{sd} and constant V_g obviates the need to solve the Poisson equation explicitly) within individual stick segments of this random stick-network system is well described by drift-diffusion theory[1]. The low bias drift-diffusion equation, $J = q\mu n \, d\varphi/ds$, when combined with current continuity equation, $dJ/ds = 0$, gives the non-dimensional potential φ_i along tube i as $d^2\varphi_i/ds^2 - C_{ij} (\varphi_i - \varphi_j) = 0$. Here, s is the length along the tube and $C_{ij} = G_0/G_1$ is the dimensionless charge-transfer coefficient between tubes i and j at their intersection point, and G_0 (~0.1 e^2/h)[5] and G_1 (= $qn\mu/\Delta x$)[1] are mutual- and self-conductance of the tubes, respectively. Here, n is carrier density, μ is mobility and Δx is grid spacing. The stick percolation networks are non-classical 2D conductor and satisfy the finite size scaling relationship [6] $I_{ON} \sim k\xi(L_C, L_S) = k/L_S(L_S/L_C)^m$ (Fig. 2d). Here, k is material specific constant and the exponent $m(D\,L_S^2)$ is universal constant that depends *only* on the areal tube density (D) and L_S (Fig. 2d). We now use this model to resolve the two puzzles of μ_{NB} discussed above:

(a) Puzzle of Randomness of Long-Channel μ_{NB}: In the literature, researchers universally use the 'top-down' definition of $\mu = (dI_D/dV_G/V_D)L_C/(L_W C_{OX})$ to characterize NB-TFTs (Fig. 3b). This μ is geometry-dependent and appears to reduce with lower density and longer L_C (Fig. 3b, blue squares)! Actually, μ for a percolative network should not be defined in this manner at all and a proper 'bottom-up' definition , $\mu_{NB} \sim (dI_D/dV_G/V_D) L_S/(L_C/L_S)^m /(L_W C_{OX})$, provides a device geometry independent value and allows a *comparison* between the mobilities from different labs[2, 7, 8] as shown in Fig. 3b and c. Not surprisingly, for high density networks (Fig. 3, red circles) with, $m \sim 1$, both the definitions give same result i. e. $\mu \sim \mu_{NB}$.

(b) Short and Long Channel Transistors and the role of C_{ij}: The second puzzle is that experiments often show that the μ_{NB} of the long channel ($L_C > L_S$) NB-TFT is almost an order of magnitude smaller than that of a short channel ($L_C < L_S$) device where the tubes bridge the source and drain directly[2, 4, 9]. Hence, we need to reanalyze the role of imperfect C_{ij} to explore differences between short and long-L_C transistors. Fig. 4 a, b and c show the simulated I_{ON} vs. L_C for short and long channel NB-TFTs for high and low values of C_{ij}. Note the increasingly abrupt reduction in I_{ON} at $L_C/L_S \sim 1$ with reduction in C_{ij} (Fig. 4a through c). The plot of the abrupt drop in current (R_{ij}) at the transition point as a function of C_{ij} for different tube-density, D in Fig. 4e shows that $R_{ij} \propto C_{ij}$ for low values of C_{ij} or poor tube-tube contact (Fig. 4b and c) but saturates to 1 for higher C_{ij} (Fig. 4a), or for good tube-tube contact, as expected. Finally, Fig. 4d plots the exponent m (see Fig. 2), for different densities are a function of C_{ij}. m increases monotonically with decreasing C_{ij} as the contribution of tube-tube resistance to the total device resistance goes up. The ratio of mobilities (μ_{NB}) for short ($L_C > L_S$) and long ($L_C < L_S$) channel NB-TFTs can be directly related to C_{ij} using Fig. 4e, resolving the puzzle discussed above. Moreover, once D is determined from SEM images and m from Fig. 2, C_{ij} can also be read out from Fig. 4d providing an experimental measure of a technology-critical parameter for design of NB-TFTs.

Conclusion: We have redefined the mobility for NB-TFTs from the 'bottom-up' perspective using the stick percolation model to allow direct comparison of NB-TFT mobilities across different labs[2, 7, 8] and with other competing technologies such as a-Si and p-Si. We have also suggested a simple experimental measure of the critical tube-tube contact parameter to allow design of optimized transistors.

References: [1] S. Kumar, et al., Phys. Rev. Lett. 95 (2005).[2] E. S. Snow, et al., Appl. Phys. Lett. 82, 2145 (2003). [3] L. Hu, et al., Nano Lett. 4, 2513 (2004). [4] S. J. Kang, et al., Nature Nano. (In Press). [5] M. S. Fuhrer, et al., Sci. 288, 494 (2000). [6] D. Stauffer, et al., *Intro. to percolation Theory* (1992). [7] C. Kocabas, et al., Nano Lett. (In Press). [8] S. H. Hur, et al., J. of the Ame. Chem. Soc. 127, 13808 (2005). [9] E. Artukovic, et al., Nano Lett. 5, 757 (2005).

978-1-4244-1101-6/07/$25.00 ©2007 IEEE

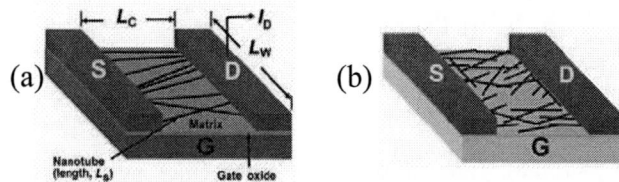

Fig. 1: Schematic diagrams for NB-TFT with (a) short channels ($L_C < L_S$, nanosticks directly bridge source S and drain D) and (b) long channels ($L_C > L_S$, electrons must percolate through the nanostick network to contribute to the drain current). L_C is channel length, L_W is channel width, L_S is stick length; S, D and G are source, drain, and gate electrodes, respectively.

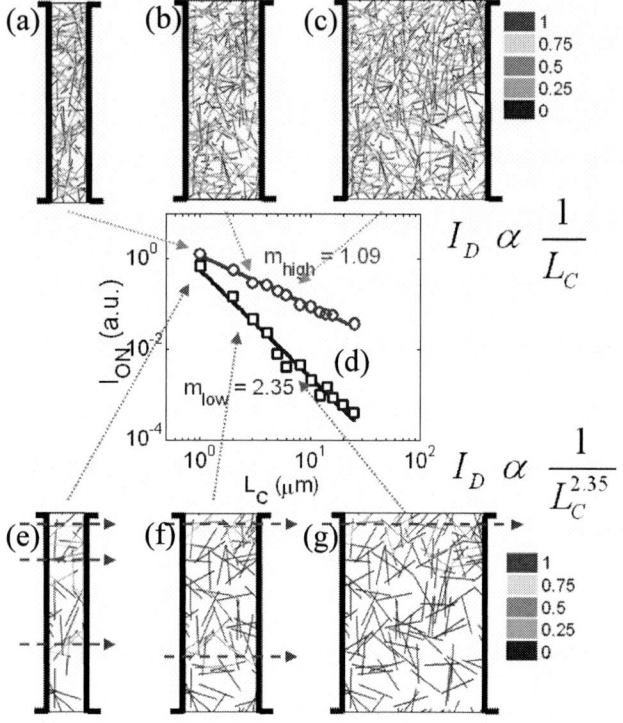

Fig. 2: Normalized current distribution for network with high (a, b, c) and low (e, f, g) density for $L_C/L_S = 1, 2, 4$, respectively. (d) I_{ON} vs. L_C plot for various tube densities (D). The symbols show experimental data from Ref. [3] and the lines show the simulations using the stick percolation model. The arrows indicate the density and L_C to which each of the samples represent. The equations show the current dependence on L_C with appropriate current exponent, m. A common color bar for all the figures is also shown. The dashed arrows in (e, f, g) show the current paths.

Acknowledgements: The authors would like to express their sincere gratitude towards S. Kang, C. Kocabas, and Prof. J. A. Rogers for helpful discussions. This work was supported by the Network of Computational Nanotechnology and the Lilly Foundation.

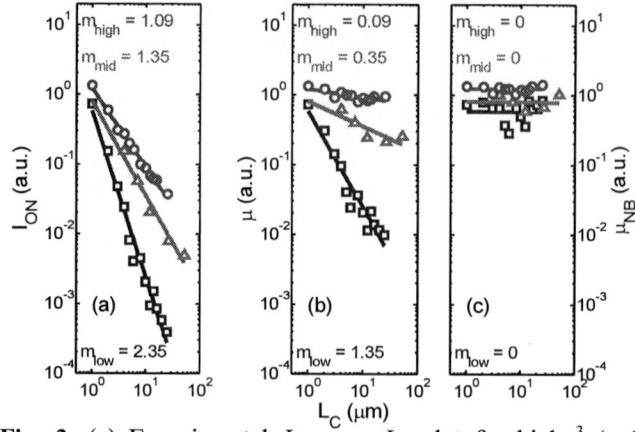

Fig. 3: (a) Experimental I_{ON} vs. L_C plot for high [3] (red), medium [4] (magenta) and low [3] (blue) density random network. The current exponent given by the slope of the lines are given by m = 1.09, 1.35, and 2.35, respectively. (b) Mobility (μ) vs. L_C plot for same networks using the conventional definition of mobility. The mobility is dependent on channel length with exponents, m = 0.09, 0.35, and 1.35, respectively. (c) The NB-TFT mobility from the 'Bottom-up' perspective vs. L_C.. Note that this mobility is independent of L_C.

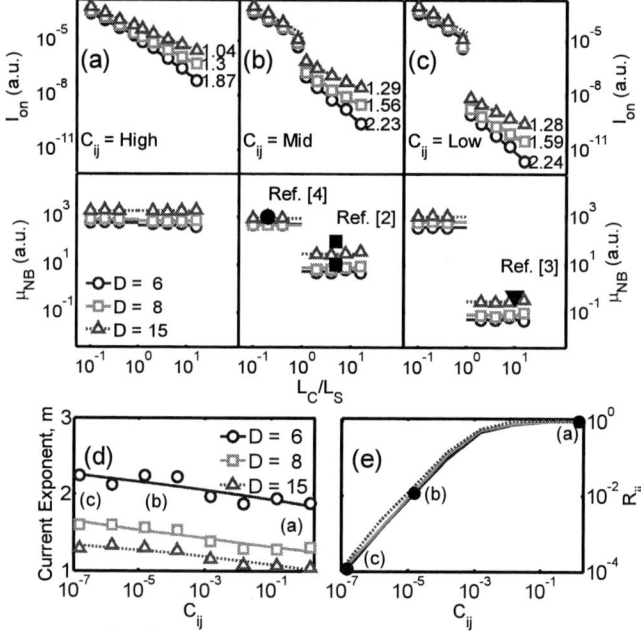

Fig. 4: Simulated I_{ON} vs. L_C/L_S plot (a) high (b) medium and (c) low tube-tube contact parameter, C_{ij}. The lower panel shows the corresponding mobilities (μ_{NB}). The black symbols in the lower panels of a, b and c map the reported mobility values from various labs as indicated by the reference numbers. Here we have assumed that all the tubes are same length (L_S) for different density (D) random tube networks. Note the abrupt transition in (b, c) at $L_C/L_S = 1$ which is a result of bad tube-tube contact. Also note that the current exponents, m, for $L_C > L_S$ change with C_{ij}. (d) Current exponent, m vs. C_{ij} for different density tube network. The exponents are relatively insensitive to the tube-tube contact, however (e) shows the normalized drop in current (R_{ij}) at the transition point $L_C/L_S = 1$. R_{ij} is insensitive to the tube density (D). The dark circles show corresponding to the figures a, b, c as indicated

Fast ZnO Thin-Film Transistor Circuits

Jie Sun[1], Devin Mourey[1], Dalong Zhao[1], Sungkyu Park[1], Shelby F. Nelson[2], David H. Levy[2], Diane Freeman[2], Peter Cowdery-Corvan[2], Lee Tutt[2], Thomas N. Jackson[1]

1. Center for Thin Film Devices and Materials Research Institute, Department of Electrical Engineering
Penn State University, University Park, PA 16802 USA
2. Eastman Kodak Company, 1999 Lake Avenue, Rochester, NY 14650 USA
Phone: 585-722-3450. E-mail: shelby.nelson@kodak.com

ZnO-based thin-film transistors (TFTs) have attracted considerable interest because of their unique characteristics [1]. TFTs fabricated using Al_2O_3 as the gate dielectric have demonstrated a mobility of 17.6 cm^2/V·s with a threshold voltage of 6 V [2]. Although there are numerous reports of ZnO TFTs, relatively few ZnO TFT circuit results have been reported. Recently, five-stage ring oscillators based on amorphous indium-gallium-zinc-oxide TFTs have been reported to operate at 2.2 kHz for a supply voltage of 30 V [3].

The dynamic properties of ZnO TFTs are of interest for use in applications such as active-matrix displays. Ring oscillators provide a simple vehicle for a first estimate of circuit dynamic performance. We report here five-stage ring-oscillator-integrated circuits fabricated using ZnO TFTs with signal propagation delays as low as 100 ns (>1 MHz oscillation frequency) for a 45 V supply voltage. These circuits also operate at a supply voltage as low as 10 V. To our knowledge, these are the fastest ZnO integrated circuits reported to date.

Our ZnO TFT circuits use a staggered, bottom-gate configuration (Figure 1a). These circuits were fabricated by first depositing 80 nm of Cr by ion-beam sputtering onto a glass substrate as the gate layer and subsequently patterning the Cr by wet etching. A 100 nm thick Al_2O_3 layer was then deposited on the Cr gate 200°C from trimethyl aluminum (TMA) and O_2. Next, a 20 nm undoped ZnO film was deposited on the Al_2O_3 from diethyl zinc and O_2. Patterning of the ZnO was done by wet etching in dilute HCl, and the aluminum oxide was etched in dilute HF. Aluminum source and drain electrodes were then deposited by thermal evaporation and patterned by lift-off.

Typical transistor characteristics were extracted from $\log(I_D)$ versus V_{GS} and $\sqrt{I_D}$ versus V_{GS}; at $V_{DS} = 30$ a field effect mobility of > 15 cm^2/V·s, a threshold voltage of 14 V, a sub-threshold slope less than 0.5 V/dec, and a current on/off ratio > 10^8 was calculated (Figure 1b).Figure 1c shows I_D versus V_{DS} characteristics for several V_{GS} for a device with channel length and width of 10 μm and 100 μm, respectively. Five-stage and seven-stage ring oscillators were fabricated with these ZnO TFTs. Five-stage ring oscillators (Figure 2a) with dimensions $L_{drive} = 5$ μm, $W_{drive} = 100$ μm, $L_{load} = 60$ μm, $W_{load} = 20$ μm, and beta ratio of 60 oscillated at 1.1 MHz at a supply voltage $V_{DD} = 45$ V (Figure 2b). The source/gate and drain/gate overlap for this circuit is 2 μm. The oscillation frequency and propagation delay as a function of V_{DD} is shown in Figure 2c. AIM-SPICE circuit-simulation results using a model fit to the discrete transistor characteristic show a reasonable match with the experimental data as shown in Figure 2c. Seven-stage ring oscillators with the same TFT load and drive transistor dimensions have ~480 kHz oscillation frequency at 35 V supply voltage; Figure 3 shows the speed-supply voltage characteristics for these oscillators. We also designed ring oscillators with different source/gate and drain/gate overlap to investigate aspects of the circuit speed. Figure 4 compares results for five-stage ring oscillators with 2 μm and 5 μm overlap. The results again are reasonably close to simulated circuits.

[1] R. L. Hoffman, B. J. Norris, and J. F. Wager, "ZnO-based transparent thin-film transistors," *Appl. Phys. Lett.,* vol. 82, p. 733 (2003).
[2] P. F. Carcia, R. S. Mclean, and M. H. Reilly, "High-performance ZnO thin-film transistors on gate dielectrics grown by atomic layer deposition," *Appl. Phys. Lett.,* vol. 88, p. 123509 (2006).[3] M. Ofuji, K. Abe, N. Kaji, R. Hayashi, M. Sano, et al., "Integrated circuits based on amorphous indium-gallium-zinc-oxide-channel thin film transistors," *Electrochem. Soc.,* vol. 602, 1596 (2006).

978-1-4244-1101-6/07/$25.00 ©2007 IEEE

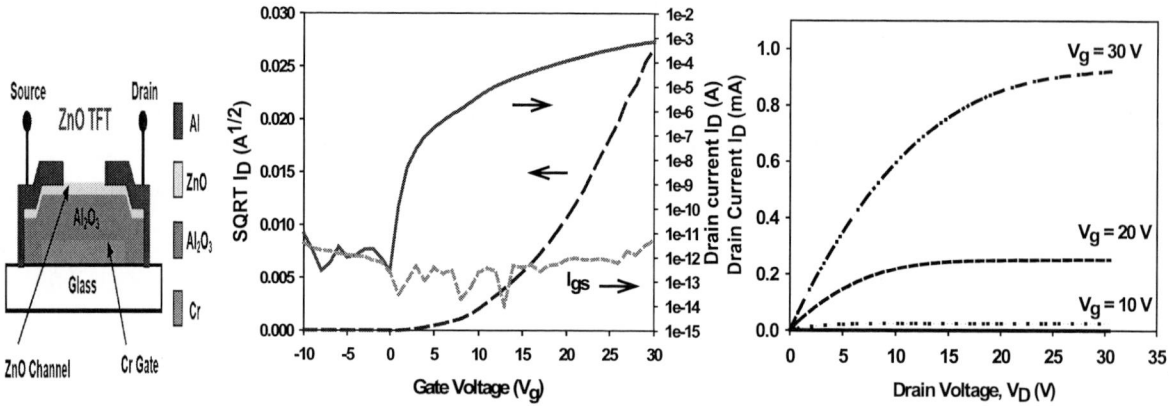

Fig. 1: (a) Bottom gate ZnO TFT structure, (b) ZnO TFT log(I_D) versus V_{GS} and $\sqrt{I_D}$ versus V_{GS} characteristics for V_{DS} = 30 V (W/L = 100 μm/10 μm, t_{ox} = 100 nm), (c) ZnO TFT I_D versus V_{DS} characteristics for several values of V_{GS} (W/L = 100 μm/10 μm, t_{ox} = 100 nm).

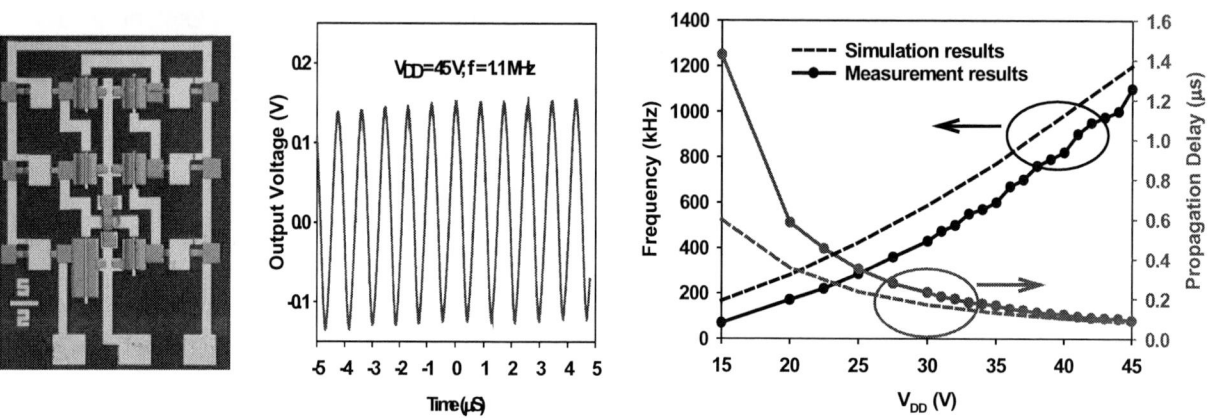

Fig. 2: (a) Photograph of a five-stage ring oscillator, (b) Output voltage of five-stage ring oscillator (L_{drive} = 5 μm, W_{drive} = 100 μm, L_{load} = 60 μm, W_{load} = 20 μm, beta ratio = 60, 2 μm source/gate and drain/gate overlap), operating at 1.1 MHz, (c) Oscillation frequency and propagation delay as a function of V_{DD} for both measurement results and AIM-SPICE simulation results.

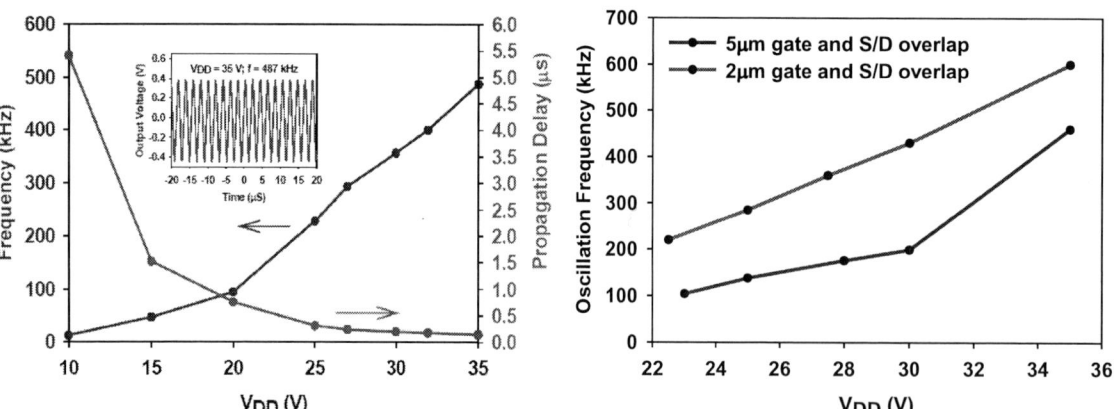

Fig. 3: Oscillation frequency and propagation delay as a function of V_{DD} for a seven-stage ring oscillator (L_{drive} = 5 μm, W_{drive} = 100 μm, L_{load} = 60 μm, W_{load} = 20 μm, beta ratio = 60, 2 μm source/gate and drain/gate overlap).

Fig. 4: Oscillation frequency as a function of V_{DD} for a five-stage ring oscillator with different source/gate and drain/gate overlap (L_{drive} = 5 μm, W_{drive} = 100 μm, L_{load} = 60 μm, W_{load} = 20 μm, beta ratio = 60).

Electrical and chemical stability of polymer transistors

R. A. Street and M. Chabinyc
Palo Alto Research Center, 3333 Coyote Hill Road, Palo Alto, CA 94304

Solution-deposited polymer thin film transistors (TFT) have mobility reaching 0.1-1 cm^2/Vsec and hence are of interest for display backplanes and related large area electronic devices. However, polymer TFTs must demonstrate long term performance stability in order to be useful. We describe the effects of both electrical stress and environmental exposure on polythiophene TFTs. Electrical stress measurements extending for nearly a year, show properties with time constants ranging from seconds to years. The threshold voltage shift increases as a power law in time and in gate voltage. However, after a few days of stress, the threshold voltage shift stabilizes because there is a thermally activated recovery mechanism. A different effect is a slow reduction in field effect mobility induced by the gate bias. At room temperature the reduction occurs with a time-constant of 1-2 years at 40V gate bias, but the effect is thermally activated and the time constant is considerably shorter at 70C. The physical mechanisms of the bias stress effects, and whether they are intrinsic to the polymer semiconductor, will be discussed.

Polymers are well known to absorb volatile molecules from the atmosphere and so chemical stability of TFTs is important to understand. Effects of exposure to the ambient, and to solvents and other volatile molecules are described. Water vapor and trace levels of ozone affect the TFT properties, but oxygen and nitrogen do not. Exposure to water and various solvents result in incorporation levels of about ~1%, and simultaneously increase the bias stress effect and reduce the mobility. We propose that molecular impurities cause an increase the local disorder and provide localized states that modify the electrical transport.

22

Solution processed OTFT circuits on plastic substrates

Sung Kyu Park[1], Sankar Subramanian[2], John Anthony[2], and Thomas N. Jackson[1]

1. Center for Thin Film Devices and Materials Research Institute,
Department of Electrical Engineering, Penn State University
2. Department of Chemistry, University of Kentucky, Lexington, KY 40506

Solution-processed organic thin film transistors (OTFTs) may allow high volume, low cost manufacturing approaches such as ink-jet printing and roll-to-roll processing. Poly (3-hexylthiophene) (P3HT) based solution-processed 5-stage ring oscillators with frequency of several kHz and 7-stage ring oscillators using soluble rubrene with frequency of 1.5 kHz (for V_{dd} = - 45 V) have been reported. We report here 7-stage and 15-stage ring oscillators using spin cast fluorinated 5,11-Bis(triethylsilylethynyl) anthradithiophene (F-TES ADT) on plastic substrates with maximum oscillation frequency of more than 22 kHz and minimum propagation delay less than 3.3 μs/stage (for 7-stage ring oscillators for V_{dd} = - 80 V). These circuits also oscillate with supply voltage as low as - 3 V and have stable operation at Vdd = -5 V with oscillation frequency of ~2.8 kHz.

For this work, transistors and circuits were fabricated on polyimide substrates using the device structure shown in Figure 1. Polyimide substrates were used for this demonstration because they provide the dimensional stability that allows small lithographically defined gate to source/drain overlaps to be simply defined. The ring oscillators use direct coupled inverters with no level shifting. Nickel (Ni) was used to form the gate electrode and silicon dioxide (SiO_2), deposited by reactive ion-beam sputtering at 80 °C, was used as the gate dielectric. Gold source and drain electrodes were deposited by thermal evaporation and patterned by lift-off. Prior to active layer deposition UV ozone cleaning was used to improve the metal/organic contact and device performance and a self-assembled monolayer of pentafluorobenzenethiol (PFBT) was formed on the Au source/drain electrodes. For the active layer, a 2 wt% F-TES ADT solution was spin-cast over the prepatterned circuit electrodes. The active layer was simply deposited by spin-casting in air (similar to photoresist application) with no high temperature post-deposition bake. On glass substrates the mobility for discrete OTFTs was typically 0.1 - 0.3 cm^2/V·s when spin-cast from toluene and 0.3 - 0.7 cm^2/V·s when spin-cast from chlorobenzene. The mobility for discrete OTFTs spin-cast from toluene on plastic substrates was typically 0.1 - 0.2 cm^2/V·s. All solution preparation and device processing steps were performed in an air ambient.

Figure 2a shows the output signal of a 7-stage spun F-TES ADT ring oscillator on a polyimide substrate with a supply bias of -80 V. For this circuit the inverter drive transistor has a channel length of 5 μm, $(W_{drive}/L_{drive})/(W_{load}/L_{load})$ = 2, and the gate to source/drain overlap is 2 μm. Figure 2b shows the oscillation frequency and propagation delay as a function of supply bias for 7-stage ring oscillators; minimum progagation delay is 3.3 μs/stage and the circuits operate for 5 V < |V_{dd}| < 80 V. Figure 3 shows the oscillation frequency as a function of supply bias for 15-stage ring oscillators with $(W_{drive}/L_{drive})/(W_{load}/L_{load})$ ratio of 5, 10, and 20. Because solution processed F-TES ADT OTFTs typically have a low subthreshold slope (as steep as 0.4 V/dec.) and threshold voltage, both 7-stage and 15-stage ring oscillators operate at supply voltage as low as - 3 V with stable, reproducible operation at -5 V. Figure 4 shows the output of a 7-stage ring oscillator on a polyimide substrate at a supply voltage of – 5 V.

1. S .K. Park, C.-C. Kuo, J. E. Anthony, and T. N. Jackson, "High-Mobility Solution-Processed OTFTs, *2005 International Electron Device Meeting Technical Digest,* pp. 113-6

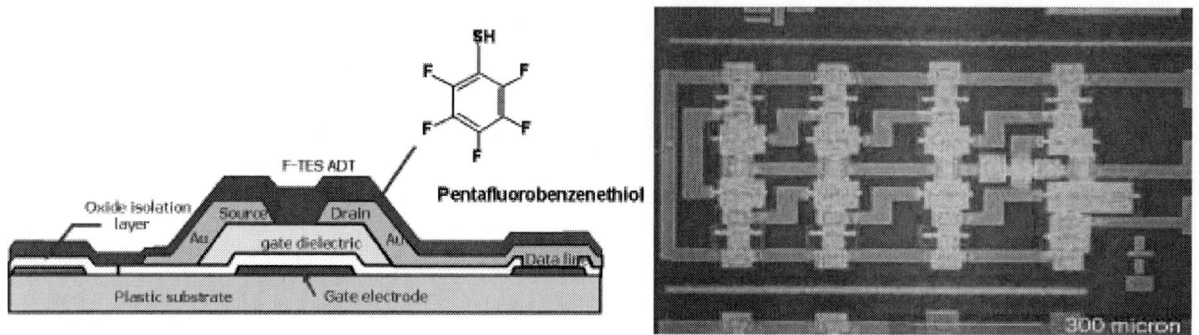

Figure 1:   Solution-processed OTFT structure and optical micrograph of 7-stage spin cast F-TES ADT ring oscillator on a polyimide substrate.

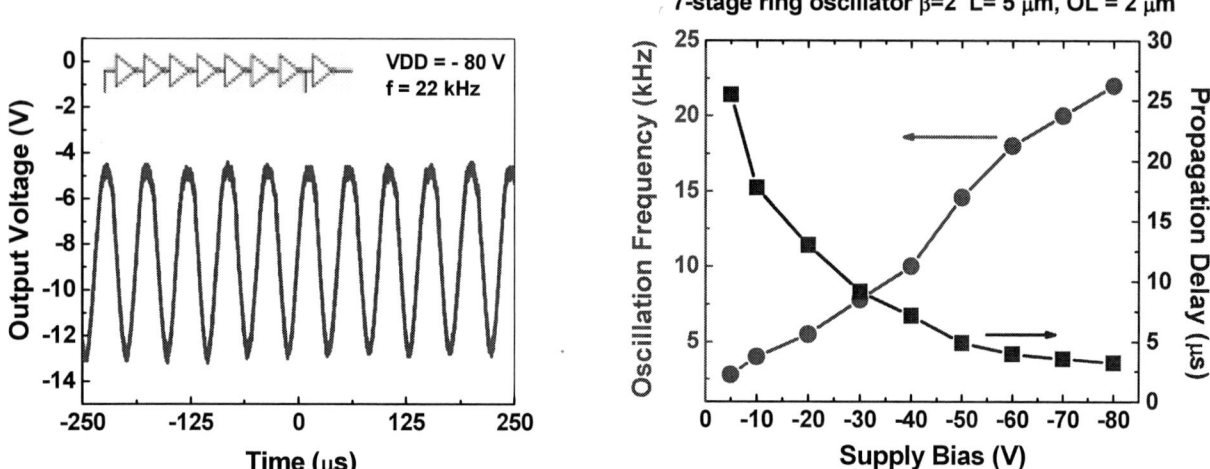

Figure 2.   (a) Output signal and (b) oscillation frequency and propagation delay of a 7-stage spin cast F-TES ADT ring oscillator on a polyimide substrate ($\beta = 2$, L = 5 μm, OL = 2 μm).

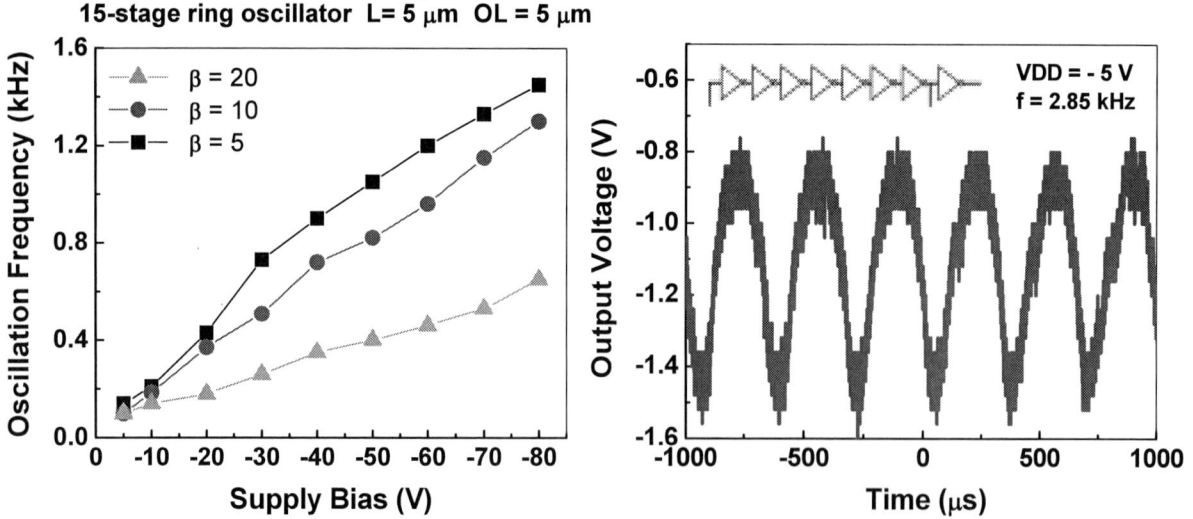

Figure 3.   Frequency versus supply voltage for 15-stage ring oscillators with different β ratios.

Figure 4.   7-stage ring oscillator output for $V_{dd} = -5$ V ($\beta = 2$, L = 5 μm, OL = 2 μm).

Session II.B (DeBartolo Hall Room 141)

Wide Bandgap Devices

Monday PM, June 18th, 2007

Session Organizer: John Zolper, DARPA/MTO and Mike Wojtowicz, Northrup Grumman
Session Chair: Mark Rosker, DARPA

1:30 PM II.B-1 Invited Paper
GaN Power Devices for Microwave/Switching Applications
D. Ueda, Semiconductor Device Research Center, Matsushita Electric Industrial Co., Ltd.,
Takatsuki, Osaka, JAPAN

2:10 PM II.B-2
**Novel 1 kV, normally-off, vertically integrated, dual-gate VJFET power switch with a
low 4.6 mOcm2 on-state resistance**
B. Nechay[1], E. Stewart[1], V. Veliadis[1], T. McNutt[1], H. Hearne[1], G. DeSalvo[1], C. Clarke[1],
and S. Scozzie[2], [1]Northrop Grumman Science and Technology Center, Linthicum,
Maryland, USA and [2]U.S. Army Research Laboratory, Adelphi, Maryland, USA

2:30 PM II.B-3
AlGaN/GaN HEMTs on Diamond Substrate
D. C. Dumka and P. Saunier, TriQuint Semiconductor, Inc., Richardson, Texas, USA

2:50 PM II.B-4 Student Paper
**10 W/mm and High PAE Field-plated AlGaN/GaN HEMTs at Ka-band with n+GaN
Source Contact Ledge**
J. S. Moon, P. Hashimoto, D. Wong, M. Hu, M. Antcliffe, C. McGuire, M. Micovic, P.
Willadsen, and D. Chow, HRL Laboratories LLC, Malibu, California, USA

3:10 PM Break

3:30 PM II.B-5 Invited Paper
Progress in GaN Performances and Reliability
P. Saunier[1], C. Lee[1], A. Balistreri[1], D. Dumka[1], J. Jimenez[1], H. Q. Tserng[1], M.Y. Kao[1],
P.C. Chao[2], K. Chu[2], A. Souzis[3], I. Eliashevich[4], S. Guo[4], J. del Alamo[5], J. Joh[5], and M.
Shur[6], [1]TriQuint Semiconductor Texas, Richardson, Texas, USA, [2]BAE Systems, Nashua,
New Hampshire, [3]II-VI, Incorporated, Pine Brook, New Jersey, USA, [4]IQE RF, Somerset,
New Jersey, USA, [5]MIT, Cambridge, Massachusetts, USA, and [6]RPI, Troy, New York,
USA

4:10 PM II.B-6 Student Paper
AlGaN/ GaN HEMTs with Large Angle Implanted Nonalloyed Ohmic Contacts
F. Recht[1], L. McCarthy[1], L. Shen[1], C. Poblenz[2], A. Corrion[2], J. S. Speck[2], and U. K.
Mishra[1], [1]Department of Electrical and Computer Engineering and [2]Department of
Materials, University of California, Santa Barbara, California, U.S.A

4:30 PM II.B-7
Self-Aligned AlGaN/GaN High Electron Mobility Transistors
V. Kumar, D-H. Kim, A. Basu, and I. Adesida, Micro and Nanotechnology Laboratory and
Department of Electrical and Computer Engineering, University of Illinois at Urbana-
Champaign, Urbana, Illinois, USA

4:50 PM II.B-8 Student Paper
**Analysis of lateral surface leakage in the vicinity of Schottky gates in AlGaN/GaN
HEMTs**
J. Kotani, M. Tajima and T. Hashizume, Research Center for Integrated Quantum
Electronics (RCIQE) and Graduate School of Information Science and Technology,
Hokkaido University, Sapporo, Hokkaido, JAPAN

5:10 PM II.B-9

Drain-to-Gate Field Engineering for Improved Frequency Response of GaN-based HEMTs

N. Pala[1,2], Z. Yang[2], A. Koudymov[2], X. Hu[1], J. Deng[1], R. Gaska[1], G. Simin[3] and M. S. Shur[2], [1]Sensor Electronic Technology, Columbia, South Carolina, USA, [2]ECSE Department, Rensselaer Polytechnic Institute, Troy, New York, USA, and [3]ECE Department, University of South Carolina, Columbia, South Carolina, USA

[Invited Paper]
GaN Power Devices for Microwave/Switching Applications

Daisuke Ueda

Semiconductor Device Research Center, Matsushita Electric Industrial Co., Ltd.
TEL:+81-72-682-7865, FAX:+81-72-685-6190, e-mail: daisuke@ieee.org

Provided with both high blocking voltage and high current handling capability, GaN-based device has been expected to be a viable alternative to GaAs, Silicon, and even to SiC ones in power electronics [1]. However, there still remains technological tasks to overcome in order to put them into practical use. Those are summarized as following three points; i.e., "Normally-off", "Collapse-free", and "Heat-release". In this paper, those points will be addressed by introducing new transistor named GIT (Gate Injection Transistor) [2] with relevant technologies.

GaN/AlGaN FETs reported so far can be operated usually as normally-on mode, which narrows the applicable field in power electronics. Though various types of normally-off FET structure have been proposed, most of them retain the trade-off between threshold voltage and available current. Fig. 1 shows the cross section of GaN/AlGaN GIT with and without positive gate bias. Under zero gate-bias condition, the channel under the gate is completely depleted due to the high potential barrier of P-AlGaN. When the p-gate is positively biased, first slightly opening the channel, and eventually minority carrier (holes) is injected to the channel where the same amount of electrons are generated. Since the effective mass of holes is approximately two order of magnitude larger than electron, injected holes play a role of slowly moving donors. Increased number of electrons drastically increase the available drain current, and thereby reduce the on resistance. It is noted that electrons injection to the gate can be suppressed like HBT by using wider bandgap AlGaN injection gate. Transfer characteristics of fabricated GIT on silicon are shown in Fig. 2, where you can see the drain current can be increased compared with that of MESFET. Observed second peak of gm is the evidence of minority carrier injection, which was also confirmed by the electroluminescence at 364nm corresponding to the bandgap of GaN. GIT is the combination of FET and HBT in lateral. Typical Gummel plot of GIT is shown in Fig. 3. The obtained current gain of GIT is approaching to the certain value, which expected to agree with the mobility ratio of electron and hole in GaN channel. The fabricated GIT achieved Ron·A of 2.6 milliohm·cm^2 with breakdown voltage of 820V, which is far lower than that of silicon limit and the lowest ever reported among those normally-off mode devices.

Recent concern to the GaN power device is the local temperature increase at the device surface caused by one-order magnitude higher product of current and voltage of GaN-based device. The simulation and micro-Raman measurement showed the channel temperature at the vicinity of gate edge exceeding 300 to 600 C depending on substrate material represented by Sapphire or SiC [3]. Anyway, such local heating may cause undesirable activation of traps or would limit the device's life by metallization failure. We developed novel passivation technique by using poly-AlN, whose thermal conductance is 300 times lager than SiN, to uniformize the surface temperature [4]. Fig. 4 shows the simulated temperature distribution obtained by the poly AlN passivation with different thicknesses. The temperature spread from the maximum to minimum is decreased as the increase of AlN film thickness. It is noted that the experimentally fabricated device with AlN passivation reduces the resultant on-resistance by the factor of two and increases drain current with 130%s, keeping complete suppression of current collapse.

GaN-based transistors distinguishes itself from the conventional ones as shown in Fig. 5. By using low cost technology like epitaxial growth over large-diameter silicon substrate, GaN device will replace those power transistors in switching as well as in microwave applications in not far distant future.

Reference
[1]. D. Ueda et al , Tech. Digests of IEDM, Dec. 2005, Page(s):377-380
[2]. Y. Uemoto, et al, Tech. Digests of IEDM, Dec. 2006, Page(s):35.2
[3]. T. Fujishima, et al., Proc. SPIE vol.6473, 647317, Feb. 2007
[4]. H. Ueno, et al, Digest Society Meeting IEICE, Sept. 2006, Kanazawa

978-1-4244-1101-6/07/$25.00 ©2007 IEEE

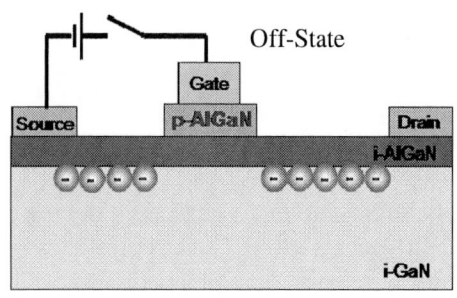 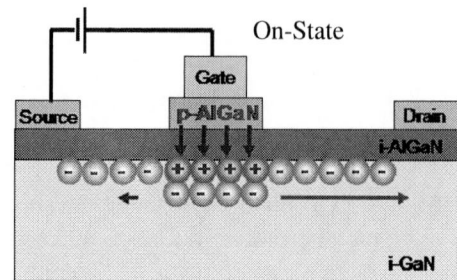

Fig.1 ALGaN/GaN SH GIT (Gate Injection Transistor)

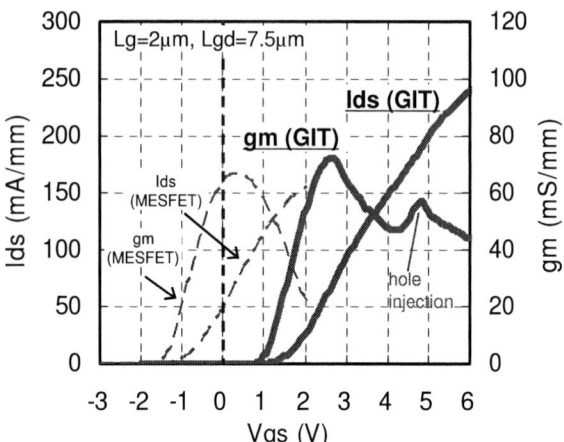

Fig.2 Vg-Id, Vg-gm Characteristics of fabricated GIT comparing with AlGaN/GaN FET

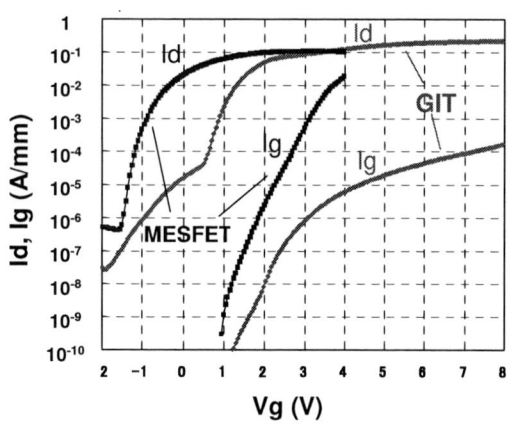

Fig.3(a) Gummel plot of fabricated GIT compared with GaN/AlGaN FET

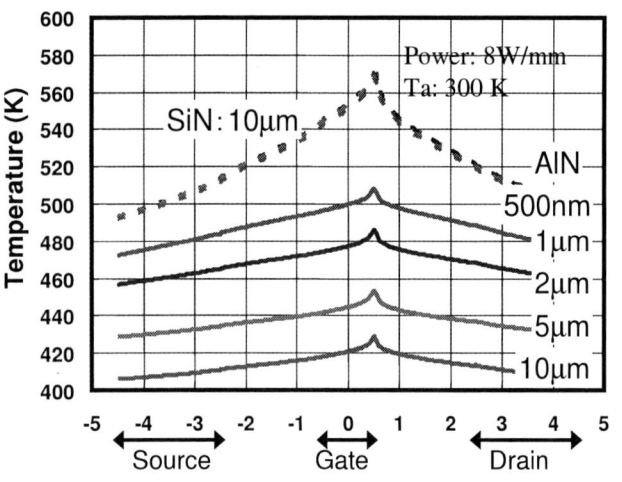

Fig.4 Simulated surface temperature distribution of AlGaN/GaN HFET on sapphire under the condition of power dissipation of 8W/mm.

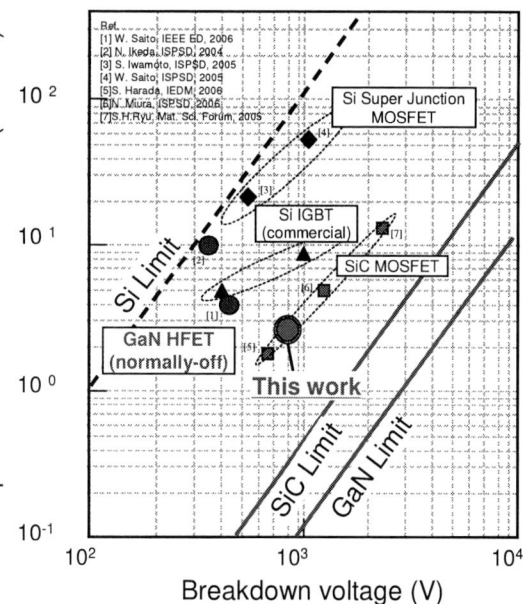

Fig.5 Comparison of GIT and other power devices reported so far in the chart of specific on-resistance vs. blocking voltage.

28

Novel 1 kV, normally-off, vertically integrated, dual-gate VJFET power switch with a low 4.6 mΩcm^2 on-state resistance

B. Nechay, E. Stewart, V. Veliadis, T. McNutt, H. Hearne, G. DeSalvo, C. Clarke, and S. Scozzie*

Northrop Grumman Science and Technology Center, MS 3B10, Linthicum, MD 21090
U.S. Army Research Laboratory, 2800 Powder Mill Road, Adelphi, MD, 20783-1197, USA
Email: bettina.nechay@ngc.com, Phone: (410)765-7744, Fax: (410)993-5530

Silicon Carbide (SiC) power transistors are ideally suited for power conditioning applications due to their high critical breakdown field and excellent thermal conductivity. The Vertical Junction Field-Effect Transistor (VJFET) possesses many advantages compared to other SiC power devices, including the lack of a critical SiC/SiO$_2$ interface, the speed of majority carrier switching, and the ability to easily scale current by paralleling multiple devices. Normally-off operation (highly desirable for power switching applications) can be achieved by adjusting the spacing between the p+ gate regions of the VJFET, Figure 1a. However, in the 1 kV range of interest, the required thin channel can result in prohibitively high on-state resistance. To remedy this problem, a low blocking voltage, normally-off device can be connected in series with a high blocking voltage normally-on VJFET (Figure 1b). Combining the two devices in either cascode or shorted-gate configuration results in a high voltage normally-off power switch with an overall lower on-resistance compared to a single normally-off VJFET. However, assembling a two device cascode switch adds cost, complexity, and introduces extra parasitics (e.g. wire-bond inductance).

In this abstract, Northrop Grumman presents a novel monolithically integrated 1 kV normally-off VJFET power switch that combines a normally-off VJFET section with a normally-on VJFET section in a single vertical structure. A schematic of the device is shown in Figure 1c. Normally-on and normally-off device sections are vertically integrated in the configuration of Figure 1b, with the drain of the normally-off device automatically connecting to the source of the normally-on device. The upper gates control the normally-off section while the lower-gates control the normally-on section. Depending on the specifics of the switching application, the device can be either controlled only by the upper gate in cascode configuration or be switched in shorted-gate configuration.

The advantages of this structure over a two device cascode switch include reduction in packaging complexity and parasitics and lower overall series resistance due the presence of a single substrate and source contact resistance. Compared to SiC MOSFET devices, this structure eliminates the need for a gate oxide, thus gaining an advantage in channel mobility and high temperature reliability. In comparison to other published JFET structures that include a lateral JFET component, fabrication of the monolithically integrated 1 kV VJFET normally-off power switch requires overall fewer processing steps, no epitaxial regrowth, and a smaller footprint of a single vertical device, which allows for high packing density and low specific on-state resistance. Also, all gate contacts are located at the surface, which eliminates the high resistivity issues associated with buried gate configurations.

Prototype vertically integrated 1 kV dual-gate normally-off VJFET power switches were successfully fabricated at Northrop Grumman on 3-inch diameter 4H-SiC wafers. The narrow upper normally-off section of the switch was reactive ion etched (RIE) and implanted with Al to form p+ upper-gate regions. The wide lower normally-on section of the switch was then RIE etched and p+ Al implanted. This was followed by implantation activation anneal. A Scanning Electron Microscope image of the device after ion-implantation is shown in Figure 2a. Passivation was then deposited and silicide contacts were formed on the backside drain, on the sources, and on the upper and lower gates and gate buses (Figure 2b). Lastly, thick field oxide was deposited, vias were etched and thick interconnect metal was deposited. The resulting switch blocks 927 V at zero volts gate-to-source bias (Figure 3a). The blocking is primarily limited by material and processing imperfections in the guard-ring edge termination. Typical on-state characteristics of the switch are shown in Figure 3b. A very low specific on-resistance of 4.6 mΩ cm^2 is measured at 2.8 V gate bias. This is significantly lower than the 7 mΩ cm^2 of a 1 kV two discrete VJFET device cascode switch. These measurements were taken in shorted gate configuration.

In conclusion, Northrop Grumman has successfully developed a novel SiC, vertically integrated, 1 kV, dual-gate VJFET normally-off power switch with a low 4.6 mΩ cm^2 on-state resistance.

978-1-4244-1101-6/07/$25.00 ©2007 IEEE

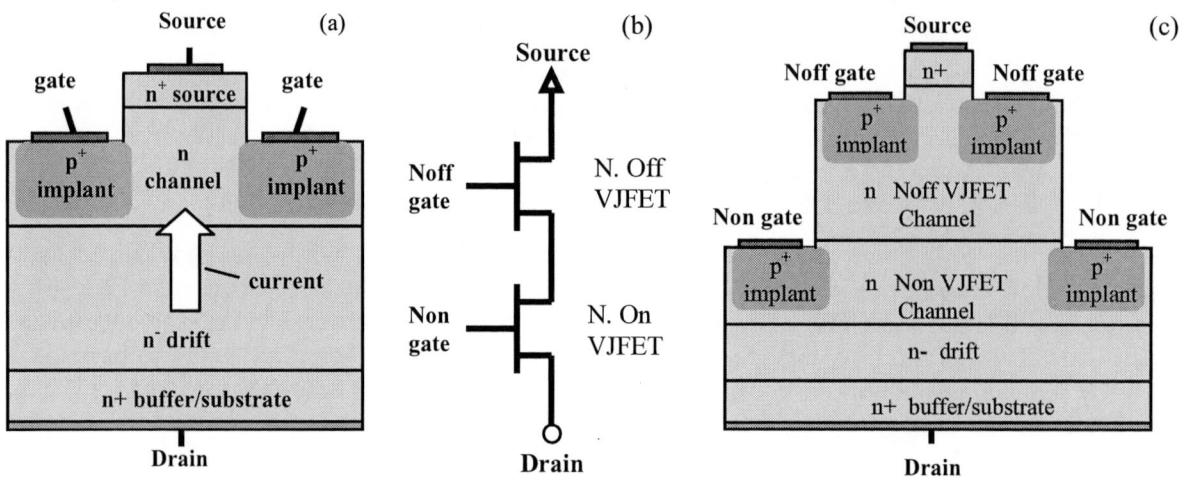

Figure 1: (a) Schematic of conventional VJFET, (b) two device normally-off switch schematic, combined in (c) a vertically integrated, dual-gate, 1 kV, normally-off VJFET power switch

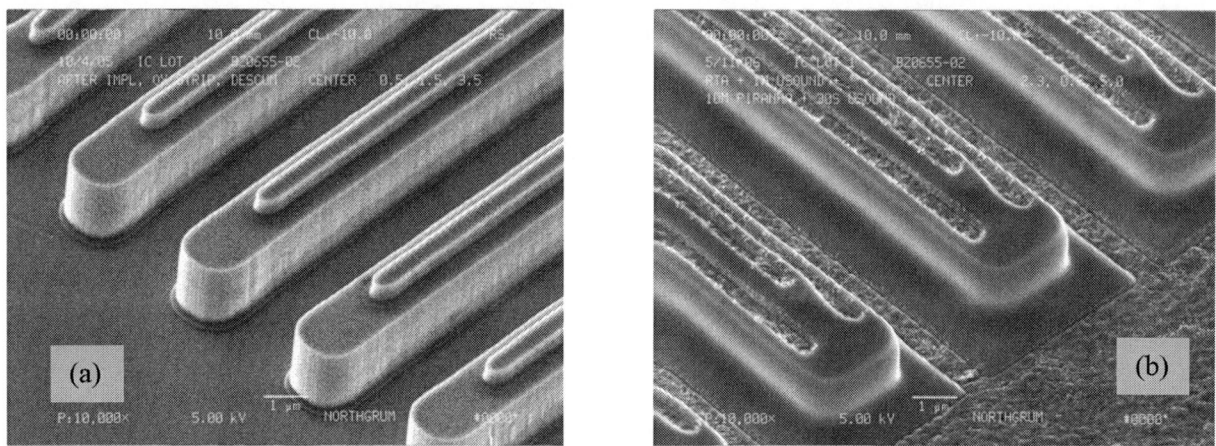

Figure 2: Integrated 1 kV, dual-gate, normally-off VJFET after (a) ion implantation and (b) silicide formation

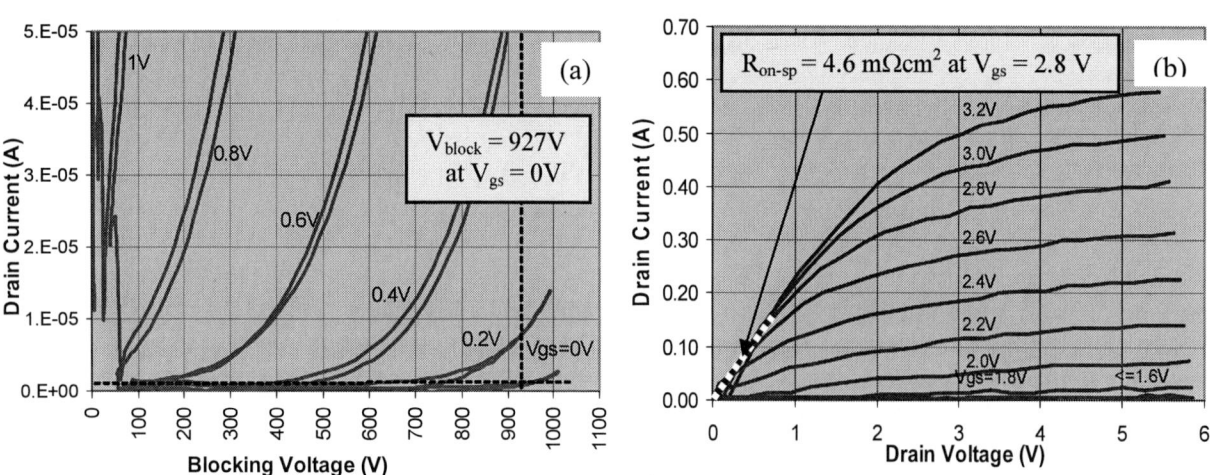

Figure 3: Integrated dual-gate VJFET shows (a) 927 V blocking voltage, and (b) 4.6 mΩ cm^2 on-state resistance

AlGaN/GaN HEMTs on Diamond Substrate

Deep C. Dumka and Paul Saunier

TriQuint Semiconductor, Inc., 500 W. Renner Road, Richardson, TX-75080, USA
Phone: +1-972-994-5609, Fax: +1-972-994-4537, E-Mail: ddumka@tqs.com

We present AlGaN/GaN high electron mobility transistors (HEMTs) fabricated on diamond substrate. Epitaxial AlGaN/GaN layers were first grown on high resistivity Si (111) substrate and transferred to polycrystalline diamond substrate, which was separately grown by chemical vapor deposition (CVD). As per our knowledge, these are the best-reported results for a transistor using GaN on diamond material.

AlGaN/GaN heterostructure is extremely promising for ultra-high power transistors [1] due to high charge density and breakdown voltage. Further, high saturation velocity and mobility of channel electrons make it attractive for high frequency applications [2]. While extremely high power transistors are being reported using GaN, the prime concern is to manage huge amount of heat generated by such devices. Single crystal SiC substrate which allows direct epitaxial growth and offers a high thermal conductivity is predominantly used for GaN HEMTs. High cost and unavailability of larger than 3-inch SiC substrates have led to cheaper substrate like Si where promising results have been reported [3] but thermal management is even more challenging due to poor thermal conductivity of Si. Diamond, being nature's most thermally conductive material, is the best substrate option for heat transfer. Thus, attempts to integrate GaN layers with diamond are expected. Further, artificially grown polycrystalline diamond may give a significant cost advantage over single crystal SiC substrate. GaN can not be grown directly on diamond. Therefore, growth of GaN films on cost-effective substrate like Si and then transferring the epi-films to diamond substrate seems a highly favorable approach.

GaN on diamond material for this work was supplied by Group4 Labs, California, USA. Group4 Labs has developed a pioneering technique to atomically attach epitaxial AlGaN/GaN layers on CVD diamond substrates. AlGaN/GaN films used here were originally grown on Si (111) substrate by metal-organic chemical vapor deposition at Nitronex, North Carolina, USA. Epi-structure on Si substrate consisted of a 1.3 μm semi-insulating (SI) GaN buffer, 10 Å AlN sub-barrier, 185 Å $Al_{0.27}Ga_{0.73}N$ Schottky layer and 15 Å GaN cap layer. To protect surface during epi-transfer to diamond substrate, 2300 Å SiN_x was deposited at 750 °C on GaN on Si wafers. Wafers were then mounted upside-down on a carrier and Si substrate was completely removed using wet etching. Revealed GaN buffer layer after Si-removal was then atomically attached to a 25-μm thick poly-crystalline CVD diamond substrate. This process required annealing at 810 °C. The carrier wafer from the frontside was subsequently removed. Fig. 1 shows the cross section of the starting material for device fabrication. Samples sizes were small ($< 2 \times 2$ cm^2).

First, 2300 Å SiN_x was completely removed using reactive ion etching (RIE). Manual lithography was used for patterning. Mesas were fabricated using BCl3-based dry-etch. TiAlNiAu stack was evaporated and rapid thermal anneal was done to realize Ohmic contacts. Source-drain spacing was 4.5 μm. Using transfer length measurements (TLM), contact resistance (R_c) of 0.48 Ω-mm and sheet resistance (R_{sheet}) of 478 Ω/Sq. were derived (Fig. 2). About 1.2 μm length gates were defined using photolithography and evaporated Pt/Au metal. DC and RF performances were measured. Drain current characteristics (Fig. 3) and transfer characteristics (Fig. 4) of 50 μm gate-width HEMTs, show excellent pinch-off performance, high current (800 mA/mm) and peak transconductance (180 mS/mm). Using Agilent network analyzer 8510C, s-parameters of 75 μm (2×37.5 μm) HEMTs were measured. Cut-off frequency (f_t) of 12.3 GHz and maximum oscillation frequency (f_{max}) of 21.8 GHz were extracted (Fig. 5) from s-parameters at drain voltage of 10 V. Also, f_t and f_{max} were measured at different drain voltages (Fig. 6) up to 20 V. Results presented here are significantly better than only published results [4] of AlGaN/GaN transistors on diamond substrate.

Development of GaN on diamond wafers used in this work was supported at Group4 Labs in part by SBIR funds from Missile Defense Agency managed by Air Force Research Laboratory (J. Blevins).

[1] Y.-F. Wu et al., *IEEE Electron Dev. Lett.*, vol. 25, p. 117 (2004)
[2] M. Micovic et al., *Int. Electron Dev. Meeting*, San Francisco, CA, USA, p. 425 (2006)
[3] D.C. Dumka et al., *Electronics Lett.*, vol. 40, p. 1023 (2004)
[4] G.H. Jessen et al., *Compound Semiconductor IC Symposium*, San Antonio, TX, USA, Late News (2006)

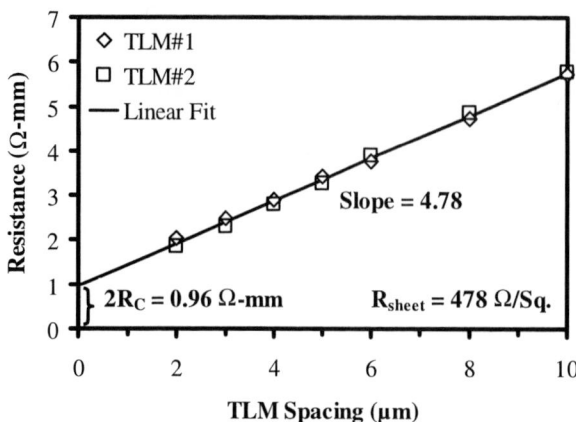

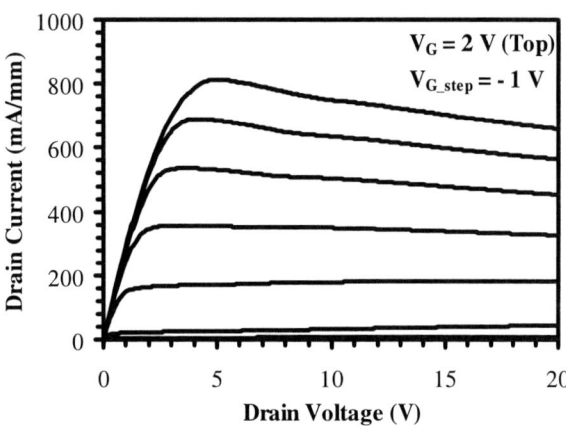

Fig. 1. *Cross-section of the starting material for fabrication of GaN HEMTs on diamond substrate.*

Fig. 2. *Measured resistance of two TLM structures on AlGaN/GaN HEMT layers on diamond. Derived R_C and R_{sheet} are shown in the figure.*

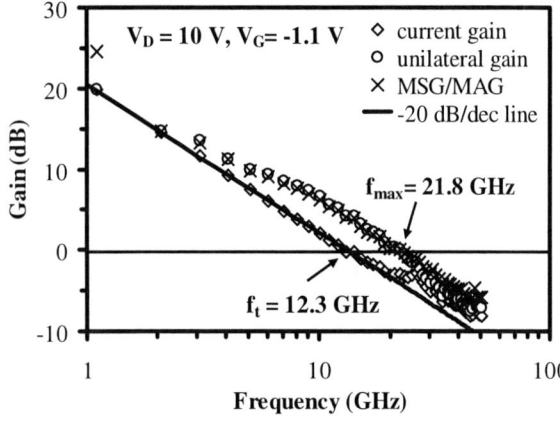

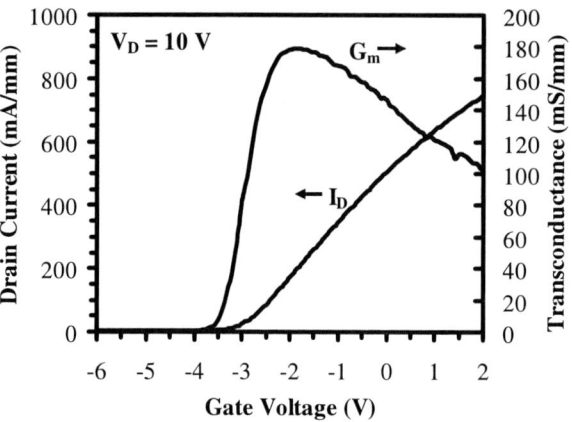

Fig. 3. *Drain current characteristics of a 50 μm gate-width AlGaN/GaN HEMTs on diamond substrate.*

Fig. 4. *Transfer characteristics of a 50 μm gate-width AlGaN/GaN HEMTs on diamond substrate measured at drain voltage of 10 V.*

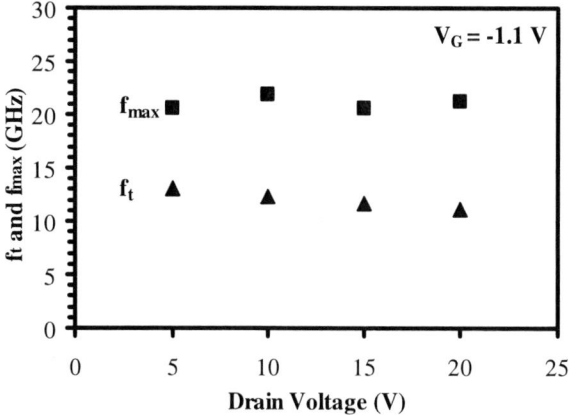

Fig. 5. *Small signal gain characteristics of a 75 μm (2×37.5 μm) gate-width AlGaN/GaN HEMT on diamond substrate.*

Fig. 6. *f_t and f_{max} vs. drain voltage of a 75 μm (2×37.5 μm) gate-width AlGaN/GaN HEMT on diamond substrate.*

10 W/mm and High PAE Field-plated AlGaN/GaN HEMTs at Ka-band with n+GaN Source Contact Ledge

J.S. Moon, P. Hashimoto, D. Wong, M. Hu, M. Antcliffe, C. McGuire, M. Micovic, P. Willadsen, and D. Chow

HRL Laboratories LLC, 3011 Malibu Canyon Road, Malibu, CA 90265, USA

Continuous progress in wide bandgap GaN/AlGaN-based HEMTs since 1993 [1] has pushed its frequency performance beyond X-band into millimeter-wave range. In particular, there has been growing interest of wide bandgap semiconductors for K-, Ka-, Q, and W-band [2] applications. While the early performances in these bands was obtained from GaN/AlGaN HEMTs fabricated with T-gates, recently demonstrated field-plated (FP) AlGaN/GaN-based HEMTs have shown dramatically improved RF power performance, where the FP devices were fabricated with reduced gate capacitance. Field-plated GaN/AlGaN HEMTs with deep-submicron gatelengths offered excellent large-signal performance with power density of 5.7 W/mm and PAE of 45 % at 30 GHz [3]. At 40 GHz, the CW power density of 10.5 W/mm was measured with PAE of 34 % [4]. The FP devices appear to enable reliable long-term operation. While the large-signal performance of AlGaN/GaN HEMTs is improved in the FP devices, their PAE, power density and linearity can be further improved if their high 2DEG sheet resistance, non-linearity in transconducance, access resistance and saturation velocity are addressed properly in device design.

We report small-signal and large-signal performance of 140 nm gatelength field-plated AlGaN/GaN HEMTs at Ka-band frequencies, in which the AlGaN/GaN HEMTs were fabricated with n+ GaN source ledge to reduce source access resistance such that the gate-to-drain feedback capacitance and breakdown voltage is not impacted. The resistance of parasitic channel is reduced by ~50 %, while the peak extrinsic transconductance is improved by 20 % from 370 mS/mm to 445 mS/mm with the n+ GaN source ledge of 0.75 µm, while the pinch-off voltage remained unchanged at -3 V. Both the saturated source-drain current, Idss (Ids at Vg = 0 V and Vds = 10 V), and maximum source-drain current, Imax (Ids at Vg = 2V and Vds = 10 V), were increased by ~23 % as the length of the n+ source ledge was increased from 0 µm to 0.75 µm. (See Figure 1) The small-signal extrinsic unity-gain cut-off frequency (f_T) increased by 10 % from 50 GHz in the case of no ledge to 55 GHz in the case of the source ledge of 0.75 µm. The AlGaN/GaN HEMTs with n+GaN source ledge exhibit improvement of maximum-stable gain by 0.5 dB at least over reference devices without n+GaN ledge. (See Figure 2) The improved frequency performance is attributed to a reduced parasitic charging delay, (Rs+Rd)*Cgd, where Cgd is a gate-to-drain feedback capacitance. Figure 3 and 4 show measured MMW power performance of field-plated AlGaN/GaN HEMTs with n+GaN source contact ledge using Maury loadpull setup at 30 GHz. At Vds = 42 V, the CW output power density of 10 W/mm is measured with PAE of 40 %. The device linear gain was 11.1 dB. At Vds = 30 V, the output power density is measured as 7.3 W/mm with PAE of 50 % and DE of 58 %, showing state-of-the-art CW large-signal performance for MMW applications.

This work was supported by he Office of Naval Research, monitored by Dr. Paul Maki.

[1] M. A. Khan et al., Appl. Phys. Lett., Vol. 63, pp 1214-1415, 1993.
[2] M. Micovic et al., IEDM Tech Digest, 2006
[3] J.S. Moon et al., EEE Electron Device Lett., Vol. 26, pp 348-350, 2005.
[4] T. Palacios et al., IEEE Electron Device Lett., Vol. 26, pp 781-783, 2005.
*Presenting and contact author: Jeong S. Moon
HRL Laboratories
3011 Malibu Canyon Rd. Malibu, CA 90265
Phone: (310) 317-5461, Fax: (310) 317-5485
Email: jsmoon@HRL.com

@ 2007 HRL Laboratories, LLC. All Rights Reserved.

Figure 1. (a) A schematic diagram of a field-plated AlGaN/GaN HEMT with an n+GaN source ledge is shown, (b) A plot of measured current-voltage curves (solid) from 2x100 μm AlGaN/GaN HEMT devices with the length of n+ ledge of 0.75 μm. The dashed and dotted lines are from devices with 0.55 μm n+ source ledge, and no n+ ledge, respectively.

Figure 2. Small-signal S-parameter characteristics of a 0.14 μm X 200 μm field-plated AlGaN/GaN HEMTs with n+GaN source ledge (line) and no n+GaN source ledge (circle), respectively. The devices were measured at Vds = 10 V, and peak gm bias.

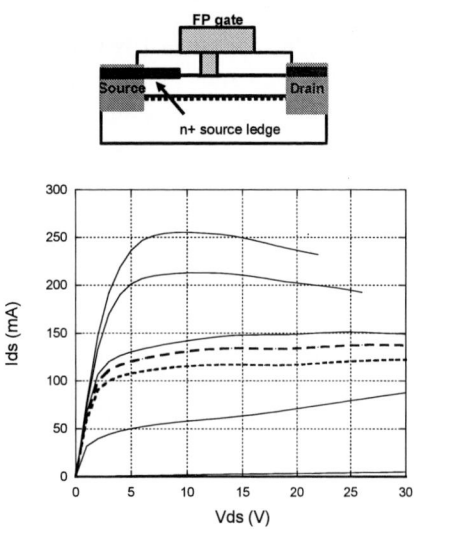

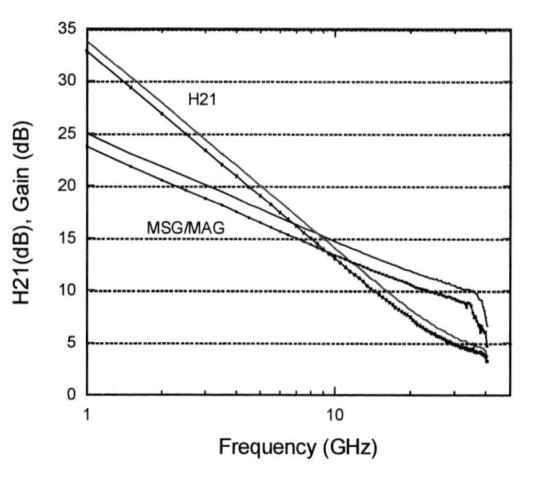

Figure 3. CW power performance of a 2 x 75 μm field-plated AlGaN/GaN HEMT with n+GaN source ledge at f = 30 GHz with Vds = 20 V (circle), 30 V (square), and 40 V (diamond). The power performance was tuned for PAE in deep class AB bias.

Figure 4. Measured power density, linear gain, PAE, and DE versus source-drain bias at f =30 GHz from 2 x 75 μm field-plated AlGaN/GaN HEMTs with n+GaN source ledge.

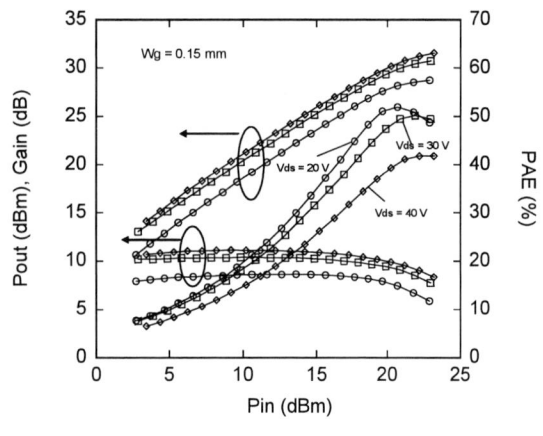

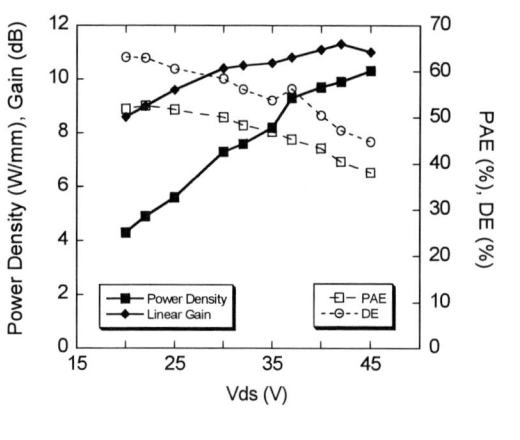

@ 2007 HRL Laboratories, LLC. All Rights Reserved.

Progress in GaN Performances and Reliability

P. Saunier[1], C. Lee[1], A. Balistreri[1], D. Dumka[1], J. Jimenez[1], H. Q. Tserng[1], M.Y. Kao[1], P.C. Chao[2], K. Chu[2], A. Souzis[3], I. Eliashevich[4], S. Guo[4], J. del Alamo[5], J. Joh[5], M. Shur[6]

1. TriQuint Semiconductor Texas, 500 Renner Road Richardson, TX 75080; psaunier@tqs.com
2. BAE Systems, Nashua, NH; 3. II-VI, Incorporated, Pine Brook, NJ; 4. IQE RF, Somerset, NJ;
5. MIT, Cambridge, MA; 6. RPI, Troy, NY

Abstract: *With the DARPA Wide Bandgap Semiconductor Technology RF Thrust Contract, TriQuint Semiconductor and its partners, BAE Systems, Lockheed Martin, IQE-RF, II-VI, Nitronex, M.I.T., and R.P.I. are achieving great progress towards the overall goal of making Gallium Nitride a revolutionary RF technology ready to be inserted in defense and commercial applications. Performance and reliability are two critical components of success (along with cost and manufacturability). In this paper we will discuss these two aspects.*

Keywords: AlGaN/GaN; GaN; MMIC; DARPA MTO

Introduction: the GaN advantages and the DARPA WBGS-RF program

The fundamental physical parameters of GaN and GaAs are well known. Much larger bandgap (3.4eV vs 1.34eV), ten times higher breakdown field (3500kV/cm), and similar or higher electron saturated velocity for GaN result in same size devices having 5- to more than 10-times the power at bias voltages of 30V to >50V [1, 2]. Critical is the excellent thermal conductivity of the recipient SiC substrate which allows effective use of the available power density.

Within this framework, the TriQuint Team of BAE Systems, II-VI Incorporated, IQE RF (formerly Electronic Materials Division of EMCORE), Lockheed Martin, M.I.T., Nitronex and R.P.I is pursuing the goal of producing reliable, reproducible, high-performance devices, and demonstrating a 100W 2-20GHz module operating at 48V with 30dB gain and 20% PAE. The critical program milestone metrics is shown on Table 1.

Table 1: Program Milestone Metrics

Metric	Unit	18-month	36-month	60-month
Unit	n/a	Device	Device	Module
Drain Bias	V	28	40	48
Size	μm	400	1250	n/a
Frequency	GHz	8-12	8-12	2-20
Pout	W	2.6	7.9	100
PAE	%	60	60	20
Gain at power	dB	10	12	30
RF yield	%	50	50	n/a
Pout uniformity	dB	1.5	1	n/a
PAE uniformity	% pts	6	3	n/a
SS gain uniformity	dB	1.5	1	n/a
Long Term Performance	hrs	1E5 @150C	1E5 @150C	1E6 @150C

Since our operating frequency goal extends to 20GHz, we have added a goal for a 400μm device to achieve 6.6W/mm at 40V with 10dB gain and 50% efficiency, with the same yield, uniformity and reliability goal as in Table 1 within 36 months. This milestone is extended to a 400μm device operating at 48V with 9dB gain and 50% PAE as a baseline device requirement to meet the 100W module goals at the end of the program.

Approach and Recent Progress

The requirements for the material come from the device performance, reliability, yield, uniformity requirements.

Uniformity starts with the substrates, and II-VI Incorporated has been developing a new substrate growth process to improve lattice tilt, micropipe density, edge quality, crystal quality and expand substrate diameter. The edge to edge lattice curvature is consistently less than 0.5^0 and usable diameter ~75mm.

IQE RF has achieved excellent uniformity sheet resistivity (standard deviation of less than 1% over 3" wafers both on 4H and 6H material) and reproducibility, wafer to wafer and run to run.

Most important also is the successful development of a suitable buffer layer: the requirements are excellent isolation and pinch-off without trapping which would induce current collapse; we have been using very successfully iron doped-buffer layers. For higher frequencies where a sharp pinch-off must be preserved with small gates, it is necessary to have electron confinement: our AlGaN buffer layers are providing very high gain.

Device structure and Process are of course key to achieving all the goals.

We have chosen a dual field-plate structure. The first field plate is the "T-cap" of the gate and the second field plate is connected to the source of the device. As reported by several other companies and universities, a significant improvement has come from the adoption of that second field plate resulting in increased gain (0.5- to 1dB) higher PAE (5 percentage points) and higher power (0.5dB).

On all the seven wafers from two recent successive lots with the same structure and process, the median 10GHz PAE of each wafer is greater than 63% at 30V. This result is shown on Figure 1 for dual field-plate devices.

978-1-4244-1101-6/07/$25.00 ©2007 IEEE

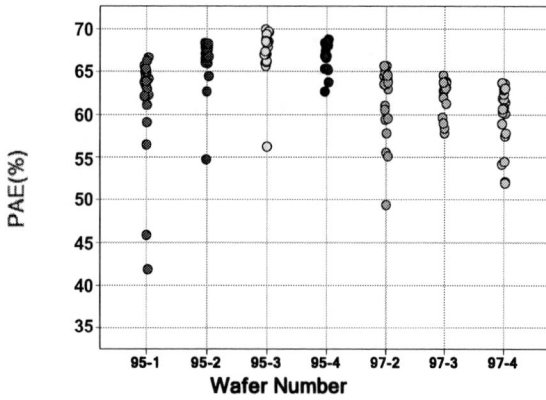

Figure 1: 10GHz PAE for all seven wafers from 2 lots

We are now able to demonstrate very good results at 40V, and we have achieved excellent preliminary 10GHz results at 48V. Figure 2 shows the compression curve of a 400μm device having 9.8W/mm power density with 11.7dB gain and 55% PAE.

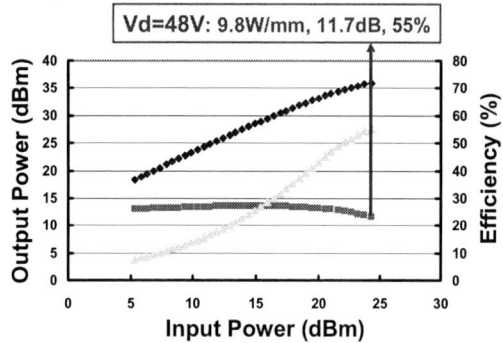

Figure 2: 400μm compression curve at $V_d = 48V$

At 20GHz and 40V we have demonstrated 7.1W/mm with 45% PAE and 7.5dB gain.

Achieving suitable reliability is the key to the viability of the GaN technology. The TriQuint Team has made outstanding progress in characterizing and understanding the failure mechanisms of GaN HEMTs. Jesus Del Alamo at MIT has greatly refined our initial findings leading to a strain related theory of degradation that is driven by electric fields [3]. Degradation can occur on the drain edge of the gate due to excessive strain given by inverse piezoelectric effect.

We have completed three, two-temperature RF life tests on baseline devices. Temperatures were chosen sufficiently far apart to account for uncertainties in the test. Devices were biased at around 5.5W/mm at 28V and the base plate temperatures were adjusted to achieve the desired junction temperatures. Devices were pre-stressed for 24-hours prior to the life test to eliminate initial transients. The results [4] show consistent behavior with estimated activation energy of

1.05eV and predicted MTTF of over 1E5 hours at 150⁰C. Figure 3 shows the results of the three RF life tests.

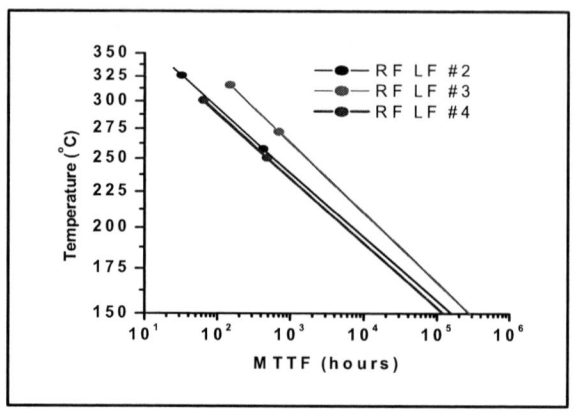

Figure 3: RF life test results

We are also aggressively pursuing modeling of the reliability (with RPI). A phenomenological model linking the current collapse observed on the RF degraded devices and defect concentration near the device channel is being developed. The core of the model is a field dependent trap induced changes in series resistances. Three independent field regions are considered: drain-edge of the gate, drain edge of the field plate and source edge of the gate. The model explains the stronger I_{max} degradation observed by MIT on Vds=0 and large Vgs when compared to RF power soak, DC and RF life test experiments. The model has been fitted to TriQuint baseline structure current collapse data and will be used to correlate powersoak DC and RF degradation.

Acknowledgements

We would like to recognize our COTR, Dr. Alfred Hung of the Army Research Laboratory in Adelphi, MD and the DARPA Program Manager, Dr. Mark Rosker. This work is performed under ARL Contract #W911QX-05-C-0087.

References

1. Y-F Wu, et al., "30W/mm GaN HEMTs by Field Plate Optimization," IEEE EDL, vol.25, pp. 117-119, Nov. 2004.

2. R. Thompson, et al., "Performance of the AlGaN HEMT Structure with a Gate Extension," IEEE Transactions on Electron Devices, vol. 51, no. 2, pp. 292-295, Feb. 2004.

3. J. Joh, Jesus Del Alamo et Al. "Mechanisms for Electrical Degradation of GaN HEMTs," IEDM, San Francisco, CA, Dec. 2006.

4. J. Jimenez, "Failure Analysis of X-band GaN FETs," CSICS ROCS, San Antonio, TX, Nov. 2006.

AlGaN/ GaN HEMTs with Large Angle Implanted Nonalloyed Ohmic Contacts

F. Recht[1], L. McCarthy[1], L. Shen[1], C. Poblenz[2], A. Corrion[2]
J. S. Speck[2] and U. K. Mishra[1]

[1]Department of Electrical and Computer Engineering [2]Department of Materials
University of California, Santa Barbara, U.S.A.
Email: recht@ece.ucsb.edu Tel:(805)893-3812

AlGaN/ GaN high electron mobility transistors (HEMTs) have a emerged as the preferred technology for next-generation radar and communications systems because of their exceptional high-power and high-frequency capability. Further improvement in device performance requires minimization of parasitic leakage currents, capacitances and source/drain (S/D) access resistance. In GaN based devices, ion implantation is one of the most promising technologies to reduce these resistances.

Nonalloyed ohmic contacts by ion ionimplanation have typically shown contact resistances in excess of 0.4 Ωmm to the two dimensional electron gas (2 DEG) [1].

In this work we report AlGaN/GaN HEMTs with nonalloyed ohmic contacts by large angle ion implantation with a contact resistance to the channel of 0.2 Ωmm.

The HEMT structures in this work were grown by MBE on SiC substrates. A 30 nm $Al_{0.28}Ga_{0.72}N$ barrier was grown on an unintentionally doped GaN buffer. Silicon ions at a dose of 5×10^{15} cm^{-2} were implanted in the S/D regions at 50 keV and room temperature. The sample was tilted ±40° relative to the ion source to reduce the resistance R_3 between the implanted area and the 2 DEG [2].

The activation annealing of the implanted species has been performed without the need of a capping layer at 1280 °C in a metal-organic chemical vapor deposition system, flowing ammonia and nitrogen at atmospheric pressure. To improve the contact resistance, the AlGaN layer was removed in the S/D contact regions. Ti/Au/Ni contacts were deposited onto the underlying implanted GaN. The edge of the S/D metallization was 0.3 mm away from the implantation edge [2].

Power measurements at 4 GHz and a drain bias of 30 V and a quiescent current of 116 mA/mm showed a gain G_T of 15 dB and an output power density P_{out} of 6.8 W/mm with a PAE of 72%. The small signal measurements on the devices (L_g=0.7 mm) showed an f_T and f_{max} of 21 GHz and 55 GHz, respectively.

These results demonstrate the first high efficiency AlGaN/GaN HEMTs with nonalloyed ion implanted ohmic contacts.

[1] Yu et al., *IEEE Electron Device Letters, Vol. 26, No. 5, May 2005*
[2] Recht et al., *IEEE Electron Device Letters, submitted Jan 2007*
[3] Recht et al, *IEEE Electron Device Letters, Vol. 27, No. 4, April 2006*

This work has been partially funded by by DARPA (Dr. C. Bozada and Dr. M. Rosker) and the ONR MINE MURI project (contract monitor: Dr. H. Dietrich and Dr. P. Maki).

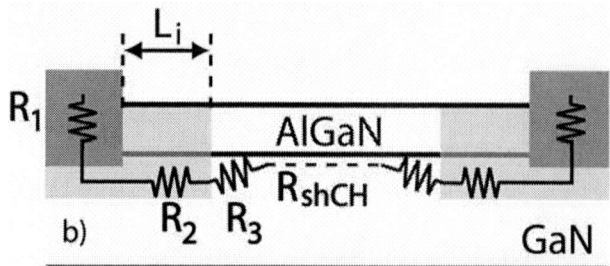

Figure 1: Resistor Model of the nonalloyed ohmic contacts for the contact resistance $R_C=R_1+R_2+R_3$. R_1 is the resistance between the metal and the implanted region, R_2 the resistance of the implanted region and R_3 the resistance between the implanted region and the channel R_{shCh} is the sheet resistance of the two dimensional electron gas and L_i is the distance between the contact metal edge and the implantation edge [2].

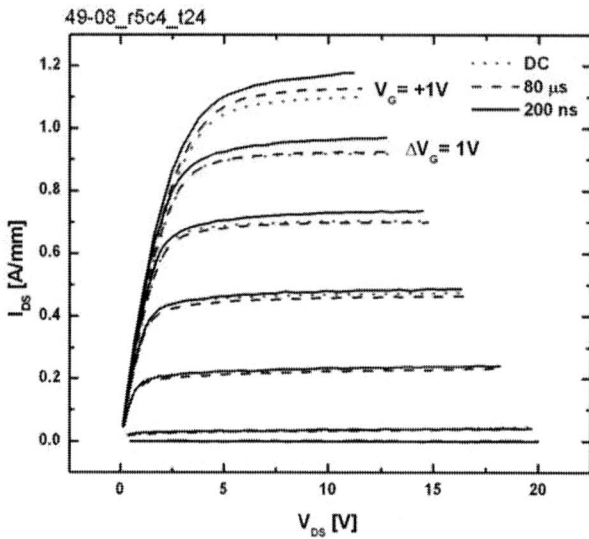

Figure 2: DC and 200-ns current–voltage curves of the implanted HEMT. The gate voltage was varied between −6 and +1 V in steps of 1 V, V_{DSMAX} = 20 V. The load line is 50 Ω.

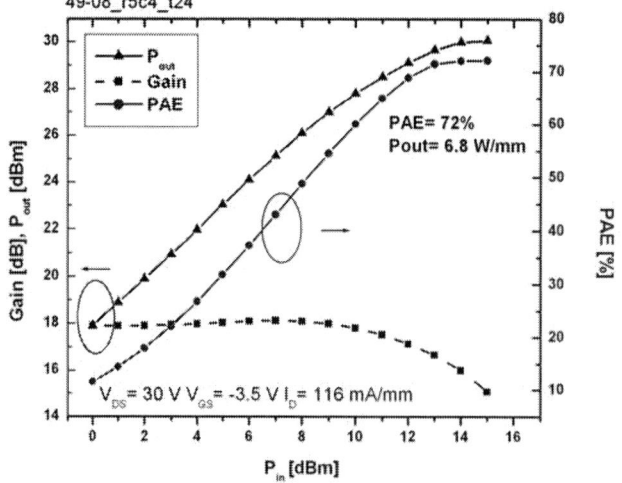

Figure 3: Load–pull measurements of implanted HEMT at 4 GHz. The device was biased with V = 30 V and I_D = 116 mA/mm.

Self-Aligned AlGaN/GaN High Electron Mobility Transistors

V. Kumar, D-H. Kim, A. Basu, and I. Adesida

Micro and Nanotechnology Laboratory and Department of Electrical and Computer Engineering, University of Illinois at Urbana-Champaign, Urbana, IL 61801

AlGaN/GaN high electron mobility transistors (HEMTs) are excellent candidates for high power and high frequency applications at room and elevated temperatures due to their superior material properties. As a result of improved material growth and processing technologies, microwave power densities have been demonstrated that are five to ten times greater than that of corresponding GaAs-based devices. Although an f_T of 180 GHz for 30 nm gate-length devices has been demonstrated, the high frequency response of these devices is still limited by the high access resistance [1]. The use of self-aligned device structure has the potential to lead to a reduction in access resistance. However, the annealing temperatures for ohmic contacts at > 800 °C makes the realization of self-aligned GaN-based HEMTs challenging. This high temperature can severely degrade the Schottky gate metal and also cause short-circuiting of the devices due to lateral flow of the ohmic contact metal. Only two reports on self-aligned GaN devices have been reported to date. Chen et al. [2] reported on self-aligned devices using regrown ohmic contacts while Lee et al. [3] utilized a double ohmic process in which the first contact was annealed at high temperature (850 °C) followed by gate definition and then by another ohmic contact annealed at 750 °C. The knee voltage of these devices was more than 4 V and the peak extrinsic transconductance was less than 150 mS/mm.

In this paper, we present high performance 0.25 μm gate-length self-aligned AlGaN/GaN HEMTs on 6H-SiC substrates using a single ohmic step. Our recently developed Mo/Al/Mo/Au-based ohmic contact requiring annealing temperatures between annealing temperatures 500 and 600 °C was utilized. Ohmic contact resistances between 0.3 - 0.5 ohm-mm have been achieved. An example of such a result is shown in Fig. 1. For both ohmic contact and device fabrication, an AlGaN HEMT structure on 6H-SiC substrates by metal organic chemical vapor deposition was utilized. It consisted of an AlN nucleation layer, 2.0 μm undoped GaN, and a 20 nm $Al_{.28}Ga_{.72}N$ barrier layer. For device fabrication, mesa isolation was achieved using Cl_2/Ar plasma in an inductively-coupled-plasma reactive ion etch system. Next, 170 nm silicon nitride was deposited using a PECVD system. Then 0.25 μm gate-footprints were patterned using e-beam lithography and etched through the silicon nitride film in a RIE system. Another, e-beam lithography was done to pattern side lobes followed by Ni/Au (300/2500 Å) e-beam evaporation. After, the gate definition silicon nitride was etched using combination of dry and wet etching process. Mo/Al/Mo/Au ohmic metallization with a total thickness of 70 nm was evaporated and annealed. The ohmic metal was self-aligned to gate as the overhangs of T-gate acted as a shadow mask i.e., the source and drain contacts separation was defined by the head of the T-gate (~ 600 nm). Finally, overlay metallization was deposited for probing pads. For comparison non-self aligned devices were also fabricated on the same wafer. Figure 2 shows a scanning electron micrograph of a self-aligned device. Figure 3 shows the typical drain current-voltage characteristics for the devices. The gate was biased from -5 V to 0 V in a step of 1 V. The devices exhibited a maximum drain current density of 1.08 A/mm at a gate bias of 0 V and a drain bias of 10 V. The knee voltage was less than 2 V showing the excellent nature of the ohmic contact. The DC transfer characteristics for this device are shown in Fig. 4. The drain was biased at 5 V. A peak extrinsic transconductance (g_m) of 321 mS/mm was measured at V_{gs} = -3.0 V. The high value of g_m is attributed to the thin AlGaN barrier layer and the low access resistance. Figure 5 shows the short-circuit current gain ($|h_{21}|$) and Mason's unilateral gain derived from on-wafer S-parameters measurements as a function of frequency for the 0.25 μm gate-length device. The values f_T and f_{max} were determined by extrapolation of the $|h_{21}|$ and MAG data at −20 dB/decade, respectively. At a drain bias of 6.5 V and a gate bias of −3.0 V, an f_T of 82 GHz and an f_{max} of 103 GHz were obtained. For comparison, the non-self-aligned devices demonstrated f_T and f_{max} of 61 GHz and 82 GHz, respectively. The source access resistances of the self-aligned and non-self-aligned devices were calculated to be 5.7 and 9.4 Ω, respectively demonstrating the improvements for the former case. Detailed device fabrication processes for this and devices with smaller gate-lengths will be presented and discussed.

[1] M. Higashiwaki et al., 64th Device Research Conf., 2006.
[2] Chen et al., Appl. Phys. Lett., vol. 73 1998.
[3] Lee et al., Elect. Lett., vol. 40 2004.

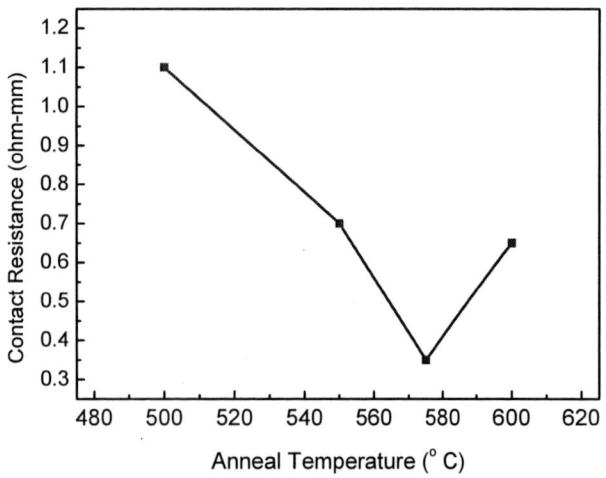

Fig. 1. Variation of contact resistance with annealing temperature.

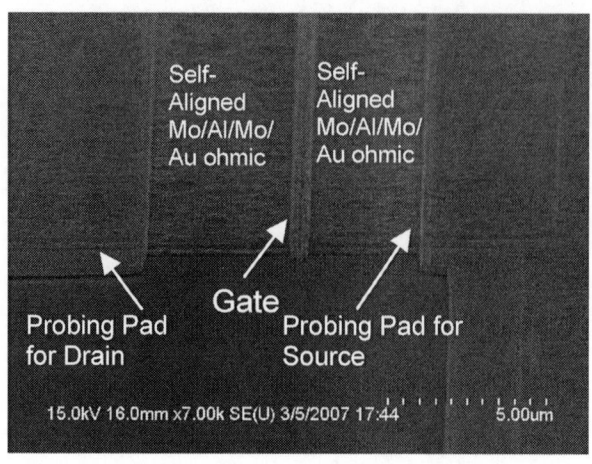

Fig. 2. SEM micrograph of a 0.25 μm × 100 μm self-aligned AlGaN/GaN HEMT.

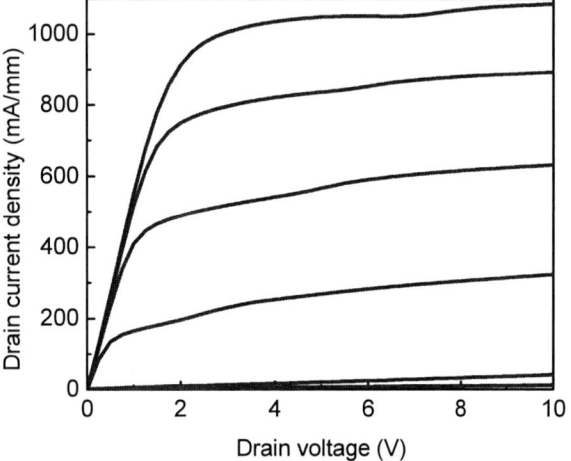

Fig. 3. Drain current-voltage characteristics of a 0.25 μm × 100 μm self-aligned AlGaN/GaN HEMT. The gate bias was swept from -5 to 0 V in a step of 1 V.

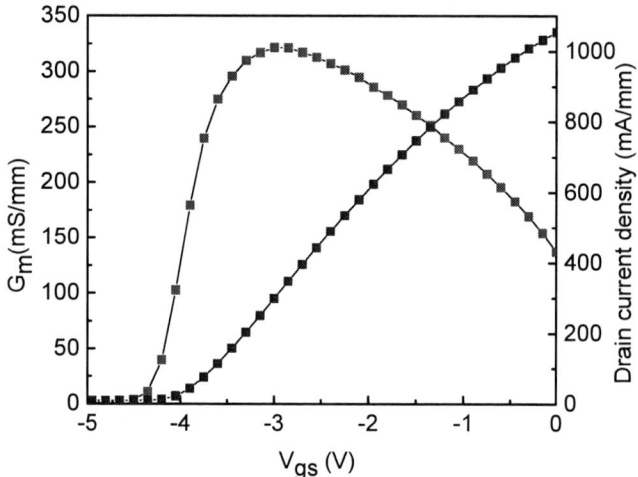

Fig. 4. Transfer characteristics of a 0.25 μm × 100 μm self-aligned AlGaN/GaN HEMT. The drain bias was 5.0 V.

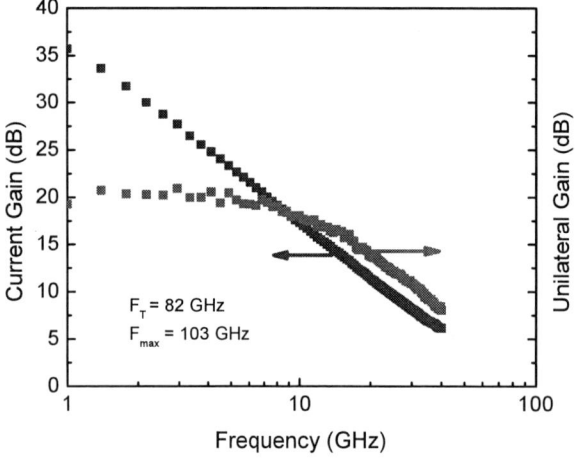

Fig. 5. Short-circuit current gain ($|h_{21}|$) and Unilateral power gain of a 0.25 μm × 100 μm self-aligned AlGaN/GaN HEMT. The device was biased at V_{ds} = 6.5 V and V_{gs} = -3V.

Analysis of lateral surface leakage in the vicinity of Schottky gates in AlGaN/GaN HEMTs

Junji Kotani, Masafumi Tajima and Tamotsu Hashizume

Research Center for Integrated Quantum Electronics (RCIQE) and
Graduate School of Information Science and Technology, Hokkaido University
phone: +81-11-706-7174; fax: +81-11-716-6004; e-mail: kotani@rciqe.hokudai.ac.jp

Schottky gates on AlGaN/GaN HEMTs are reducing their size to nanometer-scale region for the demand of further high-frequency operation. In those devices, the gate leakage includes not only one dimensional transport, normal to the Schottky interface, but also lateral electron injection from the gate edge. In the latter case, we have to pay attention to the laterally injected charges to the surface states because they can cause the instability of device operation. However, the mechanism of lateral leakage current in the vicinity of Schottky gates is not understood yet for the AlGaN/GaN HEMTs.

In this paper, we systematically characterize surface leakage current in the vicinity of Schottky gates on the AlGaN/GaN heterostructure, separating it from the normal leakage current through the Schottky interface.

The $Al_{0.27}Ga_{0.73}N$/GaN heterostructure used for this study was grown by MOVPE on a sapphire substrate. The thickness of AlGaN layer was 25nm. Typical values of the electron concentration and mobility of 2DEG at room temperature were $1.0 \times 10^{13} cm^2$ and $1380 cm^2/Vs$, respectively. The dual gate structure and the measurement circuit for characterizing surface current are schematically shown in Figs. 1 (a) and (b), respectively. In order to detect the lateral surface current (I_s) separately from vertical current (I_v), an additional Schottky gate, G2, was fabricated near the main Schottky gate G1 with a distance of L_{GG} of 200nm~10μm. The arrows in Fig. 1(b) represent the direction of electron flow. The Schottky gates and the ohmic electrodes consisted of Ni/Au and Ti/Al/Ti/Au, respectively.

Figures 2 (a) and (b) show the I_s-V_G and I_v-V_G characteristics, respectively, for various values of L_{GG}. The vertical current I_v was almost independent of L_{GG}, while the surface current I_s systematically increased with decreasing L_{GG}. The surface leakage current at V_G= -4V was plotted in Fig. 2(c), as a function of L_{GG}. It is clearly seen that I_s drastically increased when L_{GG} reduces to less than 1μm. The surface leakage behavior was not governed by a normal resistive conduction which is expected to show the $1/L_{GG}$ dependence indicated by the broken line in Fig. 2 (c).

Figure 3 shows the temperature dependence of I_s and I_v at V_G=-2V. No temperature dependence was observed in I_v, indicating that the tunneling transport through thin Schottky barrier is dominant for the vertical leakage current [1]. On the other hand, clear temperature dependence appeared in I_s. It is likely that other transport mechanism, e.g., a variable-range-hopping (VRH) conduction, governs the surface leakage current.

Figure 4 shows the I_s-V_G and I_v-V_G characteristics before and after the SiN_x passivation. The vertical current I_v almost remained unchanged before and after the passivation, reflecting that the properties of the embedded Schottky interface are not affected by the surface process. On the other hand, a pronounced reduction of I_s was observed after the SiN_x passivation.

A model for the leakage current transport near Schottky gates is shown in Fig. 5. Besides the vertical tunneling current, the lateral electron injection to the high-density surface states at the AlGaN surface may occur through electron tunneling at the edge of the Schottky gate G1. This propagates toward G2, probably assisted by the VRH conduction. It is expected that such a surface transport occurs within a limited region with a distance of ΔL from G1, as shown in Fig. 5, because some of injected electrons go down to the AlGaN/GaN interface due to the strong internal electric field induced by polarization. The distance for the drastic increase in the surface current shown in Fig. 2 (c) seems to be correlated with ΔL. The SiN_x passivation effectively reduces the density of surface states at the AlGaN surface, thereby resulting in the suppression of the surface leakage, as shown in Fig. 4.

[1] J. Kotani, T. Hashizume and H. Hasegawa, J. Vac. Sci. Technol. B **22**, 2179 (2004)

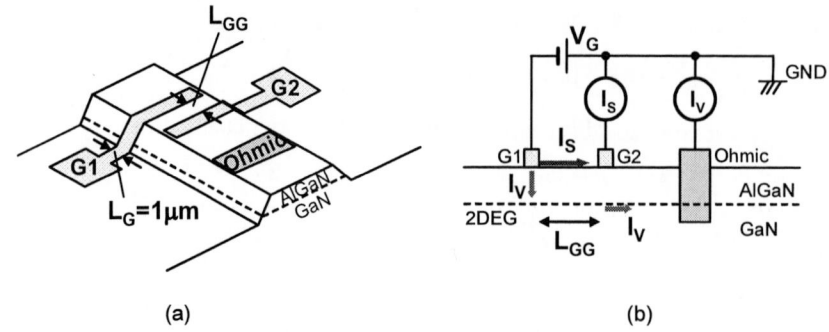

Figure 1. Schematic illustration of (a) dual gate structure and
(b) measurement circuit for characterizing surface current.

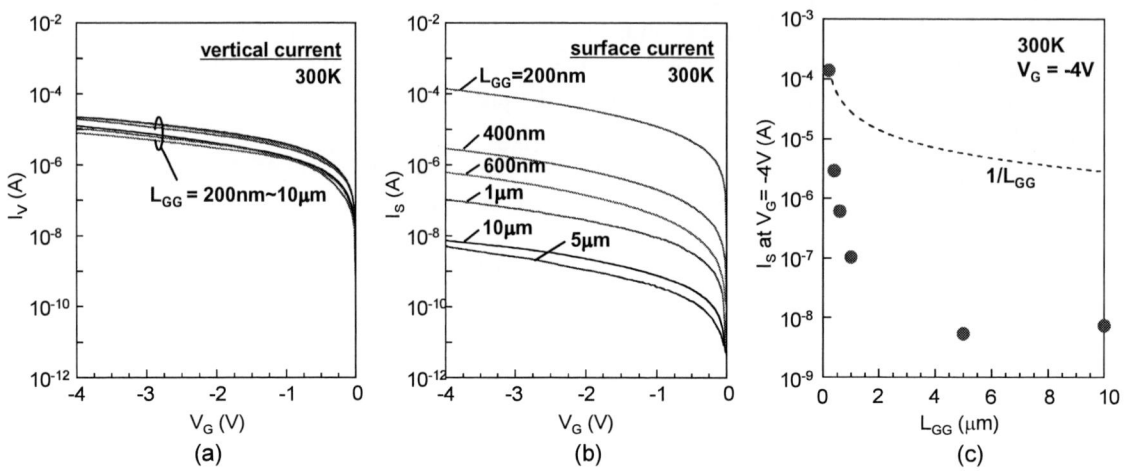

Figure 2. Measured (a) I_V-V_G and (b) I_S-V_G characteristics.
(c) L_{GG} dependence of surface leakage current at V_G = -4V.

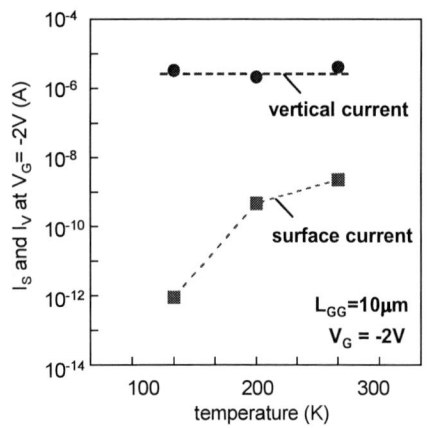

Figure 3. Temperature dependence of
I_V and I_S at V_G=-2V.

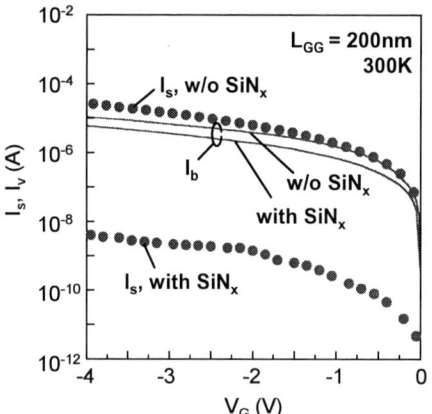

Figure 4. I_S-V_G and I_V-V_G characteristics
before and after the SiN$_x$ passivation.

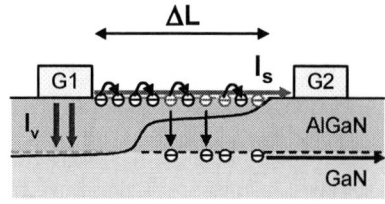

Figure 5. Lateral electron injection and a variable-range-hopping near Schottky gates

42

Drain-to-Gate Field Engineering for Improved Frequency Response of GaN-based HEMTs

N. Pala [a,b], Z. Yang [b], A. Koudymov [b], X. Hu [a], J. Deng [a], R. Gaska [a], G. Simin [c], and M. S. Shur [b]

a) Sensor Electronic Technology, 1195 Atlas Road, Columbia, SC 29209,USA, email: <u>palan@rpi.edu</u>
b) ECSE Department, Rensselaer Polytechnic Institute, Troy, NY 12180, USA
c) ECE Department, University of South Carolina, Columbia, SC 29201, USA

GaN HEMTs exhibited impressive high power and efficiency performance up to 10 GHz [1] and demonstrated potential of operating at higher frequencies [2]. However, the achieved highest cutoff frequencies are significantly lower than those expected from the gate length and peak 2D electron velocities. Recently, 2D simulations by Turin et al [3] showed that one of the major reason for a relatively low f_T is a very large extension of the space charge into the drain-to-gate spacing of GaN HEMTs with deep submicron gates. They suggested that the way to increase f_T is to achieve a more uniform field distribution in gate-drain region by providing an additional field controlling electrode (FCE) connected to the drain and optimally spaced from the gate. In this study, we fabricated a variety of devices with FCEs and investigated the effect of FCEs on extracted saturation velocity v_S , cut-off frequency and RF power performance of these devices. The results demonstrate a significant (nearly 40%) increase in the effective electron saturation velocity in the FCE devices.

The FET cut off frequency can be estimated as $f_T = v_S/2\pi L_{geff}$, where the effective gate length L_{geff} increases with the drain bias causing f_T to drop at large drain biases. Most of the reports of high-frequency performance are limited to relatively low drain biases, typically 7 to 15 V, A high power operation requires much larger biases, where both the effective electron velocity and effective channel length extension can be significantly different from those at low drain biases. In this study, we measured drain bias dependence of f_T for the fabricated devices with different FCE-to-gate separations and without FCEs up to 30 V. The decrease of f_T with increasing drain bias was found to be much less pronounced for the devices with FCEs.

For the velocity extraction, we first removed the effect of parasitics using the result of small-signal measurement of a deeply pinched-off device without FCE at zero drain bias. Then the results were analyzed based on gate length dependence of the measured f_T data. In the plot of the reciprocal cut-off frequency versus the gate length, the slope of the line gives the effective saturation velocity value, v_s, while the x-axis intercept is equal to the value of channel extension. Thus, assuming that $L_{geff} = L_g + \Delta L_g$ where ΔL_g is the channel extension beyond the gate towards the drain side, and parasitics are negligible, we estimated ΔL_g and v_s. By comparing the measurement results for the devices with gate lengths of 0.3 and 0.5 µm, we extract both electron velocity and channel extension. (The electron velocity was also extracted directly from f_T yielding the same value). The effective channel extension did not exhibit significant changes with FCE-gate separation; however the devices with FCE demonstrated the rise in the effective electron velocity. We link this result to a more uniform field distribution in field-controlled devices, as predicted by the theory in [3]. To the best of our knowledge, the values of v_s in excess of 1.34×10^7 cm/s extracted from our f_T data are record high for such devices. This observation opens a way to a novel approach in the high-frequency device design using field engineering in the gate - drain region to improve f_T vs. V_{DSmax} dependence.

RF power measurements confirmed that power performance of the devices also benefits from the more uniform electric field and improved saturation velocity. As compared to the devices without the FCE, the saturation RF power was 1.4 to 1.8 times higher. At 2 GHz, the FCE devices demonstrated the large signal gain of about 5 dB higher than conventional devices.

In conclusion, we report a novel approach in designing high frequency AlGaN/GaN HEMTs based on gate-drain field engineering utilizing a drain-connected field controlling electrode. The absence of frequency behavior degradation with drain bias as well as record high electron velocity values were obtained using gate-to-FCE separation of 0.5-0.7 µm. Thus, we demonstrated that the FCE is a powerful way to improving the high frequency, high power performance of GaN HEMTs at high drain biases.

---

[1] Y.-F. Wu, M. Moore, A. Saxler, T. Wisleder, and P. Parikh, in *Proc. 64th Device Res. Conf. Dig.*, Jun. 26–28, p. 151, 2006.

[2] M. Higashiwaki, T. Mimura and T. Matsui, *Japanese Journal of Applied Physics*, vol. 45, no. 42, p. L1111, 2006.

[3] V. Turin, M.S. Shur, and D. Veksler, *International Journal of High Speed Electronics and Systems*, In Print

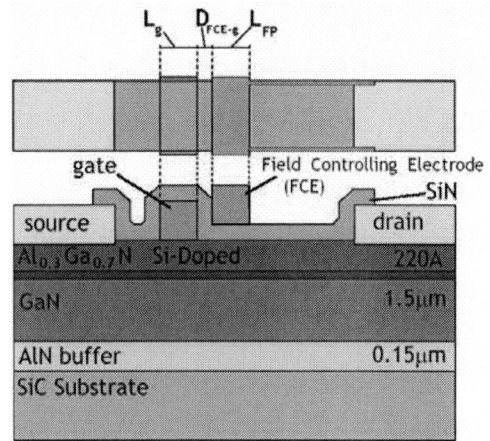

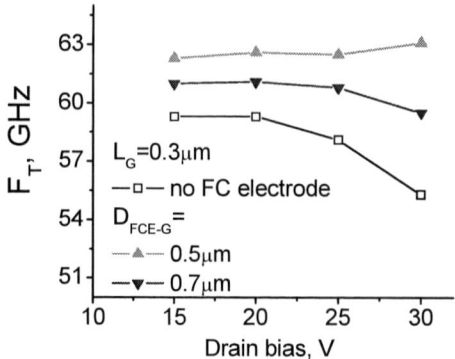

Fig. 1 Fabricated device structure. L_g=0.3/0.5μm, Field Controlling Electrode (FCE) to gate separation 0.3/0.5/0.7 μm, L_{FCE}=0.5 μm. Split-drain design was used to reduce gate-to-drain capacitance.

Fig 2. f_T vs. V_{DS} for devices with and without Field Controlling Electrode (FCE). FCE eliminates f_T degradation with drain bias.

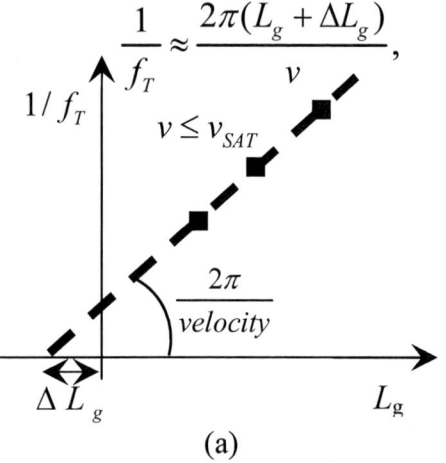

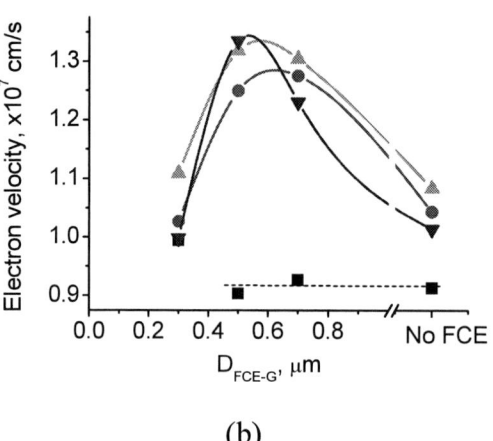

(a)

(b)

Fig. 3 (a) Saturation velocity extraction procedure and **(b)** extracted electron velocity for devices with different gate-to-FCE separations and without FCE. As seen, optimum gate to FCE separation to achieve the highest velocity is ~ 0.5 μm. Black squares: V_D = 15 V, red, green and blue – 20, 25 and 30 V, respectively. Lines are guides to the eye.

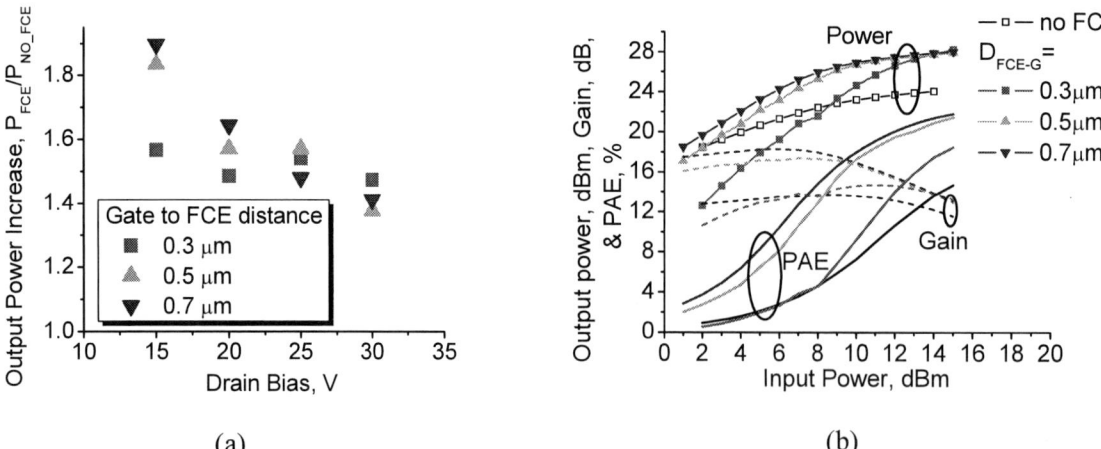

(a)

(b)

Fig. 4 (a) RF power improvement due to EFC and **(b)** 30 V RF power sweep details for devices with and without FCE.

Session II.C (DeBartolo Hall Room 155)

Silicon CMOS

Monday PM, June 18th, 2007

Session Organizer: Judy Hoyt, Massachusetts Institute of Technology and Steve Koester, IBM
Session Chair: Kerem Akarvardar, Stanford University

1:30 PM II.C-1 Invited Paper
Silicon Nanowire Field Effect Devices By Top-Down CMOS Technology
N. Balasubramanian, N. Singh, S. C. Rustagi, Kavitha, A. Agarwal, G. Zhiqiang, G. Q. Lo, and D. L. Kwong, Institute of Microelectronics, Science Park II, SINGAPORE

2:10 PM II.C-2 Student Paper
SiGe cantilever channel gate-all-around (GAA) fully depleted (FD) PMOSFET with high-? and metal gate
S.-H. Lee[1], S. Dey[1], S V. Joshi[1] , P. Majhi[2] and S. K. Banerjee[1], [1]Microelectronics Research Center, Electrical and Computer Engineering, University of Texas at Austin, Austin, Texas, USA and [2]Intel Assignee at Sematech, 2, Austin, Texas, USA

2:30 PM II.C-3
Hole Transport in Nanoscale p-type MOSFET SOI Devices with High Strain
H. M. Nayfeh[1], S. Jeng[1], S. Narasimha[1], S. Butt[1], R. Pal[2], A. Waite[2], K. Tabakman[1], J. B. Johnson[1], J. Liu[1], J. Holt[1], T. Adam[1], A. Madan[1], A. Domenicucci[1], [1]IBM Semiconductor Research and Development Center (SRDC), Hopewell Junction, New York, USA and [2]IBM Advanced Micro Devices (AMD), Hopewell Junction, New York, USA

2:50 PM II.C-4 Student Paper
3X hole mobility enhancement in epitaxially grown SiGe PMOSFETs on (110) Si substrates with high k / metal gate for hybrid orientation technology
S. Joshi[1], S. Dey[1], S.-h. Lee[1], C. Krug[2], H. Joo Na[2], P. Sivasubramani[2], P. D. Kirsch[2,3], P. Majhi[2,4], W. Wang[1], A. Campion[1] and S. K. Banerjee[1], [1]University of Texas at Austin, Austin, Texas, USA, [2]SEMATECH, Austin, Texas, USA, [3]IBM, and [4]Intel

3:10 PM Break

3:30 PM II.C-5 Student Paper
High – Mobility, Low Parasitic Resistance Si/Ge/Si Heterostructure Channel Schottky Source/Drain PMOSFETs
A. Pethe and K. Saraswat, Department of Electrical Engineering, Stanford University, Stanford, California, USA

4:10 PM II.C-6 Student Paper
Band to Band Tunneling Study in High Mobility Materials : III-V, Si, Ge and strained SiGe
D. Kim[1], T. Krishnamohan[1], L. Smith[2], H.-S. Philip Wong[1], K. C. Saraswat[1], [1]Department of Electrical Engineering, Stanford University, Stanford, California, USA and [2]Synopsys Corporation, Mountain View, California, USA

4:30 PM II.C-7 Student Paper
Process Integration and Electrical Properties of Bilayer Metal Gate/High-k MOSFETs
C.-H. Lu[1], G. M. T. Wong[1], M. Deal[2], B. M. Clemens[1], and Y. Nishi[1,2], [1]Department of Materials Science and Engineering, Stanford University, Stanford, California, USA and [2]Department of Electrical Engineering, Stanford University, Stanford, California, USA

46

Silicon Nanowire Field Effect Devices By Top-Down CMOS Technology

N.Balasubramanian, N.Singh, S.C.Rustagi, Kavitha, Ajay Agarwal, Gao Zhiqiang, G.Q.Lo, and D.L.Kwong

Institute of Microelectronics, 11 Science Park Road, Science Park II, Singapore, 117685

(email: nara@ime.a-star.edu.sg)

There has been tremendous advancement in the development of novel nano-technologies for future CMOS nanoelectronics. The challenges and opportunities have been widely discussed with the focus on the choice of materials, processes of implementation and innovative non-classical device architectures to continuously meet the scaling requirements [1, 2]. Among the non-classical device architectures, Gate All Around (GAA) FET with nanowire (NW) channel body offers the ultimate electro-static control and thus has the potential to push the gate length to few nanometers. The key challenge for NWs to be widely adopted in semiconductor industry is that they have to be formed by large scale manufacturing methods. Especially, for CMOS applications, the methods should not lead to contamination issues.

Broadly, one can classify the approaches for forming NWs as bottom-up and top-down. Several bottom-up methods for NW formation have been widely published, with Vapor-Liquid-Solid growth process [3] which uses metallic nanoclusters as the nucleation site, as a prominent candidate. Additional techniques are used to position the NWs in desired places. The potential metal-contamination presents a major concern for CMOS nanoelectronics using this approach.

On the other hand, top-down methods using lithography and etching suit CMOS applications in general and also provides a well-established CMOS based platform for creating devices for other applications. We have developed such a technology platform for Si NW fabrication [4]. The wires are patterned through lithographic method, etched, and are formed by stress-limited oxidation. With highly ordered structure, the devices show high yield of functionality, repeatability, and high adaptability to CMOS and other applications.

Fig. 1 shows the process steps for Si NW fabrication and subsequently the GAA FET devices on 8 inch SOI wafers. **Fig. 2** shows the SEM of the Si NW-FET at different process stages and the TEM cross-section (**Fig. 3**) of the NW GAA structure shows NW of ~3 nm diameter surrounded by oxide and polysilicon gate. Shown in **Fig. 4** are I-V characteristics of single SiNW channel n & p-GAAFETs with $L_g \approx 350nm$ and ~4nm grown SiO_2 as gate dielectric. Even with relatively long channel, the drive currents are higher than most of the earlier reports on nanowire devices with shorter gate lengths [5, 6]. Interestingly, the drive current with 4 nm gate oxide is only slightly higher than that of 9 nm gate oxide (not shown here). It is due to the cylindrical nature of the structure in which the electrical equivalent oxide thickness is significantly lower than the planar MOS capacitor due to diverging electric field as shown in **Fig. 5**. As shown in **Fig. 6,** the electrical equivalent thickness of gate dielectric reduces for a GAA FET for a given oxide thickness reduces with channel body thickness. It shows that gate oxide scaling requirement can be relaxed for the GAA NW architecture. Investigations on channel implant on the threshold voltage (V_t) for GAA-FETs show that V_t is insensitive to channel doping as shown in **Fig.7** (for p-MOS). For logic circuit realization one may use the channel length of n & p-MOS suitably to achieve reasonable symmetry of the inverter by matching their drive currents. We have fabricated such inverters with sharp transitions, high noise margins and low power dissipation.

We have carried out extensive studies of GAA-FETs at low temperatures to throw light on transport characteristics of such devices [4]. By virtue of the reduced channel width, sub-band splitting is very evident. Also thinner body NWs exhibit reduced sensitivity of V_t to temperature. **Fig. 8** shows the reduction in sub-threshold slope vs. T for a 6 nm wire. The minor differences between ideal and experimental data are attributed to the interface charges particularly because of the intrinsically multi-oriented wire surfaces at oxide / channel interface.

To push the technology further, we have also demonstrated SiGe-based 3-dimensional wire array channel structure [7]. Recently, we have fabricated GAA FETs consisting of Si NW surrounded by HfO2 and TaN (**Fig. 9**). The twisted NW indicates large strain induced by the metal. The strained NW FETs exhibit increased drive current.

In addition to its application to CMOS technology, NW structures shows great promise as highly sensitive bio-sensors due to their high surface to volume ratio. **Fig. 10** shows our Si NW array along a fluidic channel where bio-molecules can be specifically bound by suitable surface chemistries. In the presence of charged bio-molecule, the NW resistance changes just as the gated FET. **Fig. 11** shows sensing of the DNA down to 10 fM level of DNA concentration [8], representing high sensitivity.

---

[1] C. Hu, in *VLSI Symp. Tech. Dig.*, 2004, pp. 4-5. [2] H. S. P. Wong et al., *Proc. of IEEE*, **52**(9), p. 537 (2005). [3] Lincoln J. Lauhon, Nature, 420, p.57 (2002). [4] N. Singh et al., in *IEDM Tech. Dig.*, 2006. [5] F. L. Yang et al., in *VLSI Tech. Symp.*, 2004, pp. 196-197. [6] H. Lee et al., in *VLSI Tech. Symp.*, 2006, T_07_5. [7] L.K.Bera, et al., in IEDM Tech. Dig. 2006. [8]. G.Zhiqiang, et al., Analytical Chemsitry (in press).

978-1-4244-1101-6/07/$25.00 ©2007 IEEE

- Starting material: SOI (100) wafer
- Fin patterning and transfer into Si
- Oxidation in dry O_2 at 875°C
- Wet removal of SiO_2
- Gate oxide and poly-Si deposition
- Gate electrode definition
- S/D patterning and implants:
  - NMOS: Phos/30KeV/4E15
  - PMOS:BF_2/35KeV/4E15
- Activation 950°C/15min
- Contact and metallization
- Sintering

Fig. 1: Typical process flow for fabricating Si NW GAA transistors using top-down CMOS technology.

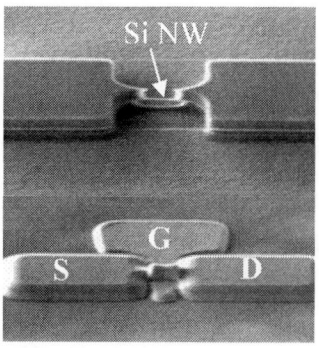

Fig. 2: Tilted view SEM of the released 200 nm long nanowire (top), and after gate electrode definition (bottom)

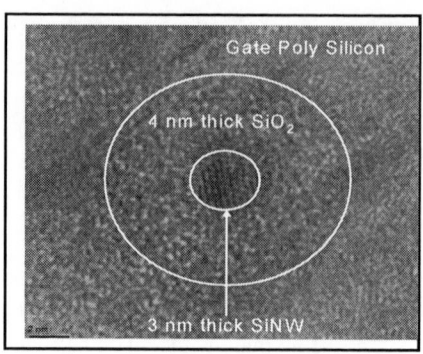

Fig. 3: TEM cross-section of a fully processed GAA device.

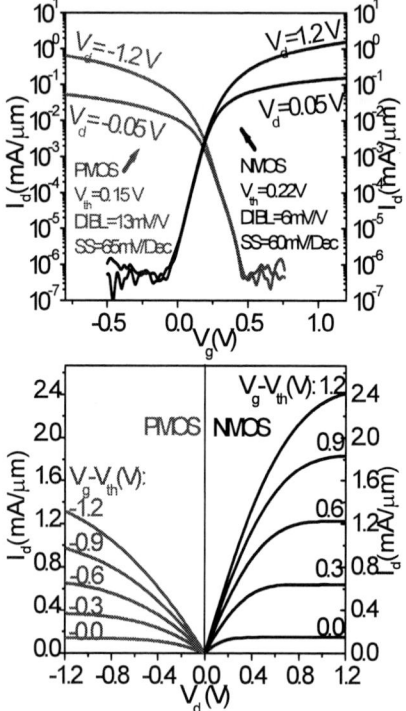

Fig.4: I-V curves of SiNW GAA FETs with 4 nm grown-oxide and poly-Si gate. NW diameter is about 4 nm.

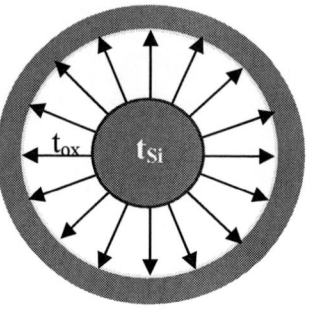

Fig.5: Schematic of diverging electric field in GAA architecture

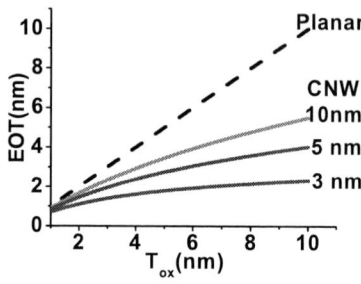

Fig.6: Electrical equivalent thickness in a cylindrical architecture compared to planar MOS

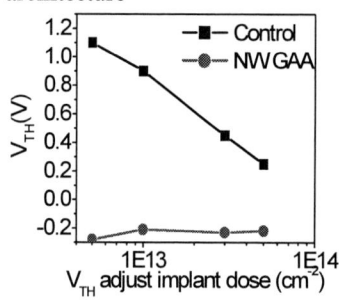

Fig.7: Effect of channel implant (P/60KeV) on V_{TH} of planar and NW GAA PMOS devices.

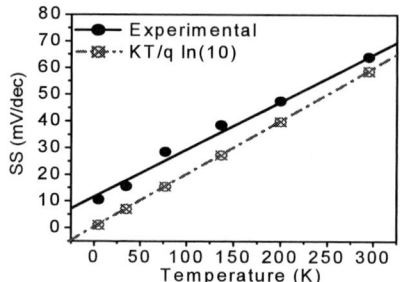

Fig.8: SS of a SiNW GAA NMOS as a function of temperature (down to 5K).

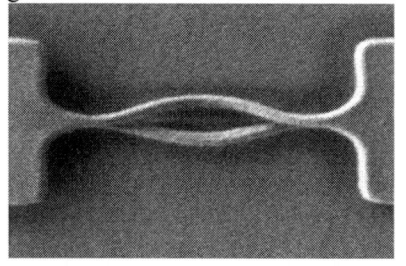

Fig. 9: Twisted twin Si NWs after high-K and metal gate depositions.

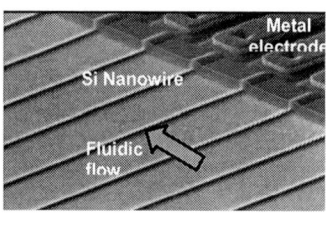

Fig.10: SEM of Si NW array with fluidic channel for bio-sensing.

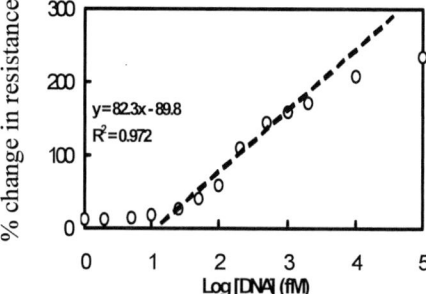

Fig.11: Charge induced Si NW resistance change, upon hybridization of different concentration of DNA.

SiGe cantilever channel gate-all-around (GAA) fully depleted (FD) PMOSFET with high-κ and metal gate

S.-H. Lee[1], S. Dey[1], S V. Joshi[1], P. Majhi[2] and S. K. Banerjee[1]

[1]Microelectronics Research Center, Electrical and Computer Engg, University of Texas at Austin,
[2]Intel Assignee at Sematech, 2706 Montopolis Drive, Austin, TX 78741
Phone: (512) 471–8658, Fax: (512) 471–5625, Email: slee2@ece.utexas.edu

Scaling the conventional CMOS transistor beyond the 45nm generation ushers in several fundamental limitations. Control of leakage currents and sustaining electrostatic integrity while maintaining historic enhancements in performance requires such ultra-thin gate-dielectrics and heavily doped bodies that a process window sufficiently large for manufacturing might not be found. While conventional SiO_2 might need to be replaced by high-κ dielectric and metal gate, it might be necessary that the conventional planar MOSFET architecture be also substituted to address the electrostatics challenges. In addition high mobility materials need to be explored to garner the additional enhancement in performance. In this paper we demonstrate a PMOSFET device architecture that integrates such a high mobility material with high-κ/metal gate in a 3D non-planar gate-all-around architecture (GAA).

The device uses a thin fully-depleted (FD) high hole mobility SiGe channel with $HfSiO_x$ and TaN wrapping all around the SiGe channel. While gate-all-around devices have been reported[1] as being very successful in suppressing subthreshold leakages and short channel effects, they have been demonstrated on Si channels and with SiO_2. We demonstrate GAA architecture with thin FD SiGe channels which can further enhance performance of the device due to high hole mobility. In addition, the high-κ/metal gate stack allows an effective oxide thickness to enhance the electrostatics and performance while maintaining a low gate leakage.

For the device fabrication, starting from a SOI wafer, a Si cantilever (~100nm thick) was first formed by selectively etching off the buried-oxide layer underneath the cantilever such that it is suspended in air between soured/drain anchors. Following this we deposited ~ 10nm epi $Si_{0.8}Ge_{0.2}$ layer all around the Si cantilever using a ultra-high-vacuum chemical vapor deposition (UHVCVD)[3] tool. For another flavor of the device following a 40nm $Si_{0.8}Ge_{0.2}$ deposition we further proceeded to oxidize the SiGe layer using a high temperature dry oxidation step such that Ge condenses [2] from all around into the Si channel and increase Ge mole fraction. Removal of the oxide layer now leads to a SiGe channel. Incorporation of Ge in such a way also induces strain in the channel that can be observed from SEM scan of the structure shown in Fig 1a. Process simulations done using Taurus also confirm the strain state of the structure (Fig 1b). Auger measurement analysis shows that the SiGe channels ~30nm thick have a uniform Ge mole-fraction(x)=0.2 (Fig 1c). Following the SiGe channel formation high-κ $HfSiO_x$ (EOT~2.5nm) and metal gate TaN was deposited using a atomic-layer deposition (ALD) system to wrap all-around the channel. Conventional implant, anneal, contact and metallization steps after this concluded the process flow. Si channel GAA devices were also fabricated for comparisons. Id-Vg characteristics of the fabricated devices are shown in Fig 2. The Ge-condensation devices were found to have a lower V_t than the other two splits which had almost similar V_t. This suggests the necessity for separate work-function engineering of the metal gate for higher Ge condensation. Fig 3 and Fig 4 compare the output characteristics and transconductance of the fabricated devices. Compared to the Si GAA channel devices the SiGe epi devices show a 2.7x and the Ge-condensation channel devices showed a 1.2x increase in drive current at the same gate overdrive. The higher enhancement and better characteristics of the epi SiGe devices than the Ge condensed devices is related to a better SiGe layer quality and lower defect density. For all these devices the currents have been normalized to the drawn width which is different from the electrical width due to the all-around gate. Enhancements in extracted mobility was seen for the SiGe epi GAA channel over the universal hole mobility (Fig 5) particularly at high fields. This could result from volume inversion induced by the all-around gate that leads to lower surface scattering. This was also confirmed from device simulations using Medici (Fig 5 inset). In conclusion, we demonstrated high mobility SiGe channel devices integrated with high-κ/metal GAA architecture. Such integration of novel material systems with novel architectures might be an attractive solution for scaling beyond 45nm nodes.

[1] S.J. Young et. al., IEEE Trans. Nanotechnology, vol. 5(3), p.186, 2006.
[2] S. Nakaharai, et al., Applied Physics Letters, vol. 83(17), p. 3516, 2003.
[3] S. Dey et. al, Jour. Elec Mat., vol. 35(8), p.1607, 2006.

978-1-4244-1101-6/07/$25.00 ©2007 IEEE

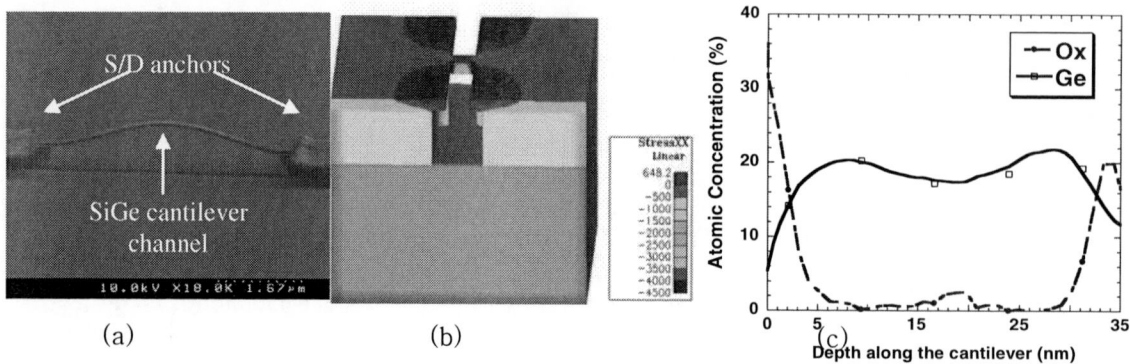

(a)   (b)   (c)

Fig 1. (a) SiGe cantilever under strain induced by Ge-condensation. (b) Process simulation using Taurus 3D confirms the strain state of SiGe cantilevers. (c) Auger profiling of Ge along the depth of the cantilever. The two oxide peaks locates the top and bottom surface of the cantilever. The cantilever thickness is ~30nm with a uniform Ge concentration (x=0.2)

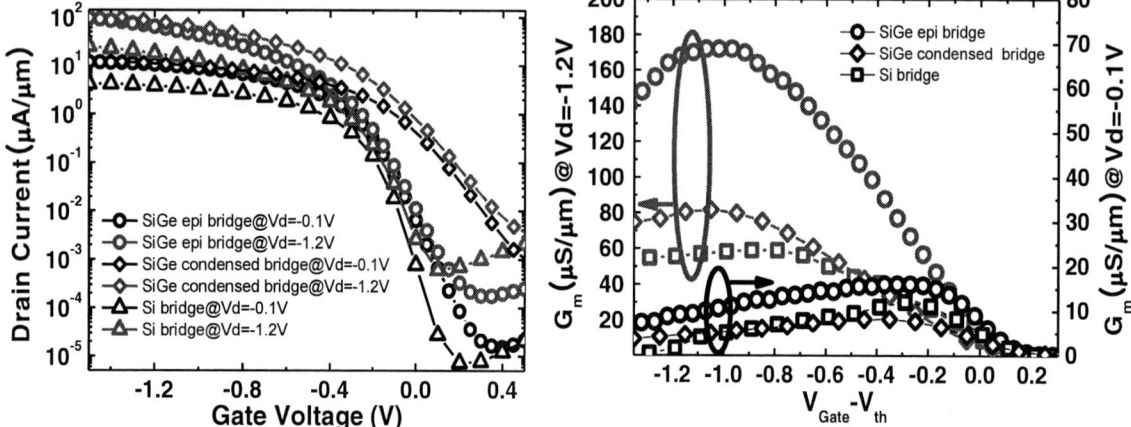

Fig 2. $I_D V_G$ characteristics of the devices are compared (L=2μm, W=0.8 μm). The Ge condensed SiGe channel is found to have a different V_t. The epi SiGe channel shows significant enhancement in drive currents.

Fig 3. Output characteristics of the epi SiGe cantilever channel and Si cantilever channel device (L=2μm, W=0.8 μm).

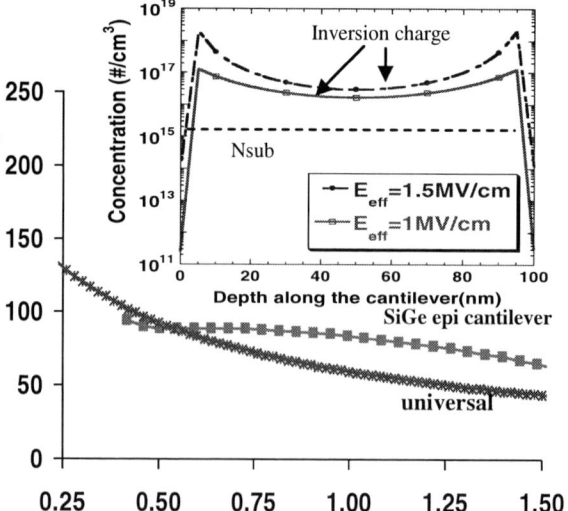

Fig 4. Output characteristics of the epi SiGe cantilever channel and Si cantilever channel device (L=2μm, W=0.8 μm).

Fig 5. SiGe epi cantilever channel shows mobility enhancement at high field due to reduced surface scattering resulting from volume inversion. Inset shows device simulation confirming volume inversion

50

Hole Transport in Nanoscale p-type MOSFET SOI Devices with High Strain

H. M. Nayfeh, S. Jeng, S. Narasimha, S. Butt, R. Pal[*], A. Waite[*], K. Tabakman, J. B. Johnson, J. Liu, J. Holt, T. Adam, A. Madan, A. Domenicucci

IBM Semiconductor Research and Development Center (SRDC), Advanced Micro Devices (AMD)[*]

contact: _nayfeh@us.ibm.com_, 2070 Route 52 MS 42J Hopewell Junction, New York 12533 Tel 845-892-2376

I. Abstract

In this paper, we quantify the relation of low lateral electric field hole mobility and channel strain to the virtual source velocity of nanoscale p-type SOI MOSFET devices with effective channel length from 35 to 50 nm and show strong correlation. The mobility is modified by the application of uniaxial compressive strain in the 1 GPa regime to the channel by employing two stressors- (1) embedded SiGe (eSiGe) at the source/drain areas and (2) compressive strain silicon nitride contact liner film. The corresponding changes in low-field mobility and saturation drain current are significant.

II. Device Fabrication

Nanoscale p-type MOSFET devices with gate length down to 35 nm where fabricated on {100} SOI wafers. Following gate stack formation that was adjusted to achieve longer than nominal gate lengths in order to facilitate the transport measurements (so that the channel resistance dominates over the parasitic resistance), cavities are formed through reactive ion etching followed by epitaxial SiGe growth. The subsequent SiGe film is under compressive strain and is fully strained according to micro-XRD measurements that show little or no relaxation and a lattice constant equal to the underlying SOI substrate (Fig.1). Shallow extension junctions and carefully designed halo channel implants for excellent short-channel electrostatic integrity are executed followed by Ni silicide for low resistance contact. Next, a compressive silicon nitride liner (CSL) film is applied to further increase the channel strain. The remaining process steps are consistent with a traditional CMOS flow. A cross-sectional TEM of the final device is shown in Fig. 2 showing the two compressive stressors.

III. Device Design

In order to compare the hole transport properties of eSiGe versus non-eSiGe devices, the device design was tailored to deliver devices with the following characteristics: (1) closely matched saturation threshold voltage ($V_{t,sat}$), measured using the constant current method, (2) comparable parasitic series source/drain resistance (R_{ext}), and (3) closely matched effective channel length (L_{eff}), all with respect to the control SOI devices. SOI e-SiGe devices have increased saturation (high drain bias) threshold voltage compared to non-eSiGe pFET (SOI pFETs) for equivalent channel doping, due to reduced floating body effect as a consequence of increased forward and reverse diode leakage currents due to reduced bandgap for compressively strained SiGe material relative to Si [1]. This is manifested by increased forward bias leakage current of ~100X and a 10X increase in reverse leakage as shown in Fig. 3. Due to this effect, the DIBL for equivalent channel doping is lower for eSiGe pFETs compared to SOI pFETs by about 50 mV/V. The improved DIBL is related to the diode currents and not to improved electrostatic properties as demonstrated in Fig. 4, which shows closely matched I_{off} versus L_{eff} characteristics. As a result in order to match the saturation threshold voltage, a halo dose lower by 15% was used for the SiGe pFETs. As can be seen this lower halo dose results in a 50 mV reduced linear threshold voltage ($V_{t,lin}$) for the SiGe pFETs, and a closely matched $V_{t,sat}$ rolloff curve as shown in Fig. 5. Finally, in order to match the extent of direct overlap of the extension junction for the two devices and as a result, deliver devices with similar parasitic resistance and effective channel length, the extension implant dose had to be increased for the SiGe FETs due to retarded Boron diffusion in SiGe versus Si [2].

IV. Results and Discussion

The incorporation of the SiGe stressor in the source/drain area of wide width devices (3μm) results in increased longitudinal uniaxial compressive strain in the channel. The composite stress due to the CSL and eSiGe stressor is ~1 GPa determined at the middle of the channel near the Si/gate oxide interface, where 0.5 GPa is due to the eSiGe stressor and 0.5 GPa is attributed to the CSL as shown from the strain simulations in Fig. 7. These stressors are degraded compared to our current baseline as reported in Ref. 5 due to the strict requirement of utilizing strained devices with closely matched electrostatics and parasitics to non-eSiGe controls for this hole transport study. The mobility in the two devices is compared using the dR_{total}/dL method, which minimizes the effect of uncertainties in L_{eff} and R_{ext} on the measured mobility [3], where R_{total} is the total device

resistance measured at low $V_{ds}=50mV$, L_{eff} is the electrically determined effective channel length and dR_{total}/dL_{eff} is determined from a linear fit of R_{total} vs. L_{eff} for a fixed overdrive voltage (V_{gs}-V_t=0.8V). The mobility enhancement is determined to be 1.76X-fold for the eSiGe pFET devices over the non-eSiGe controls with a parasitic R_{ext} that is closely matched as shown in Fig. 6. The corresponding stress versus mobility relationship, related through the lateral piezo-coefficient (Π_{xx}), by: $\mu/\mu_o=1+\Pi_{xx}*\sigma_{xx}$, where σ_{xx} is the longitudinal stress, and μ_o corresponds to an unstrained device demonstrates that if σ_{xx} =1 GPa then Π_{xx}=0.76 GPa^{-1} which is slightly increased over the linearly extrapolated Smith coefficient of 0.71 GPa^{-1} as shown in Fig. 7 [4]. We have determined Π_{xx} for a wide range of stress values utilizing numerical simulations combined with experimental data for devices with σ_{xx} as high as 1.5 GPa [5], and show that for σ_{xx} > 1 GPa, Π_{xx} increases with σ_{xx} towards 2.0 GPa^{-1}, resulting in μ/μ_o approaching 4.0, consistent with the findings in Ref. 6. The increased stress translates into a 25% increase in $I_{d,sat}$ over the non-eSiGe controls at matched L_{eff} as shown in Fig. 8. As a result, $I_{d,sat}$ can be related to fractional increase in stress and low-field mobility through the following expressions: $\delta I_{d,sat}/I_{d,sat}$=0.25*$\delta\sigma_{xx}/\sigma_{xx}$ and $\delta I_{d,sat}/I_{d,sat}$=0.33*$\delta\mu/\mu$ for devices with R_{ext} ~300 Ω*μm. Next, the virtual source velocity (v_{xo}) [7], shows a linear increase for v_{xo} with decreasing L_{eff}, with v_{xo} for the eSiGe devices approaching 9x10^{6}cm/sec for L_{eff}=35 nm, which is ~45-50% higher than the non-eSiGe control devices as shown in Fig. 9. Stress can then be related to v_{xo} through the following expression with weak dependence on L_{eff}: $\delta v_{xo}/v_{xo}$=0.45*$\delta\sigma_{xx}/\sigma_{xx}$ demonstrating that strain is a very effective method of increasing virtual source velocity in this nanoscale regime (Fig.10). Next the correlation of virtual source velocity to low-field mobility ($\delta v_{xo}/v_{xo}$=C*$\delta\mu/\mu$) is determined to be C~0.55 and is a relatively weak function of L_{eff} as shown in Fig. 11. Assuming that ballistic holes have a velocity of 1.5x10^{7}cm/sec for L_{eff}~35 nm, the ballistic coefficient, B, which is defined as the ratio of v_{xo} and the thermal velocity (v_θ) [8] of the highly strained devices appears to be B~0.60 which is 40% from the thermal limit. Using the following expression [7]: $\delta v_{xo}/v_{xo}$=[α+(1−B)(1−α+β)]*$\delta\mu/\mu$, and assuming that β (power-law coefficient relating critical length for backscattering to low-field mobility) varies from 0-0.45 and that α (power-law coefficient that relates thermal velocity to low-field mobility, $v_\theta\sim\mu^\alpha$) is less than C, then possible solutions for B=0.60 require that α<0.50 as shown in Fig. 12. This value of α implies that the mechanism for mobility enhancement for pFETs with high longitudinal uniaxial strain is a combination of lower effective mass, and reduced interband scattering differing than α=0.50 for electrons which is explained in terms of reduced effective mass only [9]. The 15% higher halo channel dose utilized for the non-eSiGe control devices was determined to impact the transport comparison by less than 5%. Furthermore, the effective electric field is sufficiently high such that Coulomb scattering mechanisms where determined to be small. Finally, SOI self-heating differences between eSiGe and the non-eSiGe control device due to reduced thermal conductivity of SiGe material in the source/drain areas was determined to be negligible, allowing for direct comparison.

V. Conclusions

The virtual source velocity is correlated to low-field mobility by a factor of 0.55 for nanoscale pFET devices with high strain in the GPa regime, and is quantified to be within 40% of the thermal limit for L_{eff} between 35 and 50 nm. These results show that techniques such as increased channel strain, surface orientation manipulation, etc. that result in increased mobility are required to achieve continued aggressive device performance.

Acknowledgment: Dr. H. M. Nayfeh would like to thank Prof. D. A. Antoniadis (MIT) for helpful discussions. **References:** [1] R. Braunstein, et al. _Phys. Rev._ Vol. 109, No. 3, p. 695, 1958 [2] J. L. Hoyt, H. M. Nayfeh, et al. _IEDM Tech. Dig_, p. 23, 2005 [3] K. Rim, et. al. _IEDM Tech. Dig._ p. 43, 2002 [4] Smith et. al., _Phys. Rev._ Vol. 94 p. 42 1954 [5] S. Narasimha, K. Onishi, H. M. Nayfeh, et al. _IEDM Tech Dig._ p. 689, 2006 [6] S.Thompson et al. _IEDM Tech. Dig._ p. 681, 2006 [7] A. Khakifirooz, D. A. Antoniadis et al._IEDM Tech Dig._ p. 667, 2006 [8] M. Lundstrom _IEEE EDL_ Vol. 22, p. 293, 2001 [9] K. Uchida, et al. _IEDM Tech. Dig._ p. 135, 2005.

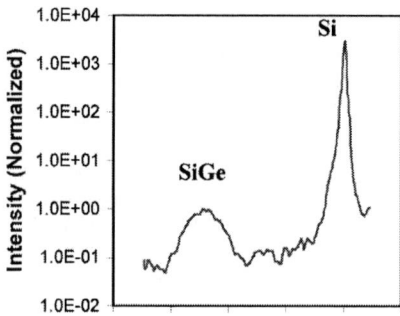

Fig. 1. High-resolution XRD rocking curve of the SiGe film grown on (100) Si substrate. The well-defined SiGe peak and fringes indicate high-quality crystal.

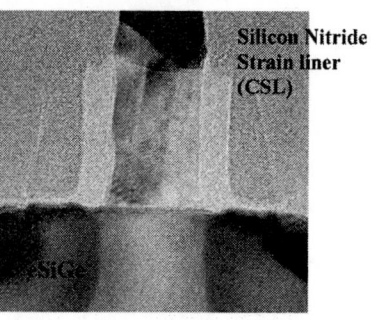

Fig. 2. High resolution cross-sectional TEM image of the eSiGe pFET device showing two stressors (1) eSiGe, and (2) Compressive silicon- nitride strain liner (CSL).

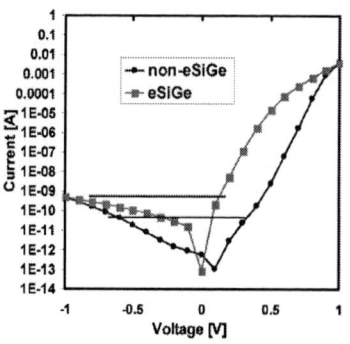

Fig. 3. Diode leakage data shows increased forward bias and reverse bias currents for the eSiGe device over non-eSiGe control. The resulting behavior results in decreased floating body voltage for the eSiGe device.

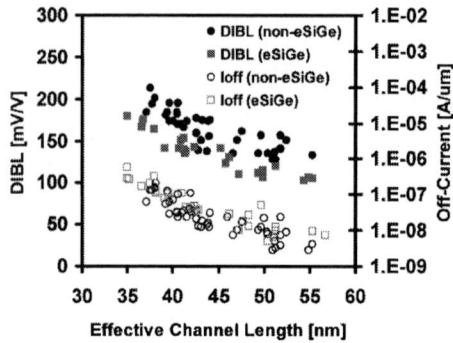

Fig. 4. Off current and DIBL versus L_{eff}. The halo ion implant dose is 15% higher for the eSiGe devices in order to compensate for the lower floating body effect. The corresponding off-current characteristics are well-matched.

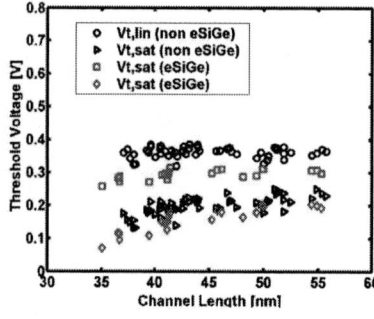

Fig. 5. Saturation (V_{ds}=1 V) and linear threshold voltage (V_{ds}=50 mV) versus L_{eff}. The curves show well-matched saturation threshold voltage for the L_{eff} range studied in this work.

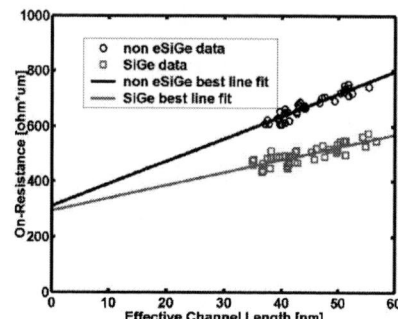

Fig. 6. The slope of on-resistance versus L_{eff} indicates a 75% increase in hole channel mobility due to the eSiGe stressor. The parasitic source/drain series resistance is well-matched.

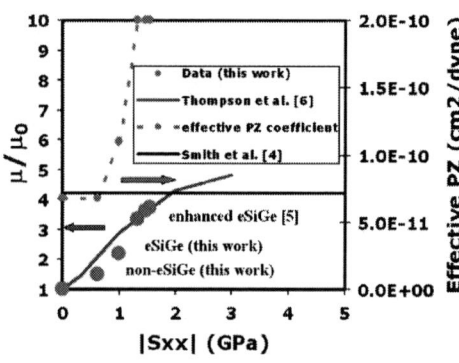

Fig. 7. Mobility enhancement versus uniaxial longitudinal compressive stress (σ_{xx}). The effective PZ coefficient is determined for stress as high as 1.5 GPa [5].

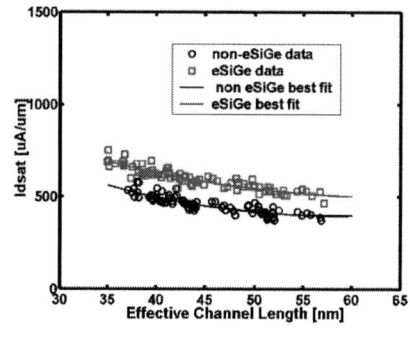

Fig. 8. Saturation drive current versus gate length. The eSiGe pFETs show a 25% increase for $I_{d,sat}$ at fixed L_{eff} due to a 100% increase in strain (2X-fold increase) at a supply voltage of V_{dd}=1V.

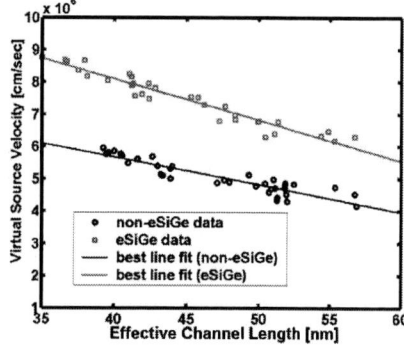

Fig. 9. Virtual source velocity versus L_{eff}. The eSiGe devices show ~45-50% increase over the non-eSiGe controls at fixed L_{eff}.

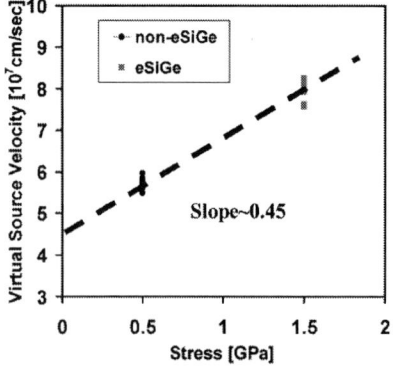

Fig. 10. Stress and virtual source velocity are strongly correlated for nanoscale devices with L_{eff} ~35 nm.

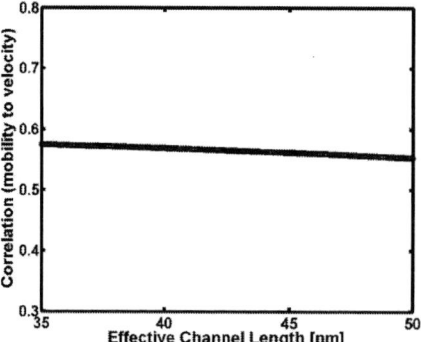

Fig. 11. Low-field mobility and virtual source velocity are strongly correlated by a factor of 0.55. The correlation has weak dependence on L_{eff}.

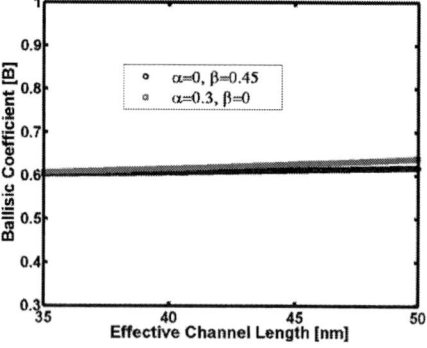

Fig. 12. Ballistic coefficient versus L_{eff} assuming B ~0.60 for L_{eff}=35 nm. The coefficient is a weak function of L_{eff}, and $\alpha < 0.50$ is required implying reduced interband scattering for holes under high strain.

3X hole mobility enhancement in epitaxially grown SiGe PMOSFETs on (110) Si substrates with high k / metal gate for hybrid orientation technology

Sachin Joshi[1*], Sagnik Dey[1], Se-hoon Lee[1], Cristiano Krug[2], Hoon Joo Na[2], Prasanna Sivasubramani[2]
Paul D. Kirsch[2,3], Prashant Majhi[2,4], Wenqian Wang[1], Alan Campion[1] and Sanjay K. Banerjee[1]
[1]University of Texas at Austin, [2]SEMATECH, [3]IBM, [4]Intel, [*]joshi@ece.utexas.edu

Demonstration of a 3X enhancement in hole mobility, in comparison to the universal Si / SiO$_2$ curve, is reported for epitaxially grown SiGe layers on (110) silicon substrates. This channel material is targeted for applications in high performance MOSFETs in a hybrid orientation technology using the (110) surface for PMOS devices. It may be envisioned that using epitaxial SiGe, while retaining the Si-like band structure at mole fractions below ~80% Ge, and using the (110) crystal surface could provide enhancements over and above the benefits accrued due to the use of Si (110) layers alone. Prior results[1] using Si$_{0.7}$Ge$_{0.3}$ indicate an enhancement of about 75% over universal Si, with a peak mobility of 210 cm^2/V-s. In this report, we demonstrate a higher Ge mole fraction (Si$_{0.4}$Ge$_{0.6}$) grown on a (110) Si wafer using a dislocation blocking technique[2] and achieve a peak hole mobility of 340 cm^2/V-s which is an enhancement of 3X over the universal Si /SiO$_2$ hole mobility. Consistent with other reports in the literature[3], a much lower epitaxial growth rate on the (110) surface was observed. Both strained and relaxed SiGe films have been fabricated on (110) Si wafers as observed from X Ray Diffraction data and the epitaxial growth process is currently under further optimization. Raman spectroscopy using a blue laser was used to confirm the presence of Si-Si and Si-Ge phonon peaks from the SiGe / strained Si cap layer grown on (110) Si.

MOSFETs were fabricated on both thick and thin epi SiGe films. An ultra thin (~ 1- 2 nm) epi Si cap grown on the SiGe layers serves to separate the Ge from the high k dielectric as well as form a SiO$_X$ interfacial layer between the SiGe channel and the high k gate dielectric. There is evidence that this cap layer is completely oxidized during the ozone based ALD high k deposition process[4]. Both epitaxial Si[5] as well as SiO$_X$ based capping layers[6] are reported to improve the interface for pure Ge devices. PMOSFETs were fabricated using a conventional 4 mask step process flow using a deposited field isolation oxide, ALD high k, metal gate electrode and implanted source / drain regions. Aluminum front and back side metallization was followed by a 20 min anneal in N2 at 400°C. This final anneal and the ALD process were the only significant thermal budget steps in the entire fabrication process. Even though thermal budgets are relatively low, the thicker SiGe films probably relax by formation of misfit dislocations. This is electrically observed as an enhanced CV response in the inversion region. For thinner, partially strained SiGe films the dislocation blocking layers are much thinner than the depletion region and do not interfere with the inversion CV response. Defect densities are currently under study. Gate leakage on both SiGe layers is comparable. Capacitance equivalent thickness is estimated to be ~2 nm. The thinner epi SiGe layers yielded well behaved high mobility PMOSFETs, while thicker films showed higher off state leakage, consistent with the enhanced CV response in inversion due to defects. Well behaved output and transfer characteristics on these devices enable a discussion of the drain current measurements. For the long channel devices under consideration, a drain current of 15 uA/um was observed for a 5 um device at a gate overdrive of 1V. Source drain junctions were created using a deep Boron implant (1x10^{15} cm^{-2}, 10 keV) in order to position the junction depletion region deeper in the silicon substrate and avoid probing the defects in the dislocation blocking layer. Three orders of magnitude I$_{ON}$ / I$_{OFF}$ and reasonable junction leakage were observed. Subthreshold swing is still degraded (166 mV / dec) and this is attributed to a higher dangling bond density for the (110) surface. Mobility was extracted on long channel MOSFETs using the split CV technique and a 3X enhancement was observed over the universal Si / SiO$_2$ hole mobility curve. Significant enhancements are also seen in comparison with data from bulk Ge (100)[6] and bulk Si (110)[7] PMOSFETs.

References:

[1] P. Liu *et al.* VLSI Symposium 2006, pp. 18.3-1-2
[2] Dey *et al.*, JEM, Vol. 25, No. 8, pp 1607 – 1612, August 2006, Joshi *et al.*, accepted, JEM
[3] Christine Ouyang *et al*, MRS 2006
[4] M.A. Quevedo-Lopez *et al.*, IEDM 2005
[5] P. Zimmerman *et al.* IEDM 2006
[6] Joshi *et al*, PCSI 34, accepted, EDL, Na *et al*, SISC 2006
[7] Pinto *et al*, MRS 2007

978-1-4244-1101-6/07/$25.00 ©2007 IEEE

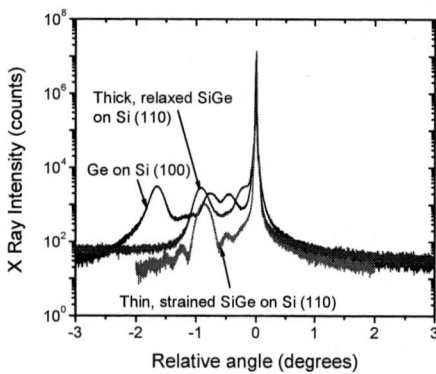

Fig. 1: XRD data from strained and relaxed SiGe layers grown on (110) Si substrates indicate Ge mole fractions much lower than similar layers grown on (100) wafers

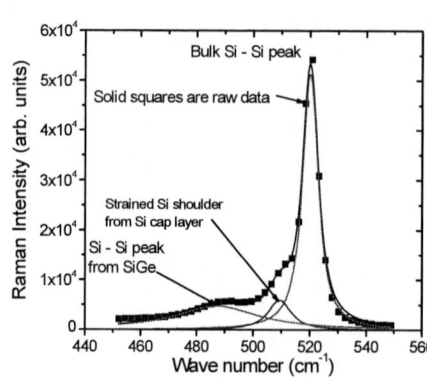

Fig. 2: Raman spectra on epi SiGe (110) layers indicate Si-Si phonon peaks from the SiGe and Si cap layers

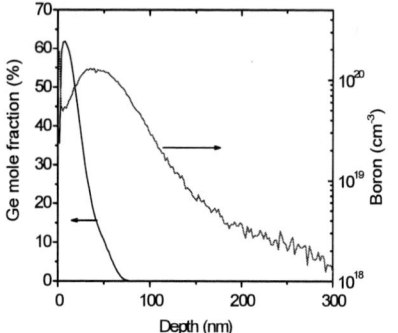

Fig. 3: SIMS profiles for Ge and Boron indicate a peak Ge concentration of 60%. The junction is designed to be deeper within the Si substrate to reduce leakage. Substrate doping was 1×10^{16} B

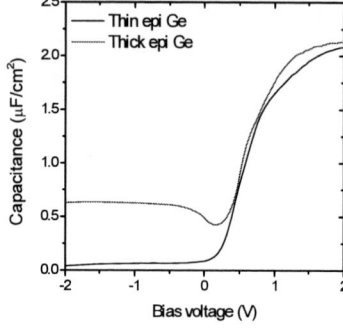

Fig. 4: High frequency CV measurements (100 kHz) indicate a much higher response in inversion / depletion for thicker SiGe layers. This is attributed to defects in the depletion region.

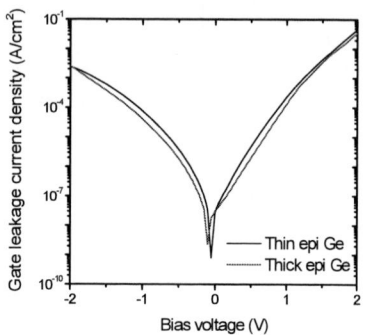

Fig. 5: Gate leakage current density for both samples is low, consistent with a relatively thick CET of 2 nm

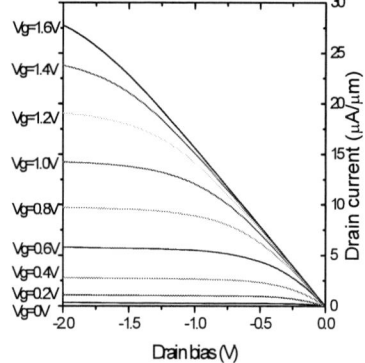

Fig. 6: Well behaved output characteristics with a drain current of 15 μA/μm for a gate overdrive of 1V.

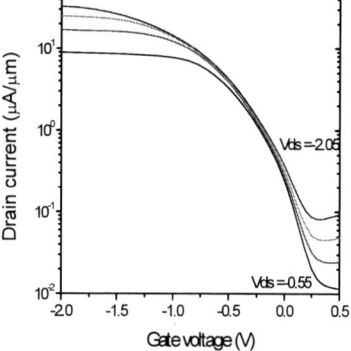

Fig. 7: Transfer characteristics are reasonable with an I_{ON}/I_{OFF} ratio of about 10^3, but also show a degraded subthreshold swing of about 160 mV / dec. This is attributed to a higher dangling bond density for the (110) surface

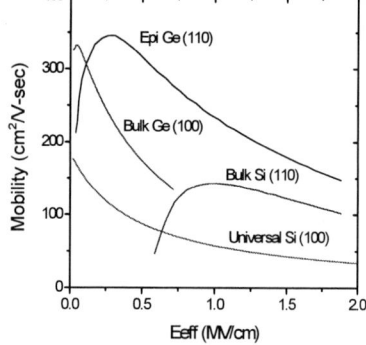

Fig. 8: Mobility data extracted using split CV technique is compared with data reported on bulk Ge (100) wafers[6] with a SiO$_X$ interface and bulk Si (110)[7] wafers. A 3X enhancement over universal Si / SiO$_2$ hole mobility is observed

High – Mobility, Low Parasitic Resistance Si/Ge/Si Heterostructure Channel Schottky Source/Drain PMOSFETs

Abhijit Pethe and Krishna Saraswat

Department of Electrical Engineering, 336 Serra Mall, Stanford CA 94305
Ph: (650)-725-3611 Fax: (650)-723-4659 E-mail: pethe@stanford.edu

Ge has been shown to provide excellent inversion hole mobility and hence is an ideal candidate to replace PMOSFET channels beyond the sub-20nm regime to maintain historic improvements in switching speeds [1]. Use of metal Source/Drain (S/D) architecture with low barrier to holes Φ_{Bp}[2] enables further improvement in the switching speeds by lowering parasitic resistance. We have built high performance PMOSFETs on Si substrates using a Si/Ge/Si heterostructure channel and NiSi Source/Drain regions. These devices exhibit ~2X improvement in mobility and orders of magnitude increase in the drive current without adversely affecting the OFF state leakage.

Recently, Si/Ge/Si heterostructures have been shown to be an excellent candidate for PMOSFETs due to their high hole mobility and reduced BTBT limited OFF current [3]. Fig. 1 shows the schematic of the device that was fabricated. Strained-Ge was epitaxially grown using germane on lightly doped n-Si substrate and capped with a thin Si using dichlorosilane. Fig. 2 shows a sample TEM of the stack. A 20nm thick low temperature SiO_2 (LTO) was deposited to form the gate insulator followed by ~400nm of polycrystalline $Si_{0.4}Ge_{0.6}$ to form the gate electrode. Gate patterning was followed with formation of LTO spacers along the gate edges. 100nm of Ni was then sputtered onto these wafers in a SCT sputter tool followed by a RTA in N_2 at 450°C for 1min to form the S/D $NiSi_xGe_{1-x}$ region. The unreacted Ni was removed in a concentrated HCl bath maintained at room temperature. 300nm of SiO_2 was then deposited at 300°C to form the intermetal dielectric. Al pads were then formed followed by anneal in forming gas at 350°C for 45min. Control NiGe/Si Schottky S/D MOSFETs were also fabricated on bulk Ge and Si.

Fig.3 depicts the I-V characteristics at the NiGe/Ge Schottky diode on n- and p-Ge. Ohmic characteristics were observed on lightly doped p-type Ge. A Φ_{Bn}=0.56eV (Φ_{Bp}=0.1eV) was extracted using low temperature measurements on Schottky diodes as shown in Fig. 4. Fig. 5 depicts the output characteristics of the heterostructure - channel device and the control Si MOSFETs. Due to the high barrier to holes at the NiSi/Si interface, Schottky diode like characteristics were observed in the linear regime in the control Si sample. Poor drive currents were obtained in the Si devices, due to this high barrier which limits the ON current. No such phenomena were observed on the Si/Ge/Si channel device indicating that a low Φ_{Bp} to the holes in the inversion channel was achieved at the NiSi-channel interface. Fig. 6 and 7 plot the transfer characteristics in the Si/Ge/Si channel devices obtained for fixed Ge- channel thickness (T_{Ge}) and cap-Si thicknesses (T_{CapSi}) contrasted against the control Si device. All the heterostructure channel devices exhibit higher drive currents. Changing the T_{Ge} leads to a V_T shifts observed in the I_S-V_G characteristics. Increasing T_{CapSi}, while maintaining the same Ge-channel thickness leads to exponential drop in the drive currents. This occurs due to the channel becoming more localized in the cap-Si layer, which has a higher Φ_{Bp} to the holes compared to the Ge channel. Fig. 8 and 9 depict the inversion hole mobilities extracted from these devices using the split-CV technique, as a function of T_{Ge} and T_{CapSi} respectively. An optimal thickness exists in both cases. Increasing T_{CapSi} relative to T_{Ge} results in the inversion charge being increasingly localized in the Si region leading to lower mobility. Due to the 4% lattice mismatch between Ge and Si, increasing T_{Ge} greater than the critical thickness results in a defective Ge layer, which exhibits lower mobility. Reducing the T_{CapSi} however leads to increased Ge concentration near the Si/SiO_2 gate interface, which causes higher D_{IT} and hence reduced mobility. Fig. 8 contrasts the switching characteristics of Schottky S/D PMOSFETs built with different channel materials. Si devices show low drive currents due to the high Φ_{Bp} at the Schottky Source-channel interface. These devices exhibit very low OFF currents due to the higher bandgap of Si, which provides a higher barrier to the back-injection of electrons. NiGe – S/D transistors with bulk Ge channels exhibit high drive currents due to the very small barrier between the holes in the channel and the NiGe. However these devices exhibit higher OFF currents due to the small barrier to electrons as well due to the small bandgap and hence increased back injection. Employing a thin Ge-channel near the gate interface of the Si MOSFET provides for high drive current due to both the low barrier at the Source-channel interface and the higher mobility of holes in Ge. The OFF state however, is limited by the large bandgap Si substrate, which has a large barrier to electrons and increased bandgap due to quantum mechanical confinemnet. Higher drive currents can hence be achieved without the huge OFF current penalty.

In conclusion, using a thin Ge layer within the inversion region of a Schottky Si – PMOSFET provides for higher hole mobility (~2X) and much higher drive currents due to almost zero barrier height to holes in the channel. Also the OFF state leakage is maintained at a low value because it is limited by the large barrier height in the wider bandgap Si and Ge quantization. The transistor hence, combines the advantages of high mobility, and low parasitic resistance and is an attractive candidate for scaling PMOSFETs into the sub-20nm regime.

References: [1] Chui et.al., IEDM 2002, pp. 437 [2] Xiong et.al., IEEE Trans. Elec. Devices, 52, 2005, pp. 1859 [3] Krishnamohan et.al., IEEE Trans. Elec. Dev., 53, 2006 pp. 990.

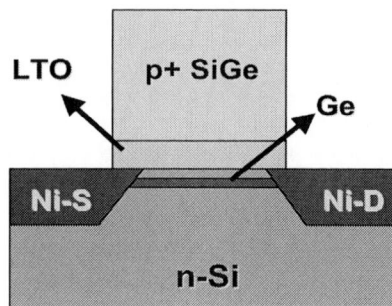

Fig. 1: Structure of the device that was fabricated. A thin Ge channel was grown pseudomorphically on the Si (100) substrates. LTO/poly-SiGe is used as gate stack.

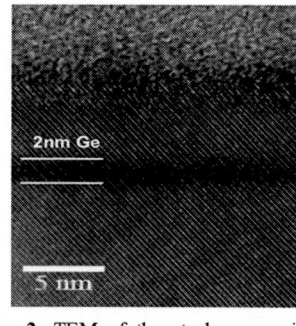

Fig. 2: TEM of the stack grown in an ASM Epi reactor. GeH₄ and DCS were used as reactive gases.

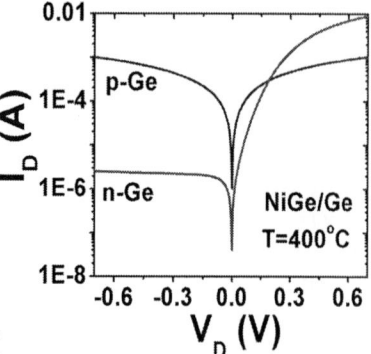

Fig 3: I_D-V_D on NiGe/Ge diodes. Ohmic characteristics observed on lightly doped p-Ge. Rectifying characteristics on n-Ge.

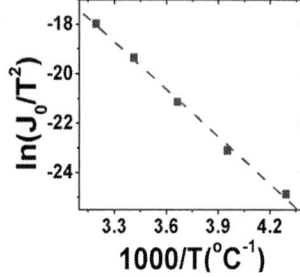

Fig 4: Arrhenius –plot for thermionic emission at the NiGe/n-Ge diode. Φ_{Bn}=0.56eV, Φ_{Bp}=0.1eV

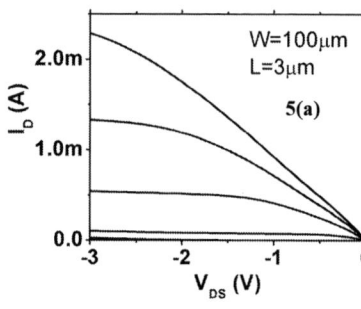

Fig. 5: Output characteristics of the Het-channel (a) and control Si – channel (b) Schottky S/D transistors. High drive currents observed in the het-channel with no Schottky-turn on in the linear regime.

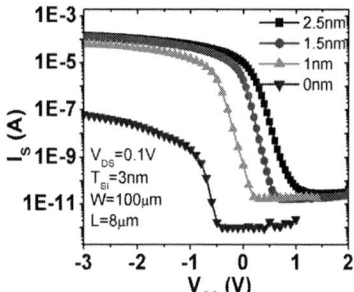

Fig. 6: Transfer Characteristics of Si/Ge/Si – channel Schottky S/D transistors for T_{CapSi}=3nm and varying T_{Ge}=2.5,2,1nm compared to control Si. The V_T shifts observed due to the VB offset and the ratio of the T_{CapSi} to the T_{Ge}.

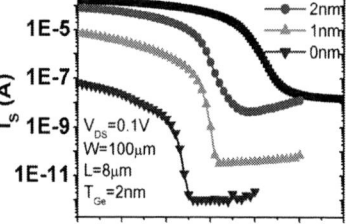

Fig. 7: Transfer Characteristics of Si/Ge/Si – channel Schottky S/D transistors for varying T_{CapSi}=4,2,1nm and fixed T_{Ge}=2nm compared to control Si. As the channel moves into the cap Si (with increasing T_{CapSi}), the barrier height to holes in the channel increases.

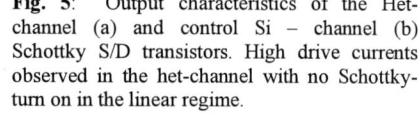

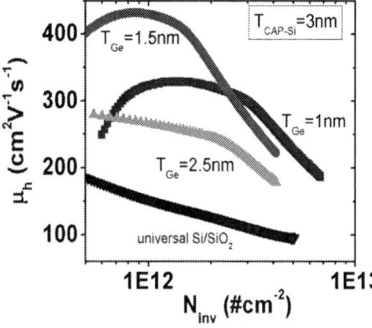

Fig. 8: Inversion hole mobility measured for a fixed T_{capSi}=3nm for varying T_{Ge}. For small T_{Ge} most of the inversion charge is located in Si and hence has less mobility. The thick Ge – layers are defective and hence have more scattering and less mobility

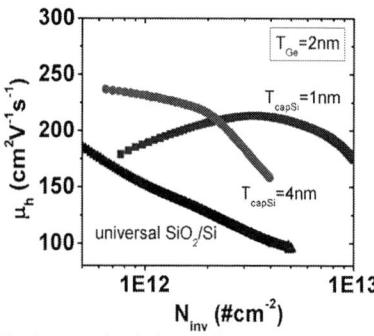

Fig. 9: Inversion hole mobility measures for a fixed T_{Ge}=2nm and varying T_{CapSi}. For small cap Si thickness, Ge is present close to Si/SiO₂ interface – high D_{IT}, lower mobility. For large T_{CapSi} mobility is reduced, as most of the inversion charge is located in the slower Si layer.

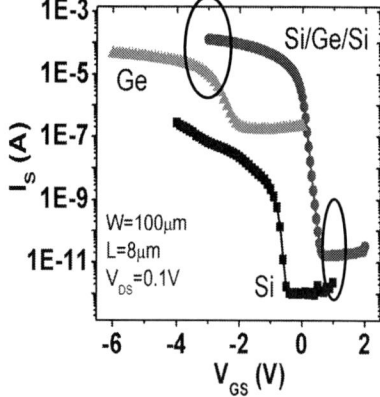

Fig. 10: Comparison between experimental Schottky S/D transistor. The Si/Ge/Si het – channel provides the advantages of higher drive due to high mobility in Ge and the lower IOFF due to the large barrier to Si substrate.

56

Band to Band Tunneling Study in High Mobility Materials : III-V, Si, Ge and strained SiGe

Donghyun Kim[1], Tejas Krishnamohan[1], Lee Smith[2], H.-S. Philip Wong[1], Krishna C. Saraswat[1]

[1]Department of Electrical Engineering, Stanford University, Stanford, CA USA, [2]Synopsys Corporation, Mountain View, CA USA

Phone : 650-725-7062, Email: dhkim81@stanford.edu

Abstract

Based on the complex bandstructure obtained by local empirical pseudopotential method (LEPM), we have developed a Band to Band Tunneling model (BTBT), which captures band structure information, all possible transitions between different valleys, energy quantization and quantized density of states. Theoretical model is verified by experimental study on tunnel diodes on various semiconductors. BTBT leakage current in high mobility (μ) channel Double Gate FET is studied. We have shown that quantum confinement effect in DGFET can suppress BTBT leakage current.

Introduction

High mobility materials are being investigated as channel materials in highly scaled MOSFETs to enhance performance [1]-[2]. The materials such as GaAs, InAs, Ge, strained Si and strained Ge have larger carrier mobility than Si, but the enhanced BTBT because of their smaller bandgap or direct band gap may limit their scalability [1]-[2]. Although it becomes important to predict the BTBT leakage of devices made with high-mobility materials, the parameters for BTBT models on these high mobility materials are not available.

Modeling Band to Band Tunneling (BTBT)

We have developed the BTBT model which takes into account full bandstructure, direct and phonon assisted indirect tunneling, quantum confinement effect, non-uniform electric field (non-local) and strain induced enhanced/suppressed tunneling in semiconductors. In this model, the tail of electron wavefunction penetrating into the bandgap is modeled and used to evaluate the interband matrix elements between conduction band states and valence band states. The final BTBT carrier generation rate (G_{BTBT}) was calculated by adding up the transition rates obtained by Fermi-Golden rule for all the possible transitions. Fig. 1 shows the possible transitions from valence bands to conduction bands such as Γ_V-Γ_C, Γ_V-L and Γ_V-Δ .

Complex Band Structure

Since BTBT process requires the movement of electron in the bandgap, it is important to predict the E-k relationship in the bandgap where k vector becomes complex number. Local empirical pseudopotential method(LEPM) is used to obtain complex bandstructure. We used simplex search method [3] to fit a model potential [4] to experimental band parameters and employed kinetic energy scaling factor [4] to have exact effective mass at bandedge. Fig. 2 shows the complex bandgap structures of Si, Ge, GaAs and InAs. In Fig. 3 G_{BTBT} rates are obtained under constant electric field (E-field) condition on these materials. Si exhibits lowest G_{BTBT} thanks to its indirect bandgap of 1.12eV. Although GaAs has larger bandgap than Si, due to its direct bandgap and to lighter light hole mass (m_{lhGaAs}=0.082m_0, m_{lhSi}=0.16m_0), it has higher G_{BTBT} than Si. InAs has highest G_{BTBT} because of its small bandgap (0.35eV) and small mass (m_{lh}=0.026m_0, $m_{\Gamma e}$=0.023m_0).

Comparison of the Model with Experimental Data

To validate our model, we collected existing experimental data on tunnel diodes and compared it to the theoretically calculated values. Kane[5] and Hurx[6] showed that G_{BTBT} can be expressed as $AF^\sigma exp(-F_0/F)$. Although most of existing data on BTBT is very out dated and hard to interpret, extracting F_0 is possible. Table 1 shows the theoretically calculated F_0 by our model, Kane's two band model and experimentally extracted values. Due to nonparabolicity of effective mass in bandgap, F_0 predicted by our model is lower than the ones by Kane's model and is closer to the experimental values. Gaul et al.[11] fabricated very abrupt junction GaAs PIN diode by MBE and their data matches very well with our prediction. Fig. 4 (a) shows G_{BTBT} obtained by experiment and our calculation. Fig 4 (b) shows the I-V of PIN diode and theoretical values given by our model and Kane's BTBT model.

The effect of tunnel orientation is studied in Fig. 5. Since in Si isotropic light hole dominates tunneling process (m_{lh}<m_Δ), both experimental and theoretical results exhibits no strong dependency of BTBT on orientation.

The experimental values of F_0 in Si vary from 19 to 35 MV/cm. When applied bias is low in a tunnel diode, the local E-field based model inevitably gives wrong results, since it doesn't count the non-uniform E-field effect. We performed the BTBT calculation on realistic tunnel diode where E-field is not uniform across the junction. Fig. 6 illustrates how much F_0 can vary, depending on the change in E-field (ΔF) while it tunnels. Theoretical values vary from 21 to 33MV/cm, which matches with the experimental result.

Strained / Quantized DGFET

We have studied off state leakage current limited by BTBT ($I_{OFF,MIN}$, Fig. 7) in DGFETs with high μ channels. Fig. 8(a) shows that compressive biaxial strain increases direct bandgap and decreases indirect bandgap in Ge. In Fig. 8 (b), $I_{OFF,MIN}$ in Ge initially decreases as applying strain, due to reduction of direct tunneling, but it eventually increases again because of large indirect tunneling. Ge with around 50% to 60% compressive strain gives lowest $I_{OFF,MIN}$. In thin body DGFET with the thickness from 3 to 10nm, the electrons and holes are quantized due to quantum confinement. Quantum confinement reduces the BTBT probability and $I_{OFF,MIN}$. In Fig 9, $I_{OFF,MIN}$ is plotted as a function of band gap (Eg) and thickness (T_{body}) for different materials. The results imply that by making body thin in a DGFET, it is possible to suppress $I_{OFF,MIN}$ over 1000x.

Conclusion

We have developed an accurate BTBT model and Local Empirical Pseudopotential method was used to simulate the complex band structure. The model is applied to tunnel diode and is in very good agreement with experiments. The BTBT model is implemented to study the BTBT leakage current in ultra thin body DGFET with high mobility channels.

[1] T. Krishnamohan et al, TED, May (2005) [2] A. Pethe et al, IEDM, (2005) [3] J.C. Lagarias et al, SIAM J. Optimization., 9 (1998) [4] K. Kim et al, PRB, 66, 045208 (2002) [5] E.O. Kane, J. Phys. Chem. Solids, 181 (1959) [6] Hurkx et al, Solid State Elect., 32 (1989) [7] Solomon et al, JAP., 95 (2004) [8] Tyagi, Solid State Elect., 11, (1968) [9] Logan et al, PR., 131 (1963) [10] Haitz, JAP., 36 (1965) [11] Gaul et al, Solid State Elect., 34 (1991) [12] Butcher et al, Solid State Elect. (1962) [13] Tyagi, Jap. J. Appl. Phys., 12 (1973) [14] Kleinknecht, Solid State Elect., 2 (1991)

978-1-4244-1101-6/07/$25.00 ©2007 IEEE

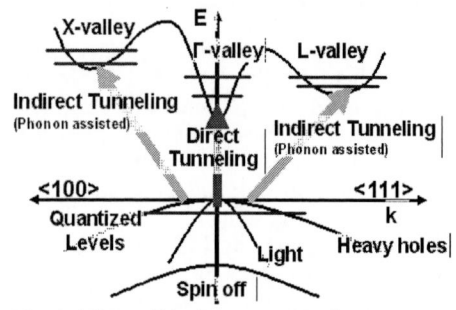

Fig. 1 All Possible (Direct and Indirect) Tunneling Paths, Γ-Γ, Γ-X and Γ-L, are captured by the model.

Mat. \ Fo		Calc. (MV/cm)	Exp. (MV/cm)	Kane (MV/cm)
Si [100]		21.6	19~35[6]-[10]	24
GaAs [111]		16.6	14.6[11]	17.5
Ge [111]	Direct	5.3	5.7[13]	5.9
	Indirect	4.7	4.6[12]	6.1
InAs[111]		1.3	1.3[14]	1.3

Table 1. Parameter F_0 of Kane/Hurkx model $AF^\sigma exp(-F_0/F)$, extracted by our BTBT model (by LEPM), Experiments and Kane(two band model for direct and EMA for indirect). The direction indicates tunnel direction.

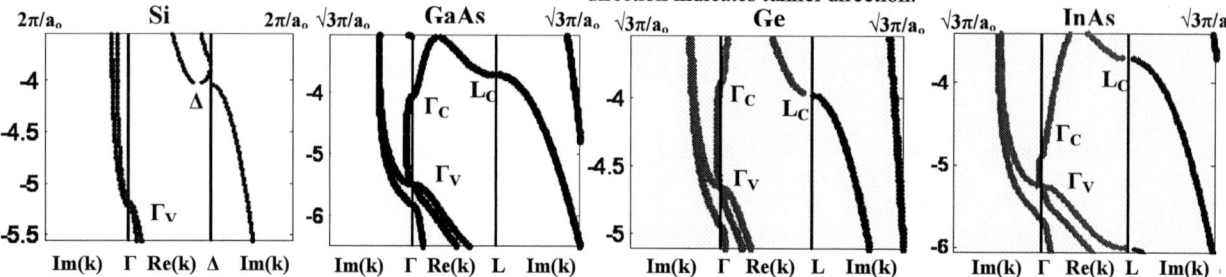

Fig. 2 Complex Band structure (E-k) of various Semiconductors. The real bands are plotted in the middle panel, and the associated imaginary bands are plotted in the left or right panels. At band edges (valleys), imaginary bands start to extend and connect conduction band and valence band. The imaginary band (Γ) has strong nonparabolicity while Δ or L don't.

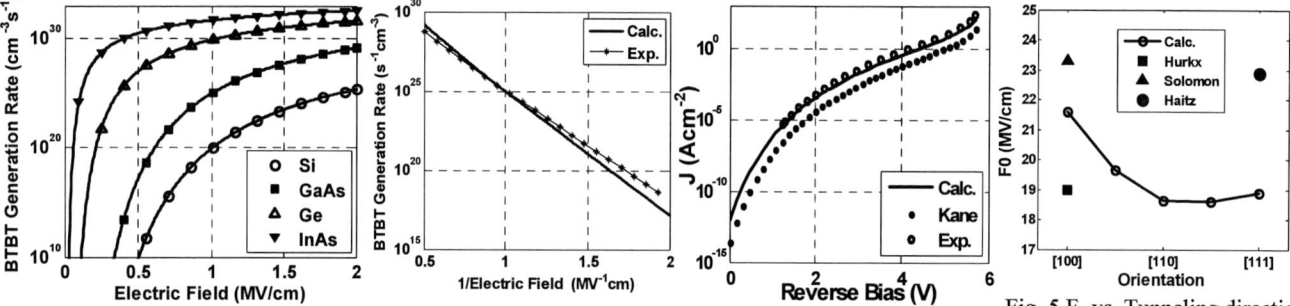

Fig. 3 BTBT generation rate vs. Electric Field. Si (indirect) has lowest BTBT rate while InAs (small bandgap & mass) is highest.

Fig. 4 (a) G_{BTBT} obtained by theory and experiments on GaAs PIN diode (Gaul[11]). (b) Current (PIN) vs. reverse bias. Solid line is calculated by our model (with impact ionization correction), Circles are by experiments and dots are by Kane's model.

Fig. 5 F_0 vs. Tunneling direction in Si. Since isotropic light hole is dominant in tunneling, there is no strong dependency on directions.

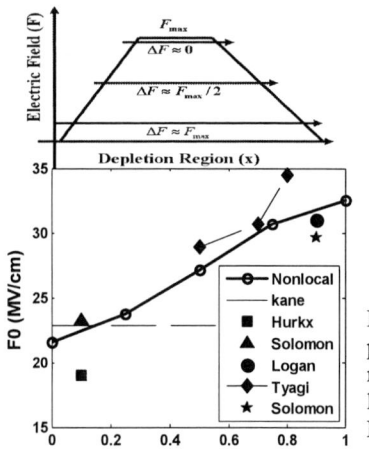

Fig. 6 F_0 vs. the change in E-field during BTBT. Simulation done by non-local model can explain the deviations in the experimental data.

Fig. 7 Typical Id-Vg for a p-MOSFET. $I_{OFF,MIN}$ is the minimum achievable leakage current in a MOSFET.

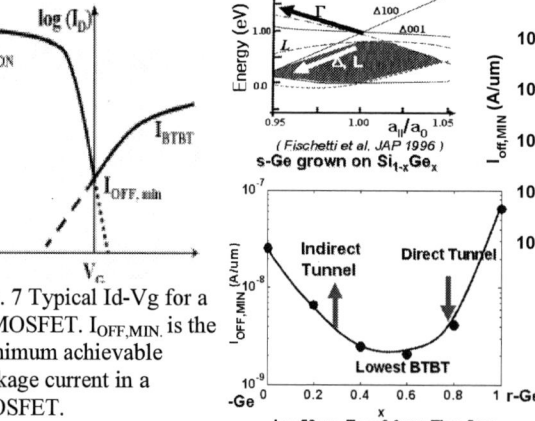

Fig. 8. (a) top. Electron bands of Ge vs. biaxial strain, (b) bottom. $I_{OFF,MIN}$ vs. biaxial strain. Compressive strain increases indirect BTBT while it decreases direct BTBT. Optimum point at around x=0.6.

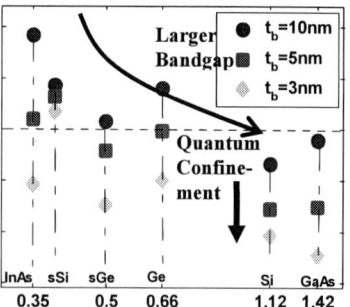

Fig. 9 . $I_{OFF,MIN}$ as a function of body thickness in DGFETs. Small bandgap materials have significantly large BTBT leakage. By reducing body thickness, BTBT can be reduced over 1000x. (Tox=0.9nm, Lg=15nm)

Process Integration and Electrical Properties of Bilayer Metal Gate/High-k MOSFETs

Ching-Huang Lu[1], Gloria M. T. Wong[1], Michael Deal[2], Bruce M. Clemens[1], and Yoshio Nishi[1,2]

[1]Department of Materials Science and Engineering, Stanford University, CA94305
[2]Department of Electrical Engineering, Stanford University, CA94305
Phone: 1(650)723-4194 Fax: 1(650)723-4659 Email: ratiug@stanford.edu

To eliminate poly depletion and to reduce gate leakage current, metal gate/high-k dielectric structures are needed for future CMOS technology. Work function engineering of metal gate electrodes and process integration with high-k dielectrics remain challenging. In this paper, we demonstrate one possible integration scheme to achieve dual work function metal gates on high-k dielectrics by utilizing bilayer metal electrodes and a selective metal wet etch process. We achieve a ~0.85eV difference in effective gate work functions and well-behaved transistor characteristics by using Al/W and Pt/W bilayer metal gates on HfO_2 dielectrics.

PVD HfO_2 dielectrics are grown by thermally oxidizing ultra-thin Hf metal thin films. The high-k/metal gate process is illustrated in Fig 1. The amount of fixed charge in the HfO_2/SiO_2 dielectrics is characterized by varying the HfO_2 or SiO_2 thickness [1]. Shown in Fig 2, the fixed charge densities of a $TiN/HfO_2/SiO_2$ gate stack are found to be $+1.33\times10^{11}/cm^2$ and $-1.68\times10^{12}/cm^2$ at the SiO_2/Si and HfO_2/SiO_2 interface respectively. The TiN work function then is extracted as 4.2eV and only slightly changes (+0.1eV) from that on SiO_2, which indicates little Fermi level pinning. To engineer the suitable work function for CMOS devices, a bilayer metal gate structure is utilized [2]. When the bottom layer is thin enough, the thick top metal layer can affect and dominate the effective gate work function due to a diffusion mechanism. This is demonstrated by the Al/W(3nm) and Pt/W(3nm) gate stacks on both SiO_2 and HfO_2 dielectrics. The extracted work functions are summarized in Fig 3. The two gate stacks show a 0.85eV difference in work function on both SiO_2 and HfO_2 dielectrics. A low work function Al/W electrode is suitable for NMOS devices and a Pt/W electrode with a high work function for PMOS applications.

To evaluate the threshold voltage control and transport properties, a gate-last process is used to fabricate Al/W and Pt/W metal gate MOSFETs on 4nm HfO_2. Fig 4 shows the I_d-V_d, I_d-V_g, and electron mobility of Al/W nFETs. Well-behaved I_d-V_d characteristics and subthreshold slope (~80mV/dec) are observed. In the I_d-V_g curve, the higher leakage current seen in the off state region might be due to reaction between Al atoms and HfO_2. Effective electron mobility of Al/W/SiO_2 FETs agrees well with universal mobility, but that of Al/W/HfO_2 devices degrades to ~60% of universal mobility. The transistor characteristics of Pt/W pFETs are shown in Fig 5. A ~70mV/dec subthreshold slope is seen in I_d-V_g and the off current is as low as $~5\times10^{-3}$ pA/μm. The effective hole mobility of Pt/W on SiO_2 agrees with the universal hole mobility but that on HfO_2 degrades to ~85% of universal mobility. The degradation of carrier mobility on HfO_2 is likely due to the fixed charge and defects in the dielectrics [3] rather than the bilayer metal gate stack. Transistor C-V characteristics of both Al/W and Pt/W stacks in Fig 6 show immunity from the gate depletion with EOT ~2nm.

Besides the ability to tune the work function, the thin bottom layers can act as buffer/etch-stop layers. In contrast to the traditional dual metal gate process [4], gate dielectrics can be protected from possible damage or contamination by employing selective etching of one top metal layer from the thin bottom metal layer. In Fig 7, three gate stacks are fabricated to demonstrate this possibility. In addition to the (1) Al/W and (2) Pt/W gate stacks, a third gate stack of (3) Pt/W is made by selective wet etching of Al from an Al/W gate stack and then depositing Pt to replace the Al just removed. The C-V characteristics of as-deposited gate stacks are shown in Fig 7a. All three gate stacks exhibit a similar V_{FB}, which indicates the effective work function in the unannealed case is controlled primarily by the bottom W layer. After a 400°C FGA, metal/metal interdiffusion occurs and results in the shift of V_{FB}'s shown in Fig 7b, giving the desired work functions. The reason why the (3) Pt/W gate stack shows a greater shift from that of the (2) Pt/W gate is likely due to the limited etch of W at the Al/W interface. The integration of dual work function metal electrodes is thus successfully demonstrated.

[1] R. Jha et al., EDL, vol.25, pp.420, 2004   [2] C-H Lu et al., EDL, vol.26, pp.445, 2005
[3] S. Saito et al., IEDM, pp.797, 2003   [4] S.B. Samavedam et al., IEDM, pp.433, 2002
This work was funded by Stanford INMP project and the devices were fabricated in SNF (NNIN).

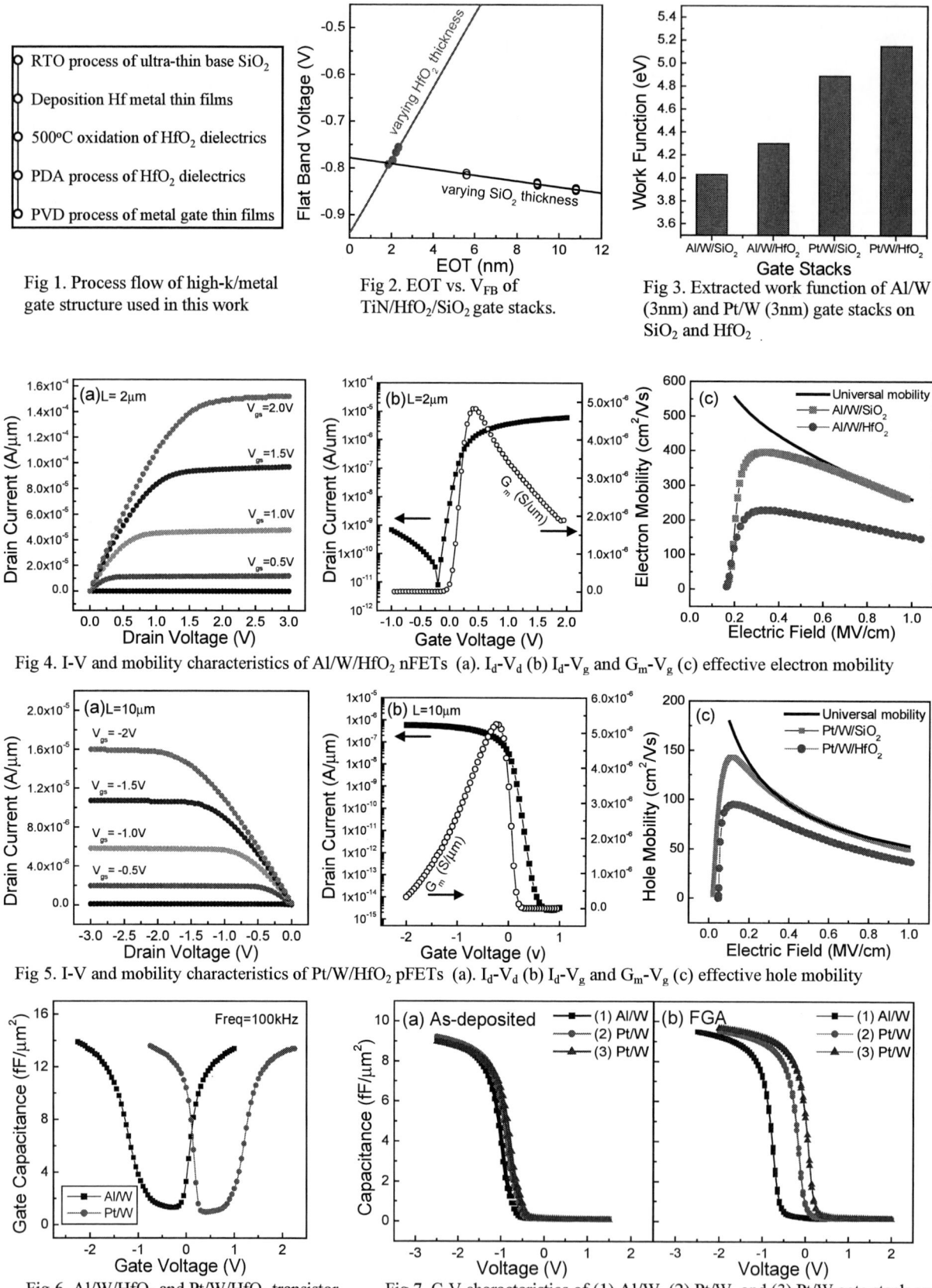

Fig 1. Process flow of high-k/metal gate structure used in this work

- RTO process of ultra-thin base SiO_2
- Deposition Hf metal thin films
- 500°C oxidation of HfO_2 dielectrics
- PDA process of HfO_2 dielectrics
- PVD process of metal gate thin films

Fig 2. EOT vs. V_{FB} of $TiN/HfO_2/SiO_2$ gate stacks.

Fig 3. Extracted work function of Al/W (3nm) and Pt/W (3nm) gate stacks on SiO_2 and HfO_2.

Fig 4. I-V and mobility characteristics of $Al/W/HfO_2$ nFETs (a). I_d-V_d (b) I_d-V_g and G_m-V_g (c) effective electron mobility

Fig 5. I-V and mobility characteristics of $Pt/W/HfO_2$ pFETs (a). I_d-V_d (b) I_d-V_g and G_m-V_g (c) effective hole mobility

Fig 6. $Al/W/HfO_2$ and $Pt/W/HfO_2$ transistor C-V characteristics showing no depletion effect with EOT=~2nm

Fig 7. C-V characteristics of (1) Al/W, (2) Pt/W, and (3) Pt/W gate stack on HfO_2. (a) As-deposited: three gates show a similar V_{FB}; (b) After FGA: V_{FB} shift due to metal/metal interdiffusion

Session III (McKenna Hall, Main Floor)

Poster Session

Monday PM, June 26th, 2006

Session Organizer: Miguel Urteaga, Teledyne

III-1
Effects of Source Access Resistance on Gate lag in AlGaN/GaN HEMTs and Current Slump Behavior
K. Horio, A. Nakajima and K. Itagaki, Faculty of Systems Engineering, Shibaura Institute of Technology, Saitama, JAPAN

III-2
Low-Voltage Organic Thin-Film Transistors with Improved Stability and Large Transconductance
H. Klauk[1], U. Zschieschang[1], R. Thomas Weitz[1], H. Meng[2], F. Sun[2], D. E. Keys[2], and C. R. Fincher[2], [1]Max Planck Institute for Solid State Research, Stuttgart, GERMANY and [2]Central Research and Development, Experimental Station, E. I. DuPont Company, Wilmington, Delaware, USA

III-3
4H-SiC RF BJTs with Long Pulse L-band Operation
F. Zhao[1], T. Shi[2], M. Mallinger[2], and B. Van Zeghbroeck[3], [1]Power Products Group, Microsemi Corporation, Bend, Oregon, USA, [2]Power Products Group, Microsemi Corporation, Santa Clara, California, USA, and [3]Department of Electrical and computer Engineering, University of Colorado, Boulder, Colorado, USA

III-4
Room-temperature lasing of type-II "W" GaSb/GaAs quantum dots embedded in InGaAs quantum well
J. Tatebayashi, A. Khoshakhlagh, G. Balakrishnan, S. H. Huang, M. Mehta, L. R. Dawson, and D. L. Huffaker, Center for High Technology Materials, University of New Mexico, Albuquerque, New Mexico, USA

III-5
SWCNT-SET fabricated by dispersion method with CMC solvent
T. Mori[1], K. Omura[1,2], S. Sato[1,3], M. Suzuki[1,4], K. Uchida[2], H. Yajima[2], and K. Ishibashi[1,4], [1]Advanced Device Laboratory, RIKEN, Saitama, JAPAN, [2]Department of Applied Chemistry, Tokyo University of Science, Tokyo, JAPAN, [3]Department of Physics, Tokyo University of Science, Tokyo, JAPAN, and [4]CREST, Japan Science and Technology (JST), Saitama, JAPAN

III-6
Microwave Noise Characterization of Enhancement-Mode AlGaN/GaN/InGaN/GaN Double-Heterojunction HEMTs
J. Liu, D. Song, Z. Cheng, W. C.–W. Tang, K. M. Lau and K. J. Chen, Department of Electronic and Computer Engineering, Hong Kong University of Science and Technology Clear Water Bay, Kowloon, HONG KONG

III-7 Student Paper
Electro-Thermally Coupled Power Optimization for Future Transistors
A. K. Chao, P. Kapur, E. Morifuji, K. C. Saraswat, and Y. Nishi, Department of Electrical Engineering, Stanford University, Stanford, California, USA

III-8
Dispersion Design of a Left-Handed Microstrip Line with Planar Double-Stub and Split-Ring Structures for Leaky Wave Radiation toward Functional RF Wireless Interconnect
M. Suhara[1,2], A. Shimizu[1], and T. Okumura[1,2], [1]Electrical Engineering, Graduate School of Engineering and [2]Electrical and Electronic Engineering, Graduate School of Science and Engineering, Tokyo Metropolitan University, Tokyo, JAPAN

III-9 Student Paper
High Power Vertical-structure GaN-based LEDs with Improved Current Spreading and Blocking Designs

T.-M. Chen[1,2], S.-J. Wang[1], K.-M. Uang[2], S.-L. Chen[1], C.-C. Tsai[1], H.-Y. Kou[1], W.-C. Lee[1], and H. Kuan[3], [1]Institute of Microelectronics, Dept. of Electrical Engineering, National Cheng Kung University, Tainan, TAIWAN, [2]Dept. of Electrical Engineering, Wu-Feng Institute of Technology, Chia-yi, TAIWAN, and [3]Optoelectronics Center of Far East University, Tainan, TAIWAN

III-10 Student Paper
Feasibility Study of Composite Dielectric Tunnel Barriers for Flash Memory

S. Verma[1], E. Pop[2], P. Kapur[1], P. Majhi[2], K. Parat[2], and K. C.Saraswat[1], [1]Center for Integrated Systems, Stanford University, Stanford, California, USA and [2]Intel Corporation,

III-11 Student Paper
Confined Optical Phonon Scattering in p-Silicon Nanowires

M. Nawaz[1], and J.-P. Leburton[2], [1]Department of Electrical and Computer Engineering, University of Illinois at Urbana-Champaign, Urbana, Illinois, USA and [2]Beckman Institute, Department of Electrical and Computer Engineering, University of Illinois at Urbana-Champaign, Urbana, Illinois, USA

III-12 Student Paper
Mid-Wavelength Infrared (MWIR) Avalanche Photodiode (APD) using InAs-GaSb type-II Strain layer Superlattice (SLS)

S. Mallick[1], K. Banerjee[1], S. Ghosh[1,] S. Krishna[2], and J. B. Rodriguez[2], [1]Lab for Photonics and Spintronics, Electrical and Computer Engineering Department, University of Illinois at Chicago, Chicago, Illinois, USA and [2]Center for High Technology Materials (ECE Dept), University of New Mexico, Albuquerque, New Mexico, USA

III-13 Student Paper
Body Thickness Optimization and Sensitivity Analysis for High Performance FinFETs

D. Lekshmanan, A. Bansal and K. Roy, School of ECE, Purdue University, West Lafayette, Indiana, USA

III-14 Student Paper
Flash Memory Fabricated with Protein-Mediated PbSe Nanocrystal Assembly as Floating Gate

S. Tang[1], C. Hun Lee[1], X. Gao[2] and S. K. Banerjee[1], [1]Microelectronics Research Center, The University of Texas at Austin, Austin, Texas, USA and [2]Texas Materials Institute, The University of Texas at Austin, Austin, Texas, USA

III-15 Student Paper
Novel Amorphous-Si AMOLED Pixels with OLED-independent Turn-on Voltage and Driving Current

B. Hekmatshoar, A. Z. Kattamis, K..e Cherenack, S. Wagner and J. C. Sturm, Princeton Institute for the Science and Technology of Materials (PRISM), Department of Electrical Engineering, Princeton University, Princeton, New Jersey, USA

III-16 Student Paper
AlGaN/GaN Bidirectional Power Switch

N. Tipirneni, B. Wang, A. Monti and G. Simin. University of South Carolina, Dept. of Electrical Engineering, Columbia, South Carolina, USA

III-17 Student Paper
n- and p-channel TaN/HfO2 MOSFETs on GaAs substrate using a germanium interfacial passivation layer

H.-S. Kim, I. Ok, F. Zhu, M. Zhang, S. Park, J. Yum, H. Zhao, and J. C. Lee, Microelectronics Research Center, Department of Electrical and Computer Engineering, The University of Texas at Austin, Austin, Texas, USA

III-18 Student Paper
Hexagonal Prism Blue Laser Diode with Low Threshold Power using Whispering Gallery Mode (WGM) Resonances

S. Kim[1] and T. D. Sands[2], [1]School of Electrical and Computer Engineering, West Lafayette, Indiana, USA and [2]Birck Nanotechnology Center, West Lafayette, Indiana, USA

III-19

Analytical Modeling of the Suspended-Gate FET and Design Insights for Digital Logic

K. Akarvardar[1], C. Eggimann[2], D.Tsamados[2], Y. Chauhan[2], G. C. Wan[1], A. M. Ionescu[2],and H. S. P. Wong[1], [1]Center for Integrated Systems, Stanford University, Stanford, California, USA and [2]Swiss Federal Institute of Technology, Lausanne, SWITZERLAND

III-20 Student Paper

Dynamic Two-Port Parameters of Ballistic Carbon Nanotube FETs: A Quantum Simulation Study

Y.n Ouyang and J. Guo, Electrical and Computer Engineering, University of Florida, Gainesville, Florida, USA

III-21 Student Paper

Schottky Drain AlGaN/GaN HEMTs for mm-wave Applications

X. Zhao, J.W. Chung, H. Tang, T. Palacios, Department of EECS and Microsystems Technology Laboratories,Massachusetts Institute of Technology, Cambridge, Massachusetts, USA

III-22

Barrier layer downscaling of InAlN/GaN HEMTS

F. Medjdoub[1], J.-F. Carlin[2], M. Gonschorek[2], E. Feltin[2], M.A. Py[2], M. Knez[3], D. Troadec[4], C. Gaquière[4], A. Chuvilin[5], U. Kaiser[5], N. Grandjean[2], and E. Kohn[1], [1]University of Ulm (EBS), Ulm, GERMANY, [2]EPFL, Lausanne, SWITZERLAND, [3]Max Planck Institute of Microstructure Physics, Weinberg, GERMANY, [4]IEMN, Villeneuve d'ascq, FRANCE, and [5]University of Ulm (ME), Ulm, GERMANY

III-23 Student Paper

Estimation of Trap Density in AlGaN/GaN HEMTs from Subthreshold Slope Study

J. W. Chung, X. Zhao, and T. Palacios Department of Electrical Engineering and Computer Science, Microsystems Technology Laboratories, Massachusetts Institute of Technology, Cambridge, Massachusetts, USA

III-24 Student Paper

High Performance ZnO Nanowire FET with ITO Contacts

Matthew A. Hollister, John D. Le, Guanghua Xiao, Xuekun Lu, and Richard A. Kiehl, Department of Electrical & Computer Engineering, The University of Minnesota, Minneapolis, Minnesota.

III-25 Student Paper

High Efficiency Oxide-Confined High-Index-Contrast Broad-Area Lasers with Reduced Threshold Current Density and Improved Near-Field Profile

Di Liang and Douglas C. Hall, Department of Electrical Engineering, University of Notre Dame, Notre Dame, Indiana, USA

III-26 Student Paper

Inversion-type enhancement-mode InP MOSFETs with ALD Al2O3, HfO2 and HfAlO nanolaminates as high-k gate dielectrics

Y.Q. Wu[1], Y. Xuan[1], P.D. Ye[1], Z. Cheng[2], A. Lochtefeld[2], [1]School of Electrical and Computer Engineering, Purdue University, West Lafayette, Indiana, USA and [2]AmberWave Systems Corp., Salem, New Hampshire, USA

III-27 Student Paper

Barrier Lowering and Widening of Schottky Barrier MOSFETs by Self-Aligned Multiple Workfunction Gate

S.-P. Yeh[1] and C.-Hsing Shih[2], [1]Institute of Electronics Engineering, National Tsing Hua University, Hsinchu, TAIWAN and [2]Department of Electrical Engineering, Yuan Ze University, Taoyuan, TAIWAN

III-28 Student Paper

Reliability of 4H-SiC DMOSFETs Evaluated by Bias Stressing

T. Okayama[1], S. D. Arthur[2], J. L. Garrett[2], and M.V. Rao[1], [1]Department of Electrical and Computer Engineering, George Mason University, Fairfax, Virginia, USA and [2]Semiconductor Technology Division, GE Global Research Center, Niskayuna, New York, USA

III-29

A 53% High Efficiency GaAs Vertically Integrated Multi-junction Laser Power Converter

D. Krut, R. Sudharsanan, T. Isshiki, R. King and N. H. Karam, Spectrolab Inc., a Boeing Company, Sylmar, California, USA

III-30 Student Paper

Native-Oxide-Confined High Index Contrast InAs Quantum-Dot Laser Diodes

J. Wang[1], D. C. Hall[1], V. Tokranov[2] and S. Oktyabrsky[2], [1]Department of Electrical Engineering, University of Notre Dame, Notre Dame, Indiana, USA and [2]School of NanoSciences and NanoEngineering, University at Albany-SUNY, Albany, New York, USA

III-31 Student Paper

Surface Treatment for Leakage Reduction in AlGaN/GaN HEMTs

R. Chu[1], L. Shen[1], N. Fichtenbaum[1], S. Keller[1], A. Corrion[2], C. Poblenz[2], J. Speck[2], and U. Mishra[1], [1]ECE Department, University of California, Santa Barbara, California, USA and [2]Materials Department, University of California, Santa Barbara, California, USA

III-32 Student Paper

AlGaN/GaN HEMT with High PAE and Breakdown Voltage Grown by Ammonia MBE

Y. Pei[1], C. Suh[1], R. Chu[1], F. Recht[1], L. Shen[1], A. Corrion[2], C. Poblenz[2], J. Speck[2] and U.K. Mishra[1], [1]Electrical and Computer Engineering Department, University of California, Santa Barbara, California, USA, and [2]Materials Department, University of California, Santa Barbara, California, USA

III-33 Student Paper

Analytical Model of Apparent Threshold Voltage Lowering Induced by Contact Resistance in Amorphous Silicon Thin Film Transistors

B. Hekmatshoar, K. Long, S. Wagner and J. C. Sturm, Princeton Institute for the Science and Technology of Materials (PRISM), Department of Electrical Engineering, Princeton University, Princeton, New Jersey, USA

III-34 Student Paper

On-Chip Clocking Scheme for Nanomagnet QCA

M. T. Alam[1], M. Niemier[2], W. Porod[1], S. Hu[2], M. Putney[2], J. DeAngelis[2], and G. H. Bernstein[1], [1]Center for Nano Science and Technology, Dept. of Electrical Engineering, University of Notre Dame, Notre Dame, Indiana, USA and [2]Dept. of Comp. Sci. and Eng., University of Notre Dame, Notre Dame, Indiana, USA

III-35

Electrical Characterization of Vertical InAs Nanowires on Si

C. Rehnstedt, T. Mårtensson, C. Thelander, L. Samuelson and L.-E. Wernersson, Solid State Physics / Nanometer Consortium, Lund University, Lund, SWEDEN

III-36 Student Paper

Band-gap engineering of enhanced spin-orbit interactions in InAs/AlGaAs heterostructures for Datta-Das spin transistor

T. Matsuda, M. Ohno and K. Yoh, Research Center for Integrated Quantum Electronics, Hokkaido University, Sapporo, Hokkaido, JAPAN

III-37

Oxide-Induced Noise in Carbon Nanotube Devices

Y.-M. Lin and P. Avouris, IBM T. J. Watson Research Center, Yorktown Heights, New York, USA

III-38 Student Paper

In-Situ Inelastic Electron Tunneling Spectroscopy of Bistable Molecular Junction Devices

H. Yoon[1], L. Cai[1], M. Maitani[2], D. L. Allara[3], and T. S. Mayer[1], [1]Department of Electrical Engineering, [2]Department of Materials Science and Engineering, and [3]Department of Chemistry, The Pennsylvania State University, University Park, Pennsylvania, USA

III-39
Electric-Field Dependence of Junction Temperature in GaN HEMTs
V. Mehrotra, K. Boutros, and B. Brar, Teledyne Scientific Company, Thousand Oaks, California, USA

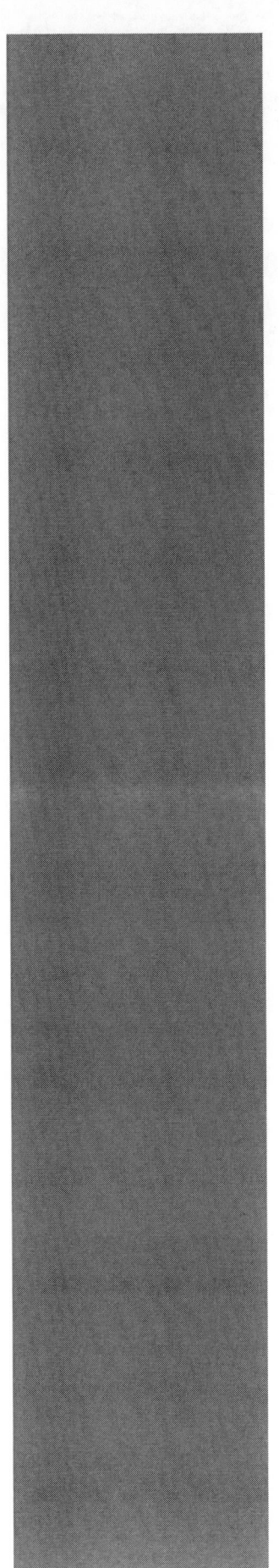

Effects of Source Access Resistance on Gate lag in AlGaN/GaN HEMTs and Current Slump Behavior

K. Horio, A. Nakajima and K. Itagaki

Faculty of Systems Engineering, Shibaura Institute of Technology
307 Fukasaku, Minuma-ku, Saitama 337-8570, Japan
(TEL: +81-48-687-5813, FAX: +81-48-687-5198, E-mail: horio@sic.shibaura-it.ac.jp)

Recently, AlGaN/GaN HEMTs have received great attention because of their potential applications to high power microwave devices [1]. However, slow current transients are often observed even if the drain voltage V_D or gate voltage V_G is changed abruptly (called drain lag or gate lag) [2]. The slow transients mean that dc I-V curves and RF I-V curves become quite different, resulting in lower RF power available than that expected from dc operation [1,2]. This is called power slump or current slump. These are serious problems, and there are many experimental works reported [1-6]. However, only a few theoretical works are made recently[5,7], where effects of a donor-type surface state (near the valence band) or effects of a bulk deep acceptor (~ 1 eV above the midgap of GaN) are studied for gate lag or pulsed I-V curves in AlGaN/GaN HEMTs. But, the type of traps and their energy levels seemed to be artificial. So, in this work, we have made simulations of AlGaN/GaN HEMTs with a semi-insulating buffer layer in which trap levels based on experiments are considered, and showed that lag phenomena and current slump could be reproduced. Particularly, we show that the gate lag is correlated with relatively high source access resistance of AlGaN/GaN HEMTs, and also give a way to reduce the current slump of the FETs.

Fig.1 shows an analyzed AlGaN/GaN HEMT structure. As a model for the semi-insulating buffer layer, we use a three-level compensation model which includes a shallow donor, a deep donor and a deep acceptor. Some experiments show that two levels ($E_C - 1.7$ eV, $E_C - 2.85$ eV) are associated with current slump in GaN-based FETs [2], so that we use energy levels of $E_C - 2.85$ eV (or $E_V + 0.6$ eV) for the deep acceptor and of $E_C - 1.7$ eV for the deep donor. Other experiments show shallower energy levels for deep donors in GaN [8,9], and hence we vary the deep donor's energy level E_{DD} as a parameter. We have calculated the drain current responses when V_D and/or V_G are changed abruptly.

Fig.2 shows calculated responses of drain current I_D when V_D is raised or lowered abruptly, where V_G is kept constant. When V_D is raised, I_D overshoots the steady-state value, because electrons are injected into the buffer layer, and deep traps there need certain time to capture these electrons. On the other hand, when V_D is lowered, I_D remains at a low value for some periods and begins to increase slowly, showing drain lag behavior. The drain currents begin to increase as the deep donors begin to emit electrons. These drain lags are also reported experimentally in AlGaN/GaN HEMTs [2,5].

We have next calculated a case when V_G is also changed. Fig.3 shows calculated turn-on characteristics when V_G is changed from the threshold voltage V_{th} to 0 V, with on-state drain voltage V_{Don} as a parameter. The characteristics are similar to those in Fig.2, and hence the change of V_D is regarded as essential in this case. However, as seen in the uppermost curve of Fig.3, some transients are observed when only V_G is changed. This indicates that gate lag occurs due to deep levels in the buffer layer. We will describe below why the gate lag arises. Fig.4 shows a comparison of (a) conduction-band-edge energy profiles, (b) electron density profiles, and (c) ionized-deep donor density $N_{DD}{}^+$ profiles between the off state (left: $V_D = 20$ V, $V_G = V_{th} = -9.24$ V) and the on state (right: $V_D = 20$ V, $V_G = 0$ V). Note that only V_G is different here. From Fig.4(a), in the on state, some potential drops are observed between source and gate (and between gate and drain), indicating that source access resistance (and drain access resistance) become important. It is understood that due to this potential drop at the source side, when V_G becomes negative and the channel is depleted, electrons do not all flow into the source and drain electrodes, but can be injected into the buffer layer as seen in Fig.4(b). These electrons are captured by deep donors, and hence $N_{DD}{}^+$ decreases in the off state, as seen in Fig.4(c). Because of this increase in negative space charges in the buffer layer, even if V_G is switched on, I_D remains at a low value until the deep donors begin to emit electrons, showing gate-lag behavior. The high access resistance in AlGaN/GaN HEMTs is considered problematic because it degrades the high-frequency performance [10].

Next, we have studied dependence of current slump on deep-level densities in the buffer layer. Fig.5 shows calculated I_D-V_D curves for different deep-acceptor density N_{DA}. In this figure, we plot by point (x) the drain current at $t = 10^{-8}$ s after V_G is switched on. This is obtained from the turn-on characteristics, and this curve corresponds to a quasi-pulsed I-V curve with pulse width of 10^{-8} s. It is seen that current reduction in pulsed I-V curves is more pronounced for higher N_{DA}, indicating current slump behavior. This is because trapping effects become more remarkable due to higher ionized (empty) deep-donor density $N_{DD}{}^+$. It is concluded that the acceptor density must be made low to minimize current slump. Finally, we have studied the dependence on off-state drain voltage V_{Doff}. As seen in Fig.5, the current slump is more pronounced for higher V_{Doff} (and the contribution of drain lag is large). This is because for higher V_D, electrons are injected deeper into the buffer layer and more electrons are captured by the deep donors, resulting in heavier current reduction for higher V_{Doff}. This tendency is also reported experimentally in AlGaN/GaN HEMTs [4].

In summary, two-dimensional transient simulations of AlGaN/GaN HEMTs have been made in which a deep donor and a deep acceptor are considered in the buffer layer. The lag phenomena and current slump could be reproduced. Particularly, it has been shown that the gate lag is correlated with relatively high source access resistance of AlGaN/GaN HEMTs, and that the drain lag could be a major cause of current slump. It is concluded that an acceptor density in the buffer layer should be made low to minimize current slump, although current cutoff behavior may be degraded when the gate length becomes short.

978-1-4244-1101-6/07/$25.00 ©2007 IEEE

[1] U. K. Mishra, P. P. Parikh and Y.-F. Wu, *Proc. IEEE*, vol.90, pp.1022-1031, 2002. [2] S. C. Binari, P. B. Klein and T. E. Kazior, *Proc. IEEE*, vol.90, pp.1048-1058, 2002. [3] G. Koley et al., *IEEE Trans. Electron Devices*, vol.50, pp.886-893, 2003. [4] A. Koudymov et al., *IEEE Electron Device Lett.*, vol.24, pp.680-682, 2003. [5] G. Meneghesso et al., *IEEE Trans. Electron Devices*, vol.51, pp.1554-1561, 2004. [6] V. Desmaris et al, *IEEE Trans. Electron Devices*, vol.53, pp.2413-2417, 2006. [7] N. Braga et al., *Proc. CSIC Symp.*, pp.287-290, 2004. [8] W. Kruppa, S. C. Binari and K. Doverspike, *Electron. Lett.*, vol.31, pp.1951-1952, 1995. [9] H. Morkoc, *Nitride Semiconductors and Devices*, Springer-Verlag, 1999. [10] T. Palacios et al, *IEEE Trans. Electron Devices*, vol.52, pp.2117-2123, 2005.

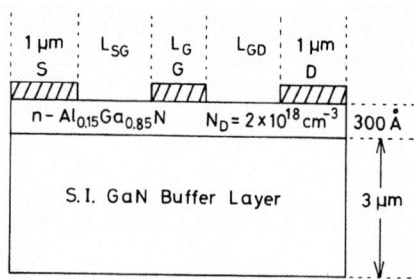

Fig.1 Modeled AlGaN/GaN HEMT analyzed in this study.

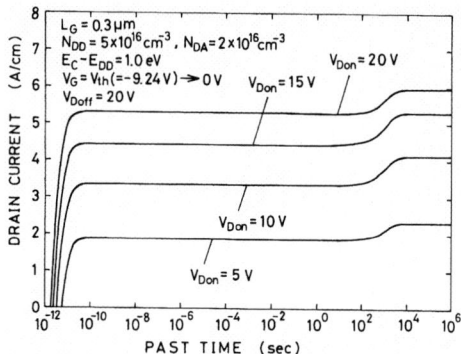

Fig.3 Calculated turn-on characteristics of AlGaN/GaN HEMT when V_G is changed from threshold voltage V_{th} to 0 V, with on-state drain voltage V_{Don} as a parameter. Off-state drain voltage V_{Doff} is 20 V. $E_C - E_{DD} = 1.0$ eV. $N_{DD} = 5\times10^{16}$ cm^{-3} and $N_{DA} = 2\times10^{16}$ cm^{-3}.

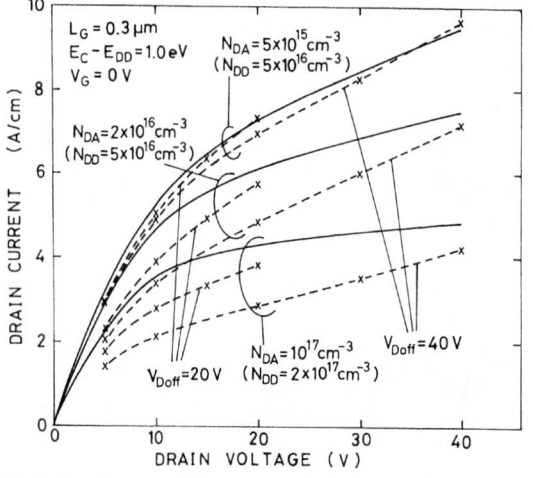

Fig.5 Steady-state *I-V* curves ($V_G = 0$ V; solid lines) and quasi-pulsed *I-V* curves (x; $t = 10^{-8}$ s) from off points for AlGaN/GaN HEMTs with different N_{DA}. $E_C - E_{DD} = 1.0$ eV.

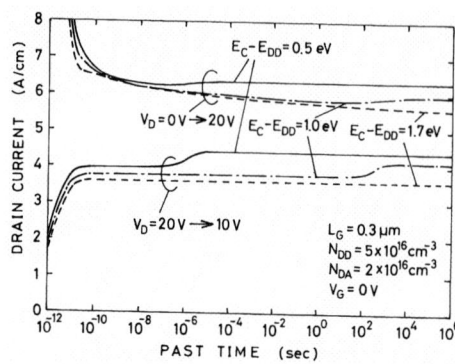

Fig.2 Comparison of drain-current responses of AlGaN/GaN HEMT as a parameter of deep donor's energy level E_{DD} when V_D is raised abruptly from 0 V to 20 V (upper) or when V_D is lowered abruptly from 20 V to 10 V (lower). $V_G = 0$ V. The deep-donor density N_{DD} is 5×10^{16} cm^{-3} and the deep-acceptor density $N_{DA} = 2\times10^{16}$ cm^{-3}.

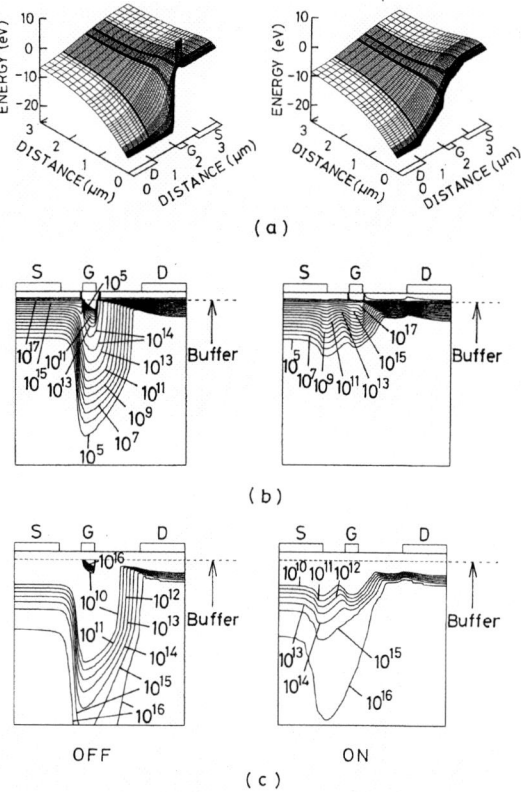

Fig.4 (a) Conduction-band-edge energy profiles, (b) electron density profiles, and (c) ionized deep-donor density N_{DD}^+ profiles when only V_G is different. The left is for $V_G = V_{th} = -9.24$ V and $V_D = 20$ V (OFF), and the right is for $V_G = 0$ V and $V_D = 20$ V (ON). $N_{DD} = 5\times10^{16}$ cm^{-3}, $N_{DA} = 2\times10^{16}$ cm^{-3} and $E_C - E_{DD} = 1.0$ eV.

Low-Voltage Organic Thin-Film Transistors with Improved Stability and Large Transconductance

Hagen Klauk, Ute Zschieschang, Ralf Thomas Weitz
Max Planck Institute for Solid State Research, Heisenbergstr. 1, 70569 Stuttgart, Germany
phone: +49 711 689-1401, fax: +49 711 689-1472, email: H.Klauk@fkf.mpg.de

Hong Meng, Fangping Sun, Dalen E. Keys, Curtis R. Fincher
Central Research and Development, Experimental Station, E. I. DuPont Company, Wilmington, DE 19880, U.S.A.

Pentacene is among the most popular organic semiconductors for organic thin-film transistors (TFTs), due to its relatively large carrier mobility [1,2]. However, the stability of pentacene TFTs under continuous dynamic operation may be insufficient for future applications. Here we compare the static and dynamic performance and the operational stability of low-voltage organic TFTs based on pentacene and a recently synthesized organic semiconductor, di(phenylvinyl)anthracene (DPVAnt).

DPVAnt was synthesized by a Suzuki coupling reaction [3], and pentacene was purchased from commercial sources. Both materials were purified by temperature-gradient sublimation. TFTs and circuits were fabricated on glass substrates using an inverted staggered TFT structure with evaporated aluminum gates, a thin gate dielectric based on a solution-processed self-assembled monolayer ($0.7 \ \mu F/cm^2$), an evaporated organic semiconductor layer, and evaporated gold top contacts [4]. All layers were patterned using manually aligned shadow masks. The device structure and the chemical structures of the semiconductors are shown in Figures 1 and 2. All electrical measurements were carried out in ambient air.

Figures 3 and 4 show the electrical characteristics of a pentacene TFT and a DPVAnt TFT before, during, and after an operational stability test. During the test a continuous square-wave signal with a period of 10 sec is applied to the gate electrode, and the drain current is monitored using a Semiconductor Parameter Analyzer. As can be seen, the performance of both TFTs changes during the test. In particular, carrier mobility and on-current of the pentacene TFT are reduced by more than an order of magnitude (mobility from $0.4 \ cm^2/Vs$ to $0.02 \ cm^2/Vs$, on-current from $2 \ \mu A$ to $80 \ nA$), while the threshold voltage appears to be unchanged. On the other hand, the mobility of the DPVAnt TFT is stable at $0.3 \ cm^2/Vs$. The threshold voltage of the DPVAnt TFT has shifted by about -0.5 V during the test, reducing the on-current by about 50% (from $1.3 \ \mu A$ to $700 \ nA$), and the subthreshold behavior of the DPVAnt TFT is somewhat degraded. Overall, the DPVAnt TFT shows better operational stability compared with the pentacene TFT.

In order to obtain TFTs with a high cut-off frequency, it is necessary to reduce the parasitic capacitances and increase the transconductance. The latter can be accomplished by reducing the channel length, but it also requires a small contact resistance. Figure 5 shows a shadow-mask-patterned pentacene TFT with a channel length of $10 \ \mu m$, a channel width of $100 \ \mu m$, a total gate capacitance of about 15 pF, and a transconductance of $4 \ \mu S$ ($40 \ \mu S/mm$) at low voltage ($V_{DS} = -1.5$ V, $V_{GS} \sim -2.0 \ .. \ -2.5$ V). For DPVAnt TFTs we have obtained a transconductance of $30 \ \mu S/mm$. Analyzing pentacene and DPVAnt TFTs with channel length ranging from $50 \ \mu m$ to $10 \ \mu m$ we have extracted the gate-bias-dependent contact resistance (sum of drain and source resistance [5]) in the linear regime and found a value of $850 \ \Omega \cdot cm$ for pentacene and $1.3 \ k\Omega \cdot cm$ for DPVAnt (see Figure 6). This is more than an order of magnitude lower than the contact resistance of photolithographically patterned bottom-contact pentacene TFTs [6].

Based on the transconductance and the gate capacitance, a cut-off frequency in the range of 30 to 40 kHz would be expected. 5-stage ring oscillators with a saturated-load design and a channel length of $10 \ \mu m$ oscillate with a minimum signal propagation delay of $200 \ \mu sec$ per stage for pentacene and $320 \ \mu sec$ for DPVAnt (see Figure 7). This is within a factor of ~10 of the cut-off frequency estimation and to our knowledge represents the smallest signal delay reported for an organic circuit operating with a supply voltage of 5 V or less. Individual inverters driven with a signal generator respond to square-wave input pulses with a period down to $100 \ \mu sec$ ($f = 10$ kHz), and from the recorded inverter output signals a time constant of $8 \ \mu sec$ has been extracted.

[1] T. W. Kelley et al., *J. Phys. Chem. B* **107**, 5877 (2003)

[2] S. Lee et al., *Appl. Phys. Lett.* **88** 162109 (2006)

[3] H. Meng et al., *J. Am. Chem. Soc.* **128** 9304 (2006)

[4] H. Klauk et al., *Nature* **445** 745 (2007)

[5] H. Klauk et al., *Solid-State Electronics* **47** 297 (2003)

[6] D. J. Gundlach et al., *J. Appl. Phys.* **100** 024509 (2006)

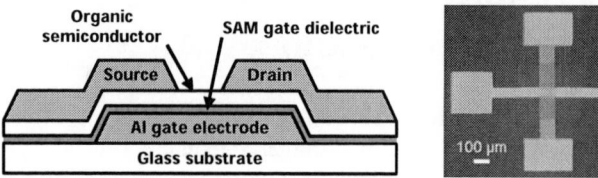
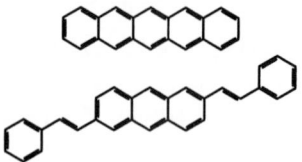

Fig. 1. Schematic cross section and photograph of the TFTs.

Fig. 2. Chemical structures of pentacene and diphenylvinyl-anthracene (DPVAnt).

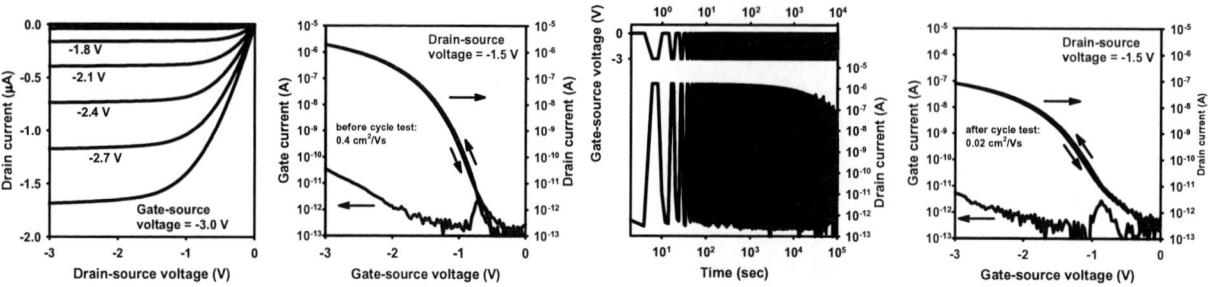

Fig. 3. Electrical characteristics of a pentacene TFT with a channel length of 20 μm and a channel width of 100 μm before and after an operational stability test with 10,000 switching cycles.

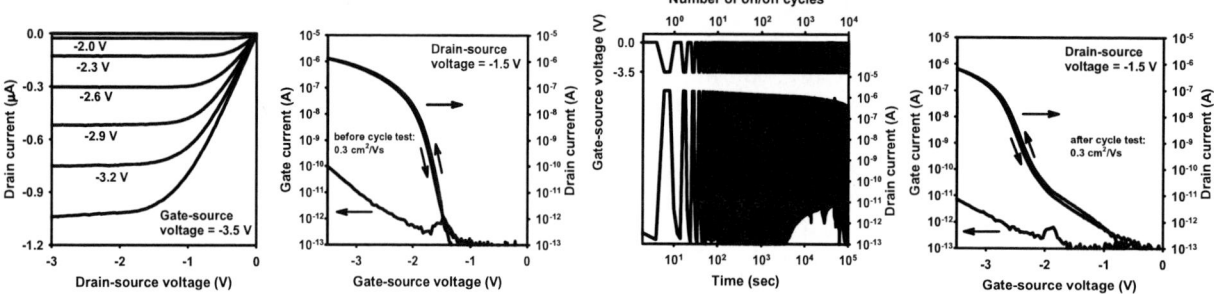

Fig. 4. Electrical characteristics of a DPVAnt TFT with a channel length of 20 μm and a channel width of 100 μm before and after an operational stability test with 10,000 switching cycles.

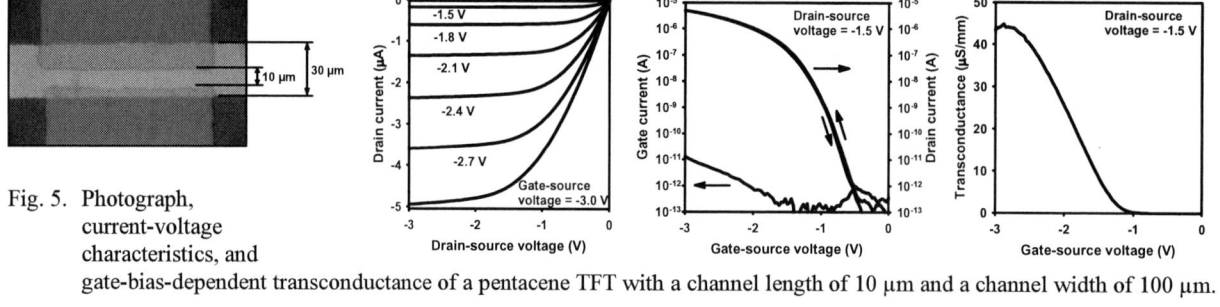

Fig. 5. Photograph, current-voltage characteristics, and gate-bias-dependent transconductance of a pentacene TFT with a channel length of 10 μm and a channel width of 100 μm.

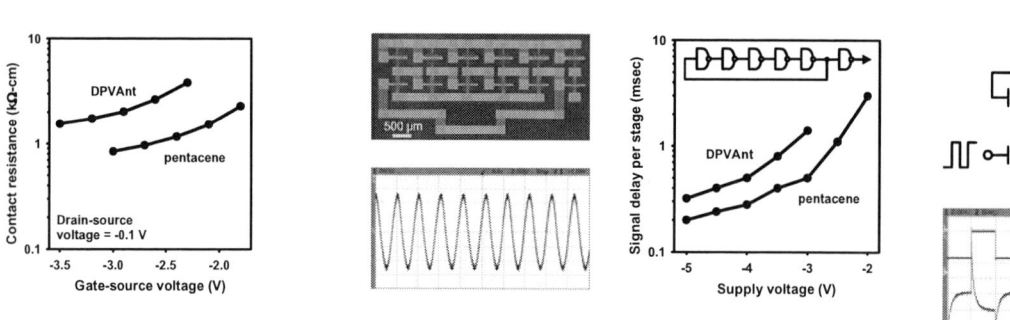

Fig. 6. TFT contact resistance.

Fig. 7. 5-stage ring oscillators (left) and inverter (right).

70

4H-SiC RF BJTs with Long Pulse L-band Operation

Feng Zhao[1], Tiefeng Shi[2], Mike Mallinger[2], and Bart Van Zeghbroeck[3]

[1]Power Products Group, Microsemi Corporation, Bend, OR 97702, USA
[2]Power Products Group, Microsemi Corporation, Santa Clara, CA 95051, USA
[3]Department of Electrical and computer Engineering, University of Colorado, Boulder, CO 80309, USA
Phone: (541)382-8028, fax: (541)383-4162, email: fzhao@microsemi.com

4H-SiC BJTs are attractive candidates for applications such as radar, avionics and communication due to their ability to handle higher RF power density [1] and higher power dissipation capability [2] compared to their silicon counterparts. Good progress has been made in the past. For long pulse UHF operation, 4H-SiC BJTs with 50 W (13.9 W/mm when normalized to total emitter length) output power and 10 dB gain at 500 MHz were reported at DRC in 2006 [1]. For higher frequency performance, the devices fabricated on a semi-insulating SiC substrate demonstrated 6 dB gain at 1.5 GHz with 1.2 W (2 W/mm) output power [3]. In this paper, we present L-band 4H-SiC RF BJTs on a conductive substrate capable of delivering an output power of 28.2 W and a power gain of 7.5 dB at 1.4 GHz under long pulse and duty cycle. The power density is 11.8 W/mm, which is comparable to the UHF devices in ref [1] while almost 6 times of reported L-band devices [3]. Furthermore, the use of SiC conductive substrate significantly reduces the material cost and simplifies the fabrication process compared to semi-insulating substrate. The RF performance makes these devices promising for long-pulse L-band radar applications.

A cross-sectional schematic of a 4H-SiC BJT structure that we report on here is shown in Fig.1. The epitaxial structures (n-p-n) were grown on a 2-inch 8° off-axis n-type 4H-SiC conductive substrate. The devices were fabricated using a double-mesa and interdigitated emitter-base finger design. A RF BJT die in a flanged package is shown in Fig. 2. The die area is 2.63×0.71 mm^2. There are six identical device cells on each die for power-scaling study. A typical I-V characteristic is shown in Fig.2. The maximum DC gain is 4 and the open-base current I_{CEO} is less than 100 μA at V_{CE} =140 V. A critical step of RF SiC BJT fabrication is the preparation of p-type ohmic contacts. It is difficult to obtain low contact resistance because of the wide band-gap energy of 4H-SiC and the high activation energy of Al (191 meV). From measurements on 18 identical on-wafer linear TLM structures across the 2-inch wafer, an average specific contact resistance of 7.8×10^{-4} Ωcm^2 was obtained as listed in Table 1 without ion-implantation. The average base sheet resistance is 20 kΩ/□, which is three times lower than previous designs [3].

Fig 3 shows the large-signal measurement results performed on a single-cell device at 1.2, 1.3 and 1.4 GHz in Class AB operation mode with 1 ms pulse width and 10 % duty cycle. The device was biased with the common-base configuration at V_{CB} =75 V and J_C = 4.2 kA/cm^2. As the output power is increased, the gain is compressed. When the frequency increases from 1.2 to 1.4 GHz and the output power is about 17 W, the power gain is larger than 7.6 dB and the difference is less than 0.5 dB. To scale the output power, we also tested devices with two cells bonded up to double the total device area, and the large-signal results are shown in Fig 4. The output power, power gain and PAE from power-scaling study were summarized in Table 2. The output power is nearly doubled while maintaining the power gains under each frequency, even though no internal matching was applied. This result indicates that the output power from the package can be further scaled up without losing the power gain when more cells are populated. Internal matching work is being undertaken and expected to further improve the bandwidth.

The authors believe that the improvement in RF performance over previous results [1, 3] is due to the optimized design, both epi structure and device lateral layout. The lower base sheet resistance and contact resistivity result in significantly lower total base resistance. The base-collector junction capacitance was minimized by reducing the total isolation-mesa area from the layout design.

---

[1] F. Zhao et al., DRC, pp 153-154 (2006). [2] A. Agarwal et al., ICSCRM, pp 1413-1416 (2005).
[3] I. Perez et al., ICSCRM, pp 1421-1424 (2005).

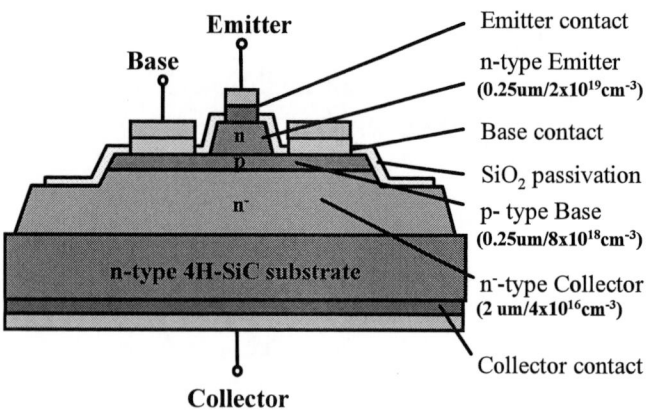

Fig. 1 The schematic cross-sectional structure of a 4H-SiC BJT on a conductive substrate.

Table 1 p-type ohmic contact resistivity from linear TLM across a 2" wafer

ρ_c (Ωcm^2)	Left	Center	Right
Row 12		3.0×10^{-4}	
Row 11		6.6×10^{-4}	
Row 10	4.9×10^{-4}	9.6×10^{-4}	4.8×10^{-4}
Row 9		7.8×10^{-4}	
Row 8		1.3×10^{-3}	
Row 7	1.3×10^{-3}	8.7×10^{-4}	8.9×10^{-4}
Row 6		8.7×10^{-4}	
Row 5		7.9×10^{-4}	
Row 4		8.2×10^{-4}	
Row 3	8.5×10^{-4}	6.9×10^{-4}	4.9×10^{-4}
Row 2		7.7×10^{-4}	
Row 1		6.4×10^{-4}	
Average		**7.8×10^{-4}**	

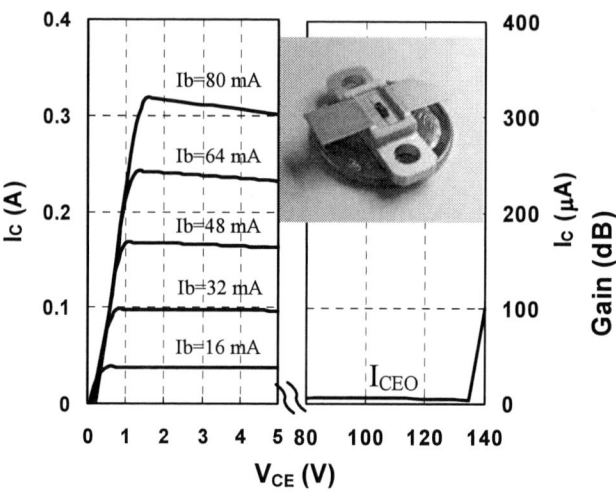

Fig. 2 Micrograph of a RF 4H-SiC BJT die in a flanged package and its I-V characteristics

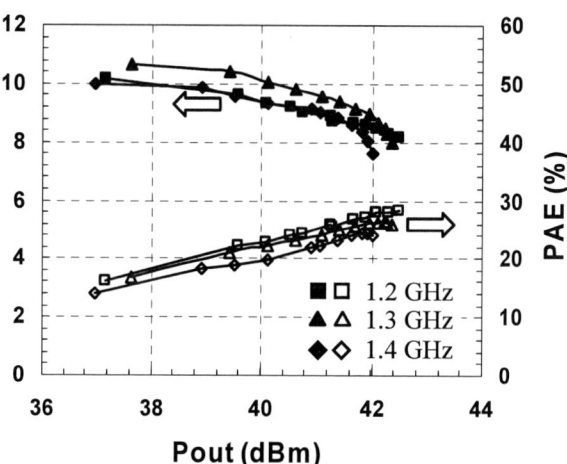

Fig. 3 Output power, power gain and PAE of a single-cell device at 1.2~1.4 GHz. The pulse width is 1 ms and duty cycle is 10 %.

Table 2 Summary of power scaling study at 1.2~1.4 GHz. The devices were biased at V_{CB}= 75 V and J_C= 4.2 kA/cm^2. The pulse width is 1 ms and duty cycle is 10 %.

Device	Freq (GHz)	Pout (W)	Gain (dB)	PAE (%)
Single-cell device	1.2	17.7	8.2	28.4
	1.3	17.2	8.0	25.9
	1.4	15.9	7.6	24.0
Two-cell device	1.2	35.0	8.0	32.6
	1.3	31.9	7.9	32.9
	1.4	28.2	7.5	28.2

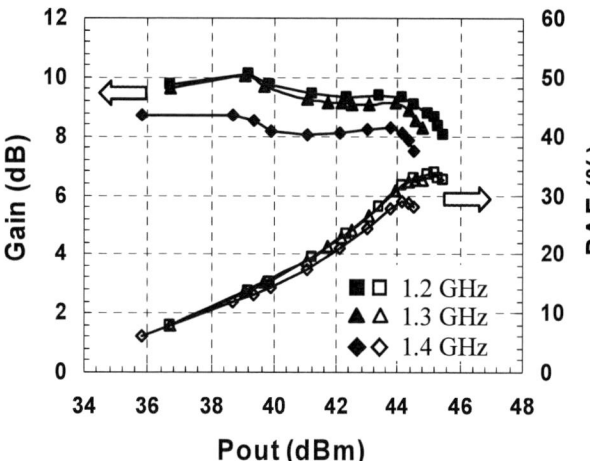

Fig. 4 Output power, power gain and PAE of a two-cell device at 1.2~1.4 GHz. The pulse width is 1 ms and duty cycle is 10 %.

Room-temperature lasing of type-II "W" GaSb/GaAs quantum dots embedded in InGaAs quantum well

J. Tatebayashi, A. Khoshakhlagh, G. Balakrishnan, S. H. Huang,
M. Mehta, L. R. Dawson, and D. L. Huffaker
Center for High Technology Materials, University of New Mexico,
1313 Goddard SE, Albuquerque, New Mexico 87106, USA.
E-mail: tatebaya@chtm.unm.edu/ Phone: +1-505-272-7945/ Fax: +1-505-272-7801

Quantum dots (QDs) in the (In)GaSb/GaAs material system are characterized by a staggered (type-II) band alignment, wide band-gap range, and large valence band offset,[1,2] along with the zero-dimensional density of states.[3] This unique collection of properties offers intriguing optoelectronic device possibilities on GaAs substrates including 1.55 μm emitters, detectors, modulators, memory and solar cells. Several groups have so far reported the formation and optical properties of type-II GaSb/GaAs QDs using the Stranski-Krastanov (SK) growth mode.[4] However, there have been no reports thus far of the lasing operation of type-II QDs, although many researchers have demonstrated the lasing of type-II quantum well (QW) devices.[5]

We report the device characteristics of type-II "W" stacked GaSb/GaAs QDs embedded in an InGaAs QW at room temperature (RT). The lasing at RT from 5 stacked GaSb QDs in InGaAs QWs is obtained at a wavelength of 1.026 μm with 2 mm cavity length. A large blueshift of the electroluminescence (EL) peak, which is typical of the type-II geometry, is observed by increasing the injection current densities. It is noted that this is the first demonstration of RT lasing of type-II QDs.

Laser structures, grown on a (100) n-GaAs substrate by molecular beam epitaxy, commence with growth of a 1.46 μm n-$Al_{0.3}Ga_{0.7}As$ cladding layer, followed by an active layer, a 1.46 μm p-$Al_{0.3}Ga_{0.7}As$ cladding layer, and a 50 nm p^+-GaAs contact layer. Si and Be are used as n- and p-type doping materials, respectively. The active layer consists of five layers of SK "W" GaSb QDs embedded in an $In_{0.13}Ga_{0.87}As$ QW separated by 23 nm GaAs spacer layers. Each QD layer is grown at 510 °C on a GaAs layer and covered with a 7 nm $In_{0.13}Ga_{0.87}As$ QW. The growth rate, V/III ratio, and nominal thickness of GaSb QDs are 0.32 monolayers (MLs)/sec, 2, and 4 ML, respectively. The dot density of the first layer is 3.0×10^{10} /cm^2 determined by atomic force microscopy (AFM). Because of the "W" configuration, the large band offsets confine electrons and holes within InGaAs QW and GaSb QDs, respectively, resulting in strong overlap of the wave function to provide high gain despite the type-II band alignment. The photoluminescence peak of the active layer is 1.18 μm.

Device characteristics of the GaSb QD lasers, such as output power-current [L-I] characteristics and EL spectra, are studied under pulsed conditions (500 ns and 0.5 % duty cycle) at RT. The laser operates at a wavelength of 1.026 μm at RT with a threshold current density (J_{th}) of 860 A/cm^2. The device length is 2 mm and reflectivity is provided by cleaved facets. The device shows 5 mW output power at an injection current density of approximately 1.1 kA/cm^2. Measurement of the EL spectra and EL peak at various injection current densities are carried out at RT. The EL emission from GaSb QDs emerges at the initial wavelength of 1.12 μm. By increasing the injection current density, a large blueshift of the EL peak can be observed due to type-II geometry effects such as band bending or band distortion caused by the charge separation of electrons and holes, along with the band filling effect of electrons confined in QW.

This work is supported by AFOSR under Gernot Pomrenke and Kitt Rheinhardt contract number FA9550-06-1-0407.

[1] G. A. Sai-Halasz et al., *Solid State Commun.* Vol. **27**, pp. 935 (1978).
[2] H. Kroemer et al., *IEEE Electron Devices Lett.* Vol. **EDL-4**, pp. 20 (1983).
[3] Y. Arakawa and H. Sakaki, *Appl. Phys. Lett.* Vol. **40**, pp. 932 (1982).
[4] For example, F. Hatami, F et al., *Appl. Phys. Lett.* Vol. **67**, pp. 656 (1995).
[5] E. Lugagne-Delpon et al., *Appl. Phys. Lett.* Vol. **60**, pp. 3087 (1992).

978-1-4244-1101-6/07/$25.00 ©2007 IEEE

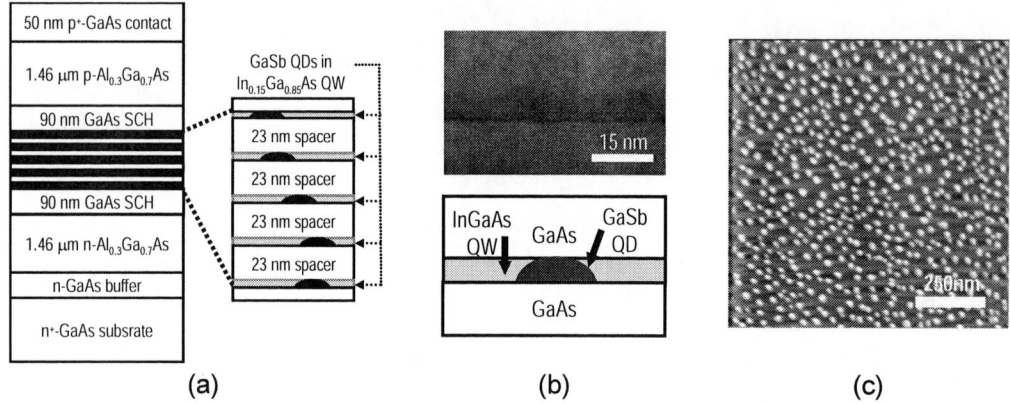

Fig. 1 (a) Schematic illustration of the fabricated laser structure. (b) Cross-sectional transmission electron microscope image and schematic illustration of single layer GaSb QDs in InGaAs QW. (c) AFM image of the first layer of surface GaSb/GaAs QDs.

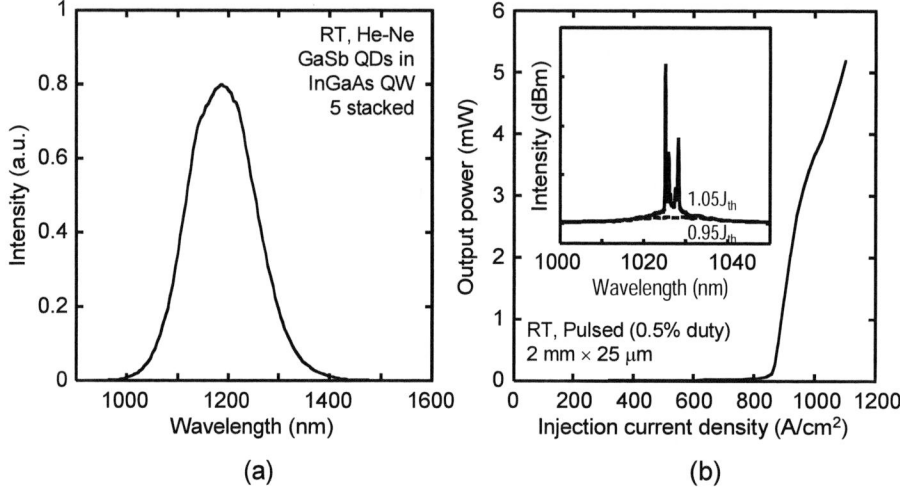

Fig. 2 (a) PL spectrum of 5 stacked GaSb/GaAs QDs embedded in $In_{0.13}Ga_{0.87}As$ QW. (b) L-I characteristics of the fabricated QD lasers with a cavity length of 2 mm at RT. Inset is the lasing spectra just above and below J_{th}.

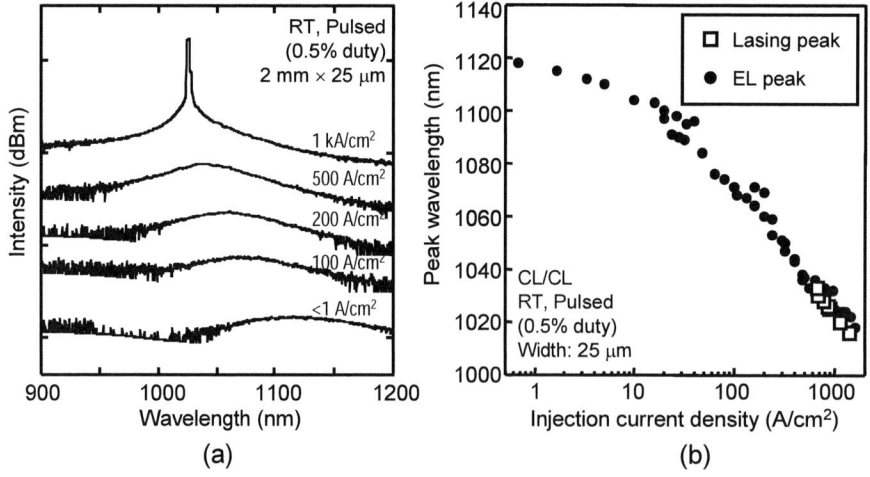

Fig. 3 (a) EL spectra of the fabricated lasers at various injection current densities ranging from below 1 A/cm^2 to 1 kA/cm^2. (b) EL peak versus injection current densities of the fabricated lasers with different cavity lengths

SWCNT-SET fabricated by dispersion method with CMC solvent

Takahiro Mori[1,*], Kazuo Omura[1,2], Shunsuke Sato[1,3], Masaki Suzuki[1,4],
Katsumi Uchida[2], Hirofumi Yajima[2], and Koji Ishibashi[1,4]

1) Advanced Device Laboratory, RIKEN, Wako, Saitama, Japan

2) Department of Applied Chemistry, Tokyo University of Science, Shinjuku-ku, Tokyo, Japan

3) Department of Physics, Tokyo University of Science, Shinjuku-ku, Tokyo, Japan

4) CREST, Japan Science and Technology (JST), Kawaguchi, Saitama, Japan

*) tmori@riken.jp, phone:+81-(0)48-462-1111, fax:+81-(0)48-462-4659

Single electron transistors (SETs) are promising candidates for fundamental components in the future nanoelectronics beyond ongoing Si technologies. The critical problem of SETs is an operation temperature. Most of the SETs operate at low temperatures, which limits practical applications, and just a few papers have reported room temperature operation of SETs[1]. Single-walled carbon nanotubes (SWCNT) are attractive material to fabricate SETs because of their extremely small diameter. An easy method to fabricate SWCNT-SETs is a dispersion method, which is bottom-up fabrication process of SWCNT-based transistors. To disperse SWCNTs on a wafer, a SWCNT suspension is used usually with a surfactant to avoid bundle formation. Popular surfactants are tritonX-100 and sodium dodecylsulfate (SDS). Adsorbed surfactant molecules on SWCNTs can affect properties of SWCNT-base devices, as reported in refs.2 and 3. In this paper, we propose another surfactant, carboxymethylcellulose (CMC). Using a CMC/water suspension to disperse SWCNTs, we have achieved high temperature operation of SET up to 80K, which is higher than the operating temperature of 40K for our SETs fabricated with tritonX-100/water suspension[4].

The SWCNT-SET device we fabricated is shown in Fig.1. The device fabrication was started with randomly dispersing SWCNTs using a CMC/water suspension onto a n^+-Si substrate with a 200nm thick SiO_2 layer. The suspension was sonicated and centrifuged. Titanium source and drain contacts were fabricated by electron-beam lithography and lift-off techniques. The n^+-Si substrate was used as a back gate.

SET characteristics were observed in transport measurements at cryogenic temperatures. Fig.2 shows temperature dependence of I_D(drain current)-V_G(gate voltage) characteristics in a range of 2.7-10K. Clear Coulomb oscillations were observed. Coulomb diamond is another evidence of SET characteristics, which can be seen in Fig.3. Estimated charging energy was about 10meV, which is almost as large as our previous SETs where tritonX-100 was used [4]. The temperature dependence of I_D-V_G characteristics in the higher temperature range is shown in Fig.4. Clear Coulomb oscillations are observable up to 80K. This is due to the higher barrier height about 16meV (for sample A in Fig.5), compared to the previously reported barrier height in SETs[4] fabricated with a tritonX-100/water suspension (about 10meV).

In conclusion, we report SET characteristics fabricated by dispersion method with a CMC/water suspension. The SET operated at higher temperatures than those fabricated with tritonX-100. The higher temperature operation was due to the higher barrier height, which suggests that adsorbed CMC molecules on SWCNT at the metal-nanotube interface affect the barrier formation.

[1] for example, H.W.Ch.Postma et al., Science **297**, 76(2003).

[2] Z.B.Zhang et al., J. Appl. Phys. **98**, 056103(2005).

[3] J.Li et al., Nanotechnology **17**, 668(2006).

[4] D.Tsuya et al, Jpn. J. Appl. Phys. **44**, 2596(2005).

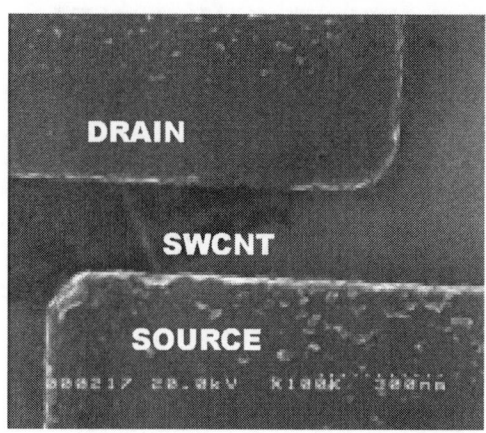

Fig.1 Typical SEM image of the SWCNT-SET devices we used.

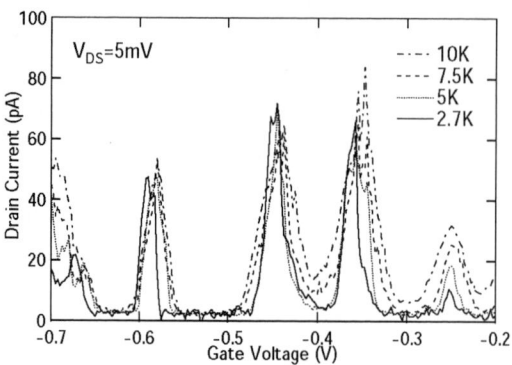

Fig.2 Temperature dependence of I_D-V_G characteristics in the range of 2.7-10K. Clear Coulomb oscillation was observed.

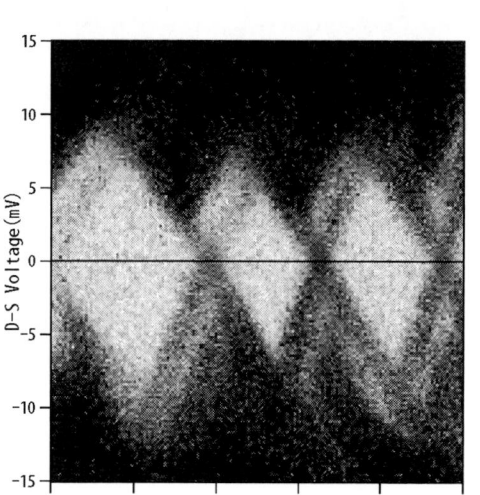

Fig.3 Mapping of the conductance as a function of V_D and V_G observed at 2.7K. White color shows low conductance region and black color shows high conductance region.

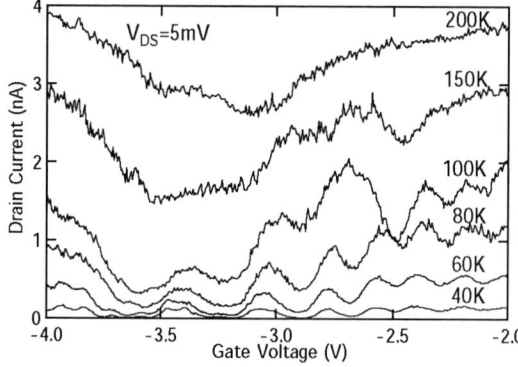

Fig.4 Temperature dependence of I_D-V_G characteristics in the range of 40-200K. Coulomb oscillation can be observed up to 80K.

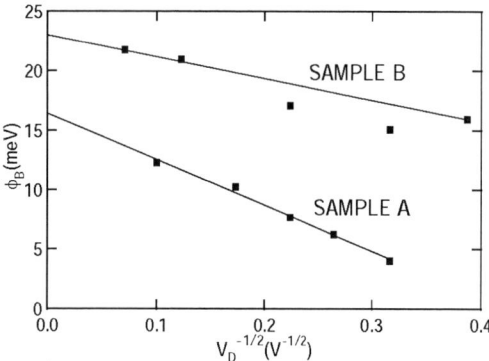

Fig.5 Drain voltage dependence of barrier height. Current characteristics shown in this presentation was those of sample A. The barrier height was linear to $V_D^{1/2}$, which suggests that the SET have MIS structure.

Microwave Noise Characterization of Enhancement-Mode AlGaN/GaN/InGaN/GaN Double-Heterojunction HEMTs

J. Liu, D. Song, Z. Cheng, W. C.–W. Tang, K. M. Lau and K. J. Chen*

Department of Electronic and Computer Engineering, Hong Kong University of Science and Technology
Clear Water Bay, Kowloon, Hong Kong
*Corresponding author: eekjchen@ust.hk Tel: (852) 235888969, Fax: (852)23581485

Wide-bandgap AlGaN/GaN HEMTs are being extensively studied for applications in RF/microwave power amplifiers and power switches. Recently, the high input power handling capabilities and high breakdown also make AlGaN/GaN HEMTs attractive for use in protection-circuit-free low-noise amplifiers, especially after excellent noise performances were reported [1, 2]. Compared to the conventional depletion-mode (D-mode) HEMTs, enhancement-mode (E-mode) HEMT enables the use of single-polarity supply voltage [3], leading to reduced circuit complexity. In this work, we report the microwave noise characterization of E-mode double-heterojunction HEMT (DH-HEMT) [4, 5]. The E-mode DH-HEMT shows reduced noise figure compared to its D-mode counterpart, mainly owing to the lower gate leakage current achieved by the Schottky barrier enhancement in fluorine-plasma treated gate region and the favorable bias conditions for the E-mode HEMT.

The sample used in this work is an AlGaN/GaN/InGaN/GaN double-heterojunction HEMT that features an InGaN notch at the back of the channel. Both D-mode and E-mode devices are fabricated using the fabrication process described in [4]. All the devices used in this work have a gate length of 1-μm. The D-mode (E-mode) HEMTs exhibit a threshold voltage of -3.8 V (+0.08 V), a peak current density of 800 mA/mm (520 mA/mm), a maximum transconductance of 225 mS/mm (210 mS/mm). The microwave noise figure measurements were carried out from 2 to 12 GHz with V_{DS} = 5 V and V_{GS} increasing from just above the threshold voltage (equivalent to increasing the drain current), using Maury Microwave's MT982B01 load-pull system with Agilent N8975A noise analyzer. At 2 GHz, the lowest minimum noise figure (NF_{min}) is 0.62 dB (at a drain current of 100 mA/mm) and 0.8 dB (at a drain current of 150 mA/mm) for the E-mode and D-mode devices, respectively. The low drain current needed for the E-mode device indicates lower power consumption. From 2 to 12 GHz, NF_{min} is consistently lower in the E-mode HEMT compared to the D-mode HEMT even though the associated gain in the E-mode HEMT is 1~2 dB lower.

The lower noise figure achieved in E-mode HEMTs can be explained by studying the gate leakage current at the bias conditions set for the noise measurement. It has been shown that the gate leakage current has significant influence over the noise figure in GaAs- and GaN-based HEMTs [6, 7]. Higher leakage current generally leads to higher NF_{min}. The E-mode HEMTs exhibits lower gate leakage current from two aspects. First, the E-mode HEMTs fabricated by the fluorine plasma treatment exhibits enhanced Schottky barrier, effectively suppressing the thermionic emission and tunneling current between the channel and the gate electrode [4]. Second, the near-zero threshold voltage of the E-mode HEMT also enable it to be biased at a voltage that results in smaller V_{GS} and V_{GD} when the device is biased for low-noise amplifier operation. For example, the gate electrode was biased at +0.5 V (with V_{DS} = 5 V) for the E-mode HEMT when the lowest NF_{min} was obtained at 2 GHz, so that V_{GS} and V_{GD} are 0.5 V and -4.5 V. The similar operation for D-mode HEMT yields a gate bias of – 2.2 V. V_{GS} and V_{GD} are -2.2 V and -7.2 V, at which higher gate leakage currents are presented.

In conclusion, the enhancement-mode III-nitride HEMTs are capable of delivering excellent microwave noise performance for applications in low-noise amplifiers with single-polarity supply voltage and low power consumption.

Acknowledgment: This work is supported by Hong Kong Research Grant Council under project 611706.
References: [1] J. S. Moon, et al. *IEEE Electron Device Lett.*, vol. 23, no. 11, pp. 637–639, 2002. [2] W. Lu, et al. *IEEE Trans. Electron Devices*, vol. 50, no. 4, pp. 1069–1074, 2003. [3] Y. Cai, et al. *IEEE Trans. Electron Devices,* vol.53, No. 9, pp. 2205-2215, 2006. [4] J. Liu, et. Al. *IEEE Tran. Electron Devices,* vol. 27, No. 1, pp. 2-10, 2006. [5] T. Palacios, et al. *IEEE Tran. Electron Devices,* vol. 27, No. 1, pp. 13-15, 2006. [6] F. Danneville, et al, *IEEE MTT-S Symp.,* pp. 373-379, 1993. [7] C. Sanabria, et al. *IEEE Electron Device Lett.,* vol. 27, pp.19-21, Jan. 2006.

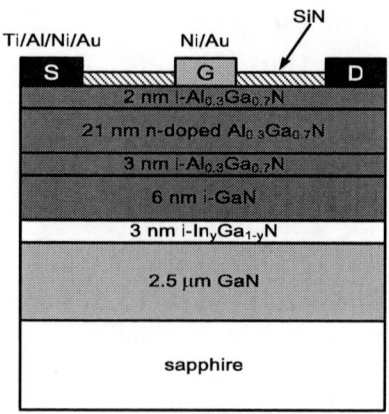

Fig .1 Cross section of the DH-HEMT used in this work.

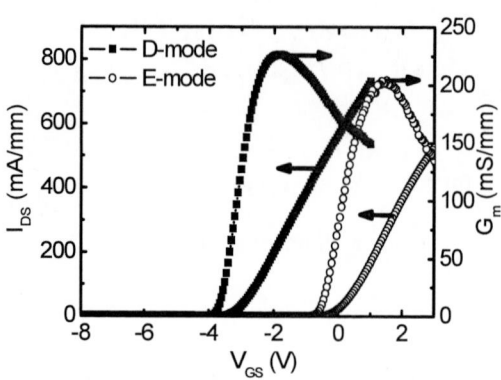

Fig .2 Transfer characteristics of the D-mode and E-mode DH-HEMTs. The D-mode HEMT has a threshold voltage of -3. 8 V and the E-mode device showed a threshold voltage of +0.08 V.

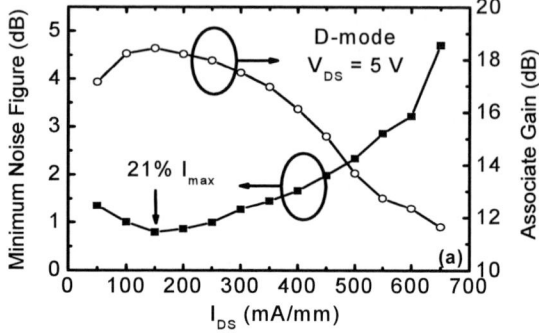

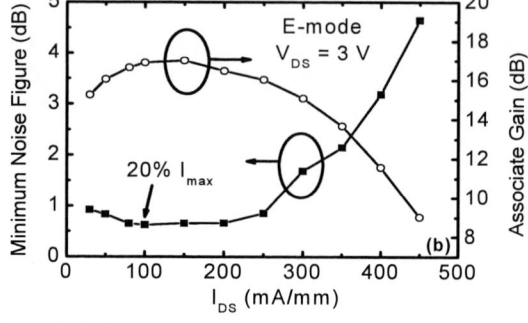

Fig. 3 Minimum noise figure and associate gain of the D-mode and E-mode devices at different drain current levels at 2 GHz. The drain bias is 5 V.

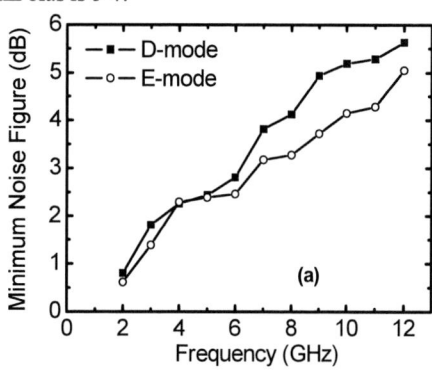

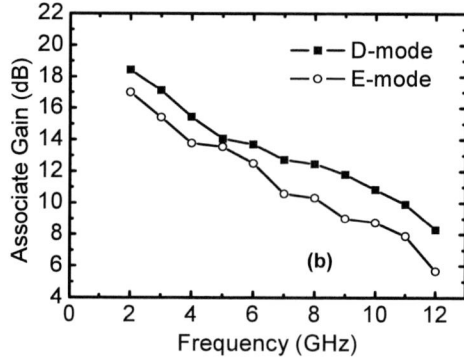

Fig. 4 (a) Minimum noise figure and (b) associate gain of D/E-Mode device at different frequencies. The bias conditions are V_{DS} = 5 V, I_{DS} = 150 mA/mm for the D-mode HEMT and V_{DS} = 5 V, I_{DS} = 100 mA/mm for the E-mode HEMT.

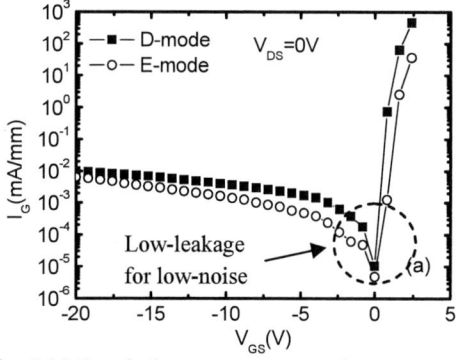

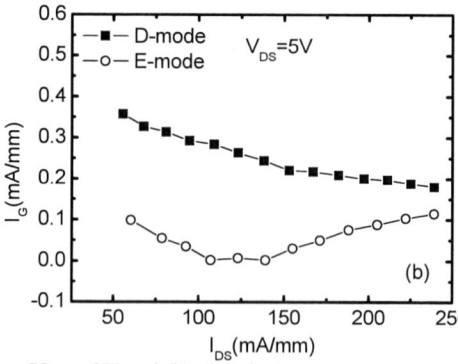

Fig. 5 (a) Gate leakage current versus the gate-source bias (V_{GS}) at V_{DS} = 0V and (b) Gate leakage current versus the drain current at V_{DS} = 5 V.

Electro-Thermally Coupled Power Optimization for Future Transistors

A. K. Chao (kuoan@stanford.edu), P. Kapur, Eiji Morifuji, K. C. Saraswat, and Y. Nishi

Department of Electrical Engineering, Stanford University, Stanford, CA 94305-4075, USA

We have developed a self-consistent, electro-thermally coupled, total power, P_{tot}, (includes dynamic, DP, and static leakage power, SDL) optimization methodology for future transistors. It calculates the best power-delay tradeoff curve, and the corresponding self-consistent, temperature-delay curves for a given transistor from its current-voltage characteristics. The methodology, by serving as a comprehensive comparison standard for different future transistor options, presents a unique and powerful tool for suitable device selection at future nodes, where no SPICE models are available. In addition, in this work, we also use the methodology to provide insight into the 1) optimum transistor design and operational parameters for minimum P_{tot}, 2) device-specific hot spot problems, 3) multi-V_t design for different functional blocks, and 4) the efficacy of novel thermal solutions (superior thermal conductivity, sub-ambient cooling).

We choose an 18nm gate length (L_g) double gate FET (DGFET) (Fig. 1) to illustrate the methodology. It is implemented as a simulator consisting of several modules (Fig. 2). In Fig. 2, first the workfunction (ϕ_m controlling V_t), V_{dd}, and T_{ox} of the transistor is chosen and the corresponding C-V and I-V curves are obtained using Medici^{TM} [1]. These curves are used by the *electrical network module* to calculate P_{tot} (and FO1 inverter delay) vs. temperature (T_{junc}). While, the *thermal network module* calculates T_{junc} as a function of P_{tot}, accounting for packaging, interposer layer, and silicon thermal resistance (R_{jtot}) as well as the ambient temperature (T_{amb}). Using the outputs from the thermal and the electrical module, the *power-temperature solver module*, self-consistently calculates the P_{tot} and T_{junc}. Once T_{junc} is known, the various power components (leakage and dynamic) and the delay are back-calculated. The process is repeated for a range of ϕ_m, V_{dd}, and T_{ox} to obtain the self-consistent P_{tot}, delay, and T_{junc}. (Fig. 2). The *global power optimization module*, subsequently calculates source-drain leakage (SDL), DP and P_{tot} vs. V_{dd} at a given delay, with ϕ_m (V_t) being the implicit parameter (Fig. 3). A clear minimum in P_{tot} with respect to V_{dd} (and implicitly ϕ_m), arising from the tradeoff between SDL and DP, is seen. Fig. 3 also illustrates the importance of using electro-thermally coupled model by highlighting the error in using a fixed T_{junc} for optimization. Other T_{ox} values (not shown in Fig. 3) reveal their own, respective, local minimum with respect to V_{dd}. A comparison of these minima yields the global P_{tot} minimum with respect to V_{dd}, V_t (ϕ_m), and T_{ox}. The process is repeated for different delays to obtain electro-thermally coupled, best power-delay curve.

At high speeds, a positive feedback between P_{tot} and T_{junc}, can result in transistor thermal runaway, manifested in our methodology as no self-consistent solution. Fig. 4 shows the V_{dd}, ϕ_m contour representing the thermal runaway wall (1% switching activity (SA) circuits, two R_{jtot}s). A better thermal solution provides a more robust design space. Fig. 5 shows the electro-thermally coupled optimum P_{tot} vs. delay curve running through the fixed T_{junc} optimum curves, clearly depicting the rise in self-consistent T_{junc} with lower delay. It also shows the corresponding optimum V_{dd}. Fig. 6 shows the electro-thermally coupled optimum power-delay curves for functional blocks differentiated by their SA (clock SA=50%, datapath SA=10%, registers SA=1%). Higher SA circuits result in larger non self-consistent errors. Fig. 7 exploits our unique ability to self-consistently evaluate T_{junc} to calculate temperature-rise at hot-spots (Higher SA circuits). It also quantifies the increase in hot spot temperature with lower delay (faster speeds). Finally, we evaluate the impact of novel thermal solutions. Fig. 8 quantifies the effect of R_{jtot} improvement on P_{tot} (and the corresponding T_{junc}) reduction, for two different delays. Above a certain thermal conductivity, further improvements do not improve P_{tot} and T_{junc} significantly, since T_{junc} starts to approach T_{amb}. Fig. 9 captures the benefit of sub-ambient cooling, showing a clear reduction in the optimum P_{tot} and T_{junc} with T_{amb} reduction. However, this P_{tot} reduction comes at the cost of extra power consumption for cooling. Fig. 10 shows both the optimum P_{tot} saving and the power expenditure for cooling as a function of T_{amb} (using 20% Carnot efficiency and [2]). The largest net power saving is obtained at a T_{amb} of 4°C (1ps delay).

In conclusion, this work uniquely fills the gap for a much needed, future transistor, low power methodology, which efficiently captures the thermal, electrical, and package interactions. We demonstrate a novel and practical framework for calculating electro-thermally coupled, minimum transistor power, and exemplify its versatility by applying it to 18nm L_g DGFET. In particular, we used it 1) to obtain optimum power (and T_{junc}) vs. delay and contrasted it with fixed T_{junc} curves; 2) to get optimum V_{dd}, V_t and T_{ox}; 3) to predict thermal runaway parameters; 4) to gain insight into hot-spot temperature rise problems; and 5) to evaluate the impact of better thermal resistance (packaging) and sub-ambient cooling.

[1] Medici version 2003 (Device simulator).12, Synopsys, CA;   [2] M. J. Ellsworth, Jr., *Electronic Cooling*, Vol. 7, No. 3, 2001.

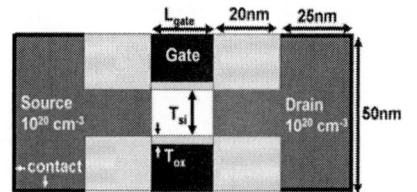

Fig. 1: Schematic of Medici-simulated double-gate (DG) FET based on ITRS 2003. Baseline DG: Lg=18nm, Tsi=7nm.

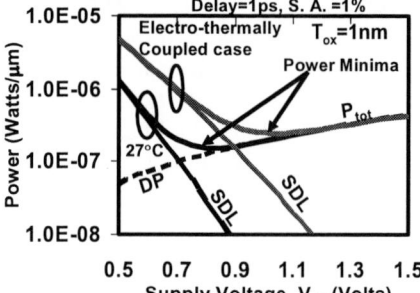

Fig. 2: Flow chart depicting self-consistent electro-thermal coupling global, power optimization methodology.

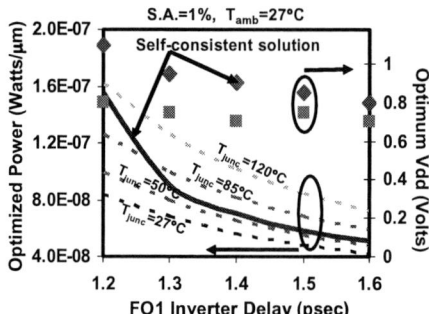

Fig. 3 (Global Power Optimization Module): Iso-performance SDL, DP (dotted curves) and their sum, Ptot (solid curves). Tox=1nm

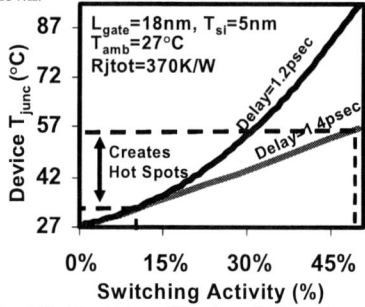

Fig. 4 Thermal runaway wall in design space. The data points show the actual optimum values for two delays

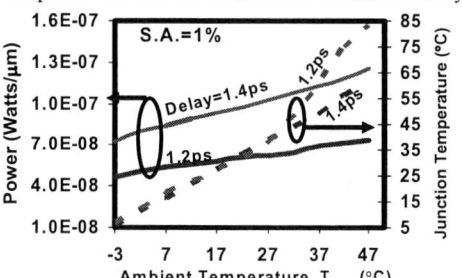

Fig. 5 Comparison between self-consistent and fixed Temp. optimized power-delay curves. Optimum V_{dd} also shown.

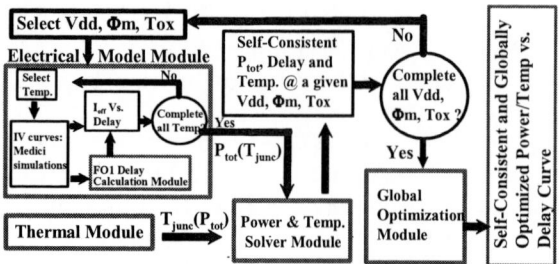

Fig. 6 Impact of switching activity (SA). Solid Curves: self-consistent optimum power-delay; dotted: Tjunc of 27°C.

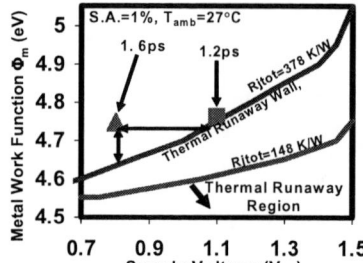

Fig. 7 Hot Spot Quantification. Self-consistent junction temperature for different SA shown for two different delays.

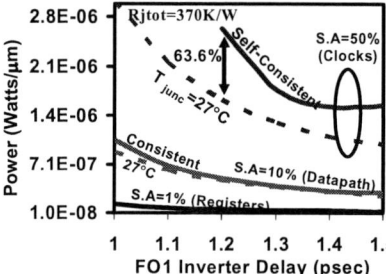

Fig. 8 Impact of improvement in thermal conductivity. Normalized value of 1 corresponds to Rjtot=370K/W

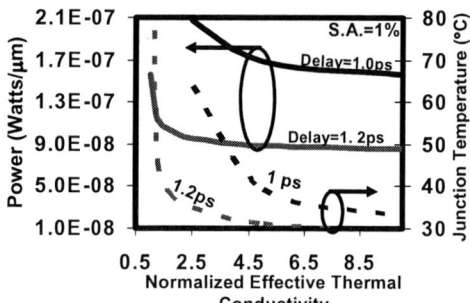

Fig. 9 Impact of Sub-Ambient Cooling. Optimal power and Tjunc under the impact of ambient cooling. Lg=18nm, and Tsi=7nm, Rjtot=370K/W.

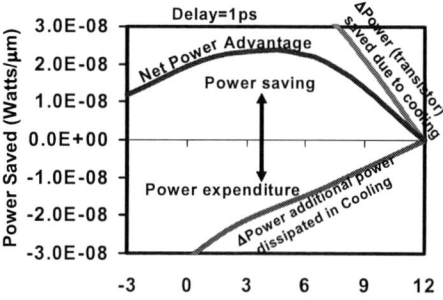

Fig. 10 Sub-ambient cooling. Shows the power trade-off during sub-ambient cooling (Carnot efficiency=20%).

Dispersion Design of a Left-Handed Microstrip Line
with Planar Double-Stub and Split-Ring Structures
for Leaky Wave Radiation toward Functional RF Wireless Interconnect

Michihiko Suhara[*,**], Akito Shimizu[*], Tsugunori Okumura[*,**]

*Electrical Engineering, Graduate School of Engineering,

**Electrical and Electronic Engineering, Graduate School of Science and Engineering,

Tokyo Metropolitan University, 1-1 Minami-ohsawa, Hachioji, Tokyo, 192-0397, JAPAN

Tel: +81-426-77-2765, Fax: +81-426-77-2756, e-mail: suhara@comp.metro-u.ac.jp

RF wireless interconnection among integrated layers and/or chips has been taken up with one of emerging technologies for future integrated system to break down global interconnect delay due to metallic wires. Although a single-input-single-output type of wireless interconnect has been reported, other functionalities with multi-input and/or multi-output have not been investigated so far.

In this paper we propose and analyze a possibility to realize functional RF interconnects by employing a double-stub and a split-ring structure integrated with microstrip transmission line (MSL) with variable leaky wave radiations for the first time on the basis of S-shaped frequency dispersion design including left-handed wave propagation[1,2] in frequency ranges of tens GHz.

Conceptual schematics for a functional wireless RF interconnect with controlling leaky wave radiation are illustrated in Fig.1. One is a case that a radiation port is varied with carrier frequency (Fig.1(a)) and the other is a case that radiation angle is varied (Fig.1(b)). Unlike electrically controlled microstrip antennas[1], we adopted and design a double-stub (DS) and a split-ring (SR) structures so that above mentioned functional leaky wave radiation can be obtained in tens GHz on the basis of left-handed transmission.

Electromagnetic field simulations and two-port S-parameter analysis are carried out for the proposed left-handed microstrip line (LHML) whose dimension is shown in Fig.2(a). The leaky wave radiates from the SR and the DS at f =14.05 GHz and 15.20 GHz, respectively. Typical two different near field patterns radiated from the optimized structure of LHML are shown in Figs.2 (b) and (c) where two given frequencies are determined to maximize the power dissipation (P_d=1-[$|S_{11}|^2$+$|S_{21}|^2$]) as shown in Fig.3. According to the Poynting power calculation, the input power at the port 1 can be divided into the power of leaky wave radiation and the power transmitted to the port 2 by a half in the present configuration. On the other hand, Figs.1 (c) and (d) show that a change of backward and forward radiations in far field patterns can be realized at f =14.15GHz and 15.30GHz, respectively.

It should be noticed that the dimension of the present device with integration of the DS and the SR structures reveals a typical S-shaped dispersion curve in a frequency range of a few GHz as shown in Fig.4. The S-shaped dispersion can be distinguished in plural frequency domains by a plus and a minus of the phase velocity v_p and the group velocity v_g of propagating wave. The left-handed transmission is defined when v_p >0, v_g <0 or v_p <0 v_g >0. On the other hand the ordinal right-handed transmission is defined when v_p >0, v_g >0 or v_p <0 v_g <0. The dimension of the device also determines two different frequency points to maximize the power dissipation as in Fig.3, namely leaky waves are resonantly occurred inside region of air lines. It can be confirmed that both two different conditions are on the left-handed dispersion region as indicated by solid -circles in Fig.4. Two other open-circles on the dispersion curve in Fig. 4 correspond to conditions giving Fig.2(d) and (c) for far field patterns. The forward and backward radiations are due to opposite handed dispersions with opposite polarities of phase constant βL each other.

Figure 5 shows that different types of S-shaped dispersion diagrams for different device structures. The S-shaped curve can have intersections on the axis of βL =0 only when the DS and the SR are located adjacent to the MSL. This is because this device structure gives particular electromagnetic field distribution so that it provides characteristics explained by an equivalent circuit with two loop elements which are inductively coupled with the thru line, and X-shaped symmetric elements in shunt branches in frequency order of several tens GHz as shown in Fig.6. According to additional studies of a scaling rule of the proposed device, the above mentioned functional leaky wave radiation can be obtained even in millimeter wave region in the same fashion when the size of the DS and the SR is miniaturized in the order of several hundreds μm.

In conclusion, toward a feasibility study of functional RF wireless interconnect, a double-stub and a split-ring structures integrated with microstrip transmission line are proposed and the leaky wave radiation properties are analyzed on the basis of designing the S-shaped frequency dispersion relation with the left-handed transmission.

[1] S.Lin et al., *IEEE Trans. MicrowaveTheory Tech.*, vol.53, no.1, p.161 (2005)

[2] S-G.Mao et al., *IEEE Trans. MicrowaveTheory Tech.*, vol.53, no.4, p.1515 (2005)

978-1-4244-1101-6/07/$25.00 ©2007 IEEE

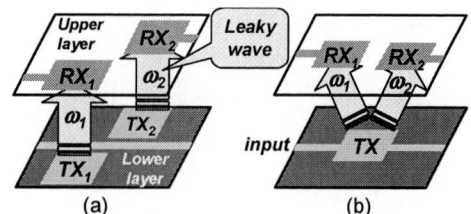

Fig.1 A conceptual schematic of reconfigurable RF wireless interconnect with variable leaky wave radiations revealing (a) emission port selectivity with and (b) angle selectivity with ω_1, ω_2.

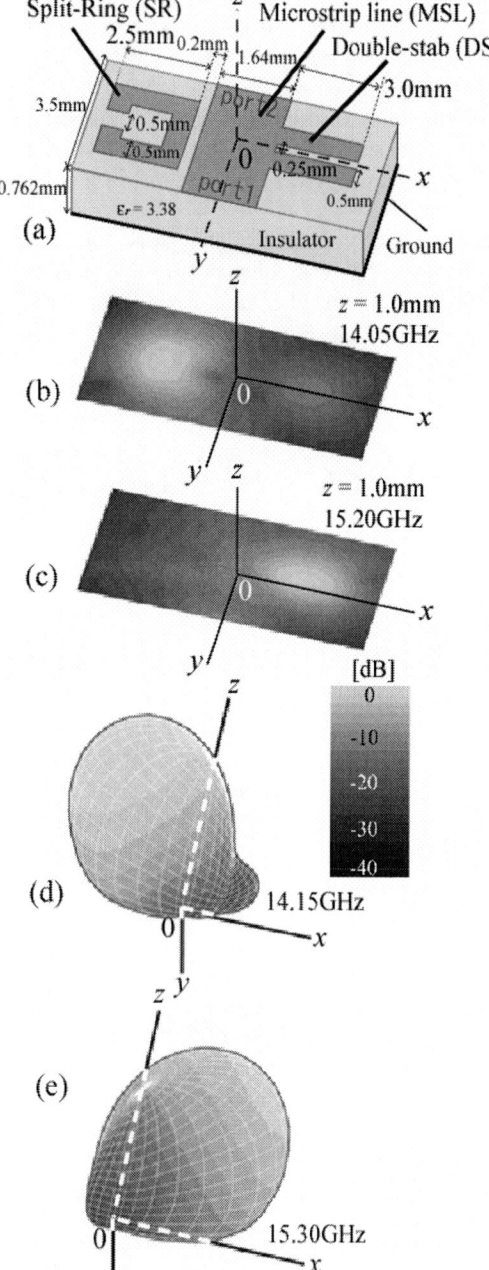

Fig.2 (a) A proposed device model with a double-stub (DS) and a sprit-ring structures adjacent to a microstrip line. Near field radiation power at 1mm high form the device at (b)14.05 GHz and (c)15.20 GHz with different emission each other. Far field patterns at (b)14.15 GHz and (c)15.30 GHz with forward and backward lobes each other.

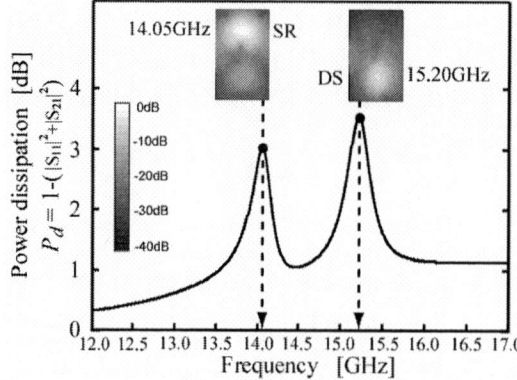

Fig.3 Power dissipation in the MSL with the DS and the SR shown in Fig.2(a). Spatially selected leaky wave are obtained around two different resonant peaks.

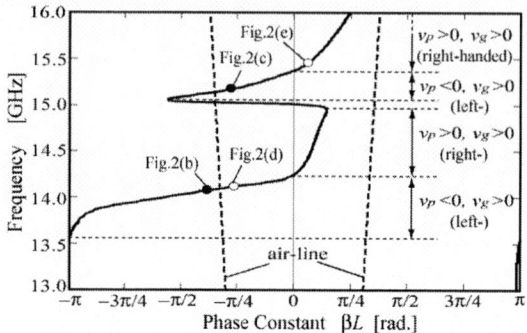

Fig.4 S-shaped dispersion curve of the MSL with DS and SR revealing the left- and the right-handed wave propagations.

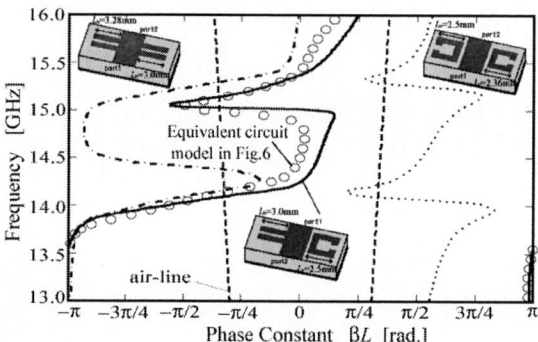

Fig.5 Different S-shaped dispersion curves for different structures.

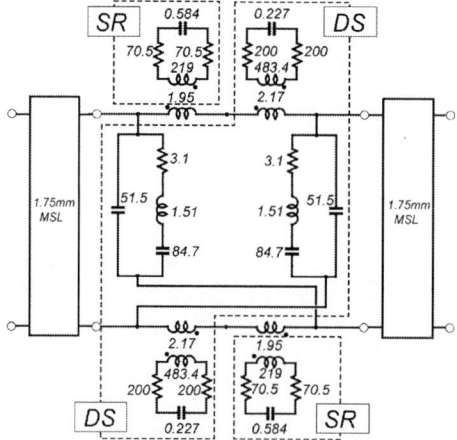

Fig.6. An equivalent circuit deembeded from the dispersion curve of the MLS with the DS and the SR. Values for R, L, C are in Ω, nH, fF.

82

High Power Vertical-structure GaN-based LEDs with Improved Current Spreading and Blocking Designs

Tron-Min Chen[1,2], Shui-Jinn Wang[1*], Kai-Ming Uang[2], Shiue-Lung Chen[1], Ching-Chung Tsai[1], Hon-Yi Kou[1], Wei-Chi Lee[1], and Hon Kuan[3]

[1]Institute of Microelectronics, Dept. of Electrical Eng., National Cheng Kung Univ., Tainan, Taiwan
[2]Dept. of Electrical Eng., Wu-Feng Institute of Technology, Chia-yi, Taiwan
[3]Optoelectronics center of Far East Univ., Tainan, Taiwan
*Phone: +886-6-2757575-62351, Fax: +886-6-2763882, E-mail: sjwang@mail.ncku.edu.tw

Abstract: Based on experimental results which reveal that the contact of Indium-Zinc-Oxide (IZO) and IZO/Ti to *n*-GaN layer is Schottky and ohmic, respectively, localized Ti deposition associated with a transparent IZO layer is proposed to serve as both current blocking and current spreading layer. In addition, an anisotropic mesa etching on the surface layer (*n*-GaN) of regular Vertical-conducting Metal-substrate GaN-based Light-Emitting Diodes (VM-LEDs) is also proposed to further decrease the resistance difference between the outside path and the inner one. The effectiveness of the proposed schemes were verified by a two-dimensional device simulator (ISE-TCAD), which indicates that significant immune of current crowding under cathode contact pad would be possible once an optimal combination of the *n*-GaN layer etching depth and width, IZO thickness, and Schottky blocking width has been achieved. In experiments, 40-mil LEDs with an anisotropic mesa etching of 400 μm in width and 2 μm in depth, 200 μm in Schottky blocking width, and a 300-nm-thick IZO layer have been successfully fabricated. Typical improvement in light output power by about 25% at an injection current of 350 mA as compared to the regular VM-LEDs has been obtained.

Recently, many attempts have been made to develop high-efficiency high-power GaN-based LEDs for solid-state lighting [1-2]. Efforts to release the problems of regular lateral structure LEDs mentioned above by means of vertical-conducting structure, transparent conduction layer (TCL), and surface texturing, etc., have been reported and encouraging results have been achieved [3-4]. The authors' group has developed a Vertical-conducting Metal-substrate GaN-based Light-Emitting Diodes named as VM-LEDs [5-6], as shown in Fig. 1(a). To effectively solve the current crowding effect (CCE) in regular VM-LEDs, we further developed a graded TCL/*n*-GaN scheme [7-8]. Though LEDs based on the graded TCL/*n*-GaN scheme has been shown providing a significant improvement in light output power (Lop) and much more uniform in current and light emission distribution as compared to those of regular VM-LEDs, the use of Excimer laser etching is still a challenge issue of expensive cost and unsuitable for batch process.

A novel scheme using patterned Ti deposition associated with a transparent IZO layer is proposed for current blocking and spreading of VM-LED. An anisotropic mesa etching on the surface layer (*n*-GaN) of the VM-LED associated with the IZO TCL is also proposed to further release CCE of the device. Figure 1(b) illustrates the basic concept behind the proposed device structure. In addition to current blocking via a Schottky IZO/n-GaN contact, once the top *n*-GaN layer was properly mesa-etched and the thickness and resistivity of the TCL layer has been optimized, the overall difference in series resistances along any two possible conducting paths could be minimized. Based on the proposed structure, distributions of current across the active region and the corresponding light emission were simulated by a device simulator ISE-TCAD [9] and results were shown in Fig. 2. Note that the thickness and doping concentration of the *n*-GaN layer are 3 μm and ~5×10^{18} cm^{-3} respectively. Although the optimum structure (including etching width, depth, and TCL thickness, etc.) design is still underway, our results reveals that, as compared to regular VM-LED, fairly good uniformities in current and light emission distributions in the active region have been obtained, indicating that the proposed device confronts a much less impact from CCE.

In experiments, an LED structure comprises a sapphire substrate, a buffer layer, a 0.5-μm-thick undoped GaN layer, a 3-μm-thick Si-doped *n*-GaN cladding layer, an undoped 5-period GaN/InGaN multiple quantum well (MQW), a Mg-doped *p*-cladding layer, and a 0.15-μm-thick Mg-doped GaN layer was used. Figure 3 illustrates the key fabrication processes flow of the proposed device. It is noted that steps (a)–(c) were employed to transfer the GaN epilayer structure from the sapphire substrate onto a metal substrate comprising a (Au/Ti/Al/Ti)/electroplated-Ni metal system [5-6]. Step (d) was the mesa etching process using an inductively coupled plasma (ICP) system. The nanofocus OM photograph shown in Fig. 4(a) reveals that the *n*-GaN layer has a smooth mesa surface with an etching width of around 400 μm and depth of around 2 μm [Fig. 4(b)]. Step (e) was the patterned Ti deposition process in which a layer Ti of 2.5 nm was deposited on the top *n*-GaN layer to form ohmic contact. Step (f) was IZO deposition process which a transparent IZO layer with thickness of 300 nm was deposited to serve as current spreading and current blocking. Finally, an ohmic electrode was deposited onto the top of fabricated device. Note that devices without anisotropic etching and without current blocking (i.e., regular VM-LEDs) were also fabricated for comparison.

Based on controlled samples, sheet resistance and resistivity of the anisotropic etching *n-*GaN/TCL structure are of around 10 Ω/□ and 8×10^{-5} Ω-cm, respectively. Based on an LED tester, typical measured light output power-current (Lop-I) and current-voltage (I-V) characteristics of proposed LED and VM-LED were shown in Fig. 5. The insets of the figure also show the photos of light emission from each device at 350 mA. The corresponding light emission patterns were obtained and analyzed by using software of Process Diffraction [10]. The solid curves at the bottom of the images denote the relative light output intensity along the dashed lines. Our preliminary results from more than 50 samples show that an average improvement

978-1-4244-1101-6/07/$25.00 ©2007 IEEE 83

in Lop by about 25% @ 350 mA has been obtained from proposed devices as compared to that of regular VM-LEDs. Slightly large forward voltage of proposed device (3.549 V @350 mA) compared to that of VM-LED (3.462 V @350 mA), which might come from the Schottky blocking.

In conclusion, a vertical GaN-based LED device with a Schottky current blocking and a mesa etching employed to improve current crowding effect and light emission distribution has been proposed and demonstrated. Theoretical calculations show that the uniformities of current and light emission distribution strongly depend on the structure and material parameters of both n-GaN and TCL layers as well as the geometries of the mesa etching and Schottky blocking. In experiments, VM-LEDs with the proposed structure have been successfully fabricated and an average improvement in light output power by about 25% at an injection current of 350 mA as compared to that of regular VM-LEDs has been obtained. It is expected that the proposed device structure using ICP mesa etching associated with TCL and Schottky current blocking would provide an efficient way in avoiding CCE in larger-area high-power LEDs and would be very advantageous for the applications of LEDs on solid-state lighting.

Acknowledgement: This work was supported by the National Science Council (NSC) of Taiwan, Republic of China, under Contract No. NSC 95-2215-E-006-014. Technical assistance in experiments from the center for Micro/Nano Technology Research, National Cheng Kung University, Tainan, Taiwan, are appreciated.

[1] Y. Arakawa et al., IEICE Trans. Electron. E83-C, p. 564, (2000).
[2] H. X. Jiang *et al*., Appl. Phys. Lett. **78**, p. 1303, (2001).
[3] B. S. Tan *et al*., Appl. Phys. Lett. **84**, p. 2757, (2004).
[4] T. Fujii *et al*., Appl. Phys. Lett. **84**, p. 855, (2004).
[5] S. J. Wang *et al*., Appl. Phys. Lett. **87**, 011111, (2005).
[6] K. M. Uang *et al*., Jpn. J. Appl. Phys., part 2 45, 3436 (2006).
[7] S. J. Wang *et al*., 64th DRC, session **II-9**, SCE, PA, 2006.
[8] T. M. Chen *et al*., Appl. Phys. Lett. **90**, 041115 (2007).
[9] ISE-TCAD Release 10.0 manual. ISE AG, Zurich (2004).
[10] http://www.mfa.kfki.hu/~labar/index.htm

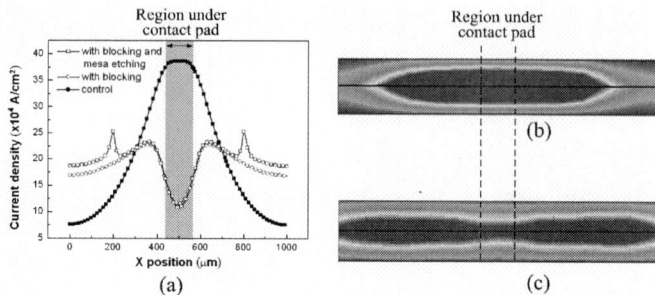

Fig. 2 (a) Current density distributions in active regions of the proposed device. Distributions of light emission in the active region of regular VM-LEDs (b) and the proposed device (c). Hues of red, orange, yellow, green, and blue indicate the relative intensity of light emission, from high to low.

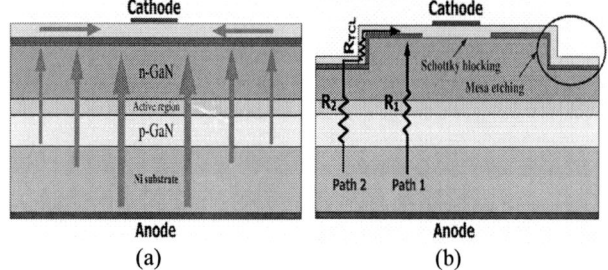

Fig. 1 Schematic cross section of (a) regular VM-LED (b) the proposed scheme.

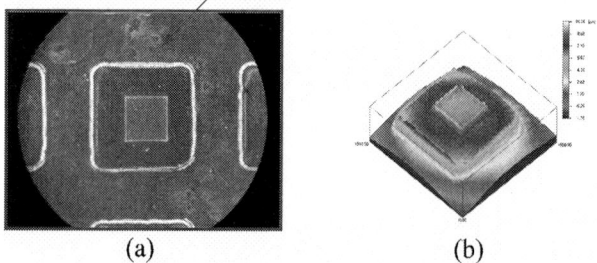

Fig. 4 Surface morphology of fabricated device measured by 3D Confocal Microscopce (nanofocus): (a) OM photograph of top-view, (b) 2-D surface profile.

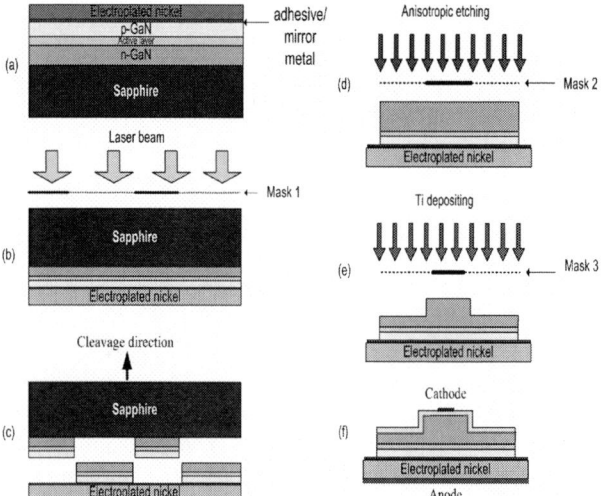

Fig.3 Key fabrication process of the proposed devices. (a) metal substrate production, (b) laser irradiation, (c) lift-off process, (d) anisotropic etching, (e) localized Ti deposition, (f) IZO deposition and cathode electrode formation.

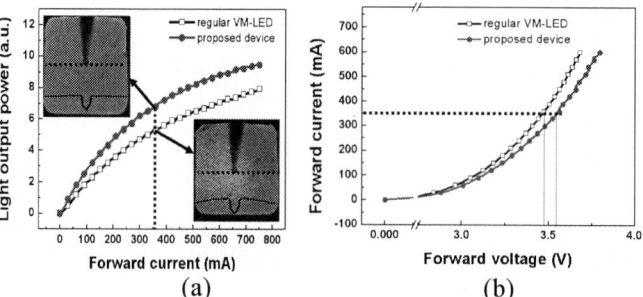

Fig. 5 (a) A comparison of Lop-I characteristics. The inset shows photos of light emission from proposed device and VM-LED at 350 mA. The solid curves at the bottom of the images denote the relative light output intensity along the dashed lines. (b) A comparisons of I-V characteristics.

Feasibility Study of Composite Dielectric Tunnel Barriers for Flash Memory

Sarves Verma[1], Eric Pop[2], Pawan Kapur[1], Prashant Majhi[2], Krishna Parat[2], Krishna C.Saraswat[1]

1 *Center for Integrated Systems, Stanford University, 20 Via Palou, MS 4075, Stanford, CA 94305*
2. *Intel Corporation, Phone: 650-725-3610, email: sarves@stanford.edu*

Lack of voltage reduction presents a serious impediment in future flash memory scaling. Replacing conventional SiO_2 tunnel dielectric with a composite dielectric material (combination of high-κ and SiO_2 layers) stack (Fig.1) potentially yields a powerful approach to achieve voltage reduction. The method relies on obtaining non-linearity in gate current vs. gate voltage (V_{gs}) characteristics, such that, low voltages (corresponding to retention and read disturb (V_{read})) render a lower current; whereas, high voltages (corresponding to program and erase) result in a higher current. Related past work involves using a composite stack with a smaller barrier material (high-κ) between two large barrier materials (SiO_2) [1] and a crested barrier approach [2]. However, in these systems, to-date, a top-down approach to optimize the design space for minimum programming voltage (V_{prog}) and equivalent oxide thickness (EOT), while meeting flash retention, read and program-disturb constrains, has not been pursued. Such an approach identifies the best high-κ option by comparisons, yielding lowest possible flash operational voltages.

Towards this end, we explore both the symmetric (low-κ/high-κ/low-κ) and the asymmetric (low-κ/high-κ) composite tunnel barriers using several high-κ materials. There are three possible tunneling mechanisms through a composite barrier depending on the V_{gs} (Fig. 2a). A corresponding simulated current density (J)-V_{gs} (Fig. 2b) indeed demonstrates a higher non-linearity for composite barrier structure compared to the conventional SiO_2 dielectric, thus, yielding a lower V_{prog}. The higher non-linearity stems from a lowering of both the high-κ barrier (modulated by voltage across low-κ) and the increased E-field with V_{gs}. By comparison, for conventional SiO_2 tunnel dielectric, the barrier height is fixed and only one factor- increase in E-field is responsible for the increase in current with V_{gs}.

Five different high-κ materials (HfO_2, La_2O_3, Y_2O_3, ZrO_2, Al_2O_3) were explored using an in-house simulator based on the transfer matrix approach [3] for current calculations. We first choose the high-κ material. Next, we fix the total EOT (high-κ + SiO_2 layers) and vary SiO_2 thickness (T_{ox}). For each T_{ox}, we simulate the J-V_{gs} curves, and obtain the V_{prog} at the required programming current density ($J_{prog} = 3 \times 10^{-2}$ A/cm^2, typical for NAND Flash). By repeating this for different EOTs, we get a family of V_{prog} vs. T_{ox} curves, each exhibiting a minimum V_{prog} (Fig. 3a). Similarly, Fig. 3b shows the voltages corresponding to the maximum allowed retention ($J_{ret} < 2 \times 10^{-16}$ A/cm^2) and read-disturb ($J_{read} < 7 \times 10^{-11}$ A/cm^2) current densities (shown as horizontal dashed lines, also see Fig. 2b), along with the actual voltages encountered during these conditions. In order to meet these constrains, the voltages obtained should be larger (in absolute value) than the horizontal dashed lines. Thus, these constrains manifest themselves in limiting the allowed T_{ox} range for certain EOTs, resulting in a domain down-selection in Fig. 3a (for V_{prog}). Combining the T_{ox} domain down-selection with the V_{prog} vs. T_{ox} plot, we obtain the optimum V_{prog} for each EOT (Fig. 4). The curve also reveals the minimum possible EOT below which no T_{ox} satisfies the Flash constrains. We repeat the process for different high-κ materials to get the lowest possible V_{prog} along with the 1) best material set, 2) the lowest EOT, and 3) the optimum T_{ox} for that EOT (Fig. 4a,b). The effect of adding constrains can be seen in Fig. 5. In general, constrains increase V_{prog} when compared to the minimum obtained in Fig. 3a, except for program disturb (corresponding to $J_{prog} < 7 \times 10^{-6}$ A/cm^2 refer Fig. 2b) which is always satisfied. Further, upon extraction of erase voltage ($J_{erase} > 7 \times 10^{-3}$ A/cm^2) (Fig. 6), the asymmetric stack was found to have a larger V_{erase} than the symmetric one leading to an important conclusion that symmetric stacks are more promising. Finally, Fig. 7 shows the maximum possible operational voltage V_{max} (maximum of V_{prog} and V_{erase} on the floating gate) vs. EOT. The optimization is across all considered high-κ materials. We find that La_2O_3 performs best for a strict read disturb criterion of 3.6 V while HfO_2 outperforms other high-κ materials for a 2.5 V read disturb. It yields the largest operational voltage of ~5–7 V constituting a ~30%-40% voltage reduction over the conventional SiO_2 based Flash cells.

Next, we experimentally corroborate these results using MOS capacitors. A J-V_{gs} comparison between pure SiO_2 tunnel dielectric and an asymmetric tunnel stack of SiO_2/HfSiON$_x$ (~6.4 nm EOT), reveals a higher non-linearity for the composite stack (Fig. 8). A good agreement between simulations and experimental results can also be observed in Fig. 8. Further, C-V measurements (Fig. 9) with minimal hysteresis confirmed a high quality composite dielectric stack. Finally, to ensure that tunneling was indeed Fowler-Nordheim (F-N) dominated, F-N slope was calculated (Fig. 10). An excellent linear fit was obtained from which both barrier height and tunneling mass were estimated.

In conclusion, we have developed a novel optimization methodology for deriving the minimum operational voltage of a composite tunnel dielectric stack for Flash memory. The methodology accounts for normal Flash operation constraints: retention, read and program disturb, and reveals that the higher J-V_{gs} non-linearity of these stacks can result in up to 40% voltage reduction. The symmetric stack is found to be more efficient than asymmetric in reducing voltage. The optimum materials along with their thickness are also identified. Experiments confirm the non-linearity in J-V_{gs} using HfSiON$_x$, a high-κ which for the first time is used in the context of Flash memory.

References :[1] Govoreanu et al., *IEEE Electron Dev. Lett.*, vol. 24, pp. 99-101, 2003 [2] K. K. Likharev, *Appl. Phys. Lett.*, vol. 73, pp. 2137-2139, 1998 [3] Y. Ando et al., *J. Appl. Phys.*, vol. 61, pp. 1497-1502, 1987

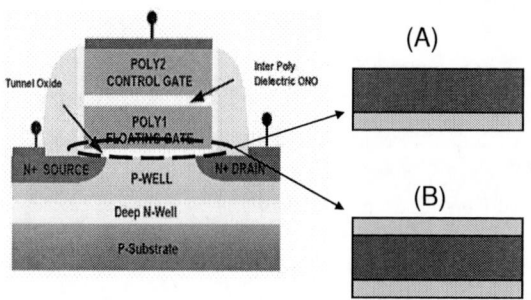

Fig. 1 Schematic of a Flash memory showing the conventional tunnel oxide. (A) Shows replacement of the tunnel stack by an asymmetric barrier while (B) Shows that by a symmetric one. The thicker layer represents high-κ material while thinner represents SiO_2

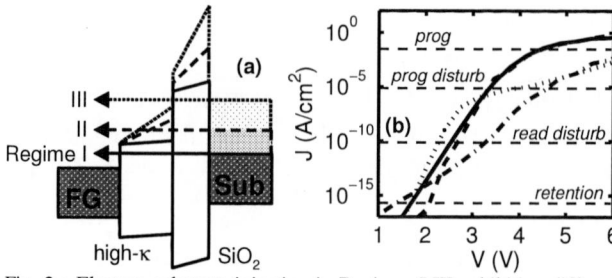

Fig. 2.a Electron substrate injection in Regimes I-III exhibiting different tunneling mechanisms. Fig. 2b Shows the J-V characteristics of asymmetric stack (solid line, $T_{ox} = 2$ nm), symmetric (dashed, $T_{ox} = 2$ nm on either side), and pure SiO_2 (dash-dot) with 6 nm total EOT. The dotted line is for a thicker $T_{ox} = 3.5$ nm in the asymmetric 6 nm EOT stack. The four horizontal dashed lines are Flash constrains mentioned in the text

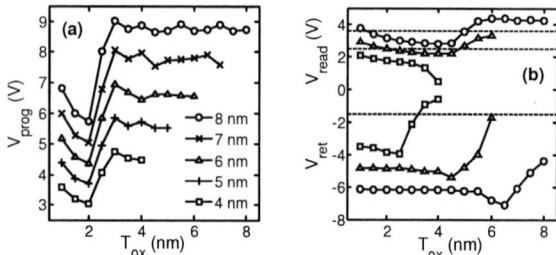

Fig. 3a V_{prog} scaling with T_{ox} for different asymmetric EOTs with HfO_2 (note the minima). Fig. 3b shows the read disturb (top) and retention voltage (bottom) scaling with T_{ox} for the 4, 6 and 8 nm EOT stacks. The top horizontal dashed lines correspond to 2.5 and 3.6 V read disturb voltage; the bottom one corresponds to -1.5V retention. Symbols for different EOTs are consistent across both figures

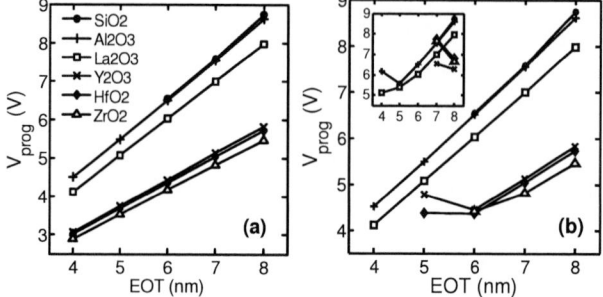

Fig. 4. Optimal V_{prog} at each EOT for all asymmetric stacks and high-κ materials. (a) Imposes only the retention constraint, while (b) adds in the $V_{read} = 2.5$ V read disturb, and considers the more restrictive 3.6 V in the inset. Symbols are used consistently across the figures.

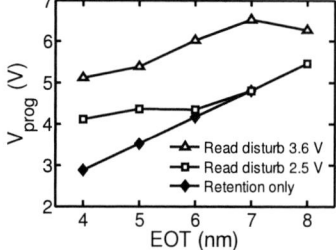

Fig. 5. Change in V_{prog} for asymmetric barriers with different constraints: retention only (no read disturb), and with different read disturb criteria V_{read} = 2.5 and 3.6 V. The values plotted represent global minima, across all high-κ materials considered here.

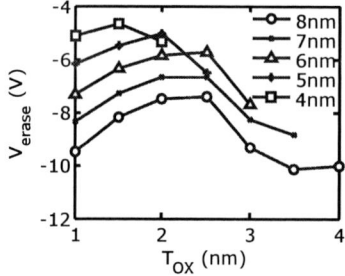

Fig. 6 Erase simulations for a representative case of SiO_2/ HfO_2/SiO_2 symmetric stack for different oxide thicknesses (within the same EOT) and with varying EOT.

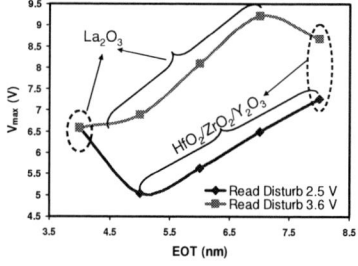

Fig. 7 Global performance optimization for all stacks in consideration irrespective of the high-κ material considered. Further, all constrains like retention, read disturb, program disturb and erase have been taken in account to find the maximum operating voltage (V_{max}) required for a Flash memory. Note only La_2O_3 sustains the stringent read disturb criterion of 3.6 V

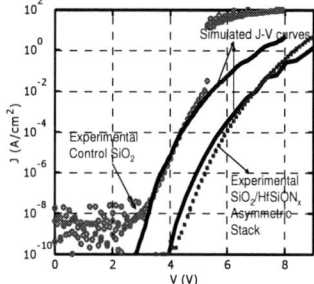

Fig. 8 Experimental J-V curves for a control oxide (SiO_2) of 5.6 nm thickness and for an asymmetric SiO_2/$HfSiON_x$ stack of 6.4 nm EOT. The dotted lines represent experimental results while the solid lines are simulated curves for pure SiO_2 stacks of same EOT. Note that the asymmetric stack has a higher non-linearity when compared to simulated pure SiO_2 of the same EOT. Simulations do not account for the breakdown in oxides. All samples are made on n-type substrates and for each sample three measurements were taken accounting for different areas.

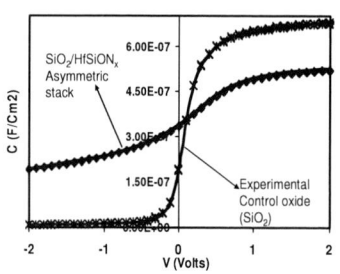

Fig. 9 C-V measurements for the same stack as in Fig. 8.Note that the high-κ /SiO_2 asymmetric stack shows no hysteresis implying a good interface with the Si substrate. Both samples were prepared under different conditions and have different doping levels. For each sample, C-V measurements were taken for 10 KHz, 100 KHz and 1 MHz frequencies.

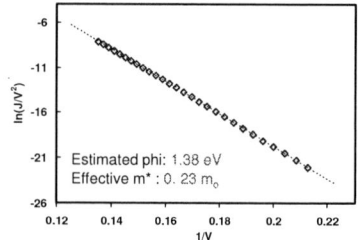

Fig. 10 Plot of Fowler-Nordheim slope for SiO_2/$HfSiON_x$ asymmetric stack (the J-V of which has been shown in Fig.7). Barrier Height (phi) and effective mass (m^*) have hence been derived.

Confined Optical Phonon Scattering in p-Silicon Nanowires

Mueen Nawaz, and Jean-Pierre Leburton[1]

[1]*Beckman Institute, Department of Electrical and Computer Engineering,*
University of Illinois at Urbana-Champaign, 405 North Mathews Avenue, Urbana, Illinois 61801

In quantum wires (QWRs), the free motion of carriers is restricted to one-dimension (1-D), whereas there is a two-fold quantum confinement in the other transverse directions. This situation limits available states for scattering, which tends to increase carrier mobility.[1] At the same time, quantum confinement enhances the interaction between carriers and lattice vibrations, which increases their scattering rates.[2] Transport properties in 1-D semiconductors is thus the result of a subtle interplay between the reduced density of states and the enhanced carrier-phonon interaction.[3]

In this paper, we investigate hole scattering by optical phonons through the deformation potential interaction, which is one of the main mechanisms for dissipation at room temperature. We consider free-standing cylindrical silicon nanowires of radius R, length L, and with the z-axis along the QWR direction (see Fig 1). We assume an infinite cylindrical well potential profile for the holes (i.e. $V = 0$ inside the wire and $V = \infty$ on the radial edges).

Solving Schrodinger's equation inside the nanowire yielded the hole wavefunctions. In Fig. 1 we plot the first few QWR subband edges where we distinguish between heavy and light hole subbands. We note that the lowest two subbands are heavy hole subbands that remain separated by an energy difference larger than $k_B T$ at room temperature for radii less than 5.2 nm. At larger radii, the separation between the subbands decreases dramatically and approaches a continuum, as is to be expected in bulk materials. Therefore for QWR's with a 5 nm radius or less, we can assume the extreme quantum limit and neglect intersubband scattering at room temperature—confining ourselves to the lowest heavy hole band.

In calculating the intrasubband hole scattering rates due to optical phonons, we used two different models for the deformation potential Hamiltonian: The plane wave model often used in bulk silicon, and a confined phonon model based on the hydrodynamic continuum approximation.

Figure 2 shows the ratio of the confined phonon emission scattering matrix elements normalized to the bulk (continuous) phonon emission scattering matrix element, as a function of q_z. In contrast with bulk phonons, the forward and backward scattering rates are actually different for the same q_z. At large q_z, all curves converge to unity and the convergence is more rapid the larger the radius of the nanowire. We observe that the normalized value of the scattering rates remains less than unity, indicating weaker scattering than obtained with the bulk (plane wave) phonon model.

Figure 3 shows the normalized scattering rates for confined phonon emission in nanowires of different radii as a function of the hole kinetic energy along the QWR. For all QWR radii, the normalized rates are smaller than unity, and decrease with smaller radii, again indicating weaker scattering is expected in the presence of confined phonons. The effect weakens at large radii, where the normalized rates approach unity, converging toward the continuous phonon values as expected.

In Fig. 4 we display the ratio between forward and backward scattering rates for different radii calculated with the confined phonon model as a function of the hole kinetic energy along the QWR. The diagram shows a stronger tendency for holes in smaller nanowires to backscatter than in wider nanowires.

In conclusion, we found considerably lower scattering rates for confined phonons than if calculated with a continuous phonon spectrum, especially at small radii. In addition, we showed confined phonons favor backward scattering over forward scattering in narrow QWRs. Our results have important consequences for the dissipation processes in nanowire based MOSFETs.

This work was supported by the Network for Computational Nanotechnology under NSF Grant No. EEC-0228390. The authors are indebted to Dr. M. Stroscio and Dr. J. Jin for his helpful discussions.

---

[1] H. Sakaki, Japan J. Appl. Phys., 19, L735, (1980).
[2] R. Kotlyar, B. Obradovic, P. Matagne, M. Stettler, and M.D. Giles, Applied Physics Letter, 84, 5270, (2004).
[3] J.-P. Leburton, Phys. Rev., 45, 11022, (1992).

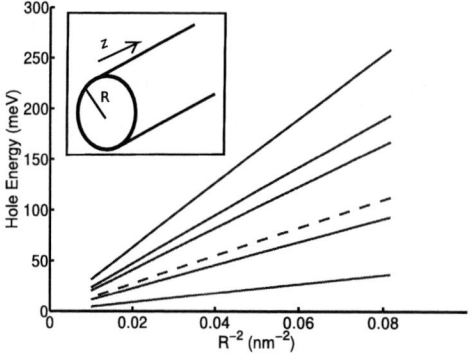

FIG. 1: Variation of the minimum heavy hole energy within a subband as a function of radius for the first six least energetic subbands. Inset: Geometry of the nanowire.

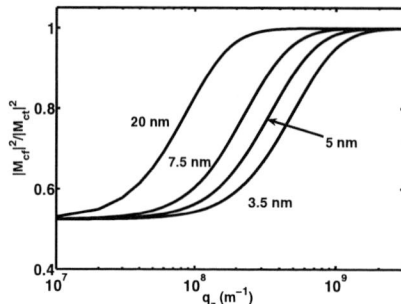

FIG. 2: Ratio of the square of the scattering matrix elements for confined phonons and continuous phonons in nanowires with different radii, as a function of q_z.

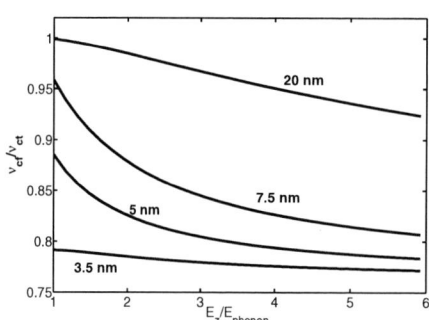

FIG. 3: Ratio of the confined phonon emission rate to the continuos phonon emission rate in nanowires of different radii.

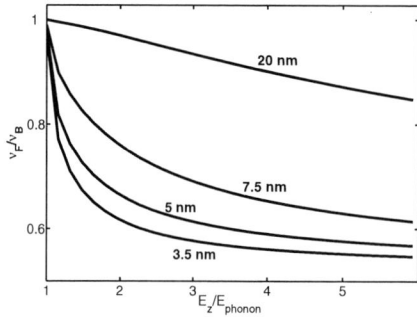

FIG. 4: Ratio of the forward emission scattering rate to the backward emission scattering rate with confined phonons as a function of hole energy for different QWR radii.

88

Mid-Wavelength Infrared (MWIR) Avalanche Photodiode (APD) using InAs-GaSb type-II Strain layer Superlattice (SLS)

Shubhrangshu Mallick[i], Koushik Banerjee[i], Siddhartha Ghosh[i], Sanjay Krishna[ii], Jean Baptist Rodriguez[ii]

[i] Lab for Photonics and Spintronics, Electrical and Computer Engineering Department, University of Illinois at Chicago, 851 S Morgan Street, 1020SEO, Chicago, IL, 60607.
[ii] Center for High Technology Materials (ECE Dept), University of New Mexico, 1313, Goddard Street SE, MSC04 2710 Albuquerque, NM, 87106.

Abstract

Modern weapon systems need to detect, recognize and track a variety of targets under a wide spectrum of atmospheric conditions. They include stationary and mobile targets against complex background and landmines. Detectors with high bandwidth and reasonable internal gain are required to detect the highly attenuated optical signal signals at long range detection. APDs are best suited for this purpose due to their internal gain-bandwidth characteristics. Since the last decade InAs-GaSb type-II SLS has been emerged as one of the competing material for the IR detection in the longer wavelength application. Due to easier growth and longer minority carrier lifetime as well as the suppression of hole ionization due to particular band structure this material system has an edge over the contemporary mostly used HgCdTe material system especially for longer wavelength applications[1,2,3]. The advances in Molecular Beam Epitaxy (MBE) have made it possible to adjust the thickness of each InAs and GaSb layers to detect the longer wavelengths (Mid-wavelength Infrared (MWIR) 3-5μm, Long-wavelength Infrared (LWIR) 8-10μm). But there has not been any report of APDs in this material system, which makes this material system more useful for longer range defense or astronomy application. We report the first p-i-n APD fabricated using InAs-GaSb SLS on GaSb substrate.

The devices were fabricated by the usual photolithography and wet-etching. The diode had a junction area of 400μm diameter. The R_0A of the passivated and unpassivated diodes diode were $2 \times 10^5 \Omega\text{-cm}^2$ and 3×10^6 at 77K which was comparable to the values reported for the diodes fabricated in HgCdTe Technology. The gains of the APDs were measured with the help of a 632nm He-Ne (1.02mW) laser. Gain of close to 600 was achieved at a reverse bias of -20V in the Giger breakdown operation of the APDs. The gain had an exponential nature as the reverse bias in increased. The avalanche gain reduces as the temperature was increased because of increase in phonon vibration at higher temperature. A detailed of the temperature variation of dark current, R_0A and the avalanche gain would be reported at the final presentation. This work is funded by DARPA HOT MWIR Program and ARL Advanced Sensors CTA.

---

[1] C. H. Grein *et. al.* Journal of Applied Physics **92**, 7311 , 2002.
[2] E.R.Youngdale *et. al.* Journal of Applied Physics **64**, 3160 , 1994.
[3] Y.Wei *et. al.* Journal of Applied Physics **94**, 4720 , 2003.

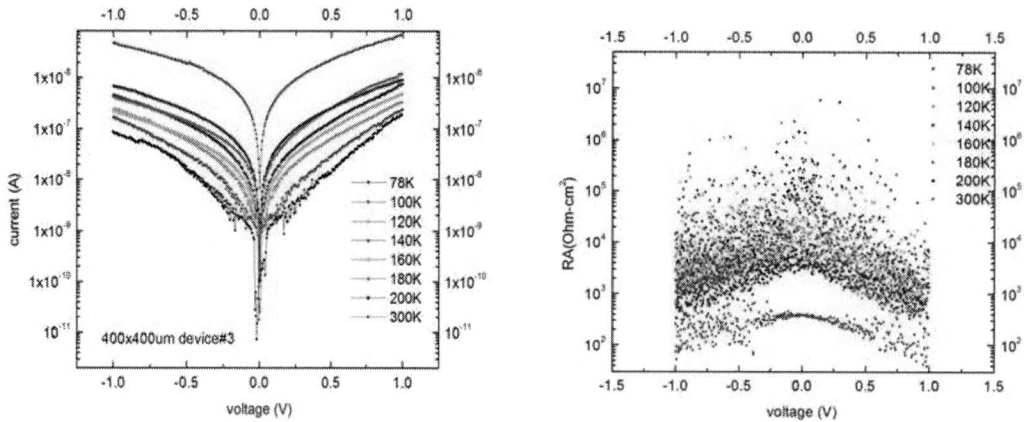

Fig.1: Dark Current and RA vs. Voltage for different temperatures

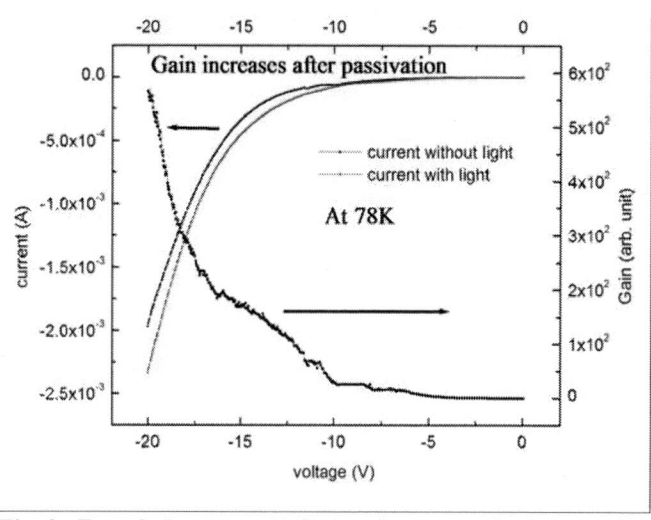

Fig.2: Breakdown and Gain characteristics st 78K

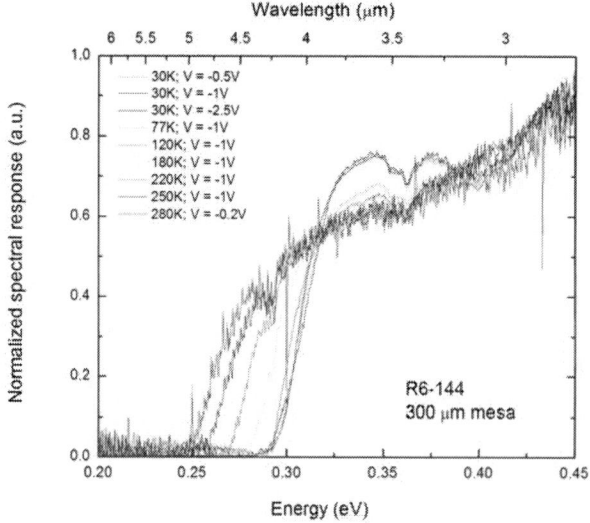

Fig.3: Spectral Response of the i-region in the p-i-n structure

Body Thickness Optimization and Sensitivity Analysis for High Performance FinFETs

Dheepa Lekshmanan, Aditya Bansal and Kaushik Roy

School of ECE, Purdue University, West Lafayette, IN – 47907. Email: dlekshma@purdue.edu

Ultra Thin Body Double Gate (UTB DG) devices are suitable for nanoelectronic circuits due to better scalability, higher on-current (Ion), improved Sub-threshold Slope (SS) and undoped body (no random dopant fluctuation). FinFETs have emerged as the best candidate of UTB DG structure because of its similarity to CMOS fabrication process. Ultra-thin body (Tsi) increases the gate control over the channel resulting in reduced short channel effects (SCE). However, thin Tsi increases the quantum confinement of charge resulting in increased threshold-voltage (Vt) [1] and hence, reduced performance. Further, very thin Tsi poses fabrication challenges and increases device characteristic mismatch due to process variations. In this work, *we optimize the Tsi in sub-50nm gate length double-gate MOS devices for designing high performance circuits under a leakage constraint. We also consider the impact of process variations in Tsi and show how delay sensitivity varies with number of fins* in a FinFET device. TAURUS drift-diffusion solver [2] is used for device simulations. Quantum confinement in ultra-thin body has been accounted for by solving 1D Schrodinger equation self-consistently with Poisson equation. Mobility model proposed in [3] has been used, which accounts for dominant scattering mechanism in UTB devices.

In UTB FETs, Vt varies with Tsi, because of quantum confinement and SCE. Hence, to achieve iso-off state leakage (@$Vgs=0V$ and $Vds=Vdd$) we tune the metal gate work function for each Tsi. Reducing Tsi improves the sub-threshold slope (Fig. 6), hence at iso-off state, Vt reduces (Fig.7) resulting in improved on-current (Fig. 2). However, for $Tsi<5nm$, Ion reduces because of reduced mobility [4] (Fig. 10). Further, reduction in Vt with Tsi (at iso-off) is also evident from C-V curves (Fig. 3). Hence, effective gate capacitance Cg (calculated by integrating the area under the C-V curve and dividing it by the supply voltage) increases with reduction in Tsi. Gate delay (Td) can be approximated as CgV/Ion. Since Cg and Ion both increase with reduction in Tsi, an optimal Tsi corresponding to minimum Td can be obtained. For $Vdd=1.0V$, minimum delay occurs at $Tsi=6nm$ (Fig. 8). For ultra low power circuits, sub-threshold operation ($Vdd<Vt$) has been proposed [5]. In sub-threshold region, the on-current (@$Vgs=Vds=Vdd<Vt$) varies exponentially with Vt. Hence on current increases exponentially with reduction in Tsi whereas Cg has similar dependence on Tsi as in super-threshold ($Vdd>Vt$). Therefore, Tsi corresponding to minimum delay reduces (Fig. 9). Note that iso-off state condition is achieved for sub-threshold operation by again tuning the metal gate work function. The above analysis shows that, *to achieve minimum delay, Tsi must be optimized based on the region of operation – super-threshold or sub-threshold.*

Since FinFET has quasi planar structure, fabrication of ultra thin silicon body is a major challenge. Process variations in silicon thickness have detrimental effect on device charactersistics. In sub-threshold region, the on-current varies exponentially with Vt (or Tsi). Hence, on-current in sub-threshold is more sensitive to Tsi variations compared to super-threshold (Fig. 11). FinFETs have similar manufacturing steps as CMOS. Therefore, random variations in Tsi (similar to line edge roughness in patterning gate length) can occur between the neighboring fins. To analyze the delay sensitivity to within-die silicon thickness, we assume the random component of Tsi follows Gaussian distribution. For calculating gate delay (CgV/Ion), where Cg is the input gate capacitance of the succeeding stage and Ion is the driving current of the current stage, we assume that Cg is constant to analyze the effect of on-current variations on the delay.

In FinFET devices, the width ($Hfin$) is in the vertical direction. Hence, it can only be increased in quanta of $Hfin$. Hence, to increase the width of the transistor, number of fins has to be increased. Typically $Hfin = 4{\sim}5\ Tsi$ to maintain the aspect ratio. For Tsi of 6nm, $Hfin$ is 30nm. Hence, for a desired transistor width of, say 300nm, 10 fins are required. Assuming that Tsi varies randomly among the fins of a single transistor, we analyze the delay sensitivity with the number of fins. In super-threshold, the delay variability (σ/μ) reduces 65% from 1 fin to 10 fin transistor whereas in sub-threshold operation, it reduces 72%.

We propose a device optimization technique to achieve high performance in double-gate MOSFETs by considering the trade-off between on-current and gate capacitance with silicon thickness. We show that the optimal silicon thickness varies based on the mode of operation (super-threshold or sub-threshold). Even though these devices have intrinsic body thickness variations, the effect of these variations reduces considerably as the number of fins increases.

Operation	Optimal Tsi (nm)	Off current Ioff(nA)	On current Ion(A)	Gate capacitance Cg(fF)	Gate Work-function(eV)	Variability of 1 fin(σ/μ)
Super-threshold	6	25	2.28e-3	0.687	4.55	0.0263
Sub-threshold	5	1	1.21e-6	0.103	4.56	0.6184

[1] Q. Chen, et. al., *IEEE SOI Conf.*, pp. 183-184, 2003.
[2] Taurus Device Simulator, Synopsys Inc, v2004.09.
[3] M.N. Darwish, et. al., *IEEE TED*, Vol. 44, No. 9, pp. 1529-1538, Sept., 1997.
[4] D. Esseni et. al., *IEEE TED*, vol. 50, no. 12, 2003, pp. 2445-2455.
[5] H. Soeleman et. al., *IEEE VLSI*, pp. 90-99, 2001.

978-1-4244-1101-6/07/$25.00 ©2007 IEEE

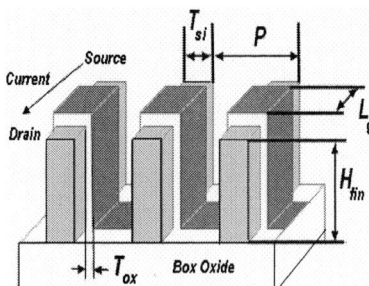

Fig 1. FinFet device structure. Lg=20nm, Tox=1nm, Hfin:Tsi=5.

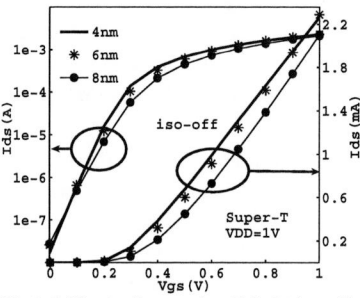

Fig 2. I-V ccts. for varying Tsi. At iso-off current, Vt reduces (SS improves) with reducing T_{si} resulting in increased Ion.

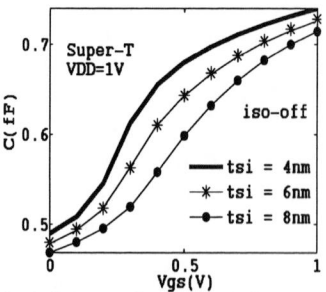

Fig 3. CV curve for varying Tsi. At iso-off current, effective gate cap. increases with reducing Tsi

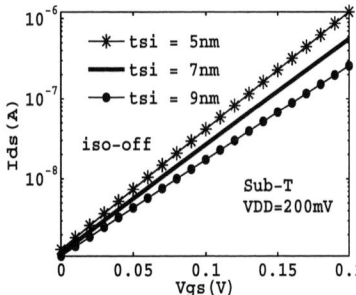

Fig 4. I-V ccts. for varying Tsi in Sub-T @iso-off current(1nA/um).

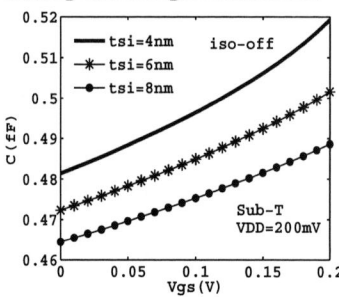

Fig 5. CV curve for varying Tsi in sub-T region @iso-off current(1nA/um).

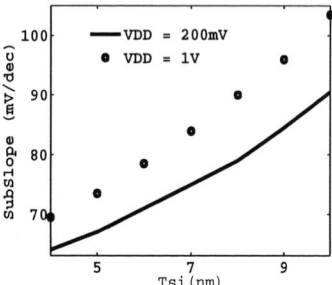

Fig 6. Sub-threshold slope (SS) vs. Tsi in super-T and sub-T operations.

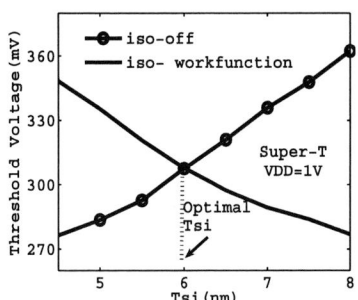

Fig 7. Vt vs. Tsi @ iso-off current and iso-workfunction.

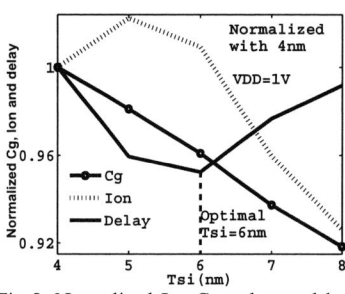

Fig 8. Normalized I_{on}, C_g and gate delay (CgV/Ion) vs. Tsi. At iso-off current minimum gate delay occurs at 6nm.

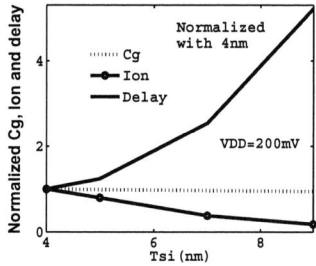

Fig 9. Norm. I_{on},Cg,gate delay(CgV/Ion) vs. Tsi. At iso-off current minimum delay occurs at lower Tsi compared to super-T (Fig: 8)

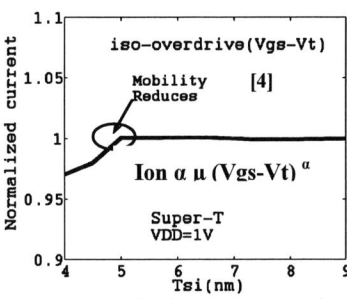

Fig 10. Normalized on-current vs. Tsi in super-T region.

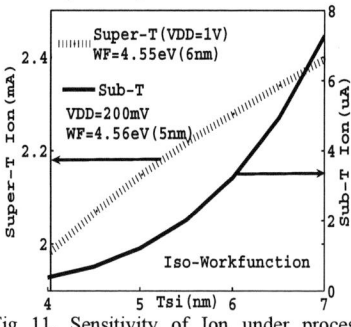

Fig 11. Sensitivity of Ion under process variations in Tsi

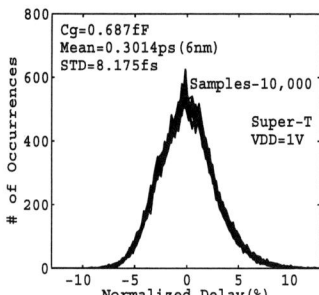

Fig 12. Gate Delay distribution in super-Threshold operation.

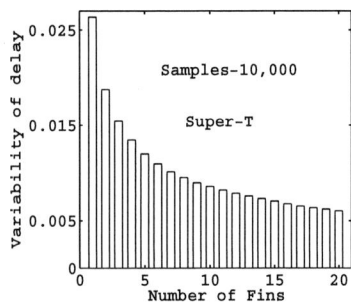

Fig 13. Variability (σ/μ) of delay vs. no. of fins. σ/μ of 1 fin is 0.0263

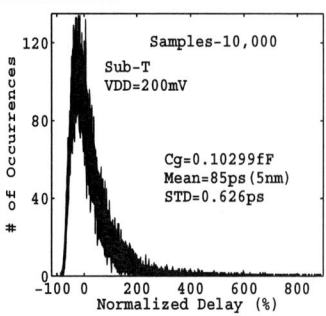

Fig 14. Gate Delay Distribution in sub-Threshold operation.

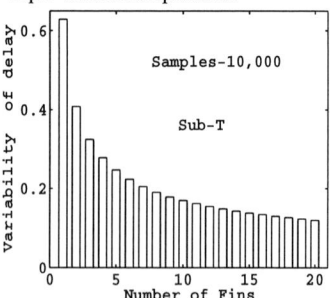

Fig 15. Variability (σ/μ) of delay vs. no. of fins in sub-T region. σ/μ of 1 fin is 0.6184

Flash Memory Fabricated with Protein-Mediated PbSe Nanocrystal Assembly as Floating Gate

Shan Tang, Chang Hun Lee, Xiaoxia Gao* and Sanjay K. Banerjee

Microelectronics Research Center, R9900, The University of Texas at Austin, Austin, TX 78758
* Texas Materials Institute, R9900, The University of Texas at Austin, Austin, TX 78758
Email: shantang@mail.utexas.edu / Phone: (512)4713706 / Fax: (512)4715625

The research on nanocrystal (NC) floating-gate flash memories in recent years has demonstrated their advantages over traditional flash memories in many areas such as better device scaling, lower power consumption and improved charge retention [1, 2]. At the same time, the control of size and distribution of NCs still remains a challenge. Using bio-nano techniques such as DNA, virus or protein as assembly tools may be feasible for self assembly and actual electronic device fabrication [3, 4]. This work is based on our previously developed technique which used chaperonin protein as a template for NC assembly on MOS capacitors [5]. Our study on floating gate flash memory fabricated with protein-mediated PbSe NCs applied to MOS transistors with greatly improved spatial uniformity will be presented in this paper.

The fabrication of NMOS transistors with PbSe NC floating gate process started with the field oxide growth and active area patterning, followed by thermal growth of 4-nm SiO_2 tunnel oxide. After pretreating the wafer surface with PTS (phenyltriethoxysilane) solution, we floated the wafer on the chaperonin protein solution with the oxide side down. Therefore NCs in colloidal suspension could be trapped inside chaperonin's central cavity through hydrophobic-hydrophobic interaction. Then the chaperonin template was removed by annealing in O_2 at 200^oC or in the air at 300^oC. A 12-nm SiO_2 tunnel oxide layer was deposited through low pressure CVD, followed by TaN sputtering and patterning to form the control gate. The remaining process was just the same as regular transistor fabrication, which included source and drain implantation and metal contact formation (Fig 1).

In our work, PbSe NCs with two different diameters (4nm and 6nm) were studied and we observed the NC densities of $4 \times 10^{12}/cm^2$ and $2 \times 10^{12}/cm^2$ from the STEM images, respectively (Fig 2). A threshold voltage shift of 1.3V was observed with ± 12V and 1s program and erase stress with 4nm NCs (Fig 3). The transient characteristics showed a larger threshold voltage window for most devices with smaller PbSe NCs under the same erase/program operation (Fig 4), which indicates overall charge storage per unit area is dominated by the NC density in this case. As we know, larger NCs can provide more energy states in a single NC and their corresponding Coulomb charging energy is lower which makes it easier for electrons to be charged into the NCs. This result shows that the factor of 2 higher density for smaller NCs, it more than compensates these two benefits of larger NCs in the effectiveness of electron trapping provided we use the same write voltage. It is also verified by the retention characteristics where a lower charge loss rate was observed for smaller NCs at 85^oC (Fig 5). The degradation of memory window is reasonable for up to 10^4 seconds which indicates good charge retention for both devices. In endurance tests, the memory window of large NC devices becomes stable after first 200 operation cycles, till 10^5 cycles, while the window for small NC ones remains open throughout the tests (Fig 6). With these device characteristics, we expect PbSe NCs as floating gate to be promising for flash memories.

References:

[1] I. Kim, etal., *IEEE IEDM Tech. Dig.*, 1998, pp. 111-114
[2] H. I. Hanafi, etal., *IEEE Trans. Electron Dev.*, Vol. 43, No.9, 1996, pp.1553-1558
[3] T. Hikono, etal., *Appl. Phys. Lett.* Vol. 88, 2006, pp. 23108.
[4] A. Miura, etal., *J. J. Appl. Phys.*, Vol. 40, 2006, pp. L1-L3.
[5] S. Tang, etal., *IEEE IEDM Tech. Dig.*, 2005, pp.181-184.

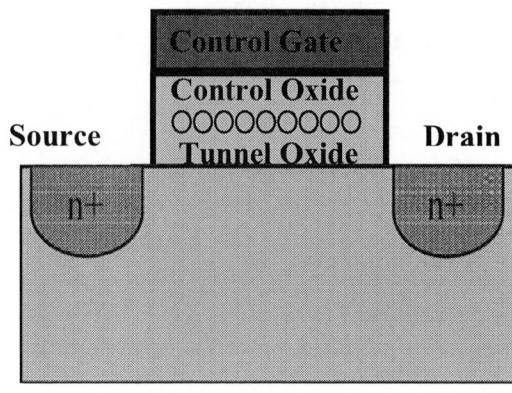

Fig. 1 Schematic cross section of the PbSe NC floating gate flash memory.

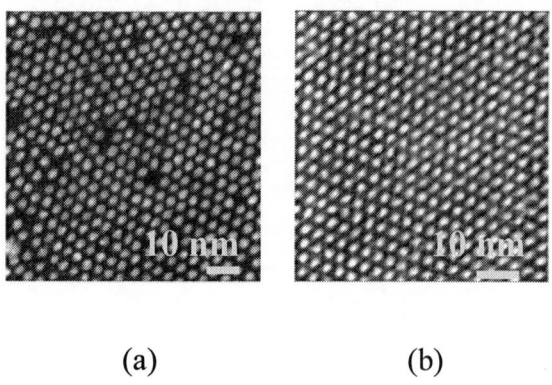

(a)   (b)

Fig. 2 STEM images of (a) 6nm PbSe NCs; (b) 4nm PbSe NCs.

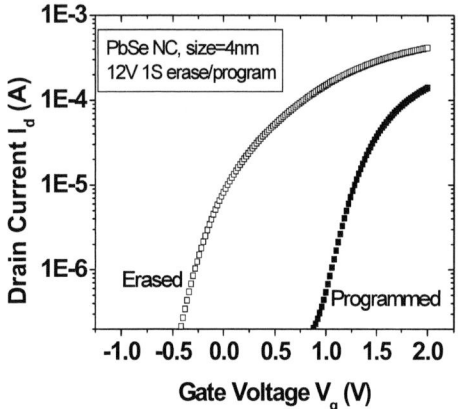

Fig. 3 I_d-V_g characteristics of memory device with 4nm NCs.

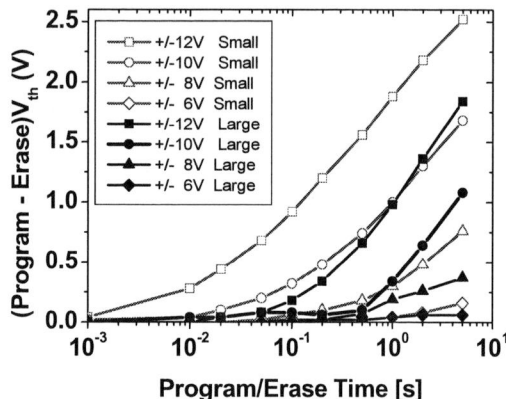

Fig. 4 Transient characteristics.

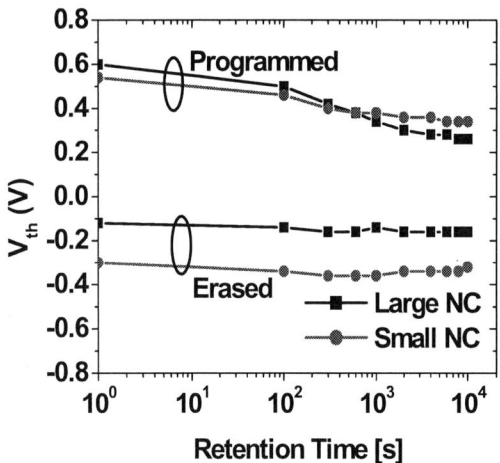

Fig. 5 Retention characteristics at 85°C; 10V ±200ms erase/program operation.

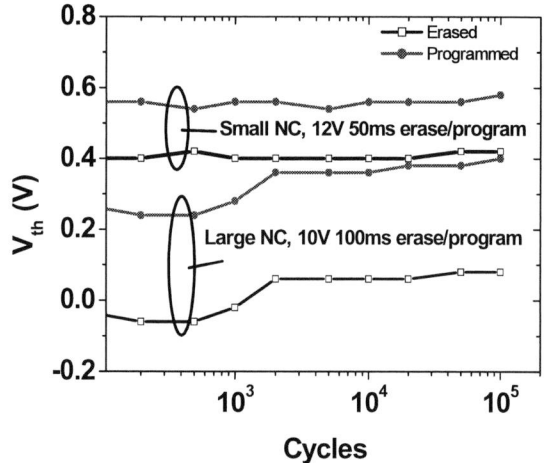

Fig. 6 Endurance characteristics.

94

Novel Amorphous-Si AMOLED Pixels with OLED-independent Turn-on Voltage and Driving Current

Bahman Hekmatshoar, Alex Z. Kattamis, Kunigunde Cherenack, Sigurd Wagner and James C. Sturm

Princeton Institute for the Science and Technology of Materials (PRISM), Department of Electrical Engineering, Princeton University, Princeton, NJ 08544
Email: hekmat@princeton.edu

In Active Matrix Organic Light Emitting Diode (AMOLED) displays, a voltage data signal is converted in the pixel to an OLED driving current (and thus brightness) by a FET. Typically p-channel FET's are used, because with conventional "bottom-anode (ITO)" OLED's, only p-channel devices allow the direct programming of the FET gate-source voltage (and thus the driving current) independent of the OLED I-V curve (Fig. 1(a)). This excludes the use of low-cost amorphous-Si (a-Si) TFT technology (as used in AMLCD's) because no p-channel a-Si device exists for fundamental reasons [1] even though a-Si devices can easily provide sufficient current. Thus more expensive poly-Si technology must be used [2]. In this abstract, we present a novel integration approach that allows conventional bottom-anode OLEDs to be integrated with n-channel a-Si TFT's in such a way that programming becomes independent of the OLED I-V curve.

The circuit schematic of a conventional AMOLED pixel with p-channel poly-Si TFT's is shown in Fig. 1(a). The TFT is processed first with temperatures up to 300°C, followed by the OLED anode (patterned ITO, connected to the TFT drain), the organic (blanket) (Fig. 2(a)) and cathode (blanket) (Fig. 2(b)). Because the best OLED's are made with the anode (ITO) deposited first, and because processing (photolithography) of the organic is not feasible, p-channel devices are required for direct programming of the pixel. With the OLED at the drain (not source) of the driving FET, the data voltage appears directly across the gate-source of the driving TFT and therefore the pixel current depends only on the characteristics of the driving FET, independent of the OLED characteristics when the FET is in saturation. If n-channel FET's (such as a-Si TFT's) are used instead with the same integration process (Fig. 1(b)), since the only accessible contact is the anode of the OLED, the OLED will be connected to the source of the driving FET, and therefore the data voltage will be divided across the gate-source of the driving FET and the OLED, based on the TFT and OLED characteristics. This is undesirable because the gate-source voltage of the driving TFT (and thus the driving current) will depend on the OLED characteristics which vary with time and in manufacturing. Also, higher programming voltages will be required for the same gate-source voltages.

To enable the use of a-Si TFT technology with a-Si TFTs as n-channel devices, and achieving an OLED-independent performance, the OLED cathode must be connected to the drain of the driving TFT as shown in Fig. 1(c) which is not possible with the conventional integration process. To achieve this, a new integration technique is required (Fig. 3). In this approach the TFT is processed first as usual, followed by the OLED anode (patterned ITO, connected to the common V_{dd} line). Next, multi-layer resist is used to form a separator around the anode (ITO) area in such a way that its projection hangs over an interconnect line which is in contact with the drain of the driving TFT. This projection shadow masks the interconnect line when the organic is evaporated in the next step at normal incidence (Fig. 3(a)). A 2-layer TPD/ALQ organic was used for this purpose. Finally the cathode is evaporated at an angle to make a contact to the mentioned interconnect line and thus the drain of the driving TFT (Fig. 3(b)). The experimental output characteristics of the a-Si TFT pixels fabricated with the new process is compared with the conventional process in Fig. 4. As a result of connecting the OLED cathode to the drain of the driving TFT in the new structure, the turn-on voltage is reduced by several volts and the driving current is increased by a 10X factor (Fig. 4(b) with respect to Fig. 4(a)). This is because in the new structure the data voltage is converted directly to the gate-source voltage of the driving TFT, rather than being split between the gate-source of the driving TFT and the OLED. This was confirmed by comparing the experimental data to SPICE modeling (Fig. 4(b)). Active matrices of the inverted pixels have also been realized with QVGA timing (Fig. 5).

This project was supported by the U.S. Display Consortium.

[1] R. A. Street, *Hydrogenated Amorphous Silicon*, New York: Cambridge University Press, 2005, ch. 10.
[2] J. Lih et. al., *Journal of the Society for Information Display*, v. 12, n. 4, 2004, pp. 367-371

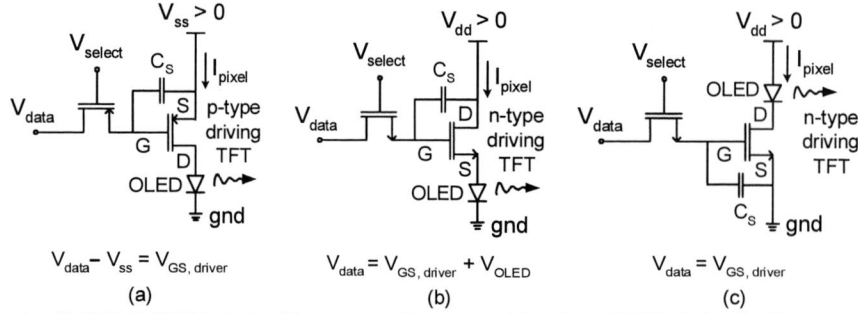

Fig.1. Circuit schematic of 2-TFT AMOLED pixels: (a) conventional structure with p-channel TFTs (poly-Si), (b) conventional structure with n-channel TFTs (a-Si) and (c) new "inverted" structure with n-channel TFTs (a-Si). In the p-channel design (a) and the inverted design (b), the data voltage programs the gate-source of the driving TFT directly, but in the design (b), the gate-source voltage of the driving TFT depends on the OLED characteristic. Realizing the inverted design (c) requires a technique for connecting the driving TFT to the OLED bottom contact (cathode).

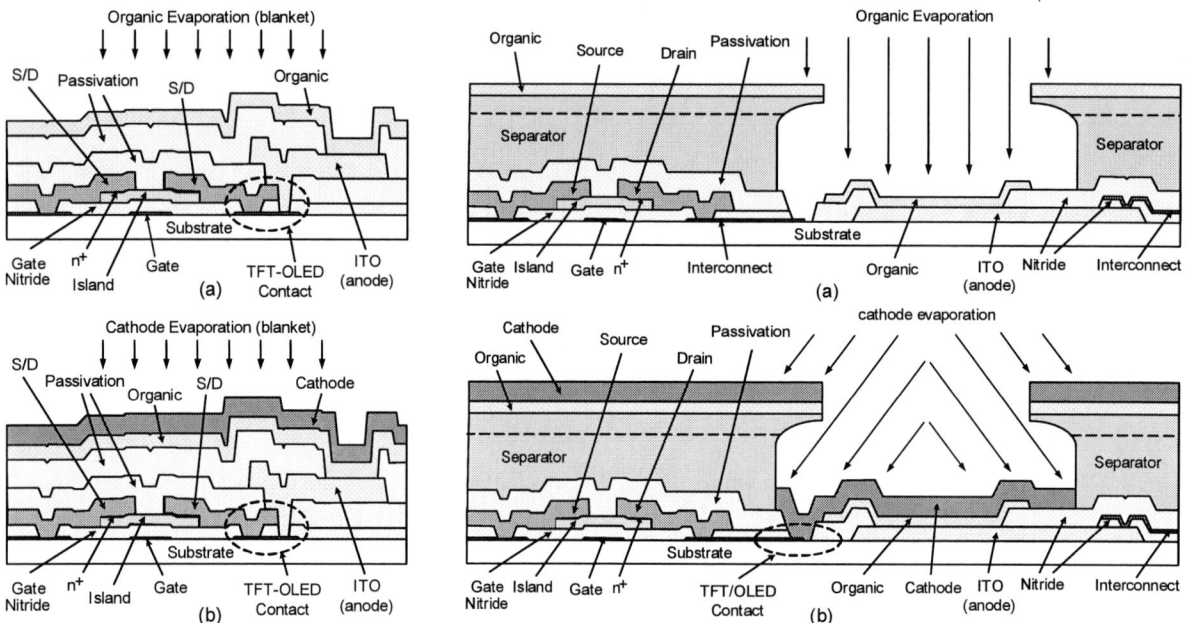

Fig.2. Schematic cross-section of the conventional AMOLED structure of Fig. 1(a), or 1(b) after the evaporation of (a) the organic and (b) cathode. The S/D electrode in contact with ITO is the drain for p-channel or the source for n-channel FETs.

Fig.3. Schematic cross-section of the inverted AMOLED structure of Fig. 1(c). Photoresist separators are used as shadow masks for evaporation of the organic layers with normal incidence (a) and evaporation of cathode at an angle (b) to form the OLEDs. Note that the drain of the driving TFT is now connected to the OLED top contact (cathode).

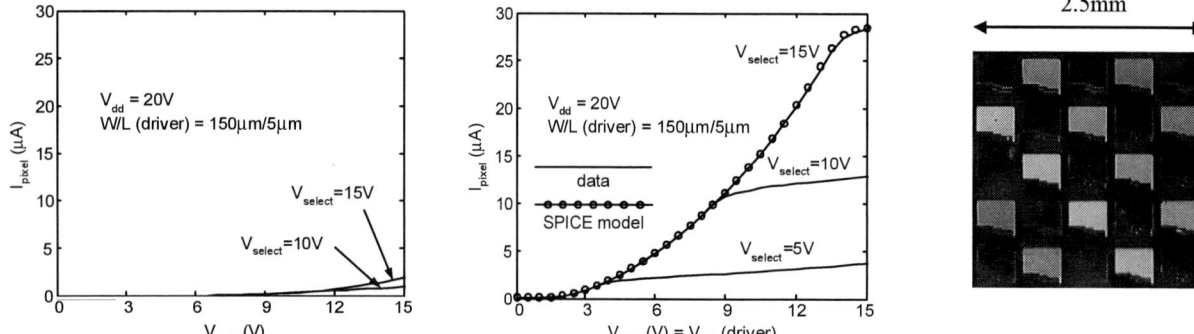

Fig.4. Experimental output characteristics of (a) a conventional a-Si AMOLED pixel of Fig. 1(b) and (b) an inverted a-Si AMOLED pixel of Fig. 1(c). A 10X improvement in the pixel current and a reduction of the turn-on voltage by several volts can be seen in the inverted pixels. This is because in the conventional design V_{data} is split between the OLED and the gate-source of the driving TFT, but in the inverted design it is converted directly to the gate-source voltage of the driving TFT.

Fig.5. image of active matrix in checkerboard test mode with QVGA timing. The pattern of emission is defined by the ITO area.

AlGaN/GaN Bidirectional Power Switch

N. Tipirneni, B. Wang, A. Monti and G. Simin.
University of South Carolina, Dept. of Electrical Engineering, 301 Main St., Columbia, SC, USA. Phone: +1 (803) 777-0751, e-mail: TIPIRNEN@engr.sc.edu

A Semiconductor Bidirectional Power Switch (BPS) which can block voltages of both polarity and conduct current in either direction is highly demanded by many modern power conversion applications like matrix converters. The functionality of bidirectional power switching is commonly achieved by connecting a pair of two-quadrant current switches back to back or voltage switches anti-parallel, because true high-power semiconductor BPS practically do not exist up to date. Multi-switch combination increases the circuit losses and complexity thus leaving many power conversion applications waiting for a truly bidirectional power switch to come [1]. The difficulty in BPS fabrication is primarily related to the fact that most of the high-power semiconductor devices have a vertical structure, with the control and the low potential electrode (cathode) generally formed on one side and the higher potential electrode (anode) on the other in order to achieve a maximum blocking voltage. This asymmetric configuration cannot be used as a BPS due to a breakdown between closely spaced control and cathode electrodes when the switch is under reverse bias. Thyristor can block voltage of either polarity owing to the back to back p-n junctions; however this feature makes this device current - unidirectional. An ideal candidate for a true BPS functionality is a lateral geometry Field Effect transistor (FET). Although the FET design has been known for many years, no high-power BPS functionality has been demonstrated because typically these devices, if made of Silicon or GaAs materials, have low breakdown voltages and/or high ON-resistances.

In the recent years, lateral geometry AlGaN/GaN Heterostructure Field Effect Transistors (HFETs) have been recognized as promising novel building blocks for power conversion and switching applications. AlGaN/GaN HFET devices with breakdown voltages (V_{BR}) above 1.5kV with an on-resistance (R_{ON}) as low as 2 mΩ.cm^2 have been demonstrated [2, 3]. Regular AlGaN/GaN HFETs demonstrated to date can be used as current bidirectional two-quadrant switches. This is because the current in a HFET can flow from source to drain or from drain to source, but the HFET can only block the current flow in one direction, i.e. from drain to source. This paper presents a novel approach to achieving BPS functionality using AlGaN/GaN HFETs and for the first time presents experimental data showing the power bidirectional capability of the devices.

We present the first detailed study of the bidirectional power switching capability of single and dual gate AlGaN/GaN BPS. One approach to achieve a symmetrical voltage blocking capability in AlGaN/GaN HFET is to place the gate electrode in the middle of the source - drain spacing. Since the gate-drain spacing must be large enough to accommodate high operating voltage, such a design would suggest a significant increase of the source-gate access resistance as compared to a regular switching HFET device, resulting in much higher ON-resistance. In order to reduce the ON-resistance of the AlGaN/GaN BPS, we have developed a new double-gated HFET with two gates, G^S and G^D, such that the first gate G^S placed close to the source, controls the switch operation when the current is flowing from drain to source and the second gate G^D placed close to the drain, controls the device when the current is flowing from source to drain. The control signals applied to the gates G^S and G^D should be applied with reference to the source and drain correspondingly depending on the input/output voltage polarity. Possible driver configurations to achieving this functionality are proposed. The HFET based BPS were fabricated over sapphire substrate. The 2D electron gas sheet resistance was around 350 $\Omega/\square$, the threshold voltage, V_T =-6V. The ohmic contacts were formed by Ti(200Å)/Al(1000Å)/Ti(500Å)/Au(1500Å) as metal combination. These were annealed at 850 °C for 1 min. in a forming gas ambient. The length of all the gates formed was 2 μm and the device width was 100 μm. The devices were passivated with SiN and field-plated to have a 2 μm extension toward the drain electrode. The devices were finally encapsulated in a 0.3 μm PECVD SiO$_2$ to achieve a symmetrical blocking voltage as high as 900V and low ON-resistance of 2.5 mΩ.cm^2. DC and pulse I-Vs of these devices demonstrate true bidirectional switching performance.

[1] P W. Wheeler, J Rodriguez, J. C. Clare, L Empringham, and A Weinstein, "Matrix Converters: A Technology Review", IEEE Transactions on Industrial Electronics, Vol. 49, no. 2, April 2002

[2] N. Tipirneni, A. Koudymov, V. Adivarahan, J. Yang, G. Simin, and M. A. Khan, *The 1.6 kV AlGaN/GaN HFETs*", IEEE Electron Device Lett., Vol. 27, Issue 9, pp. 716-718 Sept. 2006.

[3] Y. Dora, A. Chakraborty, L. McCarthy, S. Keller, S. P. DenBaars, and U. K. Mishra, "High Breakdown Voltage Achieved on AlGaN/GaN HEMTs With Integrated Slant Field Plates," IEEE Electron Device Lett., vol. 27, pp.713-715, Jan. 2006.

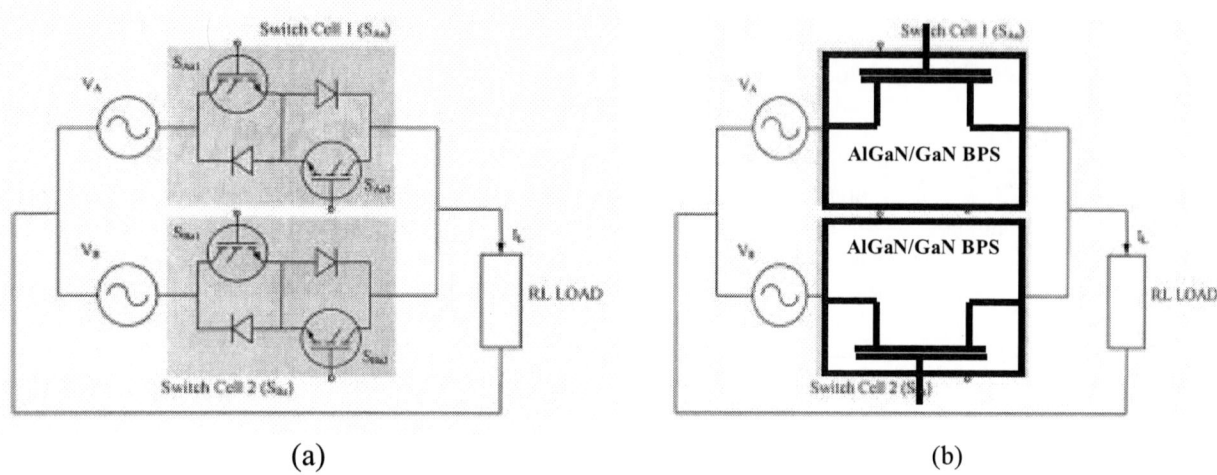

(a)   (b)

Fig 1: Illustrative two-phase into single-phase matrix converters using (a)existing two quadrant switches [1] and (b) proposed AlGaN/GaN HFET BPS.

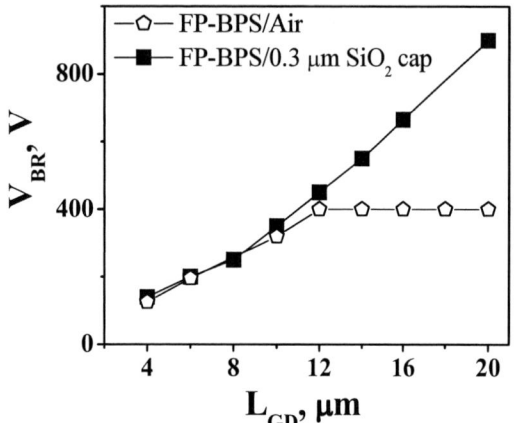

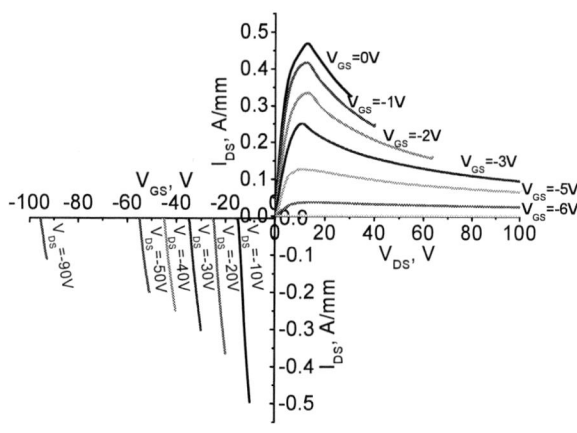

Fig 2: Breakdown voltage - gate-drain spacing dependence of a symmetrical AlGaN/GaN BPS in the air ambience and with SiO_2 encapsulation.

Fig 3: DC IVs of dual gate AlGaN/GaN BPS showing bidirectional power switching capability.

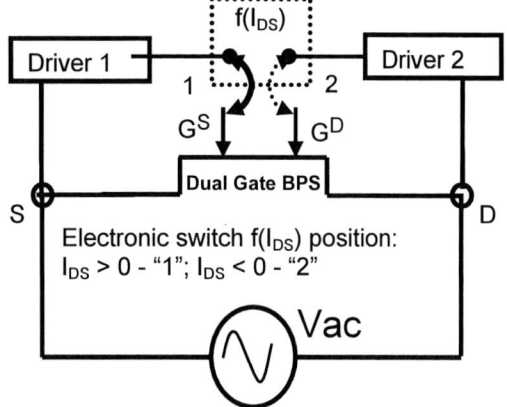

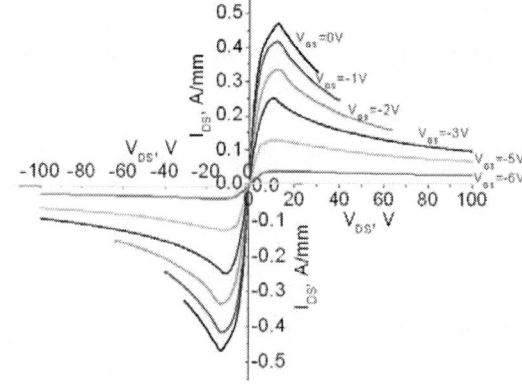

Fig. 4. Proposed illustrative switching driver for a dual gate AlGaN/GaN BPS with DC IVs shown in Fig 5.

Fig. 5. DC IVs of dual gate AlGaN/GaN BPS when driven by the driver of Fig 4.

98

n- and p-channel TaN/HfO$_2$ MOSFETs on GaAs substrate using a germanium interfacial passivation layer

Hyoung-Sub Kim, Injo Ok, Feng Zhu, M. Zhang, S. Park, J. Yum, H. Zhao, and Jack C. Lee
Microelectronics Research Center, Department of Electrical and Computer Engineering, The University of Texas, Austin, TX 78758 (Phone: 512-471-1627, Fax: 512-471-5625, email: hskim1997@mail.utexas.edu)

Using a thin germanium interfacial passivation layer (IPL), for the first time we present surface channel n- and p-MOSFETs on GaAs substrate with TaN gate electrodes and HfO$_2$ dielectric films. We used self-aligned and gate-last processes to fabricate MOSFETs on semi-insulating GaAs substrate. The electrical results from the buried channel and the surface channel-mode transistors are investigated. Both n- and p-channel transistors show excellent surface channel dc output characteristics, providing a good possibility of utilizing GaAs in CMOS technology.

The most challenging issue in GaAs devices is the lack of stable nature gate oxide like SiO$_2$ on Si substrate [1]. Recently, by employing Si or germanium (Ge) IPLs, remarkable results such as small capacitance-voltage (C-V) frequency dispersion, low D$_{it}$, and a thin equivalent oxide thickness (EOT) with low dielectric leakage currents have been achieved [2-3]. Although a Ge IPL have shown the possibility of effectively passivating GaAs surface to prevent it from Fermi level pinning based on the results from capacitor structures, there has been very little work on transistor characteristics, especially on surface channel transistors and p-MOSFETs.

Table I illustrates the MOSFET fabrication procedures for self-aligned and gate-last processes. The schematic vertical device structure of the TaN/HfO$_2$/GaAs MOSFET and ring-type transistor structure are shown in Fig. 1. Semi-insulating GaAs (100) wafers having low resistivity of $\sim 2 \times 10^8$ ohm-cm were used. In the case of gate-last process, during the RTA process for the S/D activation, a piece of GaAs wafer was placed on the sample to avoid As out-diffusion instead of using As over-pressure chamber. Thus, the real temperature exerting on the sample might be different with monitored value. The buried-channel, depletion-mode n-MOSFET was made on the MBE-grown n-type GaAs layer via the self-aligned process [4].

Atomic force microscopy result after a Ge deposition using rf sputtering (Fig. 2) shows the surface roughness (rms) of 0.28 nm from the scan area 3×3 um. In comparison to the reference sample only after surface cleaning (rms ~ 0.3 nm), the surface roughness of a Ge IPL is quite acceptable. Fig. 3 shows excellent C-V and J-V characteristics of the capacitors on n-GaAs substrate using a Ge IPL. Fig. 4(a,b) shows dc output characteristics of n-channel MOSFET (W/L = 320/20 um) fabricated via self-aligned process. The threshold voltage (V$_{th}$) ~ 0.04 V and the sub-threshold swing ~ 105 mV/decades have been obtained. The calculated effective mobility using split C-V method over entire gate voltage range exhibits a maximum electron mobility ~ 66 cm^2/Vs, which is the first report of surface channel n-MOSFET using a Ge IPL (The inset in Fig. 4b is the split C-V curves varying frequencies). Fig. 5(a,b,c) depicts dc output characteristics of the buried-channel n-MOSFET (W/L= 800/10 um, V$_{th}$ $\sim$ -0.33 V). Compared with above surface channel-mode n-MOSFET, the buried channel MOSFETs showed much larger drain current and higher mobility ($\sim$680 cm^2/Vs). Fig. 6(a,b) illustrates dc I$_d$-V$_g$ and I$_d$-V$_d$ characteristics of n-channel MOSFET (W/L = 400/10 um) from gate-last process. The V$_{th}$ and swing are 0.02 V and 171 mV/decades, respectively. In comparison to the results from the device with a longer channel length (W/L = 1000/ 50 um), the V$_{th}$ is almost similar, indicating that we might neglect the out-diffusion effect from S/D during activation and mis-alignment issue. Fig. 7(a,b) shows dc output characteristics of p-channel MOSFET (W/L= 1000/100 um) from gate-last process. The V$_{th}$ and swing are 0.26 V and 170 mV/decades, respectively. It can be concluded that from these MOSFET characteristics, a Ge IPL demonstrated its great potential for the application into the GaAs system with high-k dielectrics, although at this point the surface degradation problem during high temperature RTA still remained in both fabrication processes, which should be improved.

In summary, for the first time, surface channel n- and p-MOSFETs on GaAs substrate using a thin Ge IPL, HfO$_2$ dielectric, and a TaN gate electrode were presented. In addition, the transistor output characteristics from the depletion-mode and the surface channel n-MOSFETs via identical self-aligned process were compared.

This work is partially supported by Intel Corporation.

[1] S. Tiwari et al., IEEE EDL, vol. 9, No. 9, pp. 488, (1988)
[2] I. Ok et al., IEEE IEDM 2006, pp. 829, (2006)
[3] H.-S. Kim et al., Appl. Phys. Lett., 88, pp. 252906, (2006)
[4] H.-S. Kim et al., Appl. Phys. Lett., 89, pp. 222904, (2006)

Table I. The MOSFET fabrication procedures for self-aligned and gate-last processes

(Gate-first = Self-aligned)	(Gate-last)

• Surface preparation (HCl + (NH₄)₂S)
• Deposition of a thin Ge IPL, HfO₂, and TaN using sputtering
• Gate patterning (Photo-litho. and RIE)
• Ion implantation
  (n-channel) Si 25 KeV, 1× 10¹⁴/cm²
• Activation using RTA
  750 °C 10 sec
• S/D contact formation (Photo-litho. and RIE)
• Metal deposition
  (n-channel) AuGe/Ni/Au
• Lift-off and annealing (450 °C)
	• PECVD –SiO₂ deposition
	• Align-marks formation (Photo-litho and BOE + RIE)
	• Photo-litho. and BOE for ion implantation
	  (n-channel) Si 40 KeV, 3× 10¹⁴/cm²
	  (p-channel) Zn 90 KeV, 3× 10¹⁴/cm²
	• Removal of photo resist and PECVD-SiO₂
	• Activation using RTA (~ 950 °C 10 sec)
	• Surface preparation (HCl + (NH₄)₂S)
	• Deposition of a thin Ge IPL, HfO₂, and TaN
	• Gate patterning (Photo-litho. and RIE)
	• S/D contact formation (Photo-litho. and RIE)
	• Metal deposition
	  (n-channel) AuGe/Ni/Au
		(p-channel) Cr/Au
		• Lift-off and annealing (450 °C)

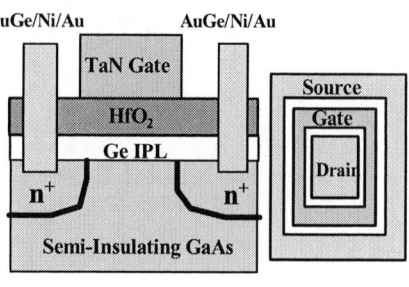

Fig. 1 The schematic vertical device structure and ring-type transistor structure.

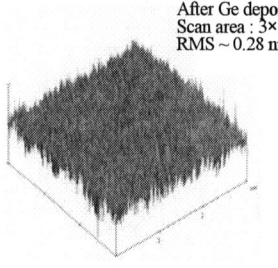

Fig. 2 AFM result after a Ge sputtering shows rms of 0.28 nm. Scan area 3 × 3 um

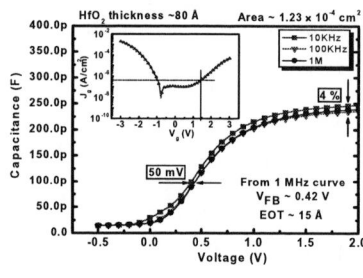

Fig. 3 C-V and J-V characteristics of MOSCAP on n-GaAs with a thin Ge IPL.

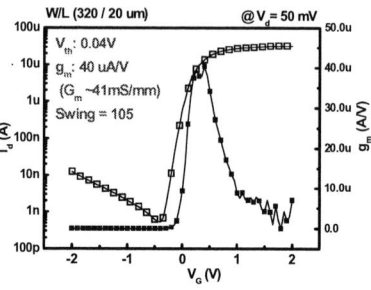

Fig. 4(a) dc I_d-V_g characteristics of n-channel MOSFET from self-aligned process.

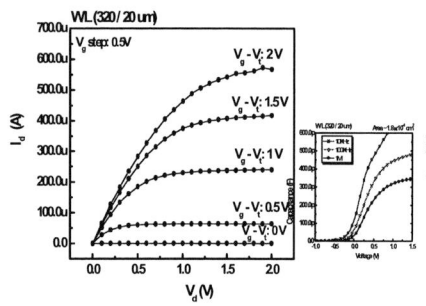

Fig. 4(b) dc I_d-V_d characteristics and split C-V curves of n-channel MOSFET from self-aligned process.

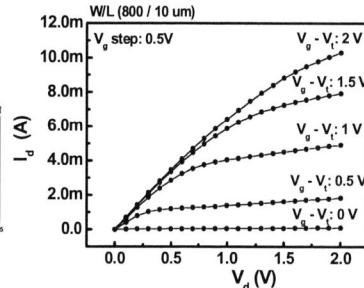

Fig. 5(a) dc I_d-V_d characteristics of depleletion-mode n-MOSFET from self-aligned process.

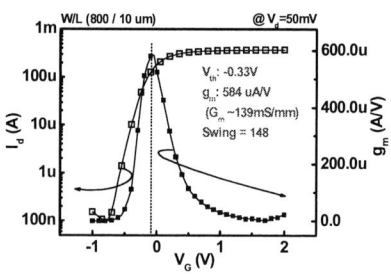

Fig. 5(b) dc I_d-V_g characteristics of depletion-mode n-MOSFET from self-aligned process.

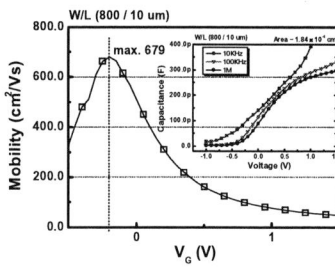

Fig. 5(c) Calculated effective mobility and split C-V curves of depletion-mode n-MOSFET from self-aligned process.

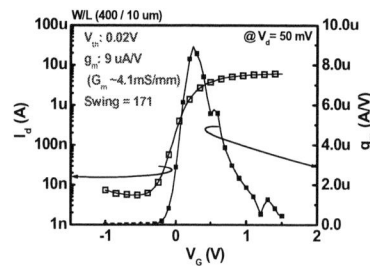

Fig. 6(a) dc I_d-V_g characteristics of n-channel MOSFET from gate-last process.

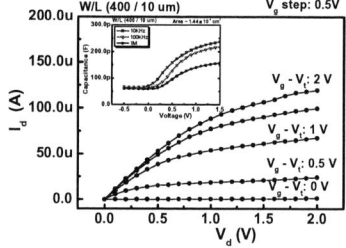

Fig. 6(b) dc I_d-V_d characteristics and split C-V curves of n-channel MOSFET from gate-last process.

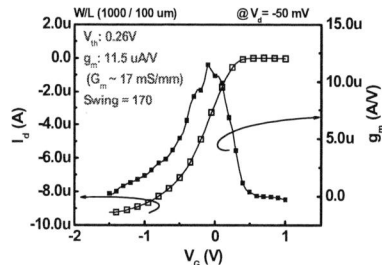

Fig. 7(a) dc I_d-V_g characteristics of p-channel MOSFET from gate-last process.

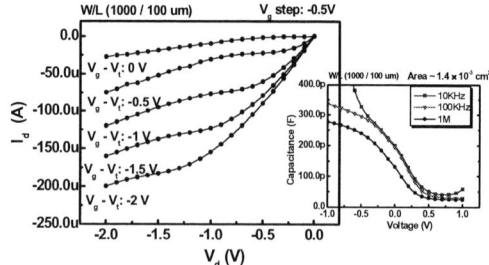

Fig. 7(b) dc I_d-V_d characteristics and split C-V curves of p-channel MOSFET from gate-last process.

Hexagonal Prism Blue Laser Diode with Low Threshold Power using Whispering Gallery Mode (WGM) Resonances.

Sangho Kim and Timothy D. Sands
School of Electrical and Computer Engineering, 465 Northwestern Ave. West Lafayette, IN 47907
Birck Nanotechnology Center, 1205 W. State St. West Lafayette, IN 47907
Phone: (765) 496-8341, Fax: (765) 496-8383, email: kim70@purdue.edu

In contrast to conventional edge-emitting lasers, lasers utilizing whispering gallery modes (WGM) and total internal reflection (TIR) offer the potential for a smaller form factor and an associated reduction in threshold current. In this work, electrically pumped room temperature emitters using hexagonal prism cavities have been fabricated. The hexagonal prism shape provides a route to atomically smooth facets through selective growth or post etching annealing treatments, exploiting the natural crystal form of the wurtzite-structured nitride heterostructures.

In this study, light-emitting diode (LED) heterostructures (sapphire substrate, 3.6µm n-type GaN, 50nm InGaN/GaN MQW active layer, and 250nm p-type GaN) was etched using optical lithography and inductively coupled plasma (ICP) dry etching to define hexagonal prism structures with a height of 0.6µm. p-type (20nm Ni/ 200nm Au) and n-type (20nm Ti/ 20nm Al) ohmic contacts were deposited by e-beam evaporation on and around the hexagonal prism features, respectively. For comparison purposes, planar LEDs were fabricated on the same substrate.

I-V characteristics of the hexagonal prism laser and planar LED both displayed typical diode behavior, but photon emission characteristics were different in many aspects. The emission intensity of the planar LED increased linearly with input power while the hexagonal prism diode displayed two distinct regimes with respect to threshold voltage. Below the threshold voltage (9V), the emission intensity at 480nm increased slowly, while above the threshold, the intensity increased rapidly along with a decrease in the full width half maximum (FWHM). In addition to lasing at 480nm, an interesting, but unexpected, peak was observed at 430nm. The origin of this short wavelength peak is not yet clear, but Tamoli et al. [1] indicated the possibility that the initial WGM resonance is limited in maximum intensity. After the first WGM mode reaches a certain intensity, a secondary WGM mode develops that takes power from first mode. External beam directionality was observed from the six apexes of the hexagon, unlike the one or two beam output of edge emitting laser diodes.

The lasing mode of the hexagonal prism diodes was verified as WGM by using focused ion beam (FIB) to notch one of the hexagon facets. The notch effectively destroyed the geometrical symmetry required for the WGM resonance mode and therefore no threshold was observed for the FIB-notched hexagonal prism. However, the spectra from the FIB-notched hexagonal prism emitters were different than those from the planar LED, with sharp peaks indicative of radial resonances and a blue shift caused by carrier renormalization.

The threshold current density and forward voltage of the hexagonal prism laser diode were 7kA/cm2 and 9V, respectively. The threshold power of this hexagonal prism laser is, to the authors' best knowledge, the lowest reported value for a GaN laser diode fabricated on a sapphire substrate without employing lateral overgrowth (LOG). WGM and TIR are believed to be the main mechanisms for reduced reflection losses and increased overall resonance efficiency.

This work was supported by the NASA Institute for Nanoelectronics and Computing (INAC) under Award No. NCC 2-1363.

[1] A. Tamboli et al. Nature Photonics vol 1, p61 (2007)

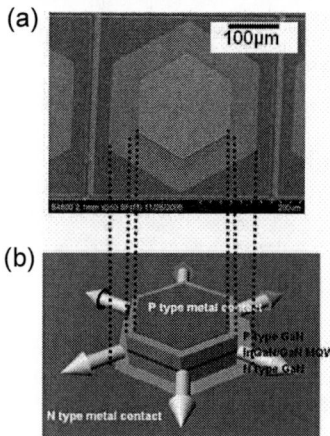

Fig. 1: (a) plan-view SEM image of hexagonal prism laser diode and (b) schematic.

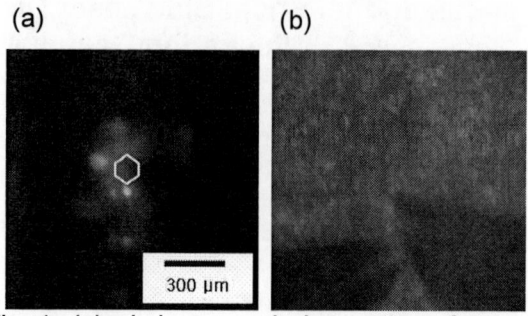

Fig 4: (a) six-beam emission pattern from six apexes of hexagonal prism laser structure (white outline indicates location of hexagonal prism structure). (b) identical image to (a) with microscope-light illumination.

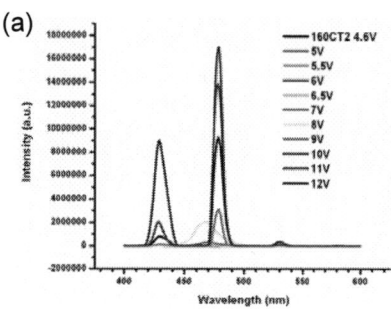

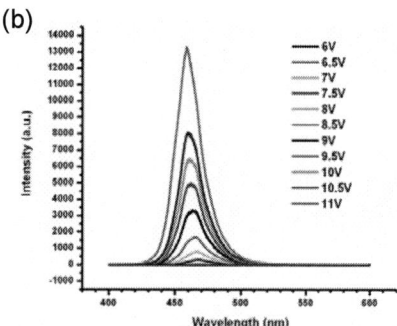

Fig. 2: Electroluminescence (EL) spectra as a function of forward bias for (a) hexagonal prism diode with primary WGM lasing peak at 480nm and secondary WGM resonance at 430nm and (b) planar LED.

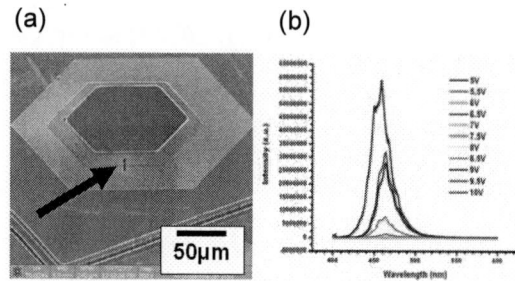

Fig 5. (a) SEM image of FIB-notched hexagonal prism structure. Notch (black arrow) disrupts WGM resonance path. (b) corresponding EL spectra demonstrate lack of lasing due to notch. Sharp peaks at 10V indicate radial resonance mode, which was not present in planar LED spectra (see Figure2b). In addition, a blue shift occurs with increasing input power, which implies excessive carrier injection and renormalization.

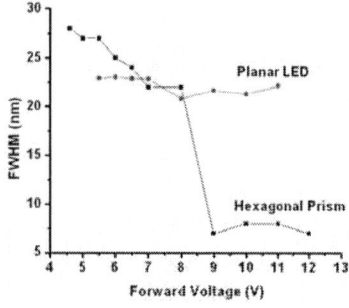

Fig. 3. FWHM as a function of forward bias for hexagonal prism diode and planar LED. Large decrease in FWHM at 9V for hexagonal prism laser diode indicates onset of lasing compared to voltage-independent FWHM of the planar LED.

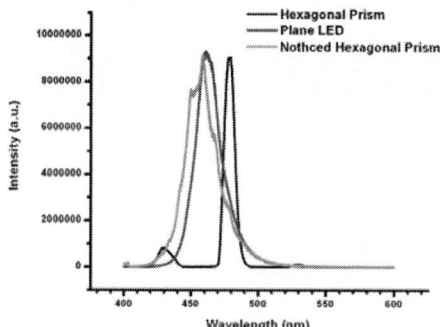

Fig 6. Comparison of EL spectra for hexagonal prism laser diode, FIB-notched hexagonal prism diode, and planar LED under 10V forward bias. Lasing peak of hexagonal prism laser diode and blue shift of FIB-notched hexagonal prism diode are clearly different than the planar LED emission spectrum.

Analytical Modeling of the Suspended-Gate FET and Design Insights for Digital Logic

K. Akarvardar[1], C. Eggimann[2], D. Tsamados[2], Y. Chauhan[2], G. C. Wan[1], A. M. Ionescu[2], and H.S.-P. Wong[1]

[1]Center for Integrated Systems, Stanford University, Stanford, 94305 CA

[2]Swiss Federal Institute of Technology Lausanne (EPFL), CH-1015 Lausanne, Switzerland

Phone: (650) 736 0778, Fax: (650) 725 7731, e-mail: kerem@stanford.edu

The scaling of the MOSFET threshold voltage is limited by the fundamental minimum value of the subthreshold swing, 60 mV/dec at room temperature. This limits the on current as the supply voltage is reduced. Suspended-gate FET (SGFET) [1] features an extremely sharp subthreshold slope and can be used to circumvent this limitation. We present a new, analytical model for the SGFET that is suitable for hand calculations and time-efficient circuit simulations. Our model expresses the pull-in, pull-out voltages and the stable travel range in terms of the structural parameters and the moving gate position as a function of the gate voltage. Starting from our model, we discuss the influence of the structural parameters on the transistor characteristics and the potential of the SGFET for logic circuits. We also introduce the SGFET SRAM cell to demonstrate the use of our model and to illustrate the interest of the SGFET for ultra-low power applications.

SGFET combines an electrostatically-actuated NEMS switch with clamped-clamped beam and an *inversion-mode* MOSFET (Fig. 1). When a gate voltage is applied, the air gap between the gate electrode and the gate oxide is reduced due to the charge-induced electrostatic attraction. The electrostatic force is equilibrated by the elastic force (that can be modeled by a linear spring constant, k [1]) as long as V_G is lower than the pull-in voltage V_{pi}, leading to a critical gap thickness, x_{lim}. For $V_G > V_{pi}$, the electrostatic component dominates the elastic component and the gate snaps down the gate oxide. When the SGFET is designed such that the mechanical pull-in occurs *before* the apparition of the inversion channel, extremely sharp on-off transitions become possible. Our SGFET model is therefore focused on this particular operation where the pull-in occurs in weak inversion. This implies $\Psi_{slim} < 2\Phi_F$, where Ψ_{slim} is the 'limit surface potential' at pull-in and Φ_F is the substrate Fermi potential.

Relationships for Ψ_{slim}, x_{lim}, V_{pi} and $x(V_G)$ are derived by using the force balance equations, depletion approximation and the equation for the limit gap height given in [2] (Table 1). The variation of the gate position as a function of the gate voltage is shown in Fig. 2, where our analytical model based on the depletion approximation is compared to the iterative solution of the force-balance equation featuring the *exact* charge equation. It is noticed that our analytical model is in satisfactory agreement with the numerical solution and the difference between the analytically and numerically calculated V_{pi} is about kT/q. In Fig. 2, it can also be remarked that the SGFET exhibits hysteresis, *i.e.*, the gate is pulled-up at a voltage $V_{po} < V_{pi}$. V_{po} is expressed by using the force equations while the gate is in down-state and neglecting the surface adhesion forces.

SGFET analytical model can be used to provide simple design guidelines. For instance a low-voltage operation requires a relatively long beam (Fig. 3a) and/or a reduced gap (Fig. 3b). On the other hand, the substrate doping should be relatively high, otherwise the negative feedback induced by the depletion capacitance risks to extend the 'stable travel range' to the whole gap height (Fig. 4) [2]. In Fig.4, as the substrate doping is increased, the MOSFET surface emulates a metallic bottom electrode and the x_{lim}/t_{gap0} ratio converges to $(2 - C_{gap0}/C_{ox})/3$, the ratio provided by a simple NEMS switch featuring a dielectric layer [2].

SGFET current-voltage characteristics can be obtained starting from *any* MOSFET model, just by replacing the oxide capacitance (C_{ox}) in the original model equations by a series combination of C_{ox} and $C_{gap} = \varepsilon_{gap}/x(V_G)$. Such a SGFET transfer characteristic, obtained by combining our analytical $x(V_G)$ equation with the MOSFET EKV model, is shown in Fig. 5. This example illustrates perfectly the interest of the SGFET for ultra-low power circuits: by selecting V_{po} just above 0 V (~250 mV in this example) and the regular MOSFET threshold just above V_{po} (~500 mV in this example), SGFET provides *many orders of magnitude reduction in off current without deteriorating the on-current, which means a tremendous improvement in the I_{on}/I_{off} ratio*.

SGFET analytical model enables an insightful design of SGFET-based circuits. As an example, the SGFET SRAM cell, featuring the complementary SGFETs connected as an inverter, is shown in Fig. 6a. This SRAM cell with considerably reduced transistor count and a very simple read scheme takes the advantage of the hysteresis in the voltage transfer characteristic (Fig. 6b). Data is stored at the output node and it is written by applying logic '0' or '1' to the input. In standby or during read operation, $V_{in} = V_{DD}/2$ is applied, allowing the output to retain its state. Remark from Fig. 6b that the width of the hysteresis window is given by $(2V_{pi} - V_{DD})$ and can be enlarged either by reducing W or increasing t_{gap0} (Fig. 3). In each static state of the SGFET SRAM cell, one of the transistors is pulled-in while the other is pulled-out. Since the pulled-out transistor has much reduced subthreshold leakage current than that of a regular MOSFET, the overall static power dissipation is reduced in complementary SGFET logic circuits as compared to their CMOS counterparts.

In summary, SGFET logic gates exhibit a significantly reduced off-state power dissipation and improved functionality as compared to CMOS gates. Our analytical SGFET model stands as an efficient tool to analyze the SGFET and explore its potential for digital circuits. Promising applications of the SGFET include header and footer switch for power management and SRAM configuration switch for FPGA.

[1] A. M. Ionescu et al., ISQED, pp. 496-501, 2002.
[2] J. I. Seeger and S. B. Crary, Proc. Int. Conf. Solid-State Sensors and Actuators, pp. 2:1133-1136, 1997.

978-1-4244-1101-6/07/$25.00 ©2007 IEEE

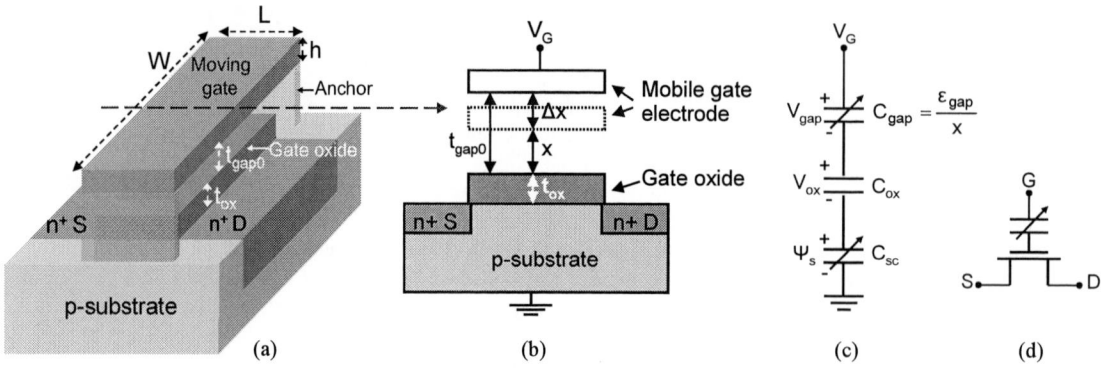

Fig. 1. N-channel SGFET: (a) 3-D structure, (b) cross-section, (c) equivalent circuit, (c) symbol.

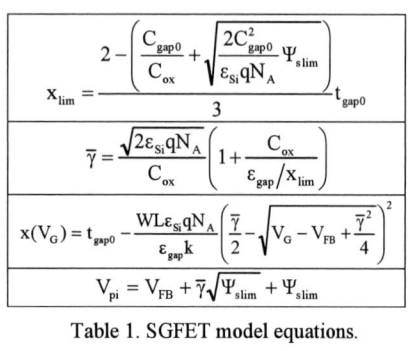

$x_{lim} = \dfrac{2 - \left(\dfrac{C_{gap0}}{C_{ox}} + \sqrt{\dfrac{2C_{gap0}^2}{\varepsilon_{Si}qN_A} \Psi_{slim}} \right)}{3} t_{gap0}$
$\overline{\gamma} = \dfrac{\sqrt{2\varepsilon_{Si}qN_A}}{C_{ox}} \left(1 + \dfrac{C_{ox}}{\varepsilon_{gap}/x_{lim}} \right)$
$x(V_G) = t_{gap0} - \dfrac{WL\varepsilon_{Si}qN_A}{\varepsilon_{gap}k} \left(\dfrac{\overline{\gamma}}{2} - \sqrt{V_G - V_{FB} + \dfrac{\overline{\gamma}^2}{4}} \right)^2$
$V_{pi} = V_{FB} + \overline{\gamma}\sqrt{\Psi_{slim}} + \Psi_{slim}$

Table 1. SGFET model equations.

Fig. 2. Variation of the normalized gap thickness as a function of the gate voltage (numerical vs. analytical model). W = 500 nm, L = 100 nm, h = 10 nm, E = 170 GPa (Si) (implying k = 4.35 N/m), t_{gap0} = 10 nm, t_{ox} = 2 nm (SiO₂), N_A = 5x10^{18} cm^{-3}, V_{FB}= 0.

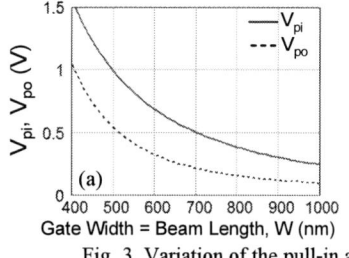

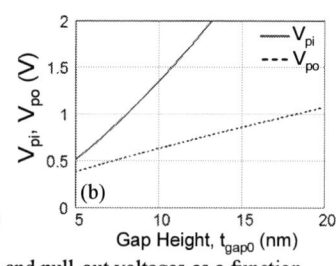

Fig. 3. Variation of the pull-in and pull-out voltages as a function of the (a) gate width (t_{gap0} = 8 nm), and (b) gap height (W = 500 nm). L = 100 nm, h = 8 nm, E = 170 GPa (Si) (implying k = 2.23 N/m), t_{ox} = 2 nm (SiO₂), N_A = 3x10^{18} cm^{-3}, Φ_G= 4.55 eV.

Fig. 4. Variation of the normalized *critical* gap thickness as a function of the substrate doping for various gate widths. L = 100 nm, h = 8 nm, E = 170 GPa (Si), t_{gap0} = 7nm, t_{ox} = 2 nm (SiO₂).

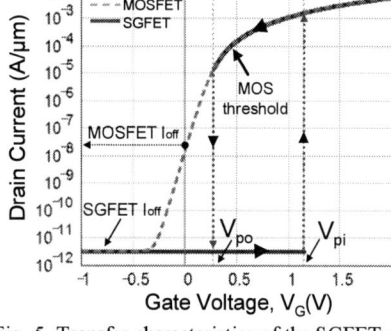

Fig. 5. Transfer characteristics of the SGFET and the MOSFET. W = 325 nm, L = 100 nm, h = 7 nm, E = 170 GPa (Si), t_{gap0} = 10 nm, t_{ox} = 2 nm (SiO₂), V_{FB} = -1.5 V, N_A = 3x10^{18} cm^{-3}, V_D = 2 V, μ = 250 cm²/Vs. Junction leakage current is assumed bias independent and equal to 1 pA (>> I_D when the gate is up).

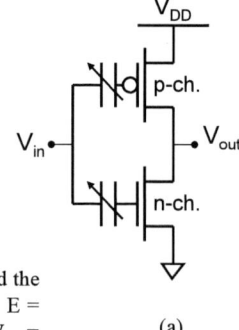

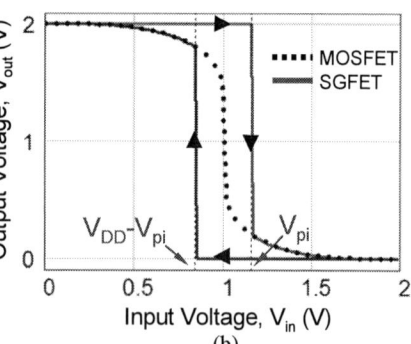

Fig. 6. (a) SGFET SRAM cell, (b) Voltage transfer characteristic. Same parameters as in Fig. 5 and fully symmetrical case (same μ, |V_t|, |V_{pi}|, |V_{po}| for both transistors).

104

Dynamic Two-Port Parameters of Ballistic Carbon Nanotube FETs:
A Quantum Simulation Study

Yijian Ouyang* and Jing Guo

Electrical and Computer Engineering, University of Florida, Gainesville, FL 32611, Tel: 352-392-0940, *yijianoy@ufl.edu

Introduction: The AC characteristics of single-walled metallic carbon nanotubes (CNT) and CNT field-effect transistor (FET) have been a topic of strong experimental and theoretical interest[1]. In most experiments, the intrinsic AC characteristics of CNTFETs, however, are blurred by large parasitic elements.[2] On the other hand, most AC simulations focused on calculating the intrinsic cut-off frequency using quasi-static (QS) approximation or simplified none quasi-static (NQS) treatment.[3] Theoretical understanding of two-port AC parameters is rare. In this work, we simulated the intrinsic dynamic conductance matrix of ballistic CNTFETs using the time-dependent non-equilibrium Green's function (NEGF) formalism.[4,5] Plasmon wave propagation along ballistic metallic CNT interconnects is also studied.

Metallic CNTs: An AC voltage source is applied to the source end of the CNT (Fig.1a). The electrostatic capacitance is infinitely large for simplicity. The CNT wire has a distributed capacitance $C=4G_0/<v>$ and inductance $L=1/4G_0<v>$[6] (G_0 is the conductance quantum per spin; $<v>$ is the average carrier velocity).

Impedance match: For ballistic electronic transport resistance only resides at the two contacts ($R_C=1/4G_0$ each) due to the mismatch between the 1 D channel and contacts. Thus, alternatively the device can be described by a loss-less transmission-line model with characteristic impedance $Z_C=1/4G_0$ and matched load at drain (Fig.1b). Therefore the magnitude of drain current doesn't change with frequency (Fig.2a).

Wave propagation: The simulation shows the NQS effect on the AC charge profile. Because the electrons cannot respond instantaneously, the imaginary part appears. The sinusoidal distribution of the real and imaginary parts of AC charge indicates plasmon wave propagation (Fig.2b). The computed velocity is equal to the Fermi velocity of the metallic band.

CNTFETs: An AC source is applied to the drain and gate of a CNTFET with a zero-height Schottky barrier (Fig.3). The energy-resolved AC charge density shows electron tunneling in band gap region and quantum interference (Fig.4), which indicates the necessity of a quantum simulation. The magnitudes of gate and drain currents are equal at cut-off frequency f_T, when AC source is applied to gate. The comparison between the full time-dependent simulation and the QS approximation indicates that the QS approach gives $f_T=5.1$ THz, which is within 18% of the full time-dependent simulation result of 6 THz (Fig.5b).

Current conservation: The summation of AC currents through source, drain and gate electrodes is zero, which indicates the AC current conservation at any frequency (Fig.5a). We assume the AC charge in the channel all images on the gate. Under this assumption, the currents through source and drain electrodes are purely conduction currents and the current through the gate electrode is the displacement current. It is necessary to include both the conduction and the displacement currents to achieve current conservation.[7]

Dynamic conductance matrix: At zero frequency, the Y matrix gives the same results as the QS simulation (Fig.6). For example the $Re(Y_{21})$ is the gate transconductance; $dIm(Y_{11})/d\omega|_{\omega=0}$ is the intrinsic gate capacitance. At high frequency, due to the finite carrier velocity, the AC charge out of the phase of AC voltage contributes to the real part of displacement current (presented by $Re(Y_{11})$ and $Re(Y_{12})$) and the imaginary part of the conduction current (presented by $Im(Y_{21})$ and $Im(Y_{22})$). The slope of $|Y_{11}|$ also shows the amount of the AC charge is decreased when frequency goes up. However this does not necessarily imply decreased currents in the intrinsic performance.

Kinetic inductance: By using the small signal π model to extract circuit parameters, we conclude the effect of kinetic inductance is almost washed away due to large drain resistance in saturation regime.

Summary: To summarize, we present a time-dependent quantum simulation approach on tight-binding basis with self-consistent AC potential. We simulated a CNT wire with infinitely large electrostatic capacitance. We find the electronic wave propagating without reflection due to impedance match. The simulated intrinsic Y matrix of CNTFET shows strong NQS characteristic. It only qualitatively agrees with the experiment because of huge parasitic elements.

References: [1] Appenzeller et al, *Appl. Phys. Lett.*, **84**, 1771(2004) [2] Bethoux et al, *IEEE Trans. Nanotechnol.*, **5**, 335(2006) [3] Yoon et al, *IEEE Trans. Electron Devices*, **53**, 2467(2006) [4] Anantram et al, *Phys. Rev. B*, **51**, 7632(1995) [5] Zhou et al, *Solid State Electronics*, **49**, 1951(2005) [6] Salahuddin et al, *IEEE Trans. Electron Devices*, **52**, 1734(2005) [7] Büttiker, *Phys. Rev. Lett.*, **70**, 4114(1993)

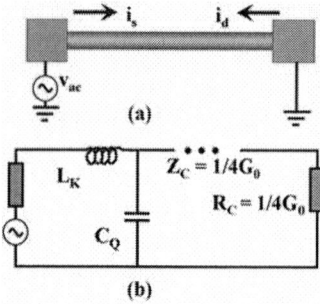

(a)

(b)

Fig.1(a) The schematic of CNT wire (b) lossless transmission-line model. The (15, 0) zigzag CNT has a length of 400 nm. The load (R_C=3.24 kΩ) matches the characteristic impedance. Input impedance seen by the ac source is 6.47 kΩ.

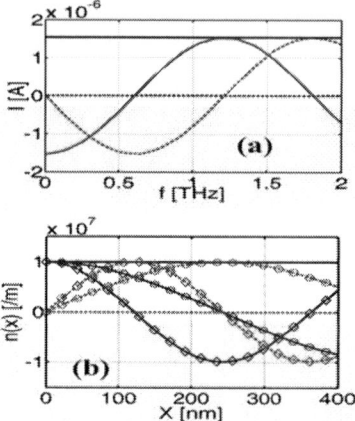

Fig.2(a) AC currents simulated by NEGF. The real parts (solid lines) and imaginary parts (dashed lines) of source (blue) and drain (red) currents are plotted. (c) The real parts (blue) and imaginary parts (red) of ac electron density vs. channel position at 0 Hz (without marks), 1 THz (with circles) and 2 THz (with squares). The ac voltage is 10 mV.

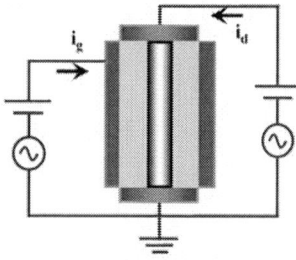

Fig.3 The schematic of modeled CNTFET. The (13,0) carbon nanotube channel is intrinsic with a diameter of ~1 nm, channel length 20 nm and a bandgap of 0.81 eV. The coaxial gate oxide thickness is 5nm and the dielectric constant is 25. The Schottky barrier height is 0. The CNTFET is DC biased at V_{GS} =V_{DS}=0.4 V.

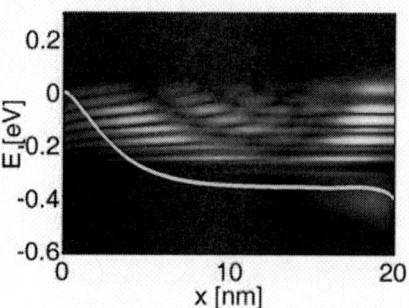

Fig.4 The energy resolved AC charge density at each channel position. The yellow curve is the conduction band edge. The interference pattern is due to the correlation between energy level E and $E+\hbar\omega$, where ω is the applied frequency.

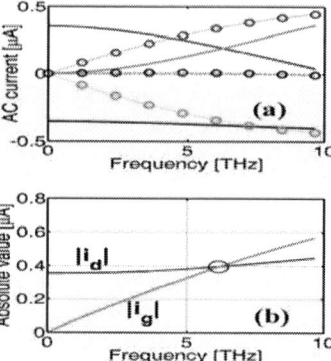

Fig.5(a) Source (blue), drain (red) and gate (green) AC currents vs. frequency with AC source at gate. The solid lines are for real parts and circled lines for the imaginary parts. The magnitude of the summation (Cyan) of three currents is also plotted (b) Extraction of cut-off frequency.

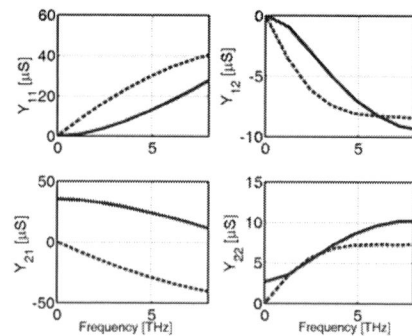

Fig.6 The dynamic conductance matrix elements as a function of the frequency. The solid lines are for the real parts, and the dashed lines for the imaginary parts.

Schottky Drain AlGaN/GaN HEMTs for mm-wave Applications

X. Zhao, J.W. Chung, H. Tang, T. Palacios
Department of EECS *and* Microsystems Technology Laboratories
Massachusetts Institute of Technology, Cambridge, MA 02139
Tel: 617-253-0717, Fax: 617-253-9622, Email: zhaoxu@mit.edu

AlGaN/GaN high electron mobility transistors (HEMTs) are some of the most promising devices for power amplification [1]. Important efforts have been taken to expand the operating frequency of these devices to mm-waves. As shown by Tasker et al. [2], the drain parasitic resistance is one of the main factors limiting the current gain cut-off frequency (f_T) and the power gain cut-off frequency (f_{max}) in high performance devices. The drain parasitic resistance consists of two components: the contact resistance between the metal and the channel and the access resistance due to the distance between the gate edge and drain contact (Fig.1). In this paper, we demonstrate a new drain contact technology based on the use of a Schottky metallization in the drain contact. This new technology has the potential to significantly reduce the contact resistance and at the same time minimize the access resistance.

The use of a Schottky contact in the drain of a HEMT has two major advantages. First, the differential resistance of a Schottky contact is given by $r_c = m k_B T/qI$, where *m* is the ideality factor. For moderate current levels, this contact resistance is much lower than what is achievable with conventional ohmic technology (Fig.2). It should be noted that the I-V characteristic of the drain contact does not need to be symmetric with respect to the voltage, as the device always operates in the first quadrant of the I-V curve. Second, due to the low thermal budget required in Schottky contact technology, these contacts can be easily self-aligned to the gate as illustrated in Fig.3 to eliminate the access resistance. Therefore, Schottky drain contact has the potential to outperform the ohmic contact both in contact resistance and access resistance.

We have fabricated standard AlGaN/GaN HEMTs following the process flow reported in ref. [3]. Using an Agilent 4155 parameter analyzer, we have measured the ohmic contact and the Schottky contact resistances as a function of the drain current (Fig.4). We have also measured the total drain parasitic resistance (ohmic contact resistance + access resistance) in devices with conventional ohmic drain contacts. In conventional AlGaN/GaN HEMTs, the total drain parasitic resistance is above 2 $\Omega\cdot$mm and increases with the drain current. In contrast, by using a Schottky drain contact, the contact resistance is ~0.2 $\Omega\cdot$mm at 500 mA/mm current level, where the ideality factor m=3 is extracted from the measurement shown in Fig.5. Even better performance can be expected upon optimization of the ideality factor. As shown in Fig. 6, our group is modeling the tunneling current in these contacts to determine their intrinsic performance. RF characterization of these devices is on the way.

We have also used the commercial software ADS to simulate the high frequency performance of Schottky drain AlGaN/GaN HEMTs. To evaluate the effects of reduction in drain parasitic resistance on f_T and f_{max}, small signal equivalent circuit simulations have been performed. The values of all the components in the equivalent circuit model are extracted from s-parameter measurement of state-of-the-art devices. As shown in Fig.7, by reducing the drain parasitic resistance from 2 $\Omega\cdot$mm to 0.2 $\Omega\cdot$mm, f_T increases from 116 GHz to 162 GHz and f_{max} increases from 162 GHz to 477 GHz. Linearity is a potential problem in these devices, but our preliminary simulations using ATLAS show that the asymmetric contacts will not significantly degrade the linearity in the output characteristics (Fig.8).

In conclusion, Schottky drain contacts have been evaluated to reduce the parasitic resistance in AlGaN/GaN HEMTs. As shown from our experimental results and simulations, Schottky drain technology can significantly improve the f_T and f_{max} of AlGaN/GaN HEMTs by reducing both the contact and access resistances. This improvement is critical to take GaN power amplifiers beyond 60 GHz.

Acknowledgement - This work was partially funded by the ONR MINE MURI Project, monitored by Dr. P. Maki and Dr. H. Dietrich. The AlGaN/GaN samples were provided by Prof. Mishra's group in UCSB.

References
[1] T. Palacios et al., IEEE Electron Device Letters, vol. 26, p. 781 (2005)
[2] P.J. Tasker et al., IEEE Electron Device Letters, vol. 10, p. 291 (1989)
[3] T. Palacios et al., Device Research Conference Digest, p. 99 (2006)

978-1-4244-1101-6/07/$25.00 ©2007 IEEE

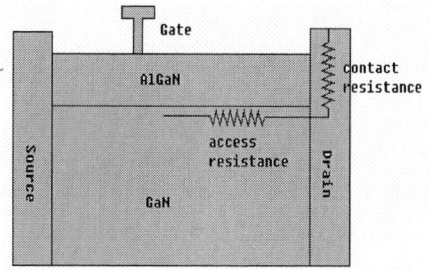

access resistance+contact resistance=drain parasitic resistance

Fig.1 Illustration of the access resistance and contact resistance in AlGaN/GaN HEMTs. The access resistance is introduced by the distance between gate edge and drain contact.

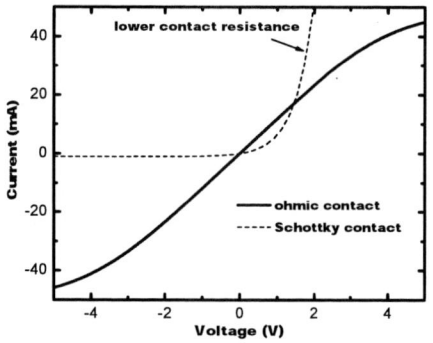

Fig.2 I-V characteristics of ohmic contact and Schottky contact. In high frequency applications, the differential resistance of Schottky contact is lower than that of ohmic contact at moderate and high current levels.

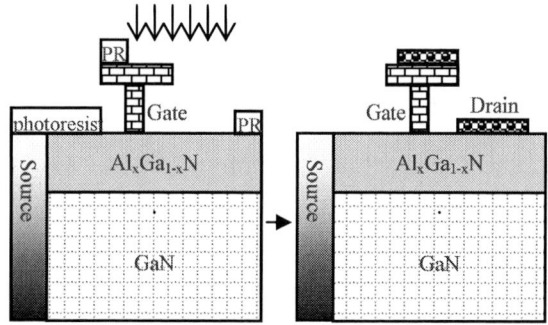

Fig.3 Illustration of the self aligned process for Schottky drain contact.

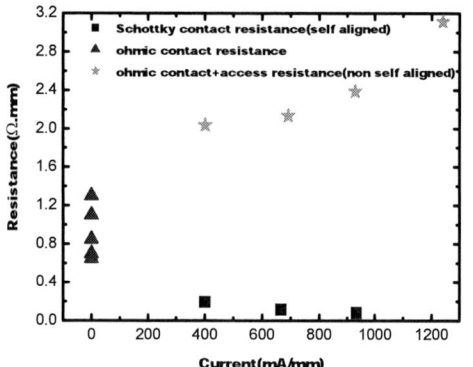

Fig.4 Measurement of the ohmic contact resistance, Schottky contact resistance and the total drain parasitic resistance in ohmic contact devices.

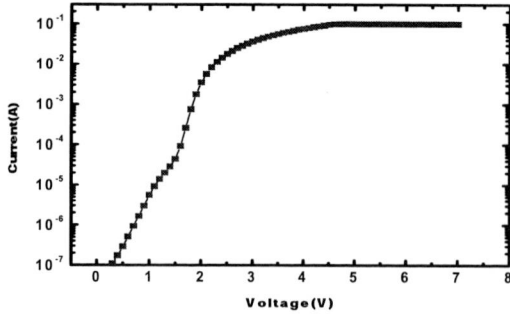

Fig.5 I-V characteristics of Schottky contact. The maximum current is set at 100 mA. An ideality factor of 3 can be extracted in this device.

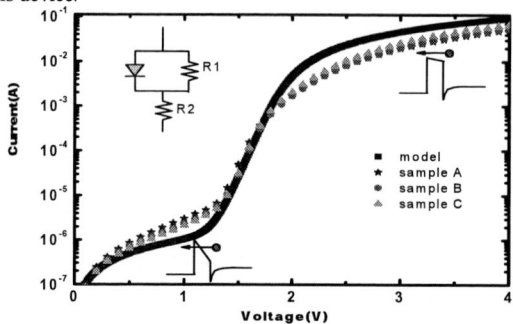

Fig.6 Tunneling currents degrade the ideality factor in Schottky contacts, which increases the differential resistance. We can model the tunneling current with the circuit shown in the inset, where R1=1 MΩ, R2=15 Ω. In this case, the contact is not self-aligned to the gate, which introduces some access resistance (R2).

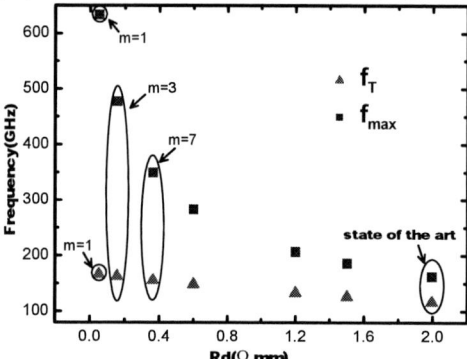

Fig.7 Small signal simulations of the f_T and f_{max} performance as a function of R_d. By reducing the drain parasitic resistance from 2Ω·mm to 0.2 Ω·mm, f_T increases from 116 GHz to 162 GHz and f_{max} increases from 162 GHz to 477 GHz. The use of Schottky drain contacts with low ideality factors (m) can significantly improve the frequency performance of these devices.

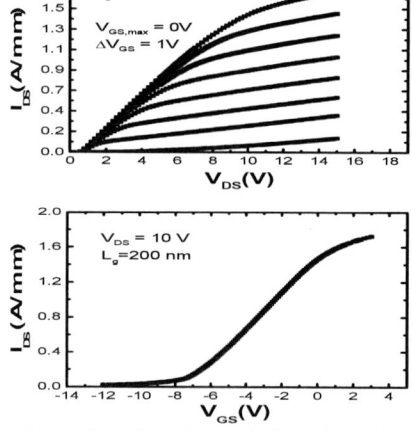

Fig.8 I_D-V_{DS} and I_D-V_{GS} characteristics of a Schottky drain AlGaN/GaN HEMT simulated in ATLAS.

Barrier layer downscaling of InAlN/GaN HEMTs

F. Medjdoub[1], J.-F. Carlin[2], M. Gonschorek[2], E. Feltin[2], M.A. Py[2], M. Knez[3], D. Troadec[4], C. Gaquière[4], A. Chuvilin[5], U. Kaiser[5], N. Grandjean[2]and E. Kohn[1], *Member IEEE*

[1]University of Ulm (EBS), Albert Einstein Allee 45, 89081 Ulm, Germany. Email: farid.medjdoub@uni-ulm.de
[2]EPFL, CH 1015 Lausanne, Switzerland
[3]Max Planck Institute of Microstructure Physics, Weinberg 2, D-06120 Halle, Germany
[4]IEMN, U.M.R.-C.N.R.S. 8520, 59652 Villeneuve d'ascq, France
[5]University of Ulm (ME), Albert Einstein Allee 11, 89081 Ulm, Germany

For high power / high frequency operation it is important to maintain a high structural aspect ratio at nm-gatelength. For V- and W-Band applications gatelengths below 100 nm are desired in conjunction with a high structural aspect ratio and thus barrier layer thicknesses of around or below 15 nm. Important prerequisits for such thin barrier devices are the need to maintain the polarization and interfacial 2DEG characteristics and to prevent interdiffusion of the contact metals with the barrier material. First hints concerning the exceptional chemical stability of this heterostructure could be obtained with lattice-matched $Al_{0.82}In_{0.18}N$/GaN HEMTs with 13 nm barrier, which had been evaluated in respect to their current handling capability and electrical and thermal robustness by the operation up to 1000 °C in vacuum [1]. In another experiment a SiN/InAlN/GaN MISHEMT (9 nm SiN/AlInN barrier and 60 nm gate length) had been reported, also pointing towards a high chemical stability of thin barrier layers [2]. Indeed first TEM studies of such InAlN heterojunctions after annealing at 1000 °C in vacuum show no detectible interdiffusion (see Fig.1), confirming the high temperature stability of Ni/Au Schottky as well as Ti/Al ohmic contacts to this system.

In this study we have investigated heterojunctions on sapphire with barrier thicknesses between 13 nm and 5 nm maintaining a high output current density (Lg = 0.2 µm). The structures used are illustrated in Fig. 2. Hall analysis shows that down to 8 nm the sheet charge density of 2.5×10^{13} cm^{-2} and a mobility above 1000 cm^2/Vs are maintained; at 5 nm surface depletion becomes noticeable, resulting in a measured Ns = 1.7×10^{13} cm^{-2}. Correspondingly, the maximum open channel current of the FETs with 13 nm and 8 nm barriers are approx. 1.85 A/mm (at Vg = +2 V), reducing to 1.35 A/mm for the 5 nm barrier MESHEMT (see Fig. 3). According to the reduction in barrier thickness, both the pinch-off voltages and the transconductances are scaled (see Fig. 4). These results confirm that no hidden parameters obscure the scaling behaviour. The peak transconductance for the 5 nm barrier device is 505 mS/mm, which is the highest reported up to date for InAlN/GaN HEMTs. The Ni/Au Schottky gate diode is highly rectifying and full pinch-off is obtained in all cases.

We have also compared a MESHEMT with 13 nm barrier with a MOSHEMT with 8 nm InAlN barrier and 5 nm Al_2O_3 gate dielectric (deposited by ALD) also resulting in 13 nm total barrier thickness. Since Al_2O_3 is a high-k dielectric with a comparable dielectric constant to InAlN, both structures (the MESHEMT and the MOSHEMT) show similar characteristics. The DC output characteristics of the two cases are shown in Fig. 5. Indeed, comparable open channel current levels, transconductances and pinch-off voltages are observed. In the MESHEMT case the gate diode starts to leak above Vg = +2 V shunting through at Vg = +4 V as can be seen near the origin of the output characteristics. The insertion of the 5 nm Al_2O_3 oxide layer suppresses this leakage up to Vg = +4 V, and the overdrive capability is increased accordingly. In reverse, the insertion of the oxide layer resulted in a decrease of the gate leakage current level by 1 order of magnitude and an increase of the gate to drain breakdown voltage of 25 % of the planar and unpassivated devices.

For gate length scaling, unpassivated devices with a gatelength of 500 nm, 200 nm, 100 nm and 80 nm have been fabricated on the 13 nm barrier material. For all gatelengths the same open channel current densities are reached, confirming that all devices operate at saturated velocity with the same sheet charge density, the $F_T L_g$ product being approx. 11 GHzµm.

The results indicate that InAlN/GaN HEMT device structures can be reliably designed and fabricated with barrier layer thicknesses approaching the tunnelling thickness and approaching enhancement mode characteristics. Using Al_2O_3 as high-k gate dielectric also high aspect ratio MOSHEMTs could be designed with comparable characteristics to MESHEMTs of the same barrier thickness. This is an ongoing study and results on further barrier downscaling experiments will be presented.

Acknowledgements: This work is carried out under a European project (UltraGaN), contract #6903.

---

[1] F. Medjdoub, *et al. IEDM Tech. Dig, p. 927, 2006*
[2] M. Higashiwaki, *et al.* J. J. Appl. Phys, Vol. 45, No. 32, p. L843, 2006

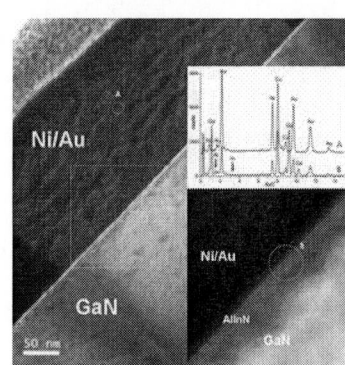

Fig. 1: TEM images and EDX spectra of the Schottky contact upon cooling from 1000°C

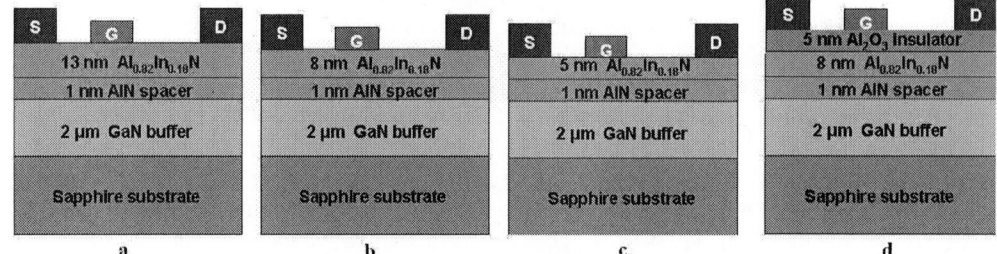

Fig. 2: Schematic cross section of the studied structures based on AlInN/GaN heterostructure with an AlInN layer thickness of 13 nm, b) 8 nm, c) 5 nm, d) 8 nm covered by 5 nm Al_2O_3 dielectric layer

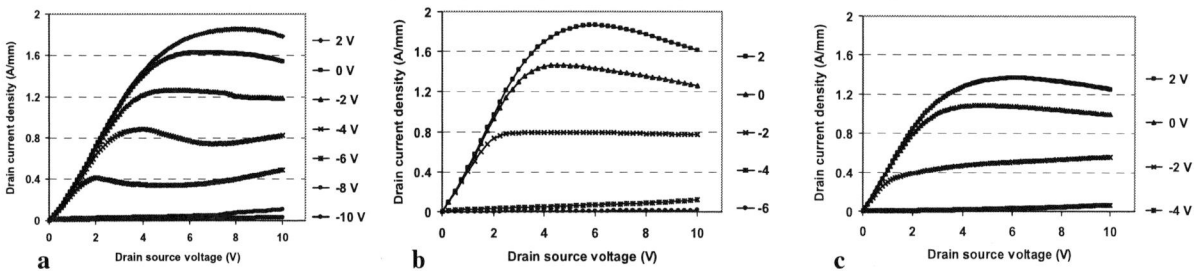

Fig. 3: DC output characteristics of 0.2×50 μm AlInN/GaN HEMTs with an AlInN layer thickness of a) 13 nm, b) 8 nm, c) 5 nm

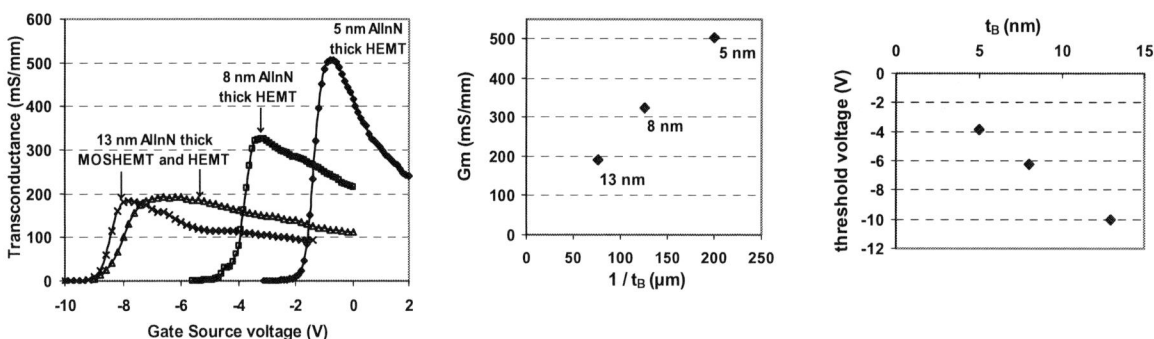

Fig. 4: Transconductance and threshold voltage scaling of 0.2×50 μm AlInN/GaN HEMTs

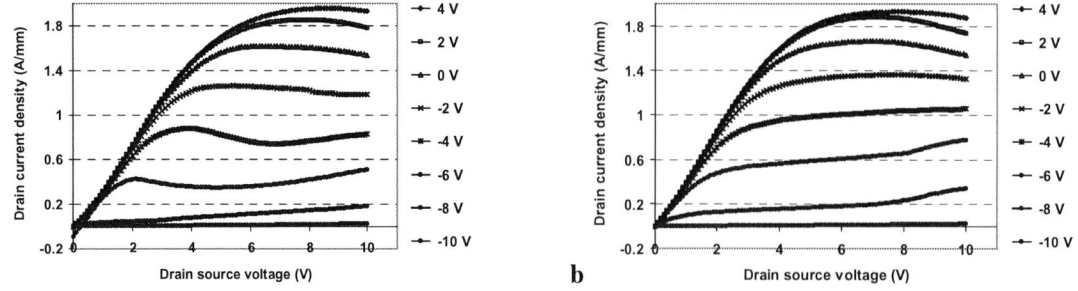

Fig. 5: DC output characteristics at room temperature of 0.2×50 μm AlInN/GaN a) HEMTs b) MOSHEMTs. V_{GS} swept from -10 to 4 V

Estimation of Trap Density in AlGaN/GaN HEMTs from Subthreshold Slope Study

J. W. Chung, X. Zhao, and T. Palacios

Department of Electrical Engineering and Computer Science,
Microsystems Technology Laboratories, Massachusetts Institute of Technology, Cambridge, MA 02139
E-mail: wilchung@mit.edu / Phone: (617)913-8592 / Fax: 617-253-9622

AlGaN/GaN high electron mobility transistors (HEMTs) have shown outstanding performance on high power and high frequency applications. However, AlGaN/GaN HEMTs suffer from much higher subthreshold slope than the theoretically limit of 60 mV/decade, an issue which has been only scarcely addressed. Subthreshold slope (S) is very important not only to assure excellent pinch-off and low dissipated power in digital applications but also for achieving good power added efficiency in analog applications. Most crucially, subthreshold slope can be used to quantify trap density in the gate modulated region of AlGaN/GaN HEMTs. The presence of traps in this region degrades gate modulation efficiency which is closely related to the subthreshold slope and ultimately to high frequency performance. In this paper, we first demonstrate effects of thermal annealing on subthreshold slope in AlGaN/GaN HEMTs. Then, based on the temperature dependence of subthreshold slope, we introduce a new method to estimate interface trap density in these devices.

All the samples used in this work were standard AlGaN/GaN HEMTs with 0.6 μm gate length and 150 μm gate width. Layers consist of 25 nm $Al_{0.33}Ga_{0.67}N$ followed by 7 Å AlN interlayer and 1.8 μm GaN buffer. Samples are grown by metal organic chemical vapor deposition (MOCVD) on SiC substrates and passivated. Individual dies are cut from the sample and annealed from 25°C to 200°C and from 200°C back to 25°C 4 times using a Temptronic TP03010B Thermo Chuck System. Measurements are performed every 25°C. We performed all the current and voltage measurements using an Agilent 4155C Semiconductor Parameter Analyzer.

Figure 1 shows the temperature dependence of S in two different transistors with different initial values of S. Notably, when the initial subthreshold slope is very high (typically in excess of 500 mV/dec, Figure 1(b)), reproducible results are only achieved after the first heating sequence. In our best transistors, the subthreshold slope is as low as 122 mV/decade (Figure 2), the drain induced barrier lowering (DIBL) over the drain bias range of 0.1 to 5 V is 82 mV/V, and on/off ratio is more than 10^4. To the best of our knowledge, these are the best values reported in passivated GaN devices. This device had a gate leakage current of 1.42 μA at V_{GS}=-5 V and V_{DS}=5 V.

The gate leakage has a strong effect on the subthreshold slope. Although this effect is not clear in unannealed devices, we have measured a strong linear dependence after temperature annealing, as shown in Figure 3. By extrapolation from Figure 3(b), we can extract the expected subthreshold slope when the gate leakage current is zero (130 mV/dec) which roughly corresponds to our best measured value. However, this value is still almost twice the value of the ideal subthreshold slope, 60 mV/dec, and we associated this difference to interface traps.

The interface trap density in these devices can be extracted from the change of S with temperature (T). By analogy with MOSFETs (Figure 4), the equation for subthreshold slope in HEMTs is given by,

$$S = \frac{kT}{q}\ln(10)(1+\frac{C_Q+C_{it}}{C_i}) = \frac{kT}{q}\ln(10)(1+\varsigma) \rightarrow \frac{\partial S}{\partial T} = \frac{k}{q}\ln(10)(1+\varsigma) \tag{1}$$

where C_i is AlGaN layer capacitance, C_Q is quantum capacitance, and C_{it} is associated with the interface trap density. Here, ς is a non-ideality factor related to the interface trap density. ς can be calculated from the equation (1) after measuring the slope from the S vs. T curve. As shown in Figure 5, ς varies between 2 and 4. Slight variations of ς might be caused by differences in gate leakage current and its behavior with temperature. From the value of ς, we estimated an interface trap density of $3\times10^{12}\sim 8\times10^{12}$ $cm^{-2}eV^{-1}$, assuming C_Q is negligible. This value is similar to previously reported data obtained from low-frequency noise data[1] or from gate-drain conductance and capacitance dispersion studies[2].

In conclusion, we have studied the change in subthreshold slope in AlGaN/GaN HEMTs with temperature. From this experiment, we proposed for the first time a relatively simple way to estimate interface trap density. The understanding of this interface trap is critical to optimize the gate modulation efficiency of these transistors and maximize their high frequency performance.

Acknowledgements.- This work was partially funded by the ONR MINE MURI project (Drs. H. Dietrich and P. Maki). The samples were provided by Prof. Mishra's group in UCSB.

---

[1] D. Kotchetkov et al., "Carrier-Density Fluctuation Noise and the Interface Trap Density in GaN/AlGaN HFETs", Mat. Res. Soc. Symp. Proc. Vol. 680E (2001).
[2] E. J. Miller et al., "Trap characterization by gate-drain conductance and capacitance dispersion studies of an AlGaN/GaN heterostructure field-effect transistor", J. Appl. Phys., Vol. 87 (2000).

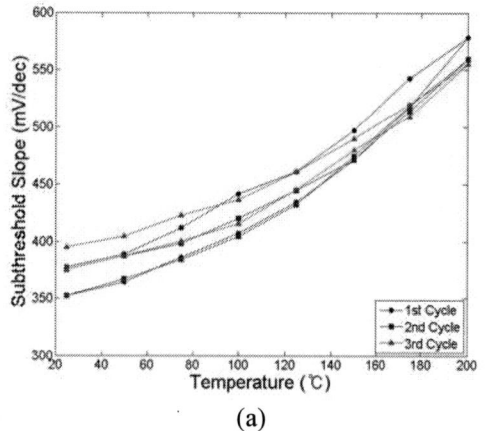 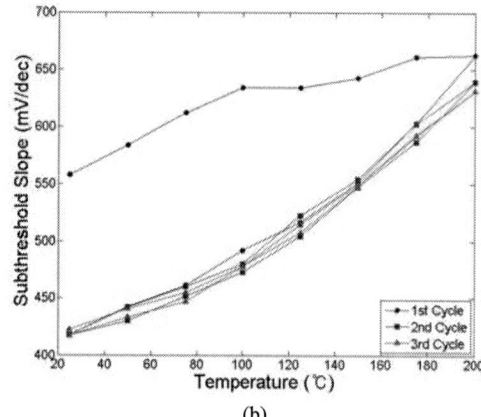

(a)　　　　　　　　　　　　　　　　　(b)

Figure 1. Dependence of subthreshold slope on temperature for two different transistors. (a) Transistors with an initial subthreshold slope below 500 mV/dec typically show a very reproducible linear dependence with temperature. (b) However, transistors with higher initial subthreshold slope reach stable operation only after the first heating sequence.

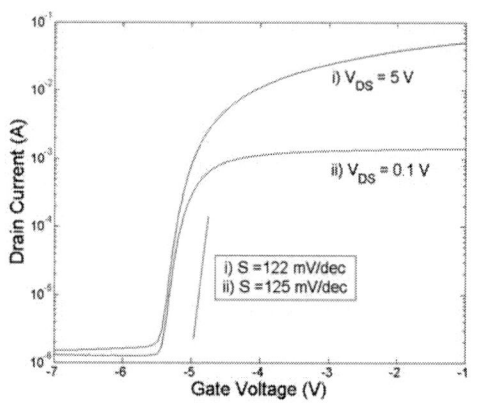

Figure 2. Subthreshold characteristics of one of our best devices. (DIBL=82 mV/V, On/off ratio>10^4, and S=122 mV/dec)

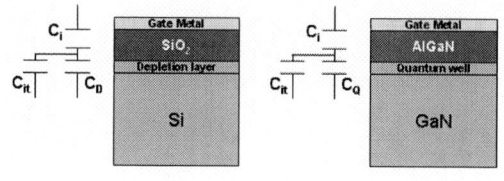

(a)　　　　　　　　(b)

Figure 4. Analogy between Si MOSFET (a) and AlGaN/GaN HEMT (b) structure.

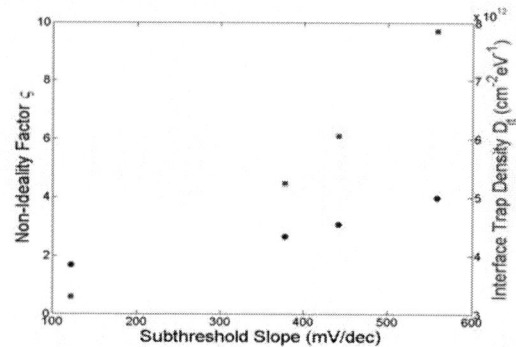

Figure 5. Calculated non-ideality factor and interface trap density.

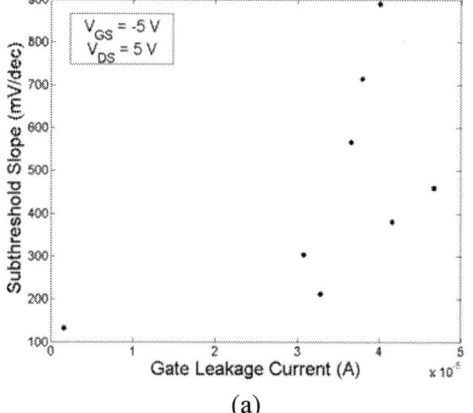 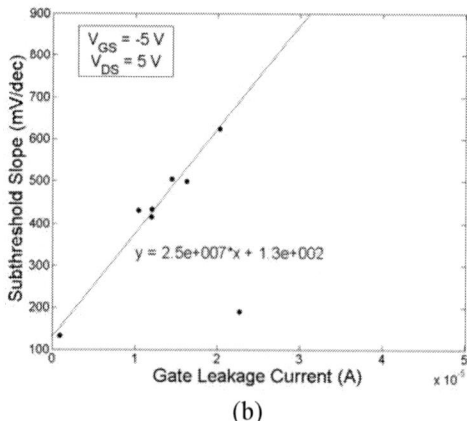

(a)　　　　　　　　　　　　　　　　　(b)

Figure 3. Dependence of subthreshold slope on gate leakage current before (a) and after (b) thermal annealing. After the annealing, the transistors showed a much more stable performance and the subthreshold slope is clearly proportional to the gate leakage current.

High Performance ZnO Nanowire FET with ITO Contacts

Matthew A. Hollister, John D. Le, Guanghua Xiao, Xuekun Lu, and Richard A. Kiehl

Department of Electrical & Computer Engineering

Minneapolis, Minnesota 55455

Abstract:

Nanowire FETs based on a ZnO channel, a SiO_2 gate-dielectric and ITO source-drain contacts are reported. The 55 mS/mm transconductance and other performance parameters are the best reported for any ZnO-based FET. The results demonstrate that high-performance ZnO NW FETs can be fabricated by conventional processes without special gate dielectrics or surface layers.

Introduction:

ZnO exhibits many attractive properties [1]. It has a wide bandgap (3.4 eV), a large exciton binding energy (60 meV) and one of the highest Hall mobilities (200 cm^2/V-s) of the oxide semiconductors. ZnO-related materials also exhibit ferroelectric and ferromagnetic properties, which are of interest for novel applications such as spin-based electronics. These properties, coupled with techniques to synthesize high quality single-crystal nanowires (NWs) in ZnO and related compounds, open the door to a promising family of future devices. The development of a high-performance ZnO NW FET would provide an important member of this family.

In contrast to more conventional FET materials, the conductivity of semiconducting oxides like ZnO is highly sensitive to the concentration of oxygen vacancies, which act as electron donors. In a ZnO NW FET, the concentration of oxygen vacancies is sensitive to chemical adsorption on the NW surface and to reactions between ZnO and the source-drain contacts, which are usually metallic (e.g., Al or Ti/Au) [2]. Although the sensitivity to surface reactions might be useful for chemical sensors, these reactions can affect the channel conductance and contact resistance, thereby hampering amplification and switching applications. In previous work on ZnO NW FETs, special organic gate dielectrics [3] or organic surface layers [4] were needed to optimize the transconductance g_m, presumably because of their affect on channel conductance or contact resistance via reactions at the nanowire surface. In this paper, we report on ZnO NW FETs that exhibit record performance parameters without the need for such surface modification layers.

Fabrication:

Our device is based on a ZnO NW, a SiO_2 gate-dielectric and indium-tin oxide (ITO) contacts (Fig. 1). While ITO is an established contact for large-area TFT devices, it has heretofore not been used for small-area NW contacts. Optical lithography and lift-off were used to define Cr/Au probe pads and alignment marks on the SiO_2 surface of a Si substrate. ZnO nanowires grown by solid-vapor transport were then dispersed by spin deposition from a suspension in IPA. Candidate ZnO FET nanowires were targeted via SEM imaging for e-beam patterning of source-drain regions. Contacts were formed by sputter deposition of ITO and liftoff (Fig. 2). No annealing step was used. The resistivity of the resultant ITO film was ~ 1 mΩ-cm.

Current-Voltage Characteristics:

The ZnO NW FETs exhibited excellent characteristics and record performance parameters. The I_D-V_{DS}, I_D-V_{GS} and g_m-V_{GS} characteristics (Fig. 3) closely follow the ideal long-gate FET model (constant-mobility model) over most of the bias range. R_{SD}^{min}, I_D^{max}, and g_m^{max} are 122 kΩ, 20 μA, and 4.2 μS (55 mS/mm), respectively. The corresponding effective mobility μ_{eff} is 104 cm^2/V-s. which is near the Hall mobility of bulk material. The values of R_{SD}^{min} I_D^{max} g_m^{max} are the best reported for a ZnO NWFET.

Conclusion:

These results demonstrate that high-performance ZnO NW FETs can be fabricated by conventional processes without special dielectrics or surface layers. They show that sputtered ITO provides low-resistance source-drain contacts to small nanowire regions without annealing. The fabrication approach used here is of particular interest for applications where simple, low-temperature processes are required.

---

[1] Y.W. Heo et al., *Mat. Sci. Eng.* **R 47**, 1–47, 2004.

[2] U. Ozgur et al., *J. Appl. Phys.* **98**, 041301, 2005.

[3] S. Ju, et al., *Nano Lett.* **5**, 2281-2286, 2005.

[4] W. I. Park et al., *Appl. Phys. Lett.* **85**, 5052-5054, 2004.

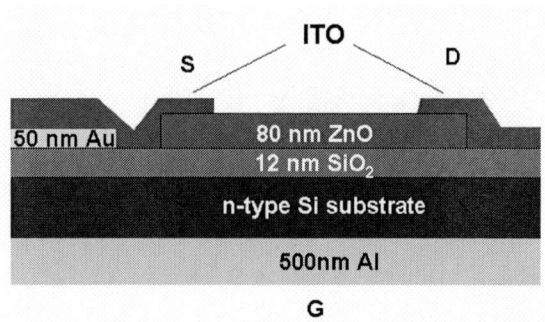

Fig. 1 – Schematic diagram of the ZnO NW FET. The devices were fabricated by spin deposition of ZnO NWs onto the SiO$_2$ surface of a Si substrate. The ITO source-drain contacts were formed by magnetron sputtering of a 40-nm ITO film from an In$_2$O$_3$/SnO$_2$ (10% Sn) sputter target without annealing. The Cr/Au probe pads are also shown.

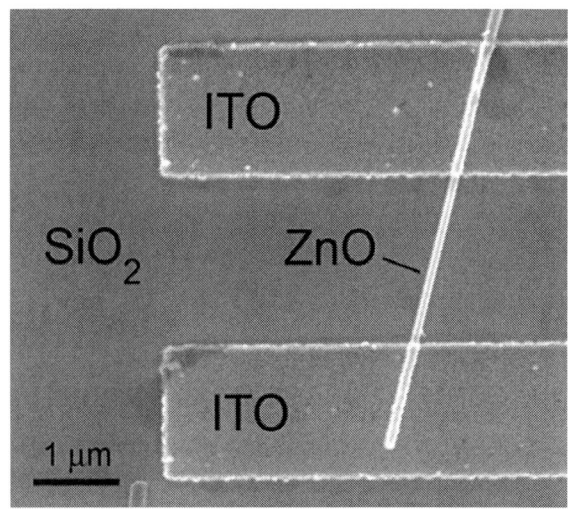

Fig. 2 - Scanning electron micrograph of a device fabricated by SEM-targeting of candidate ZnO NWs and e-beam patterning of ITO source-drain contacts.

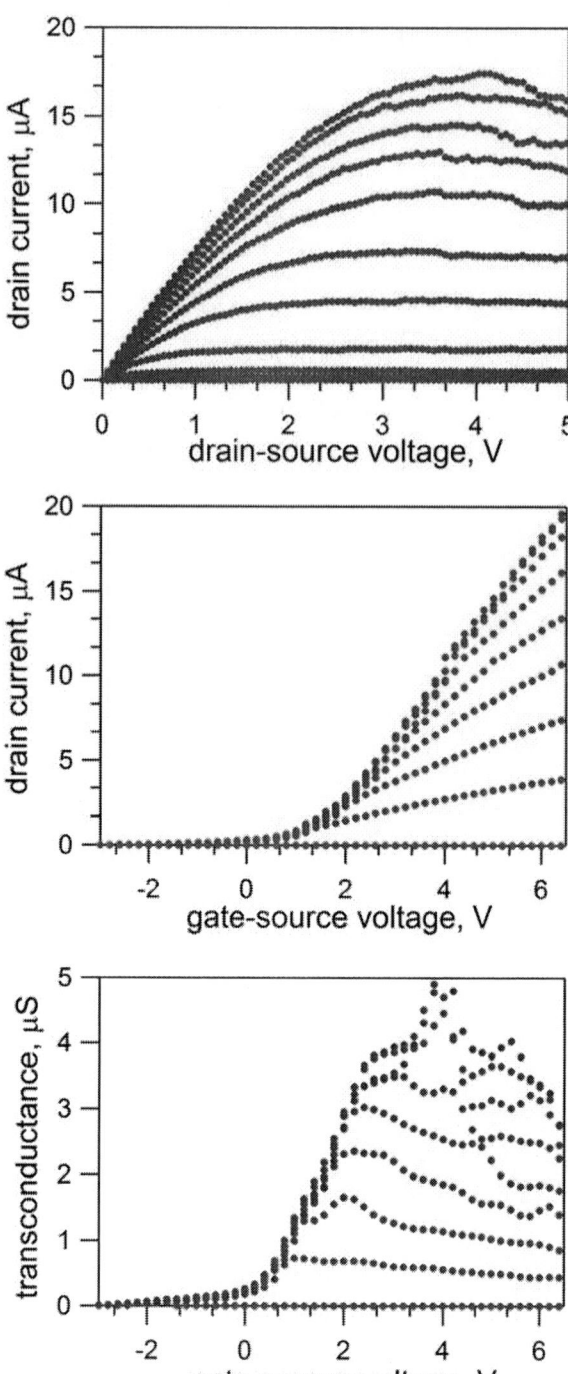

Fig. 3 – Current-voltage characteristics of a ZnO NW FET. For I_D-V_{DS}, V_{GS} varies from -3.0 to 7.0 V in 1.0 V steps. For I_D-V_{GS} and g_m-V_{GS}, V_{DS} varies from 0.0 to 4.0 in 0.5 V steps. The nanowire diameter and source-drain spacing of this device are 76 nm and 1.72 μm, respectively. The characteristics were measured at room temperature in the dark.

High Efficiency Oxide-Confined High-Index-Contrast Broad-Area Lasers with Reduced Threshold Current Density and Improved Near-Field Profile

Di Liang and Douglas C. Hall

Department of Electrical Engineering, University of Notre Dame, Notre Dame, IN 46556

Phone: (574) 631-8631, Fax: (574) 631-4393, Email: dhall@nd.edu

With the advance of crystal growth technique, high power diode lasers have become a highly favorable light source for solid-state laser pumping, direct material processing and medical applications. A gain-guided or weak index-guided broad-area laser structure with an aperture size of w=50 μm-200 μm is normally employed to maximize the output power while minimizing the waveguide scattering loss and non-radiative recombination. The poor lateral electrical and optical confinement, however, make current spreading unavoidable and lead to filamentation, a nonlinear effect, limiting the maximum output power achievable while keeping the facet load below the threshold of catastrophic optical damage.

We have recently developed a simple, self-aligned ridge waveguide laser fabrication process for high-performance GaAs-based narrow-stripe devices [1]. A high-index-contrast (HIC) structure is formed by deep etching (through the waveguide core layer), followed by non-selective O_2-enhanced wet thermal oxidation [2] to grow a uniform thickness layer of high-quality native oxide (200~300 nm) at the sidewalls of the etch-exposed high Al-composition cladding layers and low Al-composition core layer and active region. As shown by the schematic inset in Fig. 1, the oxide-confined waveguide sidewall totally eliminates the lateral current spreading while simultaneously providing a strong lateral index step of Δn~1.7. All devices in this work are fabricated in an 808 nm single quantum well graded-index separate-confinement heterostructure (GRINSCH) with no facet coating or heatsinking employed, and are tested p-side up with 1% duty cycle current pulses at room temperature. A high average slope efficiency of 1.32 W/A (corresponding to a differential quantum efficiency of η_d=86%) is demonstrated (Fig. 1) in a 490 μm long bar length for aperture sizes w ranging from 40 μm to 120 μm. Conventional weak index-guided broad-area devices are also made from the same material to comparatively study the benefit of good lateral electrical and optical confinement in our HIC oxide-confined devices. Fig. 2 compares the laser threshold current densities versus aperture size for similar cavity length structures, and shows a reduction of up to 1.74X at w=40 μm.

Good thermal dissipation is also found in oxide-confined devices primarily because the deeply etched structure and thin native oxide place the high-conductivity deposited p-side metallization in close proximity to the active region sidewall. This is evident from the data of Fig. 3 which shows that the heat dissipation enables a steady output power increase without thermal roll-over up to 5.4X above threshold (I_{th}=74 mA) in a w=40 μm oxide-confined device under a fast-dc sweep, while a shorter weak index-guided device with the same aperture size starts rolling over at 4X threshold and ceases lasing. Fig. 4 shows near-field images for the same two devices under cw operation. The near-field profile in Fig. 4(a) clearly exhibits filamentation due to the spatial-hole burning effect for the conventional device at injection currents of 120 mA and 130 mA. Fig. 4(b), however, shows a much more uniform near-field profile, indicating that a better beam quality can be achieved in HIC broad-area laser structures.

[1] D. Liang, J. Wang, and D. C. Hall, *Electron. Lett.* **42**, 349-350 (2006).

[2] Y. Luo and D. C. Hall, *IEEE Journal of Selected Quant. Electron.* **11**, 1284-1291, (2005).

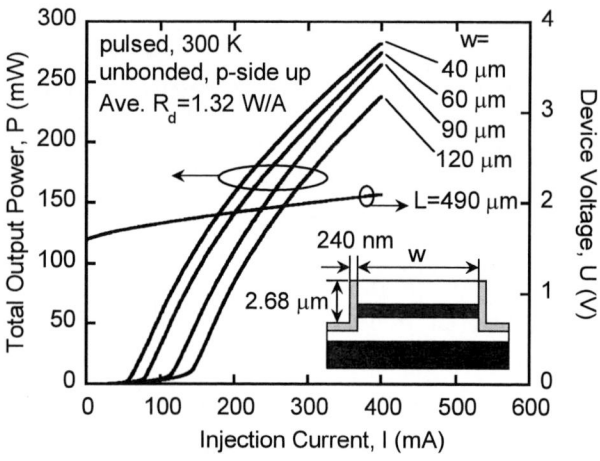

Fig. 1. Pulsed light-current (LI) characteristic and current-voltage relationship of high-index-contrast (HIC) broad-area lasers with aperture size varying from 40 μm to 120 μm and a cavity length of 490 μm, showing an average slope efficiency of 1.32 W/A.

Fig. 2. Threshold current density vs. laser stripe width for HIC and conventional weak index-guided devices with similar cavity length. Up to 1.74X threshold current density reduction is achieved at w=40 μm.

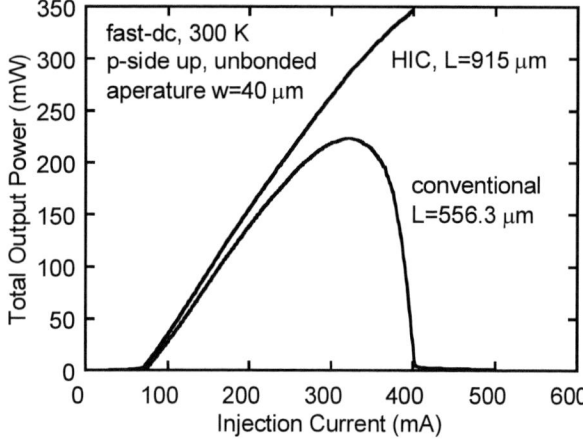

Fig. 3. Fast-dc LI characteristic of a HIC laser and a weak index-guided laser with the same 40 μm aperture size.

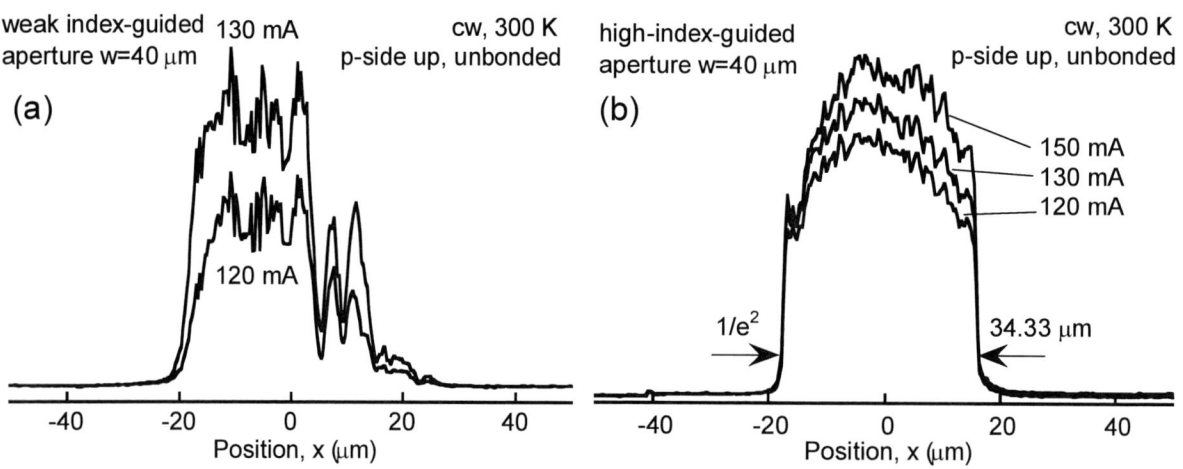

Fig. 4. Continuous wave (cw), 300 K near-field patterns of (a) weak index-guided and (b) HIC devices.

Student Paper

Inversion-type enhancement-mode InP MOSFETs with ALD Al_2O_3, HfO_2 and HfAlO nanolaminates as high-k gate dielectrics

Y.Q. Wu [1], Y. Xuan [1], P.D. Ye [1,*], Z. Cheng [2], A. Lochtefeld [2]

[1] *School of Electrical and Computer Engineering, Purdue University, West Lafayette, IN 47906*
[2] *AmberWave Systems Corp., 13 Garabedian Drive, Salem, NH 03079*
* Tel: 765-494-7611, Fax: 765-494-0676, E-mail: yep@purdue.edu

With recent announcements from Intel and IBM regarding implementation of atomic-layer-deposition (ALD) high-k gate dielectrics and metal gates in high-volume manufacturing for upcoming complementary metal-oxide-semiconductor (CMOS) integrated circuits (ICs), the potential for novel channel materials for future CMOS ICs is growing. By eliminating SiO_2 as the gate dielectric for the channel surface, a key advantage of Si over compound semiconductors is minimized; there is a growing hope that the ALD high-k dielectrics developed for Si may also be applicable to compound semiconductors. Although high-performance depletion-mode (D-mode) GaAs MOSFETs have been demonstrated by various research groups, the reported *inversion-type* enhancement-mode (E-mode) GaAs MOSFETs suffer from relatively low drain currents.[1-3]

In this paper, we report on fabricating inversion-type E-mode n-channel InP MOSFETs using ALD Al_2O_3, HfO_2, and HfAlO nanolaminates as high-k gate dielectrics and demonstrating more than *a factor of 1000* increase in maximum drain current, compared to inversion-type E-mode GaAs MOSFETs.[1-3] InP is widely believed to be a more forgiving material with respect to Fermi level pinning and has a higher electron saturation velocity (2×10^7 cm/s) as well. Detailed Monte-Carlo simulations of deeply scaled n-MOS devices indicate that an InP channel could enable high-field transconductance ~60% higher than either Si, Ge, or GaAs at equivalent channel length [4]. Fig. 1 and Fig. 2 show the schematic cross section of the device structure and the process flow. Table 1 summarizes the device performance of the same gate length devices with different gate dielectrics in terms of maximum drain current, peak transconductance, and drain current drifts. ALD Al_2O_3 shows better interface properties than HfAlO and HfO_2, though its k value is about half of HfO_2 and HfAlO. Our detailed analysis on device characteristics in this abstract is focused on ALD Al_2O_3 only due to the limited space. A well-behaved I-V characteristic of an E-mode InP NMOSFET is demonstrated in Fig. 3 with maximum drain current of 78 mA/mm and Al_2O_3 thickness of 8 nm. Fig. 4 shows the effective gate length (L_{eff}) and series resistance (R_{SD}) extracted by plotting channel resistance R_{Ch} vs. mask gate length L_{Mask} which is important to determine the intrinsic device performance and accurately extract the effective mobility. To evaluate the output characteristics more accurately, the intrinsic transfer characteristics is calculated by substracting R_{SD} and using L_{eff} instead of L_{Mask} and is compared with the extrinsic counterparts as shown in Fig. 5. The threshold voltage is determined to be 0.5 V for Al_2O_3 (8nm) and 1.3V for HfAlO (8nm), respectively. Fig. 6 shows the sub-threshold slope (S.S.) and DIBL characteristics of 280 mV/dec. and 50 mV for 8 nm Al_2O_3 devices. As shown in Fig. 7, the "split-CV" method is used and the extracted mobility has a peak value of 650 cm^2/Vs around a normal electric field of 0.22 MV/cm. Fig. 8 shows a representative *C-V* characteristic of an Al_2O_3 (8nm)/n-InP MOS capacitor with a clear transition from accumulation to depletion for HF *C-V* and the inversion features for LF *C-V* and quasi-static *C-V*, which demonstrates channel inversion operation. It verifies that the conventional Fermi-level pinning phenomenon reported in the literature is overcome in this ALD high-k/InP material system with a mid-gap D_{it} of 2-3 $\times 10^{12}$ /cm^2-eV determined by HF - LF method. More detailed analysis on HfO_2 and HfAlO devices with different surface treatments, compared to Al_2O_3, is ongoing.

We demonstrate here the use of ALD high-k dielectrics for the fabrication of E-mode InP MOSFETs exhibiting well-behaved transistor characteristics. These results suggest new opportunities for evaluating and applying InP as a novel high-mobility channel material for future ultimate CMOS applications.

[1] Y. Xuan et al., *Applied Physics Letters* **88**, 263518 (2006).
[2] F. Ren et al., *Solid-State Electron.* **41**, 1751 (1997).
[3] S. Oktyabrsky et al., *Materials Science and Engineering* B **135**, 272 (2006).
[4] M.V. Fischetti and S.E. Laux, *IEEE Trans. Electron Devices* **38**, 650 (1991).

978-1-4244-1101-6/07/$25.00 ©2007 IEEE

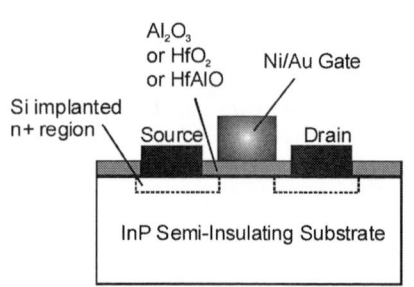

Fig. 1 Schematic view of an E-mode n-channel InP MOSFET with ALD Al_2O_3, HfO_2 or HfAlO as gate dielectrics.

- Surface clean and pretreatment $(NH_4)_2S$
- Deposition of 30nm Al_2O_3 using ALD
- Ion Implantation (Si 35Kev, $1 \times 10^{14}/cm^2$)
- Activation using RTA 720 ºC for 10sec
- For regrown oxide, etch away oxide using BHF and regrow 8nm Al_2O_3 or HfO_2 or HfAlO and PDA
- S/D region patterning and metal deposition AuGe/Ni/Au and RTA
- Gate region patterning and metal deposition Ni/Au

Fig. 2 Fabrication process flow for E-mode high-k/InP MOSFETs. 30 nm thick Al_2O_3 is used as an encapsulation layer for dopant activation process, while thin regrown Al_2O_3, HfO_2 and HfAlO is used as high-k gate dielectrics.

V_g=5V V_d=2V L_g=2µm	Al_2O_3 30nm As-growtn	Al_2O_3 8nm Re-growth	HfAlO 8nm Re-growth	HfO_2 8nm Re-growth
I_{dss} (mA/mm)	32	67	55	<50
g_{max} (mS/mm)	8.5	22	20	~20
I_{dss} drift percent (10^4 sec)	12%	5%	35%	~35%

Table 1 Drain current, transconductance, drain current drift vs. the same gate length InP MOSFETs with the different Al_2O_3, HfO_2 and HfAlO as dielectrics.

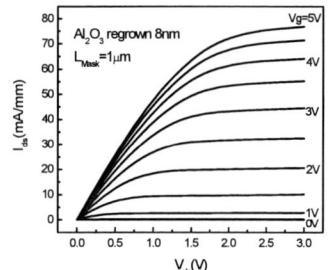

Fig. 3 Drain current vs. drain bias as a function of gate bias for 1 µm InP MOSFET with 8 nm regrown Al_2O_3 as gate dielectric.

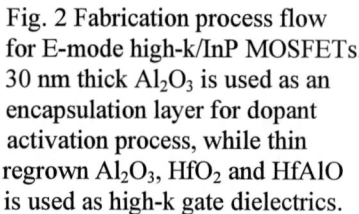
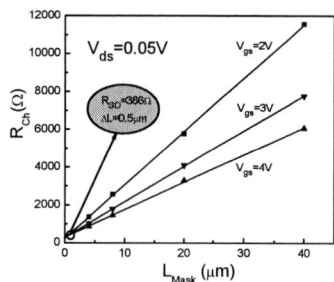

Fig.4 Measured channel resistance vs. different mask gate length as a function of gate bias. Three dashed fitting lines are used to determine R_{SD} and ΔL. Al_2O_3 thickness is 30 nm

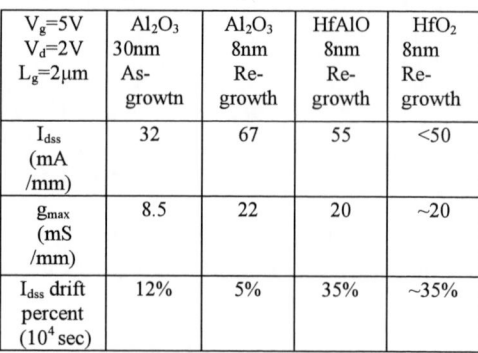

Fig. 5 Extrinsic (empty) and intrinsic (solid) drain current and trans-conductance versus gate bias. The dashed lines are eye-guided to determine the threshold voltage of the devices.

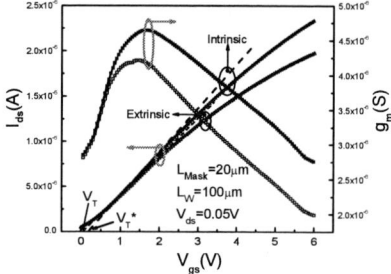

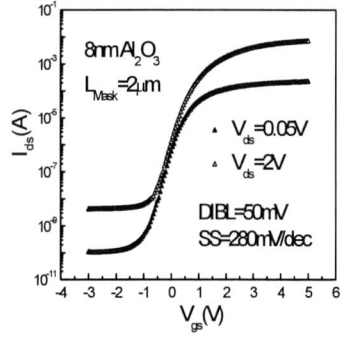

Fig.6 Drain current versus drain bias as a function of gate bias for 2µm InP MOSFET with 8nm Al_2O_3 oxide.

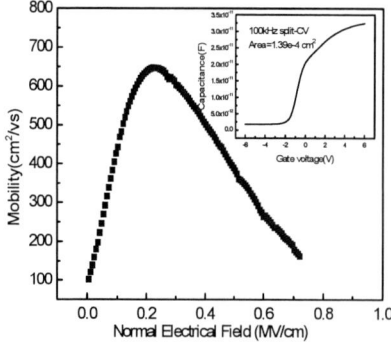

Fig.7 Effective mobility versus normal electric field for the InP MOSFET with 30 nm Al_2O_3 as gate oxide. Inset is 100kHz split-CV measurement of the MOSFET.

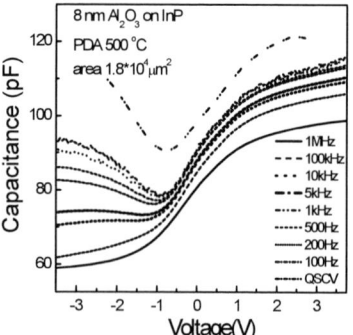

Fig.8 C-V measurements on 8 nm Al_2O_3/n-InP MOSCAP from quasi-static up to 1MHz.

Barrier Lowering and Widening of Schottky Barrier MOSFETs by Self-Aligned Multiple Workfunction Gate

Sheng-Pin Yeh[1] and Chun-Hsing Shih[2]

[1]Institute of Electronics Engineering, National Tsing Hua University, Hsinchu 30013, Taiwan
Phone: +886-3-5715131, Email: d919009@oz.nthu.edu.tw

[2]Department of Electrical Engineering, Yuan Ze University, Taoyuan 32003, Taiwan
Phone: +886-3-4638800, Email: chshih@saturn.yzu.edu.tw

As the MOSFET devices scale into nanoscale regime, extensively minimizing source/drain depth limits the driving current due to the enhanced series resistance. By eliminating the implanted ultra-shallow junctions, the metallic Schottky Barrier MOSFETs (SBMOS) become attracting candidates in deep sub-50 nm regime [1]. However, unique Schottky barrier still constrains the on and off currents for its typically higher contact resistance and ambipolar conduction behavior [2]. Although the Dopant Segregation (DS) provides an available approach to modify the Schottky barrier for practical CMOS technologies [3], still, device characteristics are highly dependent on the gate offset [4] and variations of segregation profiles [5]. Most importantly, as gate insulator is shrinking, the performance of DS SBMOS will be severely degraded. Fig.1 shows the numerical [6] results of drain current with various EOT at several CMOS technologies. Both for on- and off- state operations, the barrier engineering from DS technique gradually loses its effect as gate control became stronger. Here, an ideal 5 nm, uniformly heavy segregation layer of 1×10^{19} cm^{-3} is used.

We propose a Multiple Workfunction Gate (MWG) structure for SBMOS to meet the challenges mentioned above. Fig. 2 gives the schematic view of MWG Schottky Barrier MOSFETs (MWG SBMOS), where High (H) and Low (L) indicate different workfunction values are utilized at center and side gate regions, respectively. The MWG structure can be easily realized by a tilt angle implantation with a fully CMOS process. Fig. 3 illustrates the key steps of a feasible fabrication flow with simulated gate profile. As the in-situ polysilicon deposited and the gate photoresist patterned, a tilt angle implantation is performed to make the heavier side regions. After that, the normal CMOS process is resumed with polysilicon gate etching.

Fig. 4 sketches typical energy band diagrams of MWG SBMOS at on-state, V_{gs}= 0V, and off-state with comparisons of SBMOS and DS SBMOS. Avoid from variations of dopant segregation and gate-insulator scaling issues, the self-aligned Multiple Workfunction Gate can lower the distribution of Schottky barrier for electron to achieve high driving current. For off-state, MWG structure can widen the shape of Schottky barrier for hole, and thus to effectively suppress ambipolar conduction behavior.

Fig. 5 plots the dependences of on- and off-state drain currents on various EOT, (a) 2.5 nm, (b) 1.5 nm, and (c) 1.0 nm, with several CMOS technologies for different types of SBMOS, where $L = 90$ nm, $V_{ds} = 1.0$ V. And, Schottky Barrier Height = 0.4 eV, an ideal 5 nm, uniformly segregation layer of 1×10^{19} cm^{-3} and a feasible workfunction difference $\Delta \phi_{\text{H-L}} = 0.55$ eV are adopted for comparisons. For the 2.5 nm devices, comparable behaviors can be observed both for DS and MWG SBMOS. But for the thinner gate-insulator (EOT = 1.5 and 1.0 nm) devices, the MWG SBMOS exhibits superior characteristics over DS counterparts, such as on-current, subthreshold swing and off-current. Contrary to DS SBMOS, the thinner gate-insulator is, the more performance is enhanced.

Instead of adjusting the Schottky barrier from junction engineering, the self-aligned MWG can effective modify both the tunneling barriers of hole and electron to achieve high driving and low off-state currents by a simple tilt angle implantation. As gate insulator continues to scale along roadmap, the self-aligned MWG structure is a more promising alternative for next generation SBMOS.

This work is financially supported by National Science Council (NSC 95-2221-E-155-082).

[1] ITRS, International Technology Roadmap for Semiconductors (2003).
[2] M. Zhang et al., IEEE Electron Device Lett., vol. 28, p. 223 (2007).
[3] A. Kinoshita et al., Symp.VLSI Tech. Dig., p. 168 (2004).
[4] D. Connelly et al., IEDM Tech. Dig., p. 972 (2005).
[5] Q. T. Zhao et al., International Workshop on Junction Technology, p.147 (2006).
[6] MEDICI and TSUPREM-4, User's Manual, Synopsys (2005).

978-1-4244-1101-6/07/$25.00 ©2007 IEEE

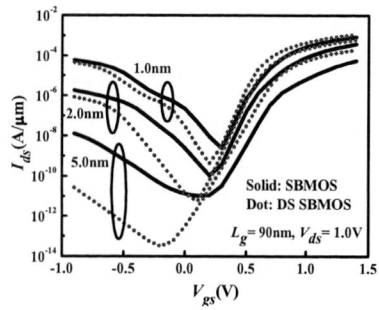 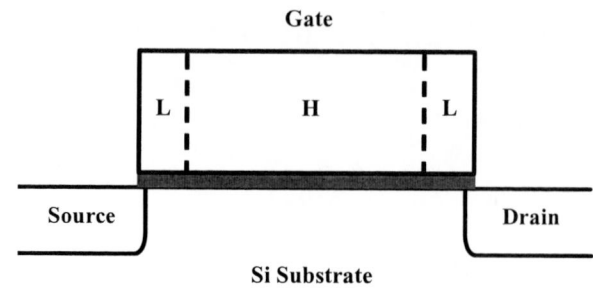

Fig. 1: Drain current for SBMOS with various EOT. Fig. 2: Schematic view of MWG SBMOS.

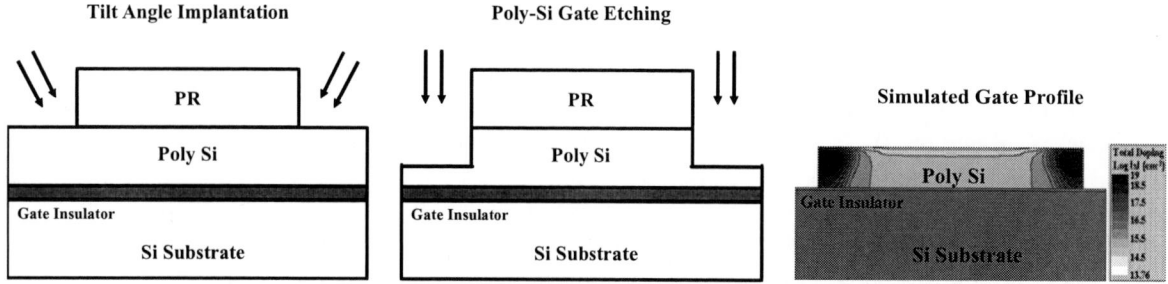

Fig. 3: A feasible process flow for MWG Schottky Barrier MOSFETs with simulated gate profile.

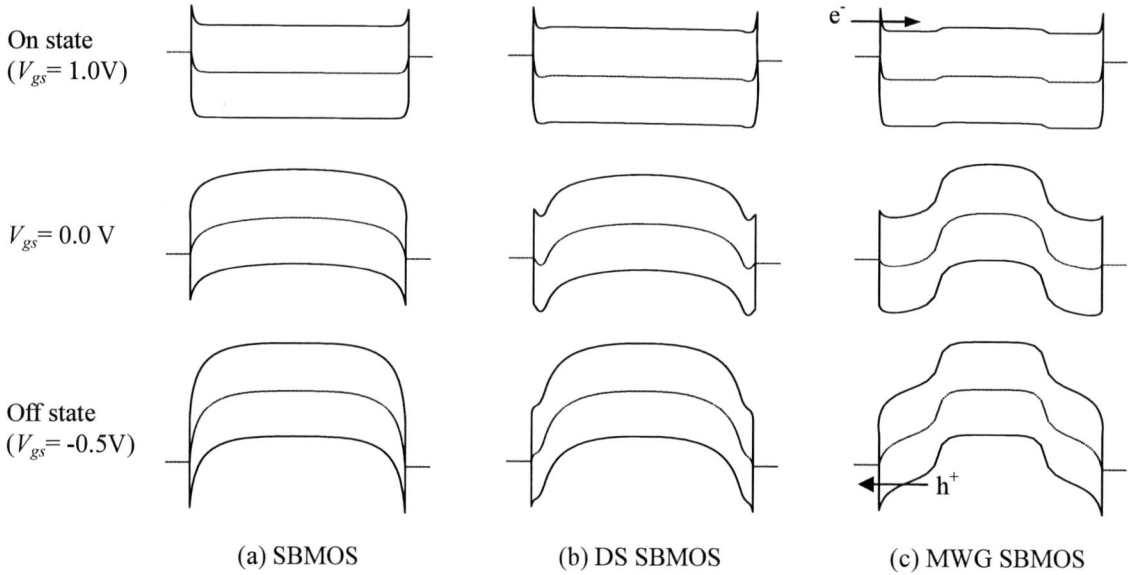

Fig. 4: Sketches of typically simulated energy band diagrams for different types of SBMOS (L= 90 nm, EOT= 1.5 nm, and V_{ds}= 0.05V).

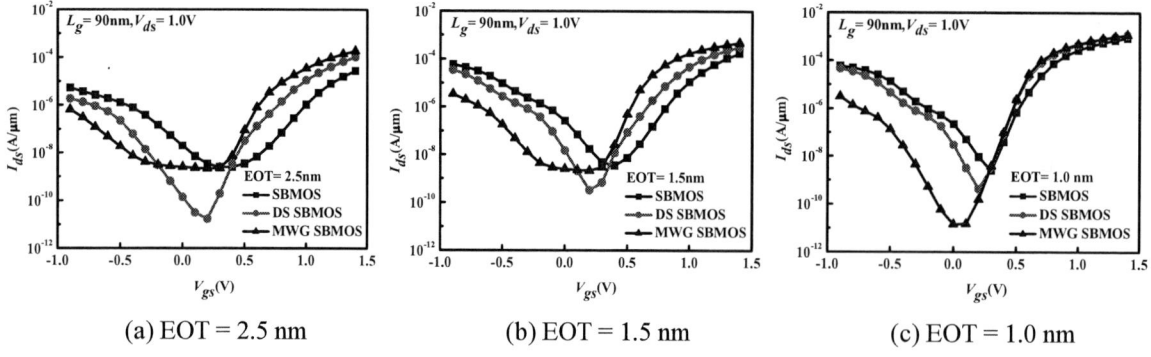

Fig. 5: Results of drain currents for different types of Schottky Barrier MOSFETs.

120

Reliability of 4H-SiC DMOSFETs Evaluated by Bias Stressing

T. Okayama[1], S. D. Arthur[2], J. L. Garrett[2], and M.V. Rao[1]

[1] Department of Electrical and Computer Engineering, George Mason University, Fairfax, VA 22030
[2] Semiconductor Technology Division, GE Global Research Center, Niskayuna, NY 12309
Tel.: +1-518-387-7111; fax: +1-518-387-5997; E-mail address: arthurs@crd.ge.com

In this work, the threshold voltage (V_{TH}) of n-channel 4H-SiC double-implanted metal-oxide-semiconductor field effect transistors (DMOSFETs) was measured after different gate-bias-stress durations to determine if the bias-stress induces a shift in the V_{TH}. If measurement sweep time to acquire transfer curves at the conclusion of bias stress is long, the electrons may be captured by the bulk SiO_2 traps or the SiC/SiO_2 interface traps or emitted from these traps during the sweep, which may result in altered transfer curves and consequently an erroneous extraction of V_{TH}. Recent measurements by Gurfinkel et al. [1], driven by the interest in faster measurement, have shown a clear dependence between the magnitude of the instability and the speed of the measurement. Hence, in this work we pursued a different approach to determine the bias-stress-induced device instability, which is to study the drain current (i_{DS}) transient recorded during the stress. The V_{TH} shifts and the i_{DS} transients of n-channel 4H-SiC DMOSFETs (half cell structure as depicted in Fig. 1) were measured for different gate-bias-stress durations in the range 100 s – 5500s, and at elevated temperatures. The bias stress cycle used in this work is shown in Fig. 2.

Transfer curves, measured immediately after different positive- and negative-stress durations during sweep-down and sweep-up, respectively, at 30 °C, are shown in Fig. 3. The V_{TH} in this work is defined as the v_{GS}-axis intercept of a linear fit to the 90% and the 60% values of I_{DS} at $V_{GS} = 15$ V. The V_{TH} shift (ΔV_{TH}) is calculated as the difference between the V_{TH} extrapolated from the transfer curves taken after the positive and the negative bias-stress durations. The ΔV_{TH} is plotted against the stress duration in Fig. 4. The ΔV_{TH} appears to be linear with log time and is attributed to charge tunneling into the bulk SiO_2 traps or SiC/oxide interface traps during the positive bias-stress, which later is emitted by the traps during the negative bias-stress [2].

The i_{DS} transient, recorded during the positive bias-stress, appears to be linearly decreasing with log time (Fig. 5). During the bias-stress, the devices are in linear region ($V_{DS} = 0.1$ V) and the linear equation of the MOSFET suggests that the Δi_{DS} is proportional to $-\Delta V_{TH}$. Therefore, a linear decrease in i_{DS} with the bias stress time corresponds to a linear increase in V_{TH}, which agrees with Fig. 4 results. Thus, the i_{DS} transient technique is simpler and more accurate than the ΔV_{TH} technique, to characterize the bias stress induced instability in SiC DMOSFETs. The i_{DS} transient and the ΔV_{TH} measurements were also performed at different elevated temperatures in the range 70 °C – 130 °C (Fig. 6). Unlike as seen at 30 °C, the i_{DS} at 130 °C increased with an increasing positive bias-stress time and the V_{TH} decreased with the stress time. The i_{DS} transient at 90 °C initially shows a decay but at around 20 s starts showing an increase in i_{DS}. This reversal in i_{DS} and ΔV_{TH} trend could be due to the presence of mobile ions in the gate oxide, which are active at high temperatures.

In conclusion, the i_{DS} transient technique is simple and useful to characterize the bias-stress induced instability in DMOSFETs. At low temperatures, the V_{TH} shift and i_{DS} decay can be attributed to the capture of electrons by the SiO_2 and SiC/SiO_2 interface traps during positive gate-bias-stress and the release of captured electrons during the negative gate-bias-stress. A prolonged positive gate-bias-stress results in the trapping of more electrons, causing a larger increase in V_{TH}, but the negative gate-bias-stress of the same duration as the positive gate-bias-stress releases the trapped electrons fully, offsetting the increase in V_{TH}. At high temperatures, the mobile ion hypothesis suggests the positively charged mobile ions are repelled away from the gate towards the SiC/SiO_2 interface when positive bias is being applied to the gate. The number of mobile ions at the interface increases with the bias-stress duration. The positive ion charge at the interface results in an increased electron concentration in the channel and consequently an increase in i_{DS}.

[1] M. Gurfinkel et al., IRPS, 2007
[2] A.J. Lelis et al, MRS Proceedings, 2006.

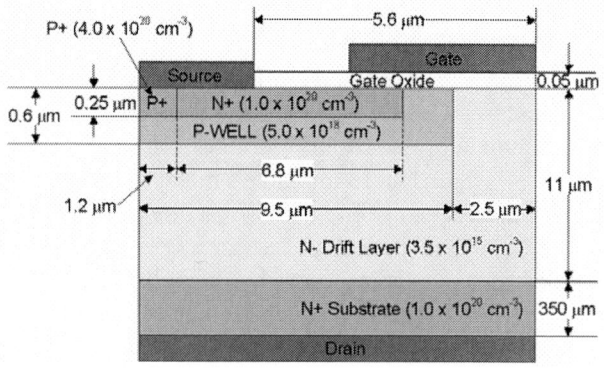

Figure 1: Schematic of the simplified DMOS half cell structure

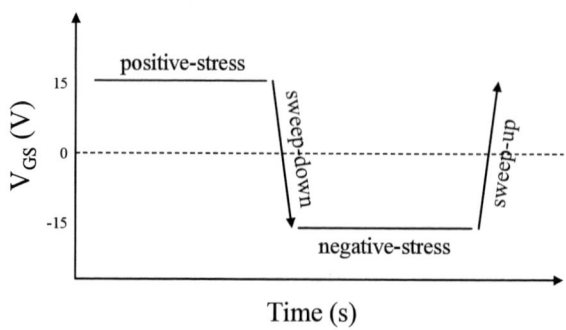

Figure 2: Bias stress cycle

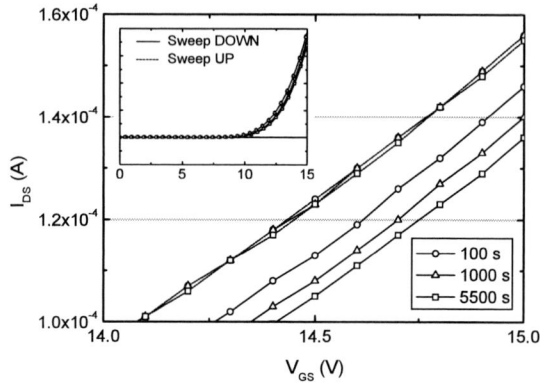

Figure 3: Transfer curves (i_{DS}-v_{GS}) at 30 °C, measured immediately after different durations of positive and negative gate stress

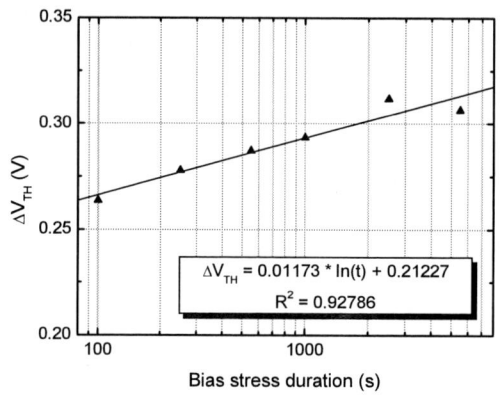

Figure 4: V_{TH} shift versus stress time, at 30 °C

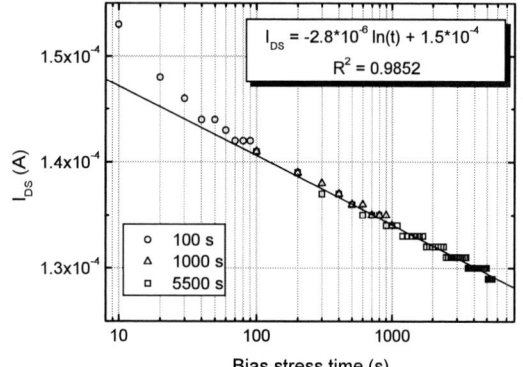

Figure 5: The i_{DS} transients, recorded during three different duration positive bias stress steps

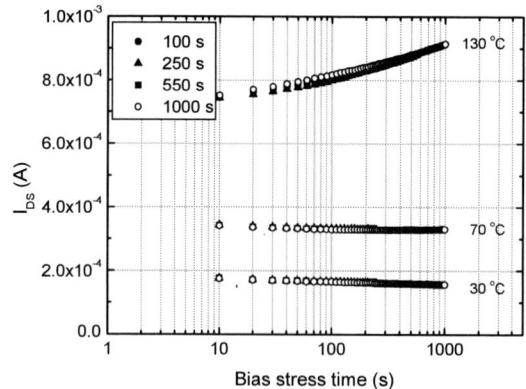

Figure 6: The i_{DS} versus stress time (recorded during positive-stress step), at different temperatures

A 53% High Efficiency GaAs Vertically Integrated Multi-junction Laser Power Converter

D. Krut, R. Sudharsanan, T. Isshiki, R. King and N.H. Karam

Spectrolab Inc., a Boeing Company, 12500, Gladstone Ave, Sylmar, California, USA, 91342

Email: dkrut@spectrolab.com /Phone: (818) 898 -2827/Fax: (818) 838-7474

Power-by-light schemes to provide electrical power to remote opto-electronic modules typically involve the use of multi-segmented photovoltaic (PV) cells or commonly known as laser power converters (LPC) illuminated by an optical fiber to produce output voltages between 2 and 12 volts, with up to several hundred milliwatts of electrical power. One of the most common laser power converters is GaAs based device operating the 810-850 nm wavelength range. Typically these photovoltaic devices are based on a GaAs single junction cell grown on semi-insulating GaAs substrates and they are further segmented and connected in series depending on the output voltage required. Theoretically, these cells should have efficiencies near 60% but currently these cells have efficiencies only in the range of 40 to 45% [1-4]. The reduction in conversion efficiency is attributable to the considerable optical and electrical sheet and interconnection losses of this approach. Optical losses due to high grid metal coverage and cell isolation trench can reduce efficiency by nearly 10%. With relatively high current density for optical devices of over $30 A/cm^2$ laser power converters need high level metal grid coverage.

In order to eliminate the losses in the multi-segmented cells and to improve the efficiencies further we have designed a vertical multi junction device as shown in Figure 1. The vertical multi-junction approach has several advantages over the multi-segmented approach. The advantages are; a) reduction in optical losses due to trench formation, b) lower metal obscuration losses, due to lower current density in the device, c) reduction of sheet conduction losses, and d) simple process to fabricate the devices. For a 2 V device, the optical loss due to trench is totally eliminated and furthermore the reduction in current density by half reduced the obscuration losses due to metal grids by as much as 50%.

The device structures were grown on n-type Ge substrates by metal organic vapor phase epitaxy (MOVPE). Figure 2 shows the grid design and the die dimensions of a 2V GaAs laser power converter. The devices were tested using a 1W 810 nm laser. Figure 3 shows the I-V curve data for a GaAs 2V device. We observed 52.7% efficiency, a considerably higher efficiency than the regular multi-segmented 2V GaAs laser power converter and this is the highest efficiency reported so far for a photovoltaic device.

In conclusion, we have successfully designed and fabricated a new vertical multi-junction 2V laser power converter with an efficiency of over 52%. This efficiency was obtained with high level of direct laser illumination at 810 nm.

[1] L.C.Olsen, D.Huber, G.Durnham, W.Addis "High efficiency monochromatic GaAs solar cells" *Proceedings of 22-nd IEEE PVSC*, (IEEE, New York, 1991) p. 419.

[2] R.Pena, C.Algora "The influence of monolithic series connection on the efficiency of GaAs photovoltaic converters for monochromatic illumination", *IEEE Trans. on Electron Devices* **48**, 196 (2001).

[3] S.Van Riesen, U.Schubert, A.W.Bett "GaAs photovoltaic cells for laser power beaming at high power densities" *Proceedings of 17 EU-PVSEC*, Munich (2001) p. 182.

[4] D.Krut, R.Sudharsanan, W.Nishikawa, T.Isshiki, N.H.Karam "Monolithic multi-cell GaAs laser power converter with very high current density", *Proceedings of 29th IEEE PVSC*, New Orleans (IEEE, 2002) p 908.

978-1-4244-1101-6/07/$25.00 ©2007 IEEE

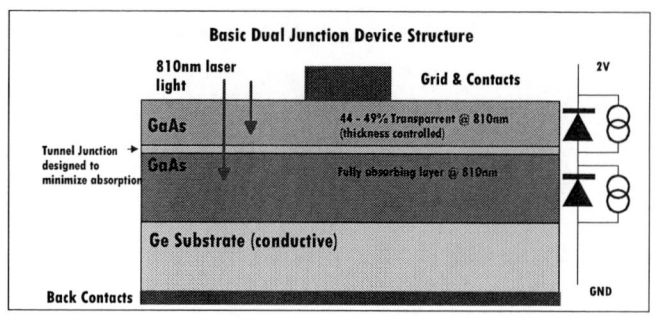

Figure 1: Dual junction LPC device structure

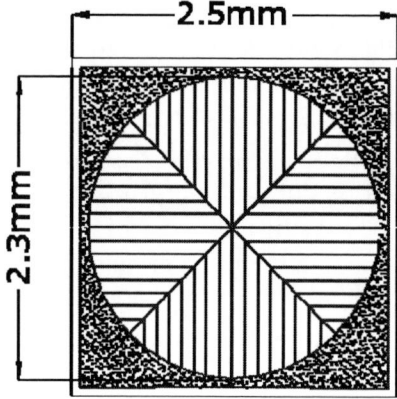

Figure 2: LPC device grid pattern design

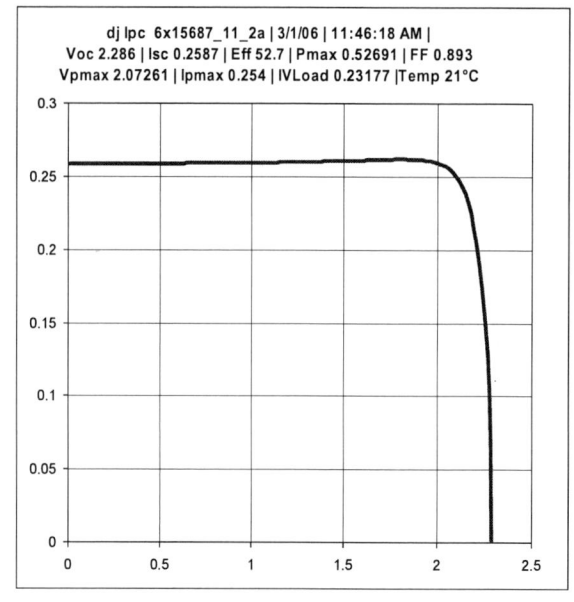

Figure 3: I-V curve data of a 52.7 % efficient 2V GaAs laser power converter

Native-Oxide-Confined High Index Contrast InAs Quantum-Dot Laser Diodes

Jusong Wang,[a] Douglas C. Hall,[a] Vadim Tokranov[b] and Serge Oktyabrsky[b]

a) Department of Electrical Engineering, University of Notre Dame, Notre Dame, IN, 46556
Email: jwang1@nd.edu, phone: (574)631-5498, Fax: (574)631-4393,

b) School of NanoSciences and NanoEngineering, University at Albany-SUNY, Albany, New York 12203
Email: soktyabrsky@uamail.albany.edu

The use of InAs quantum dot (QD) gain media provides superior performance for several laser diode (LD) parameters with the possibility to fabricate long-wavelength lasers on GaAs [1]. QD LDs utilizing a deeply-etched ridge waveguide (RWG) structure [2]-[4], compared with a conventional shallow-etched structure, can provide high electrical confinement in the QD active region to greatly reduce current spreading, leading in turn to decreased threshold current density and increased external quantum efficiency [4]. Deeply-etched RWG structures provide high index contrast (HIC) confinement for the optical mode as well, reducing the bend loss for curved resonators such as the 100 µm radius ring lasers reported in [2]. We at Notre Dame have recently developed a novel non-selective oxidation process for AlGaAs/GaAs heterostructures [5], enabling HIC RWG quantum well (QW) LDs with exceptional performance [6] and extremely small (r≤10 µm) bend radius [7], [8] to be realized.

In this work, we have fabricated HIC RWG LDs with a gain medium consisting of 7-layers of InAs shape-engineered QDs in GaAs QWs separated with 29 nm of $Al_{0.2}Ga_{0.8}As$ similar to Ref. [1]. Three different kinds of RWG structures are fabricated. For the three sample types (A, B, and C), stripe width ranging from 100 µm down to 4 µm are defined, followed by dry etching via reactive ion etching in $BCl_3/Cl_2/Ar$. The etching depth for the conventional shallow-etched sample A is ~ 0.9 µm, stopping above the QD active region. Samples B and C are etched through the QD active region to a depth of ~2.4 µm. Subsequently, samples A and C are wet oxidized at 450 °C for 30 min with 7000 ppm O_2 added to grow a ~ 450 nm native oxide. Sample B has a ~400 nm thick SiO_2 layer deposited on the etched sidewall by PECVD. The SEM cross-sectional images of samples A, B, C at this stage are shown in Fig. 1. Following standard lapping, metallization and cleaving processes, the LDs are tested under pulsed condition (10 kHz, 1% duty cycle) stripe-up at room temperature. The emission wavelength is λ~1.2 µm.

Fig. 2 shows a comparison of the threshold current density (J_{th}) for these three sets of LDs with similar cavity length (L~1 mm). Sample C (deep etch + native oxide) exhibits the lowest J_{th} of ~300 A/cm^2 for all stripe widths ≥8 µm. For sample A (shallow etch + native oxide) and B (deep etch + PECVD SiO_2), the minimum J_{th} values are ~500 A/cm^2 for the broad-area (w≥50 µm) LDs, 1.66X higher than for sample C. For the narrowest stripe widths of ~4 µm, J_{th}=480 A/cm^2 for sample C is 45% lower than that of sample B (853 A/cm^2), and only one sixth of that of sample A (3122 A/cm^2). For further comparison of the two HIC structures B and C, Fig. 3 plots the output power vs. current for broad area (50 µm-wide) and narrow stripe (~7-8 µm-wide) devices, Fig. 4 the external differential quantum efficiency η_d vs. stripe width, and Fig. 5 the internal loss α_i vs. stripe width as derived from analysis of inverse η_d data vs. L (not shown). Fig. 4 shows that the efficiency of sample C is 1.4X higher that that of sample B for LD widths ≥ 20 µm, and 1.7X higher for w~4 µm. Clearly, sample C, fabricated by combination of deep dry etching and non-selective oxidation, exhibits superior laser performance over samples A and B. This can be attributed to more effective optical and electrical confinement and, especially, the reduced sidewall scattering loss (as is evident from Fig. 5) achieved through smoothing of sidewall roughness during non-selective oxidation [9]. Fig. 5 shows a reduction of internal loss by 5.5X (from 63.6 cm^{-1} for sample B to 11.6 cm^{-1} for sample C) for the w~7 µm wide LDs due to the non-selective oxidation process.

In conclusion, with a native-oxide-confined HIC structure enabled by non-selective wet thermal oxidation, we have fabricated InAs QD LDs with improved performance (>40% decrease in J_{th} and >40% increase in η_d) over HIC devices utilizing a deposited dielectric layer, thus demonstrating a simple, promising method for future QD LD fabrication.

[1] V. Tokranov, M. Yakimov, A. Katsnelson, M. Lamberti, S. Oktyabrsky, *Applied Physics Letters*, vol. 83, p. 833, (2003)
[2] Y. Barbarin, S. Anantathanasarn, E.A.J.M. Bente, Y.S. Oei, M.K. Smit *et al*, *IEEE Photonics Technology Letters*, vol. 18, p. 2644, (2006)
[3] D. Ouyang, N.N. Ledentsov, S. Bognar, F. Hopfer, R.L. Sellin, *et al.*, *Superconductor Science and Technology*, vol. 19, p. 43, (2004)
[4] D. Ouyang, N.N. Ledentsov, D. Bimberg, A.R. Kovsh, A.E. Zhukov, *et al.*, *Semiconductor Science and Technology*, vol. 18, p. 53, (2003)
[5] Y. Luo and D. C. Hall, *IEEE Journal on Selected Topics in Quantum Electronics*, vol. 11, p. 1284, (2005)
[6] D. Liang, J. Wang and D. C. Hall, *Electronics Letters*, vol. 42, p. 349, (2006)
[7] J. Wang, D. Liang and D. C. Hall, IEEE 20th Int. Semiconductor Laser Conf., paper P5 (Kohala Coast, Hawaii, 2006).
[8] D. Liang, J. Wang and D. C. Hall, *IEEE Photonics Technology Letters, vol. 19, in press (2007)*.
[9] D. Liang, D. C. Hall and G. M Peake, *LEOS 2005, paper TuY4 (Sydney, Australia, 2005)*.

978-1-4244-1101-6/07/$25.00 ©2007 IEEE

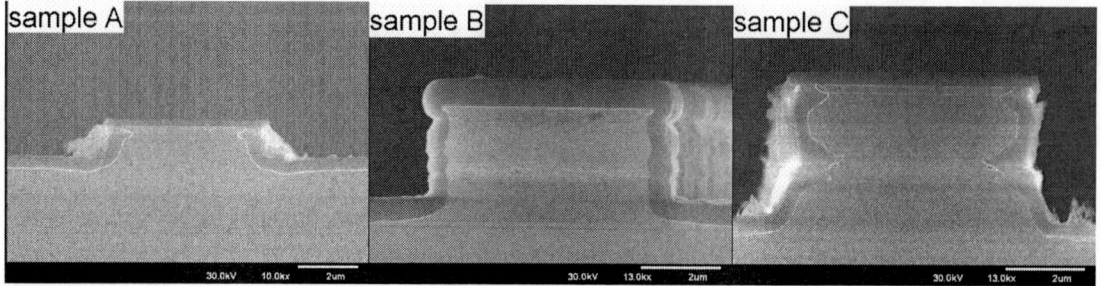

Fig. 1. Cross-section images of QD waveguide structures fabricated via (sample A) shallow etch + non-selective oxidation, (sample B) deep etch + PECVD deposited SiO_2, and (sample C) deep etch + non-selective oxidation. Samples A and C are shown as capped with 300 nm thick SiN_x oxidation mask.

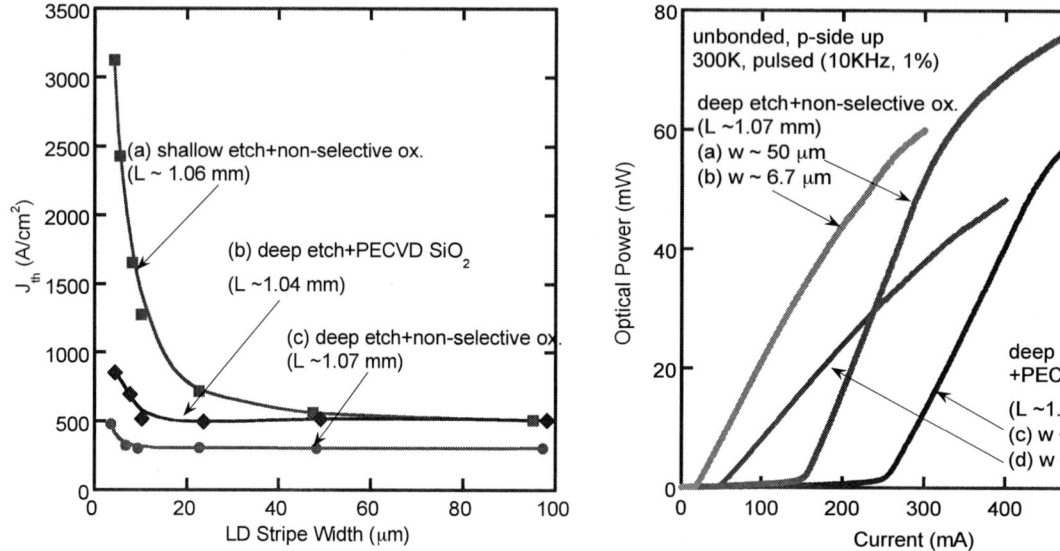

Fig. 2. J_{th} vs. LD Cavity Length for devices with (a) shallow etch+non-selective oxidation (sample A), (b) deep etch + PECVD SiO_2 (sample B), and (c) deep etch + non-selective oxidation (sample C).

Fig. 3. Output powe vs. current for QD LDs with deep etch+non-selective oxidation for (a) w~50 μm and (b) w~6.7 μm, and for deep etch+PECVD SiO_2 for (c) w~50 μm and (d) w~7.6 μm.

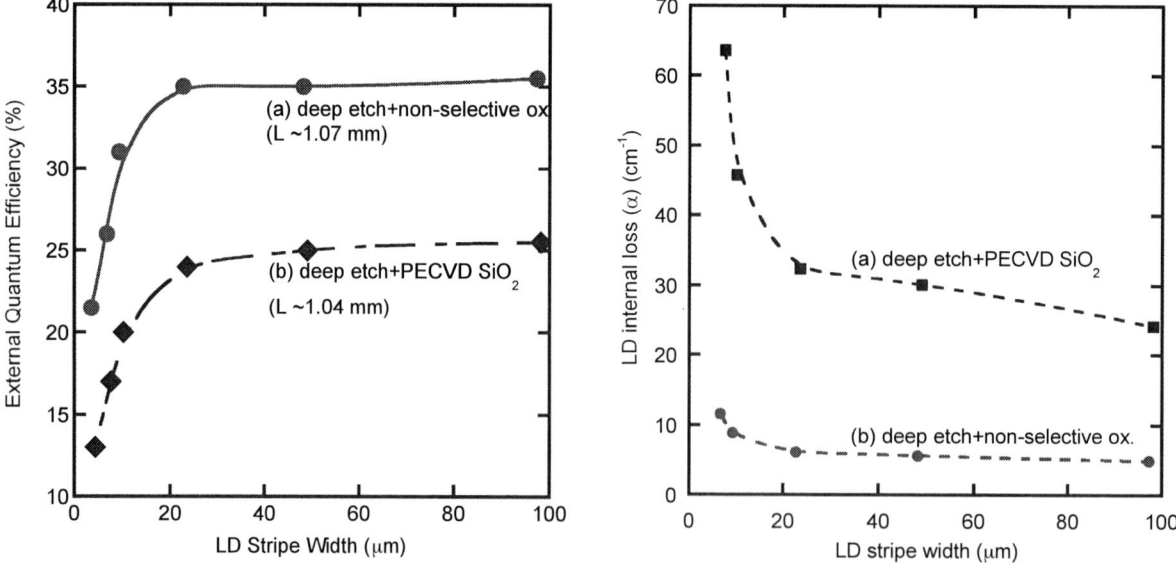

Fig. 4. External quantum efficiency vs. stripe width for QD LDs with (a) deep etch+ PECVD SiO_2 and (b) deep etch + non-selective oxidation.

Fig. 5. Internal loss α_i vs. stripe width for QD LDs with (a) deep etch+PECVD SiO_2 and (b) deep etch+non-selective oxidation.

126

Surface Treatment for Leakage Reduction in AlGaN/GaN HEMTs

Rongming Chu[1], Likun Shen[1], Nick Fichtenbaum[1], Stacia Keller[1], Andrea Corrion[2], Christiane Poblenz[2], James Speck[2], and Umesh Mishra[1]

[1]*ECE Department, University of California, Santa Barbara, CA 93106*
[2]*Materials Department, University of California, Santa Barbara, CA 93106*

Development of AlGaN/GaN HEMTs has been largely advanced in recent years, leading to record microwave power performance from solid-state devices. With the AlGaN/GaN HEMTs emerging as a viable technology for high frequency low noise amplifiers and power amplifiers, there are still several problems to be solved. One key problem is the relatively high gate leakage current, which may cause extra noise and reliability problems. In this report, we present a technology which reproducibly reduces gate leakage by two orders without introducing any negative effect on the device performance.

A CF_4 plasma treatment prior to Schottky metal evaporation has been used to reduce the reverse leakage current of Schottky contacts on both GaN and AlGaN/GaN. The initial data showed that the reverse leakage decreases with increasing CF_4 treatment time, as shown in Fig. 1. However, prolonged CF_4 plasma treatment decreases the 2DEG density and mobility (Fig. 2). We developed a two-step plasma treatment for effective leakage reduction without affecting the 2DEG. For the first step, a low power BCl_3 plasma was employed to remove the native oxides and clean the AlGaN surface. After this, a brief CF_4 plasma exposure was used to reduce the leakage. Fig. 3 shows that the two-step plasma treatment leads to a two orders of magnitude reduction of reverse leakage current for MOCVD grown AlGaN/GaN HEMTs. The effectiveness of short CF_4 plasma exposure indicates that the leakage reduction is caused by a surface effect, rather than the negative charge implantation into the bulk. This viewpoint is further supported by the increase of leakage current with a 100 sec BCl_3/Cl_2 etching after the 15 min CF_4 treatment (Fig. 4). The two-step plasma treatment is also effective in reducing the reverse leakage of Schottky contacts on MOCVD grown GaN MESFET samples, while keeping the forward current unaffected (Fig. 5). CV characterization suggests that the plasma treatment does not change the barrier height (Fig. 6). This leakage reduction technology can be applied to samples grown by MOCVD and NH_3 MBE. However, we did not observe consistent leakage reduction for samples grown by plasma MBE. This could be explained by the leaky screw dislocations of plasma MBE samples, which are usually grown under N-lean conditions.

Un-passivated AlGaN/GaN HEMTs are plagued with DC-RF dispersion. This dispersion can be effectively reduced by either the SiN_x passivation or the deep gate recessing. Fig. 7 shows that the deeply recessed HEMT structure can fully take advantage of the surface treatment technology, with two orders of magnitude reduction in leakage current. Without using any field plate, the deeply recessed HEMTs with the two-step plasma treatment exhibited a breakdown voltage of 100 V for a gate-drain spacing of 2 μm. Microwave power measurement of these devices shows an output power density of 9.5 W/mm at 40 V bias, with the associated PAE of 60%.

This work is supported by ONR MINE project monitored by Harry Dietrich and Paul Maki.

978-1-4244-1101-6/07/$25.00 ©2007 IEEE

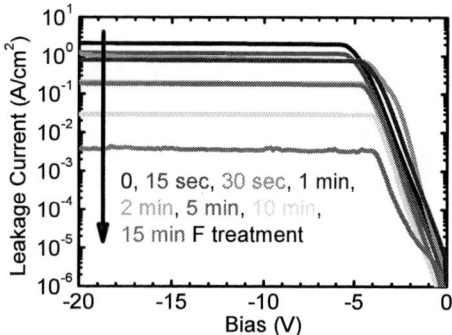

Fig. 1 Reverse leakage characteristics of AlGaN/GaN Schottky diodes with different CF_4 plasma treatment time.

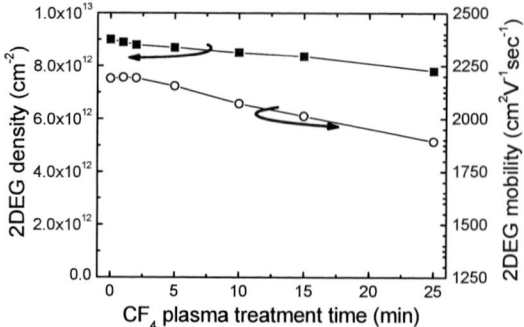

Fig. 2 2DEG Hall density and mobility of AlGaN/GaN HEMTs with different CF_4 plasma treatment time.

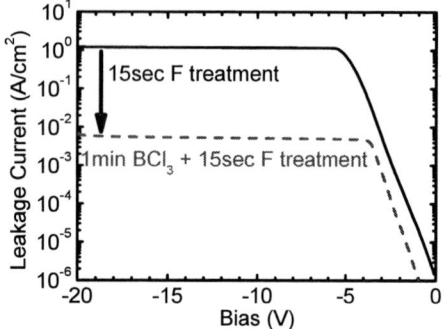

Fig. 3 Comparison of reverse leakage characteristics of AlGaN/GaN Schottky diodes with CF_4 and BCl_3+CF_4 treatment.

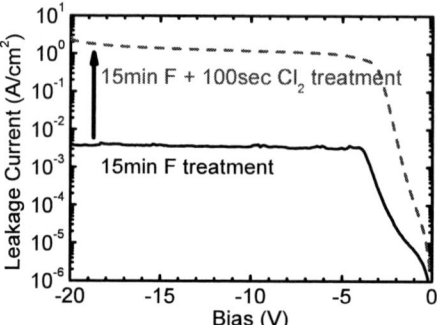

Fig. 4 Comparison of reverse leakage characteristics of AlGaN/GaN Schottky diodes with CF_4 and CF_4+Cl_2 treatment.

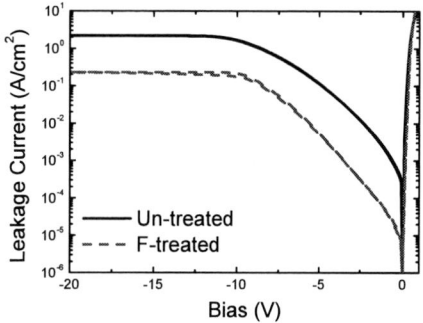

Fig. 5 Leakage characteristics of GaN Schottky diodes with and without CF_4 plasma treatment time.

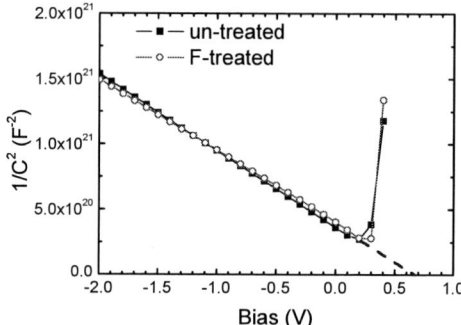

Fig. 6 $1/C^2$ vs. bias for GaN Schottky diodes with and without CF_4 plasma treatment time, showing the same barrier height.

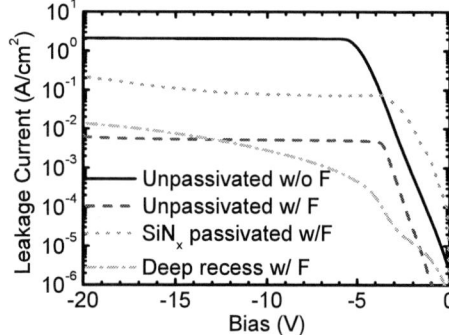

Fig. 7 Comparison of Schottky diode reverse leakage current characteristics of different AlGaN/GaN HEMT structures.

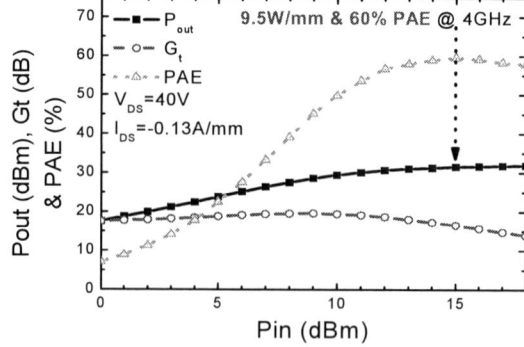

Fig. 8 Power performance of deeply recessed AlGaN/GaN HEMTs with CF_4 treatment.

128

AlGaN/GaN HEMT with High PAE and Breakdown Voltage Grown by Ammonia MBE

Y. Pei[1], C. Suh[1], R. Chu[1], F. Recht[1], L. Shen[1], A. Corrion[2], C. Poblenz[2], J. Speck[2] and U.K. Mishra[1]

[1]Electrical and Computer Engineering Department, University of California, Santa Barbara, CA 93106

[2]Materials Department, University of California, Santa Barbara, CA 93106

AlGaN/GaN high electron mobility transistors (HEMTs) have significantly advanced during the past years leading to exceptional high power and high frequency performance. In those devices, efficiency becomes the key point, especially for high power applications. At 35 GHz, output power of 7 W/mm and power-added efficiency (PAE) of 52% has been reported on deep submicron HEMTs grown by MOCVD [1]. At 18 GHz, PAE of 20%, and power density of 5.1 W/mm has been reported on HEMTs grown by ammonia MBE [2].

In this paper we report on high PAE, high breakdown-voltage HEMTs grown by ammonia MBE. The epi-structure is shown in figure 1. First, an AlN nucleation layer was grown by plasma-assisted MBE on SiC. Then GaN buffer was grown by Ammonia MBE, followed by 30 nm ammonia $Al_{0.3}Ga_{0.7}N$. A sheet charge of 1×10^{13} cm^{-2} with a mobility of 1500 cm^{2}V^{-1}s^{-1} was obtained from Hall measurements.

For the fabrication of this HEMT, Ti/Al/Ni/Au source/drain metallization scheme was used, followed by 30s 870 °C annealing in N_2 atmosphere to achieve a 0.3 ohm·mm contact resistance. Device isolation was formed by Cl_2 RIE-etch. 160 nm SiN_x was deposited by PECVD as a passivation layer. For optical-gate devices, 0.6 μm gates were defined by the stepper. A fluorine etch was used to remove SiN_x in the gate trench before gate metal evaporation. For deep submicron HEMTs, a 50 nm sacrificial Germanium layer was evaporated on top of SiN_x layer. Then 130 to 250 nm foot-gates were defined by electron beam lithography. The Germanium and SiN_x were removed by fluorine etch in the foot gate trench, followed by a 12-16 nm AlGaN recess etch. A second lithography was done to define a 0.6 μm top gate. The gate contact was formed by evaporating Ni/Au/Ni. After lift-off, Germanium was removed by H_2O_2 to have an air gap between the top gate and the SiN_x layer. This helped to decrease gate-source and gate-drain capacitances. Top view of SEM image is shown in figure 2.

Figure 3 shows the I-V curve of the deep submicron ammonia AlGaN/GaN HEMT. A peak transconductance of 350 ms/mm was obtained. The device showed very high breakdown voltage and very low leakage even for deep submicron HEMT after passivation. At V_{GD}=80 V, without any field plates, the gate leakage was as low as 0.8 mA/mm (figure 4). At 4 GHz, power measurement was carried on a 0.6 μm gate-length device. Power density of 10.6 W/mm and PAE of 63% was achieved at 48 V drain bias (figure 5). Figure 6 shows power density, PAE and linear gain as a function of drain bias for this device. For a 130 nm gate-length device, f_T of 68 GHz and f_{max} of 110 GHz were measured at V_{ds}=16 V (figure 7). At 30 GHz, this deep submicron HEMT showed an excellent PAE of 43%, 9 dB linear gain and 3.4 W/mm power density at a drain bias of 30 V (figure 8).

We demonstrated the ammonia MBE HEMTs with excellent PAE and breakdown voltage. The low leakage and excellent PAE show the great potential for ammonia MBE HEMT on applications in K band and beyond.

This work has been funded by the ONR MINE project, monitored by Dr. H. Dietrich and Dr. P. Maki.

References.

[1]M. Rosker et al. CS Mantech, Apri 25, 2006

[2]D. Ducatteau et al. Elect. Dev. Lett., 2006, vol. 27, pp. 7-9

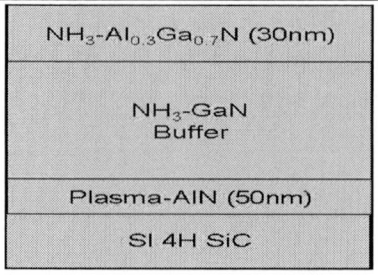

Fig 1: Epi-structure of an ammonia MBE HEMT

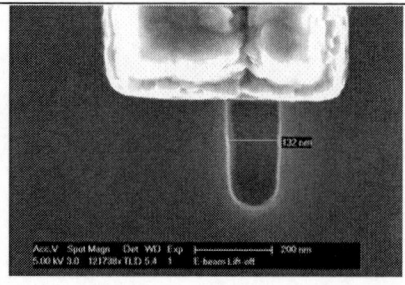

Fig 2: SEM image of foot gate trench and top T-gate

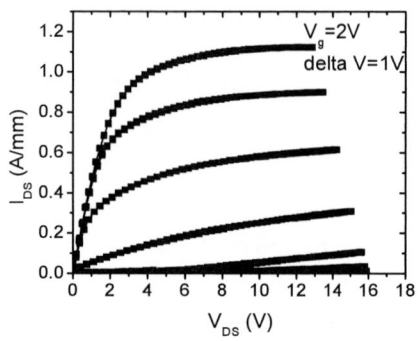

Fig 3: DC performance of an ammonia MBE deep submicron HEMT

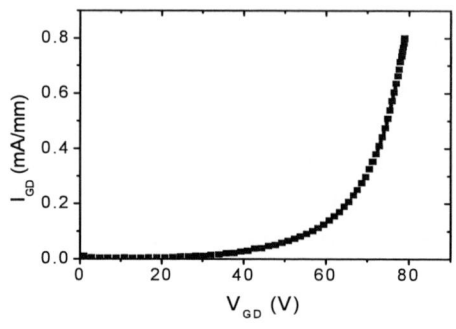

Fig 4: Gate-drain leakage for a L_g=160 nm deep submicron HEMT

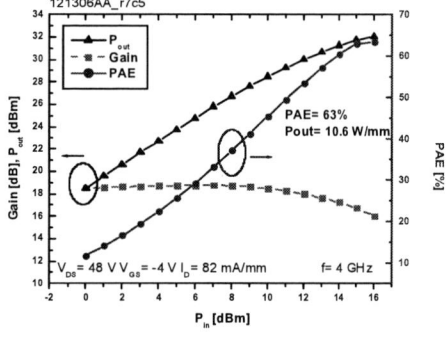

Fig 5: Measured power density, PAE and Gain at 4 GHz, 48 V drain bias for the 0.6 μm gate length HEMT

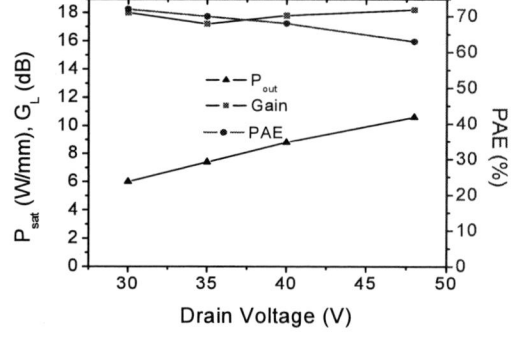

Fig 6: Measured power density, PAE and linear gain versus drain bias at 4 GHz for the 0.6 μm gate length HEMT

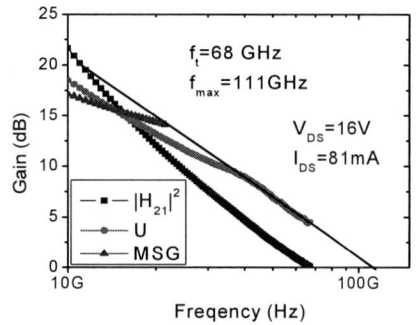

Fig 7: Small signal gain of L_g=130 nm W_g=2×75μm GaN HEMT at V_{ds}=16 V

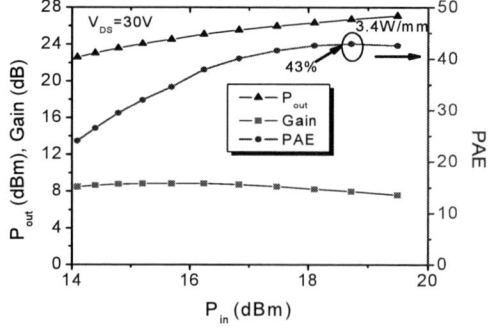

Fig 8: Measured power density, PAE and Gain at 30 GHz, 30 V drain bias for the 130 nm gate length HEMT

Analytical Model of Apparent Threshold Voltage Lowering Induced by Contact Resistance in Amorphous Silicon Thin Film Transistors

Bahman Hekmatshoar, Ke Long[*], Sigurd Wagner and James C. Sturm

Princeton Institute for the Science and Technology of Materials (PRISM), Department of Electrical Engineering, Princeton University, Princeton, NJ 08544
[*]Present address: *Flexible Display Center, Arizona State University, Tempe, AZ 85284*
Email: hekmat@princeton.edu

The interest in amorphous Si (*a*-Si) thin-film transistors (TFT's) has increased recently due to active matrix organic light emitting diode (AMOLED) display applications. Unlike active matrix liquid crystal display (AMLCD) applications where the TFT only charges a capacitor, the DC driving requirement in AMOLED's makes the source/drain series resistance and modeling of current in saturation very important. It is well-known from Si VLSI MOS technology that the presence of the drain/source contact resistance lowers the driving current and as a result the "apparent" mobility extracted from the electrical characteristics is lower than the "true" mobility especially at short channel lengths [1]. Amorphous Si TFT's have a much larger series resistance than VLSI FET's because in addition to the metal/n$^+$ contact, *a*-Si which is a low conductivity material also contributes to the contact resistance. This is because in the TFT structure, the metal/n$^+$ contact is on top of the a-Si film while the channel is at the bottom. In this abstract, we (i) show that this series resistance causes a large lowering of the "apparent" threshold voltage when it is extracted by conventional methods, and (ii) develop an analytical model to explain this effect. The model is supported by experimental data at different channel lengths and series resistances.

Conventionally, the apparent mobility and threshold voltage of a FET are extracted by plotting the square root of the saturation current versus the gate voltage and a least square fit (LSF) calculation is performed to approximate the square root of the drain current with a straight line. Fig. 1 shows that the values of apparent threshold voltage extracted by this method are lower at shorter channel lengths and higher contact resistances. To explain this behavior, we first assume that the contact resistance in a-Si TFT's is a constant voltage-invariant resistance as in MOSFET's. Also, in our model, we assume that a-Si TFT's are described by MOS I-V characteristics, which is a relatively good approximation for hydrogenated a-Si TFT's. As shown in Fig. 2, an LSF to the saturation regime of the MOS-based characteristic results in an apparent threshold voltage and mobility which are lower than their true values. However, the actual a-Si TFT problem is more complicated, since the contact resistance is not constant, and has a gate-voltage dependent series component that results from the presence of a-Si beneath the metallurgical junction [2] as shown in Fig. 3 for a typical inverted-staggered a-Si TFT. The extracted values of contact resistance for our test TFT's are plotted in Fig. 4. It is observed that both components of the contact resistance increase with decreasing the overlap between the gate and drain/source. By direct application of the MOS equation (the model given in Fig. 2) it can be shown that the gate-voltage dependent component of the contact resistance lowers the apparent mobility but does not change the apparent threshold voltage. By adding this effect, the apparent mobility and threshold voltage can be evaluated analytically. This model is given in Fig. 5 and compared with the experimental I-V curves for two different contact resistance values. Fig. 6 shows the variation of the apparent threshold voltage as a function of contact resistance for different channel lengths. In Fig. 7, the model is compared with the experimental I-V curves to show that changing the drain bias changes the extracted apparent threshold voltage and mobility by changing the saturation to linear transition point and therefore changing the range over which the LSF is calculated. Finally, the variation of the apparent threshold voltage is plotted versus the channel length in Fig. 8 and compared to the experimental data at two different drain biases.

In summary, we have shown that in a-Si TFT's the apparent threshold voltage extracted by conventional methods is lowered by the presence of the source/drain contact resistance, especially at short channel lengths and the analytical model presented to explain this effect is in good agreement with the experimental data. This model is particularly useful for AMOLED applications where the contact resistance has a crucial role in determining the driving current and thus the brightness of the pixels.

[1] P. R. Gray and R. G. Meyer, *Analysis and Design of Analog Integrated Circuits*, 3rd Edition, John Wiley & Sons, New York, NY 1993
[2] S. Luan and G. W. Neudeck, *J. Appl. Phys.* vol. 72, no. 2, July 1992, pp. 766-772

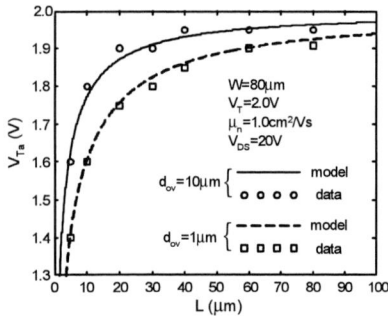

Fig.1. Apparent threshold voltage (V_{Ta}) extracted from test TFT's using the conventional method, for several channel lengths (L) and two gate to drain/source overlap lengths, d_{ov} (see Fig. 3). The shorter d_{ov} corresponds to the higher contact resistance. It is observed that V_{Ta} is lower at shorter L and also lower for higher contact resistance (shorter d_{ov}).

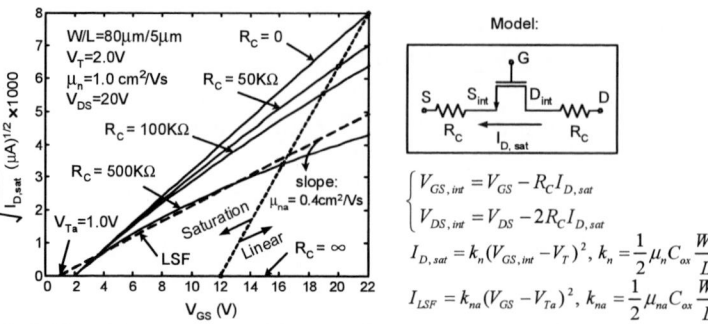

Fig.2. Plot of the square root of the saturation current in a FET for several values of contact resistance R_C, when the current is calculated by a simple FET model plus R_C at the source and drain. Using a least square fit (LSF) to estimate the current results in an apparent threshold voltage (and mobility) that is lower than its true value, i.e. $V_{Ta} < V_T$ (and $\mu_{na} < \mu_n$), as shown for $R_C = 500K\Omega$. The LSF is calculated (and the result is valid) only where V_{GS} is low enough (for a given V_{DS}) to ensure saturation. The saturation to linear transition point occurs at $V_{DS, int} = V_{GS, int} - V_T$.

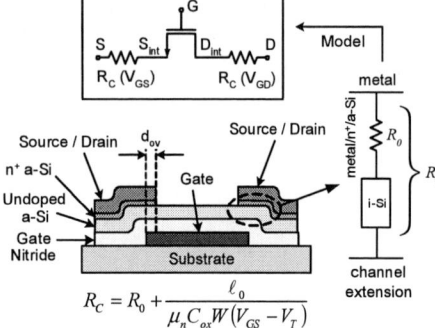

$$R_C = R_0 + \frac{\ell_0}{\mu_n C_{ox} W (V_{GS} - V_T)}$$

Fig.3. Cross-section of an inverted-staggered a-Si TFT. The contact resistance R_C is composed of a metal/n$^+$Si/a-Si metallurgical junction and an a-Si region beneath that junction. The latter component makes R_C gate-voltage dependent [2].

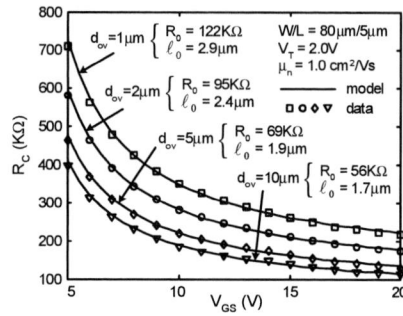

Fig.4. Gate voltage dependence of the contact resistance R_C for various gate to drain/source overlap d_{ov}. The R_C values are extracted from the I-V curve of the test TFTs following the approach introduced in [2].

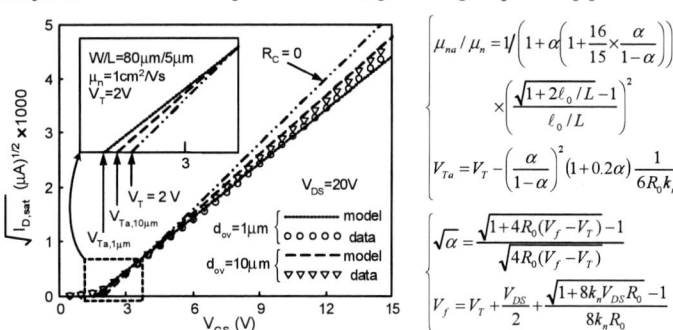

Fig.5. Comparison of our LSF-based model with the experimental data. The apparent threshold voltage V_{Ta} (and μ_{na}) is lower for smaller d_{ov} (higher R_0). The model predicts that V_{Ta} is not affected by the gate-voltage dependent component of R_C (though μ_{na} is). In the model, V_f is the gate voltage at which the saturation to linear transition occurs and α is the ratio $R_0 I_{D, sat} / (V_{GS} - V_T)$ at the transition point, $V_{GS} = V_f$.

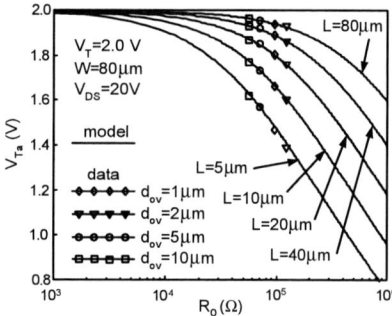

Fig.6. Variation of V_{Ta} versus R_0 for several values of channel length, L. The extracted V_{Ta} is lower at shorter L and higher R_0, as predicted by the model.

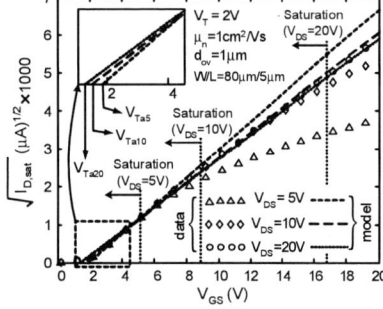

Fig.7. Variation of V_{Ta} for different drain biases. Reducing V_{DS} shrinks the saturation regime over which the LSF is calculated. The extracted V_{Ta} (and μ_{na}) increases with reducing the drain bias as predicted by the model.

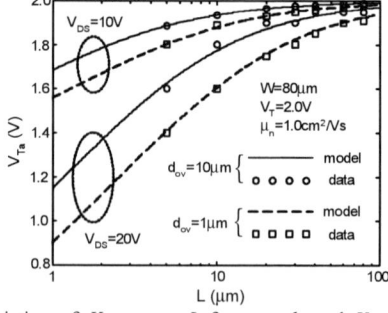

Fig.8. Variation of V_{Ta} versus L for two d_{ov} and V_{DS} values and comparison with the model. The extracted V_{Ta} is lower at shorter L, higher R_0 and higher V_{DS}, as predicted by the model.

On-Chip Clocking Scheme for Nanomagnet QCA

Mohmmad T. Alam[1], Michael Niemier[2], Wolfgang Porod[1], Sharon Hu[2], Michael Putney[2], Jarett DeAngelis[2], and Gary H. Bernstein[1]. *[1]Center for Nano Science and Technology, Dept. of Electrical Engineering, U. of Notre Dame. [2]Dept. of Comp. Sci. and Eng., U. of Notre Dame. Address: 275 Fitzpatrick Hall, University of Notre Dame, Notre Dame, IN 46556. Phone: 1-574-631-6269; fax: 1-574-631-4393; email: malam1@nd.edu.*

Quantum-dot Cellular Automata (QCA) has been demonstrated using aluminum tunnel junction single-electron transistor technology at mK temperatures[1], and molecular QCA is under development for operation at room temperature (RT)[2]. All of the basic building blocks needed for QCA have been experimentally demonstrated. Our work on nanomagnet-based QCA (NMQCA) holds the most promise for achieving viable RT operation in the near term[3]. One requirement of the QCA architecture is low-power clock structures, which is the subject of this paper.

Experiments have shown that a series of nanomagnets placed side by side, with small gaps between them, can be used to carry digital data. The long axis is the "easy" axis of magnetization (i.e. low coercivity), and in a ground state, the magnets polarize in either up or down directions, which can be used to represent digital logic values. Fig. 1 shows such an arrangement where an external magnetic clocking field is applied along the "hard" (shorter, and high coercivity) axes. When the field is increased, polarization is aligned along the hard axes of the nanomagnets, forcing the line of magnets to a "null" state, and when it is released, the line relaxes adiabatically into the ground state set by the end magnet whose coercivity is made higher than the wire magnets via a narrower geometry. Alternatively, an electrical input structure, or some other method can be used. Previously, we have employed external magnetic fields to polarize magnets along their hard axes. Although an external field was used successfully for magnetic bubble technology, it is not a viable scheme for multiple clock phases at high frequencies. Consequently, we have designed and simulated an on-chip clocking scheme that should be able to create a confined, high-speed, and sufficiently strong nulling field.

One clocking structure has been simulated using the Maxwell® 2D electromagnetic simulator. A cross-sectional view of a line, or "wire," of nanomagnets along with the clocking structure is shown in Fig. 2. The magnetic fields generated due to currents in the clocking wires are strengthened in the nanomagnets by a ferrite yoke. The dimensions of the nanomagnets are 60 nm X 90 nm X 20 nm, with 15 nm gaps between them. As we have used a 2D simulator, each nanomagnet in Fig. 2 is 60 nm in the x direction and 20 nm in the y direction. Thus, both x and y directions are hard axes for the nanomagnets.

Digital data is propagated through the QCA wire using multiphase clocking. First, assume that there is a fixed input (up or down state of a hard nanomagnet) at the left side of the structure. During the first phase of the clock, both current-carrying wires are excited. As a result, all the nanomagnets experience a high magnetic field in the x-direction (Fig. 3) that polarizes the magnets in the x-direction, nulling the state of the magnets. The field in the y direction (B_y) is small compared to B_x and, hence, is not shown. In the second clock phase, the right wire and the wire to its right (not shown) are excited, while the left one is de-excited. Now, only the nanomagnets on the right side experience a high magnetic field in the x direction (Fig. 4), and are nulled. The left nanomagnets relax to the ground state, and the state (up or down) of the individual nanomagnets are determined by the input on the left. Thus, the data is propagated from left to right up to the first half of the structure. In the next clocking phases, wires to the right are excited sequentially two at a time, and the data propagates from left to right. A four-phase clock is sufficient to facilitate data propagation.

Micromagnetic simulations using OOMMF demonstrate that we should be able to clock the nanomagnets to perform logic functions. For example, Fig. 5 illustrates a short QCA wire segment in three different states. Fig. 5a shows a 3-cell wire segment after a 50 mT clock field is applied to the wire. Fig. 5b illustrates the same wire segment with the polarization of the first cell changed. A 15 mT field is applied along the x-axis of Fig. 5b (analogous to the middle row of Fig. 1) and is then reduced to 0 mT. The result is shown in Fig. 5c. The rest of the wire has changed in accordance with the new input. Fields as small as 3-5 mT have been shown to drive longer lines of similarly sized magnets to the correct ground state.

The combined power consumption of the clock circuit and the nanomagnet switching power is very small, in the range of 1.5 μW per clock wire. The clocking circuit proposed in this paper should be applicable to NMQCA majority gate logic as well. Thus, this technology has the potential to allow for non-volatile, ultra-low power digital logic with minimal CMOS support.

---

[1] A. O. Orlov et al., *Science*, **277**, pp. 928-930 (1997).

[2] K. Sarveswaran et al., *Langmuir*, **22**, 11279-11283 (2006).

[3] A. Imre et al., *Science*, **311**, pp. 205-208 (2006).

978-1-4244-1101-6/07/$25.00 ©2007 IEEE

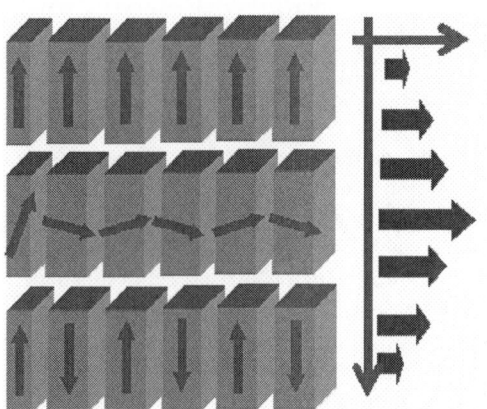

Fig 1. NMQCA wire showing polarization of nanomagnets as the external magnetic field is increased and then decreased. The resulting antiferromagnetic coupling follows the state of the input magnet on the left end.

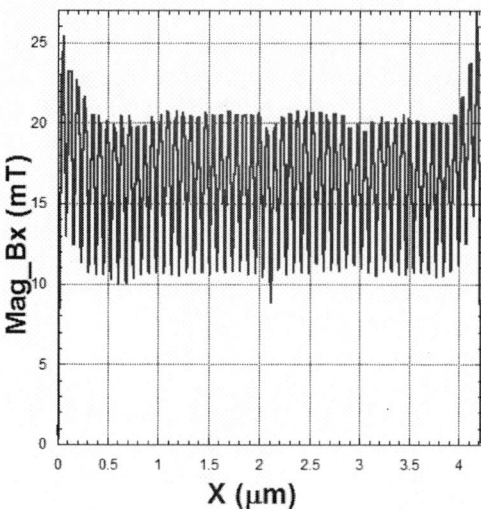

Fig. 3: Magnitude of B_x vs. x along y = 0.35 μm line, when both the wires are excited with current density of 10^6 A/cm². y = 0.35 μm line goes through the center of the nanomagnets.

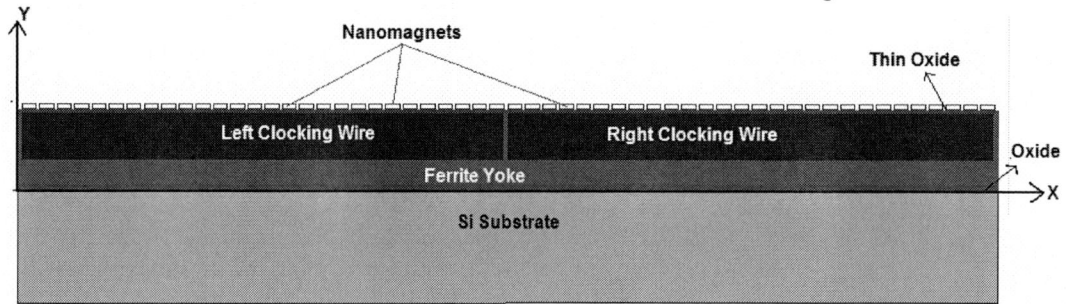

Fig. 2: The cross sectional view of two blocks in the data line. The clock wires are made of copper. The oxide layers between the Si substrate and the ferrite yoke and between the copper wires and nanomagnets provide electrical insulation.

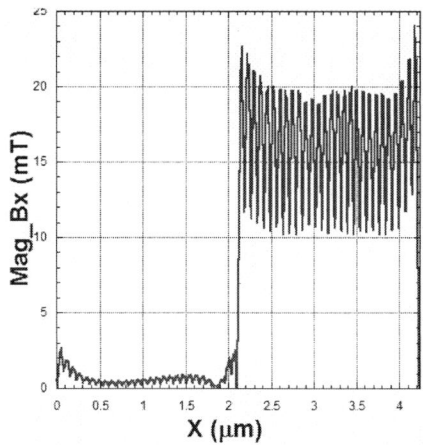

Fig. 4: Magnitude of B_x vs. x along the y = 0.35 μm line, when the right wire is excited (current density of 10^6 A/cm²) and left wire is not.

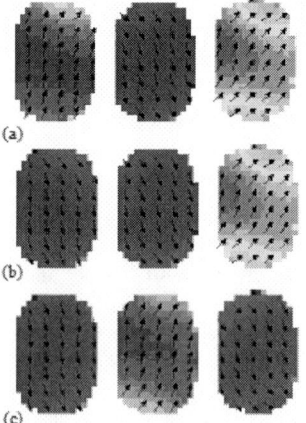

Fig. 5. OOMMF simulations of 60 nm X 90 nm nanomagnets. (a) Wire segment after input applied, (b) new input applied to wire (simulated electrical input structure), (c) 15 mT field applied to magnets, field relaxed to 0 mT -- magnets take on new (and correct) polarizations.

134

Electrical Characterization of Vertical InAs Nanowires on Si

C. Rehnstedt, T. Mårtensson, C. Thelander, L. Samuelson and L.-E. Wernersson

Solid State Physics / Nanometer Consortium, Lund University, Box 118, S-22100 Sweden
Email: Lars-Erik.Wernersson@ftf.lth.se, Phone: +46 46 2227678

Growth of III-V nanowires (NWs) on silicon enables integration of III-V materials with mainstream silicon technology. Many III-Vs, including InAs, have material properties, such as increased combinations for bandgap engineering, low contact resistance, and enhanced transport properties that enable high-speed and low power devices. Vertical wrap-gate transistors [1] in turn impose a new potential, but also challenges, for transistor processing and scaling.

InAs NWs (n-type, $2-5 \times 10^{17}$ cm^{-3}) were grown on Si(111) substrates with different doping levels using metal-organic vapor phase epitaxy (MOVPE). No Au seed particles were used for the NW growth, but instead we used self-assembled organic coatings that create an oxide template, which guides NW nucleation [2]. The wires were simultaneously grown on all samples at a growth temperature of 570 °C with tri-methyl indium and arsine as precursors. All wires had a length of 1.5 μm, while the diameter and area density varied with the Si substrate doping. The NWs grown on highly doped n-type (phosphorus, 10^{19} cm^{-3}) substrates had a diameter of 70 nm and a density of 0.5 μm^{-2}. NWs on low doped n-type (phosphorus, 10^{15} cm^{-3}) had a diameter of 170 nm and a density of 0.2 μm^{-2}, while NWs with a diameter of 250 nm and a density of 0.12 μm^{-2} were grown on p-type (boron, 10^{15} cm^{-3}) substrates. NWs were grown uniformly over the substrate (Fig. 1).

Using optical lithography, 100 μm^2 areas were covered with resist while etching away the NWs on the surrounding substrate. After removing the resist, the remaining NWs were covered by 135 nm SiN$_x$ deposited by plasma enhanced chemical vapor deposition (PECVD). The SiN$_x$ on the NW tops was then etched with CF$_4$ reactive ion etching (RIE) using a photoresist mask. Ti/Au top contacts were finally deposited by evaporation and patterned by optical lithography (Fig. 2, 3).

Electrical characterization of the InAs NWs and the InAs/Si heterojunctions was performed on all three differently doped Si substrates. The wires were measured in parallel, contacting 50, 20 and 10 wires per device, respectively, for highly n-doped, n-doped, and p-doped Si. The IV characteristics (Fig. 4) for each doping level were measured with the Si substrate grounded, while probing the top contact to the InAs NW. The IV of the n-type sample indicates a quasi-linear behavior, while the p-type sample has a more non-linear IV. Temperature dependent measurements were done for temperatures from 25°C to 125°C (Fig. 5, 6) and the data were analyzed in Arrhenius plots (Fig. 7). The exact values of the observed barrier heights depend on the details in the model used, which may differ between the samples. To capture the essence in the transport, we modeled the thermionic part of the transport as $J=C \times \exp(-q\Phi_B/kT)$, where C is a constant, which allows us to compare the barriers formed at the interface between the differently doped substrates and the NWs. The results demonstrate that a low and almost negligible barrier is formed for InAs NWs grown on n-type Si, while a barrier is formed on p-type Si. These findings, suggest an alignment of the InAs conduction band close to the Si conduction band. However, the presence of a few hundred meV discontinuity at the conduction band for the n-type samples may not be ruled out, since the growth of the InAs may change the carrier concentration at the interface. This strongly affects the measurements. Finally, the resistance of single NWs broken off and placed on an oxidized surface and contacted with Ti/Au contacts in both ends, showed a resistivity of 59 $\mu\Omega$m. This corresponds to a resistance of approximately 460 Ω for the vertical devices of the highly doped n-type Si substrate, a value that agrees well with the deduced value for the vertical device (534 Ω). Also the current densities agree well, 9.4 mA/μm^2 for the vertical NWs as compared to 14.5 mA/μm^2 for NWs broken off.

This work shows that InAs NWs can be integrated with Si and that low barriers may be formed at the InAs/Si interface. It also shows that the barrier height and the resistance at the interface may be controlled by means of the substrate doping.

Acknowledgments: This work was supported by the Swedish Research Council, the Swedish Foundation for Strategic Research, the EU program NODE 015783, and Knut and Alice Wallenberg Foundation.

[1] T. Bryllert et.al., DRC 2005 pp. 157, and T. Bryllert et. al., IEEE EDL, 27, 323, 2006
[2] T. Mårtensson et al., unpublished results

978-1-4244-1101-6/07/$25.00 ©2007 IEEE

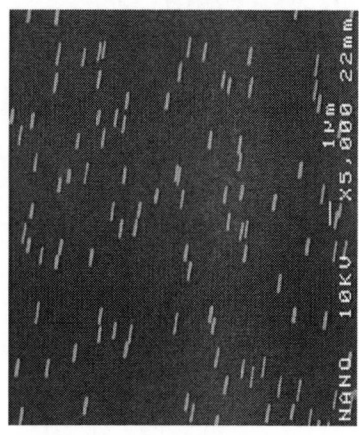

Fig. 1 SEM image of InAs NWs on Si p substrate.

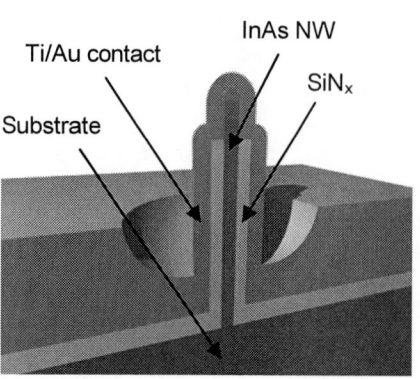

Fig. 2 Schematic cross-section of fabricated structure.

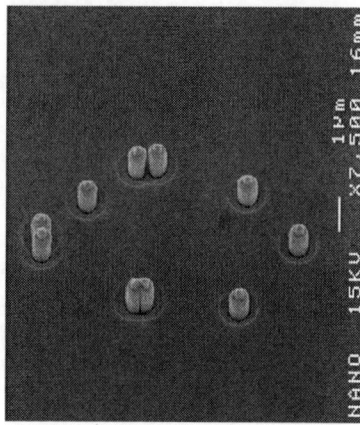

Fig. 3 SEM image of fabricated structure.

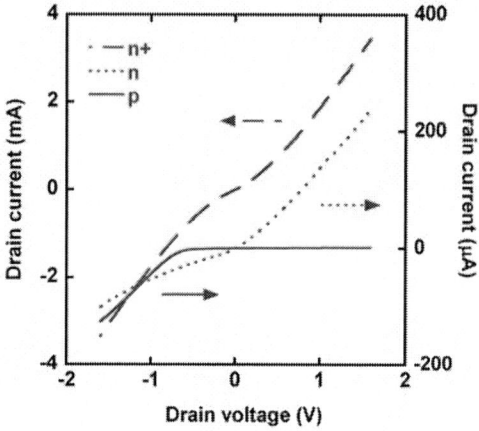

Fig. 4 Room temperature IV characteristics of InAs NWs on differently doped Si substrates.

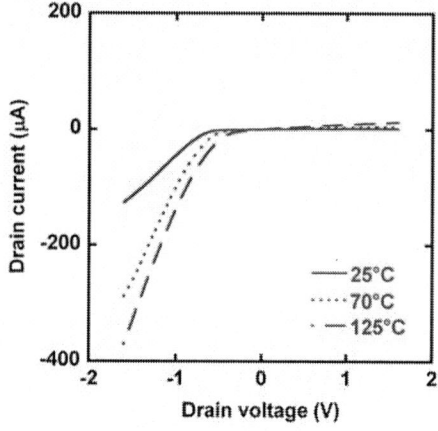

Fig. 5 IV characteristics for different temperatures of InAs NWs on Si p substrate.

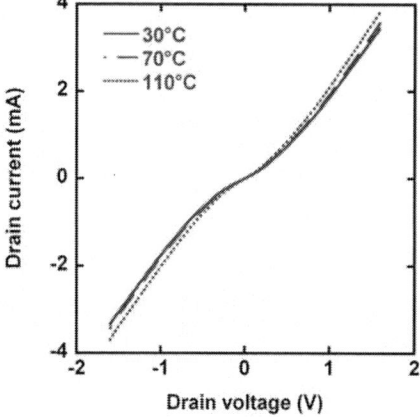

Fig. 6 IV characteristics for different temperatures of InAs NWs on Si n$^+$ substrate.

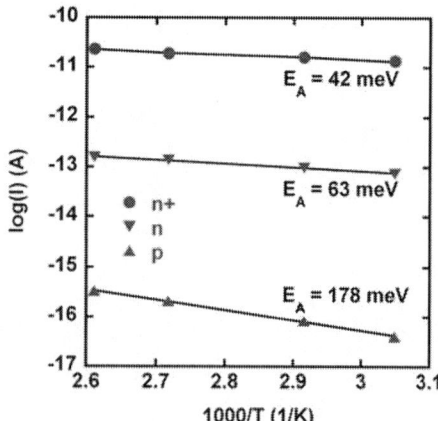

Fig. 7 Arrhenius plots for 20 mV drain voltage.

136

Band-gap engineering of enhanced spin-orbit interactions in InAs/AlGaAs heterostructures for Datta-Das spin transistor

Takashi Matsuda, Munekazu Ohno and Kanji Yoh

Research Center for Integrated Quantum Electronics, Hokkaido University, Sapporo, 060-8628 Japan
Phone: +81-11-716-7174   e-mail: matsuda@rciqe.hokudai.ac.jp

Much effort has been done on spin injection and MR measurements after the proposal of spin-FET by Datta & Das[1] (Fig.1). The remaining issue of gate controlled spin rotation in the quantum well has not been focused together with heterostructure design in conjunction with increased spin-orbit interaction. We will show that increased spin-orbit interaction is possible by band-gap engineering in InAs/AlGaAs heterostructure grown on InP, thereby spin 2π rotation in less than 70nm in channel length becomes possible. Importance of conduction band discontinuities in a quantum well was pointed out as the source of the energy shift by Pfeffer and Zawadzki[2] in InGaAs/InAlAs heterostructures grown on InP, as opposed to the naive interpretation that the spin-orbit interaction is proportional to the average electric field in the quantum well (QW) [3]. We have designed and fabricated psuedomorphic InGaAs channel structures with high indium content and verified enhanced spin-orbit interaction theoretically and experimentally. The device structure is designed in such a way as to bring tha peak of the wavefunction lies on top of the band discontinuity on front side between main QW and sub-channel QW. The enhancement of spin-orbit interaction was verified experimentally and confirmed theoretically by k•p calculation a la Zawadzki.

Sample1 has $In_{0.9}Ga_{0.1}As$ channel of 50A thickness and sample2 does InAs channel of 40A thickness (Fig.2a, b). The mobility and carrier concentration of sample1 and sample2 are 76000cm^2/Vs, 1.6×10^{12}cm^{-2} and 58000 cm^2/Vs, 2.0×10^{12}cm^{-2}, respectively at 3.8K. Shubnikov de Haas (SdH) oscillations were measured and Fourier transformation was carried out to estimate Rashba coefficient α which indicates the magnitude of the spin splitting as $\Delta E = 2\alpha |k_{//}|$ (Fig.3). Measured Rashba coefficient was 52×10^{-12}eVm for sample1 and 50×10^{-12}eVm for sample2. These values turned out to be extremely large compared with the value, 30×10^{-12}eVm, usually reported for InAs-based HEMTs in conventional simple quantum well structure whose wavefunction peak lies in the center of the well (Fig.4) and our control sample of $In_{0.81}Ga_{0.19}As$ HEMT Rashba coefficient of 21×10^{-12}eVm by SdH analysis (Fig.2c). This result is promising to provide a channel as short as 65nm for spin FET to the length become much shorter than spin relaxation length along the channel.

The k•p approximation was carried out to theoretically verify such a enhanced Rashba coefficient observed experimentally. Fig.5 shows the estimated Rashba coefficient of each sample and the contribution of the electric field in the well, electric field on the conduction band offset and the interaction between conduction band and valence bands to it. It can be seen that the component of electric field on the conduction band offset drastically increased in sample1 and 2, which leads the total contribution of δ-functional electric field to dominate the spin-orbit interaction strength. Unlike Zawadzki's simple quantum well structure, our result indicate that the δ-functional electric field at the band discontinuity of main well and sub-channel in front side, but not the effect of valence band and split-off band as in case [2]. This was made possible by the wavefucntion engineering so that the peak position of the wave function lies on the band discontinuity edge.

In conclusion, we have proposed and verified InAs (or InGaAs) channel HEMT structure with an enhanced spin-orbit interaction. The result suggests a new direction to achieve short channel Datta-Das spin transistor (Fig.6) which is a key to make the device work by minimizing spin relaxation in the channel (Fig.7).

[1] S.Datta and B.Das, Appl.Phys.Lett.56, 665 (1990)

[2] P.Pfeffer and W.Zawdzki, Phys. Rev. B, 59, R5312 (1999)

[3] E.A.de Andrada e Silva et al.,Phys. Rev. B, 50, 8523 (1994)

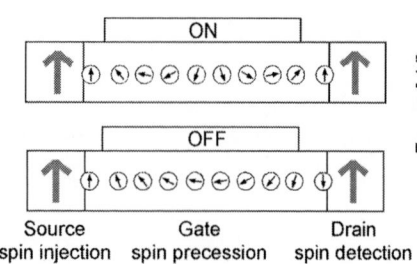

Source — spin injection | Gate — spin precession | Drain — spin detection

Fig1. Schematic diagram of the operation of spin-FET prorosed by Datta & Das

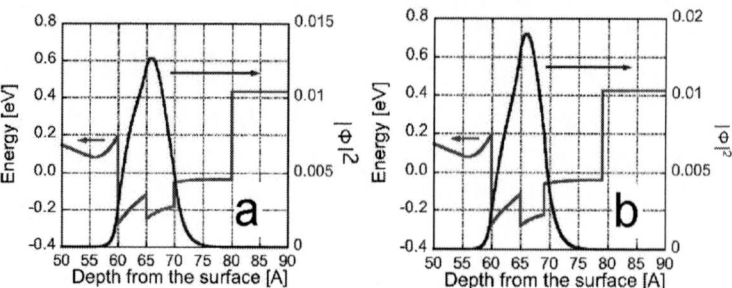

Fig.2 Conduction band structure and the distribution of squared value of wave function in a: sample1, b: sample2, c: sample3.

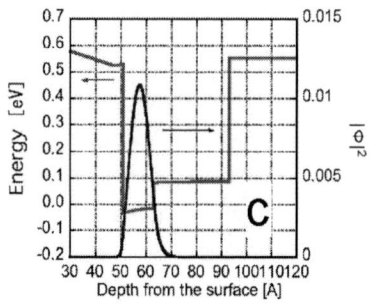

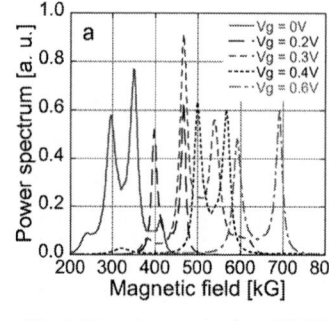

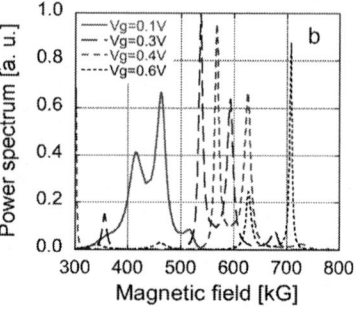

Fig.3 Fourier analysis of SdH oscillations. a: sample1, b: sample2

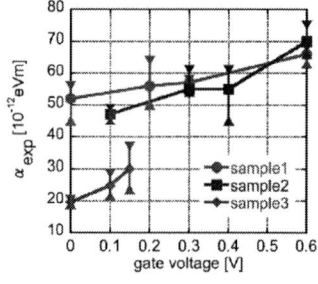

Fig.4 Rashba coefficient estima--ted from the SdH oscillations.

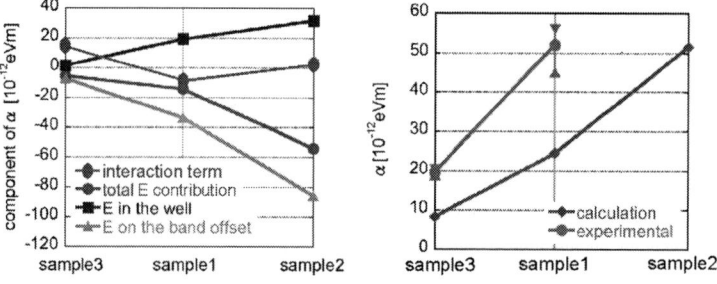

Fig.5 a: Components of Rashba coefficient. E stands for electric field and interaction term indicate the interaction between conduction band and valence bands. b: Comparison of Rashba coefficient obtained by calculation and experiment.

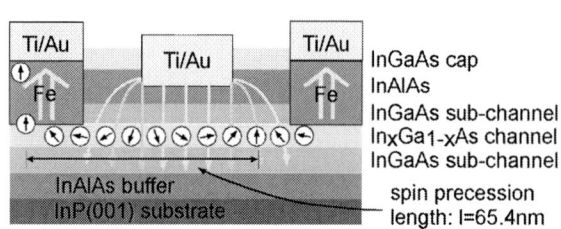

Fig.6 Spin precession length become 65nm when $\alpha = 50 \times 10^{-12}$ eVm.

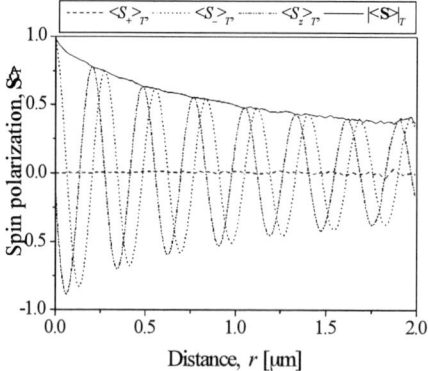

Fig.7 Monte Carlo simulation of spin relaxation length calculated with $b = 11 \times 10^{-12}$ eVm and $a = 15 \times 10^{-12}$ eVm

Oxide-Induced Noise in Carbon Nanotube Devices

Yu-Ming Lin and Phaedon Avouris
IBM T. J. Watson Research Center, Yorktown Heights, NY 10598

Single-walled carbon nanotubes are important building blocks for nanoelectronic devices and circuits. Recent advances in device fabrication and optimization have enabled high performance CNT devices rivaling state-of-the-art Si counterparts, showing outstanding dc electrical characteristics approaching theoretical limits[1-4]. However, most CNT-based devices also possess much more pronounced current fluctuations than conventional bulk devices, presenting one serious concern for practical applications[5-6]. In CNT devices, the electrical noise is dominated by the 1/f noise, and trapped charges in the oxide have been suggested as one possible source[6]. Nevertheless, the contribution of the oxide substrate to the 1/f noise has not been determined, the knowledge of which is critical in controlling and lowering the noise level in SWNT devices. In this report, we employ suspended nanotubes to directly study the impact of the oxide on the electrical properties and noise characteristics of carbon nanotube devices, showing that the 1/f noise can be reduced significantly in a substrate-free environment.

Fig. 1 shows the device structure used in our study, where two devices are fabricated side-by-side on the same nanotube. As shown in Fig. 1, the left device consists of a fully-suspended nanotube between the electrodes. To fabricate such structures, Pd electrodes are first deposited on arc-discharge CNTs dispersed on a SiO_2(10 nm)/Si substrate. The suspended nanotube is subsequently produced by etching the oxide and Si substrate in buffered HF and KOH solutions, respectively, to a depth of 200nm. Both semiconducting and metallic nanotube devices have been fabricated and studied using the same device configuration.

Fig. 2 compares the subthreshold characteristics (I-V_g) of semiconducting CNT devices before and after the etching process. The doped Si substrate is used as the back gate and the drain voltage is -0.5 V. Prior to the etching, the two side-by-side devices possess similar I-V_g characteristics with an on-current of 4μA and an inverse subthreshold slope of $S \sim 130$ mV/dec (closed circle in Fig. 2). After the etching, the I-V_g characteristics of the nanotube segment that remains supported by the substrate are nearly unchanged (solid line in Fig. 2), indicating that the etching process does not affect the nanotube and substrate properties. In contrast, the suspended nanotube device shows a lower on-current and a shallower I-V_g slope (dash line in Fig. 2) due to the weaker gate control from the back gate. As also shown in Fig. 2, there is less, if any, hysteresis in the suspended nanotube device compared to the substrate-supported one.

Fig. 3 plots the I-V_g curves of metallic CNT devices measured at $V_d = 0.1$V for the suspended nanotube (circle) and the substrate-supported nanotube (square), exhibiting a similar current level independent of back gate. The inset of Fig. 3 shows the linear I-V_d behavior of the suspended nanotube device up to 1V. To illustrate the impact of the oxide on the 1/f noise, we compare the noise power spectra S_I of both devices before and after the etching. In CNT devices, the 1/f noise is proportional to the square of the dc current I, and can be expressed by $S_I = A_N \times I^2/f$, where A_N is referred to as the 1/f noise amplitude. For the control device where the nanotube channel remains lying on the substrate after the etching process, the noise level is unchanged before and after the treatment, as shown by the noise power spectra, normalized by I^2, in Fig. 4. In comparison, Fig. 5 shows that the 1/f noise level is reduced by almost an order of magnitude in the CNT device when the nanotube becomes suspended. These results clearly illustrate the dominant role played by the oxide substrate with respect to the 1/f noise behavior in carbon nanotube devices. Fig. 6 shows the 1/f noise amplitude A_N as a function of temperature for the two types of devices. We note that the noise amplitude in the suspended CNT device exhibits a much weaker temperature dependence than that in the substrate-supported device, indicating that the current fluctuations are indeed dominated by the thermally-activated charge traps in the oxide.

In conclusion, we have fabricated suspended nanotube devices to elucidate the impact of the oxide substrate on the electrical transport properties of CNT devices. We find that while the dc electrical properties of carbon nanotubes seem to be relatively unaffected by the presence of the oxide substrate, the oxide plays a dominant role in determining the 1/f noise level. We have demonstrated that the noise level can be lowered significantly, up to an order of magnitude, by removing the oxide substrate.

---

[1]Lin et al., IEEE Trans. Nanotech. **4**, 481 (2005)
[2]Ph. Avouris, Physics World **20**, 41 (2007)
[3]Dai et al., Nano **1**, 1 (2006)

[4]Appenzeller et al., IEEE Trans. Elec. Devices 52, 2568 (2005)
[5]Collins et al., Appl. Phys. Lett. **76**, 894 (2004)
[6]Lin et al., Nano Lett. **6**, 930 (2006)

978-1-4244-1101-6/07/$25.00 ©2007 IEEE

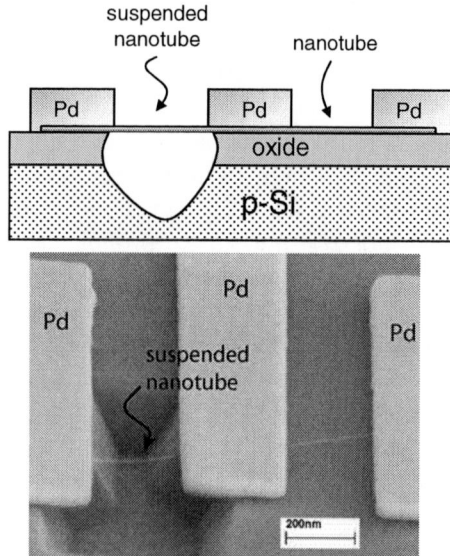

Fig. 1: (top) Schematic cross-section of two side-by-side nanotube devices, one of which consists of a fully suspended nanotube segment. (bottom) SEM image of the two devices fabricated on the same nanotube.

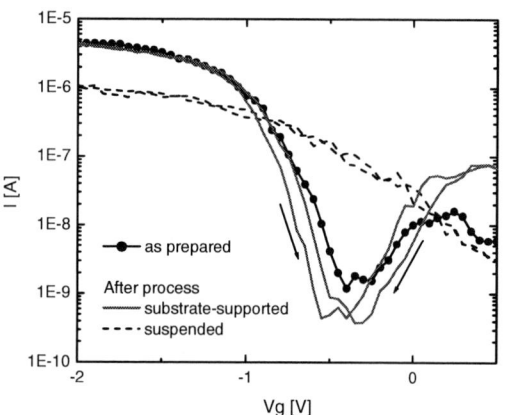

Fig. 2: Subthreshold characteristics of semiconducting CNT devices before (close circle) and after (dash and solid lines) the etching process.

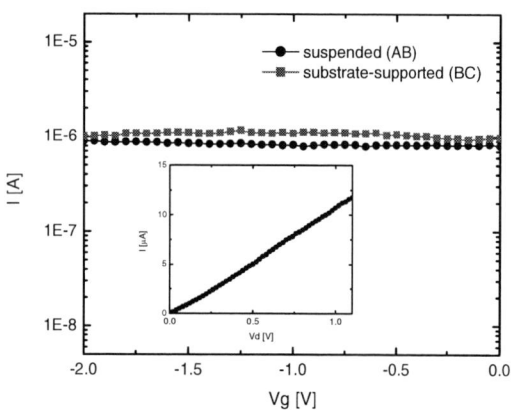

Fig. 3: Electrical characteristics of metallic CNT devices after the etching. The inset shows the linear I-V behavior of the suspended nanotube.

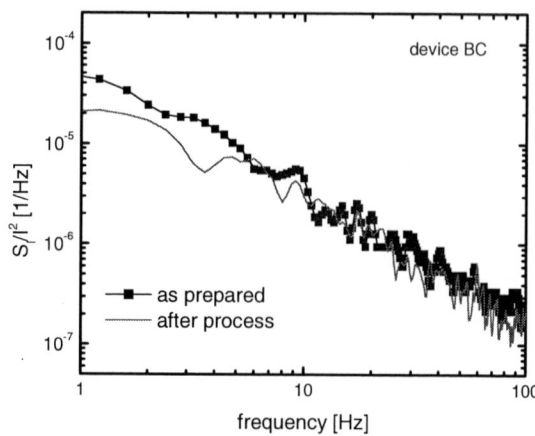

Fig. 4: Measured noise power spectra, normalized by the current square, of the metallic CNT device before and after the etching. In this device, the nanotube remains on the substrate after the etching.

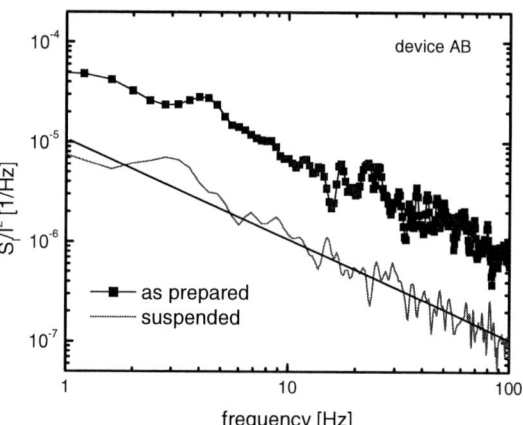

Fig. 5: Comparison of the noise power spectra of the nanotube device before and after the nanotube becomes suspended.

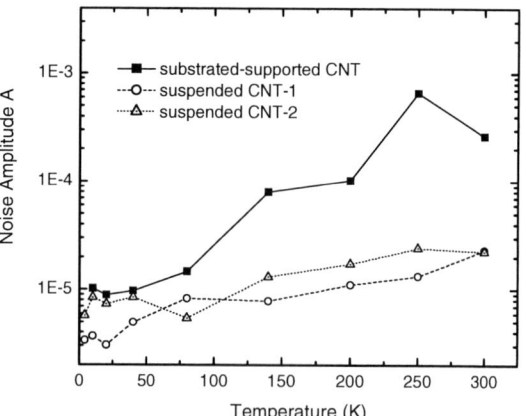

Fig. 6: Measured temperature dependence of the 1/f noise amplitude A of a substrate-supported nanotube device and two suspended nanotube devices.

140

In-Situ Inelastic Electron Tunneling Spectroscopy of Bistable Molecular Junction Devices

Heayoung Yoon[1], Lintao Cai[1], Masato Maitani[2], David L. Allara[3], and Theresa S. Mayer[1]

[1]*Departments of Electrical Engineering,* [2]*Materials Science and Engineering, and* [3]*Chemistry*
The Pennsylvania State University, University Park, PA 16802, USA

Inelastic electron tunneling spectroscopy (IETS) is a valuable *in-situ* spectroscopic analysis technique that can be applied in conjunction with electrical transport measurements to characterize and identify the molecular species under study by measuring its vibrational fingerprint. IETS was used previously to confirm that the measured properties of nanoscale molecular junctions are due to the intended molecule rather than process induced artifacts.[1] Here we use IETS to probe the vibrational modes excited in oligoaniline molecular junctions after switching between high and low conductance states by applying a critical threshold voltage. These spectroscopic measurements allow us to distinguish between the two molecular conformations that give rise to the observed conductance change and clarify the mechanism responsible for switching.

Recently several studies reported that self assembled monolayers (SAMs) and individual oligoaniline molecules exhibit reversible bistable switching behavior.[2,3] Temperature dependent current-voltage (I-V) and IET measurements were conducted on thiol-substituted oligoaniline dimers (OMAn) assembled in a nanoscale crossed-wire molecular junction devices. In these devices, the bottom contact was formed by self assembling the OMAn molecules directly onto 70 nm wide thermally evaporated Au electrodes, while the top contact was formed by physically contacting the SAM with a single 200 nm diameter Au nanowire. This high yield fabrication process produces junctions containing ~1000 OMAn molecules that are linked by Au-S chemical bonds to the top and bottom Au electrodes. Immediately after fabrication, the molecular devices were loaded into a variable temperature high vacuum chamber for I-V(T) and IET measurements.

The junctions showed reproducible bistable switching between low and high conductance states with the threshold voltages of ± 1.0~1.5 V at room temperature and ± 1.5~2.0 V at 4.2 K. The on-state current at 1 V is ~5 μA with an on-off ratio of ~ 5 for these Au-OMAn-Au; higher on-state currents and on-off ratios can be obtained by replacing Au with Pd or Pt metal contacts. Once switched to the high state, the junction remains in this state for > 24 hrs at room temperature. The extremely long high state retention times could be explained by a molecular transition between two energetically stable conformational states.

IET spectra were obtained at 4.2 K using a lock-in amplifier to collect directly the second harmonic signal, which is proportional to d^2I/dV^2. Prominent vibrations were observed at 88 mV (C-H op), 142 mV (C-H ip), and 204 mV (C=C str) in low state and 150 mV (C-H ip), 170 mV (C-N str) and 198 mV (C=C str) in high state. The intensity and peak position of our experimental IET spectra show very good agreement with IET spectra determined by density functional theory (DFT) calculations using neutral and dication molecular confirmations in the low and high conductance states. The predominant changes in the peak intensity associated with the transition from low and high states are observed around 170 mV C-N stretch modes, which are related to electron delocalization around –N atom. These results show that IETS can be used to study the mechanisms responsible for changes in electrical transport properties in molecular junctions.

The authors would like to acknowledge the Tour group (Rice University) for providing the molecules used in this study and A. Troisi (University of Warwick) and M. Ratner (Northwestern University) for theoretical calculations.

[1] A. Troisi and M. Ratner, *Small* 2 **2**, 172 (2006)
[2] L. Cai, T. Mayer et al., *Nano Lett.*, **12**, 2365 (2005)
[3] F. Chen, C. Nuckolls, and S. Lindsay, *Chem. Phys.* **324**, 236 (2006)

978-1-4244-1101-6/07/$25.00 ©2007 IEEE

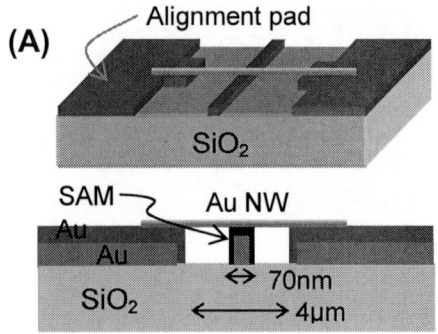

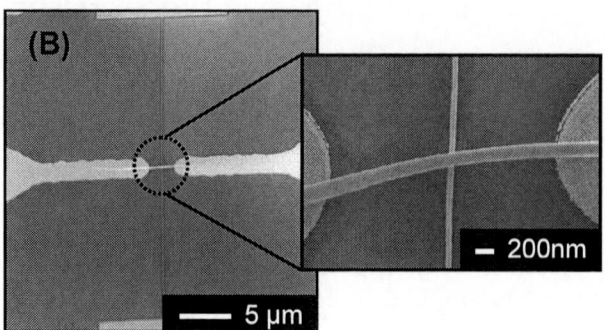

Figure 1: Crossed-wire molecular junction device **(A)** Schematics. Au electroplating performed to cover the SAM on the Au alignment pads before NW alignment. **(B)** Field Emission Scanning Electron Microscope (FESEM) mages showing the device structure comprised of lithographically defined 70-nm wide bottom electrode and 200-nm diameter NW top electrode. Inset also shows smooth Au NW surface.

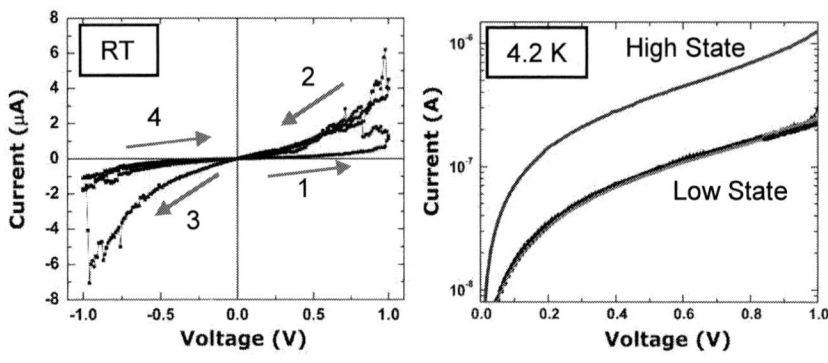

Figure 2: I-V characteristics at RT and 4.2 K **(Left)** Low current state(1) switched to high current state(2) at V_{th} (1-1.5V @RT and 1.5-2V @4.2 K). High state can be switched again to low state with negative V_{th}(3,4) **(Right)** I_{low} / I_{high} ratio was 8 in the device. Low current state remained until applying V_{th}.

Figure 3: Configuration of thiol-substituted oligo(N-methylated-anilines) in high and low state

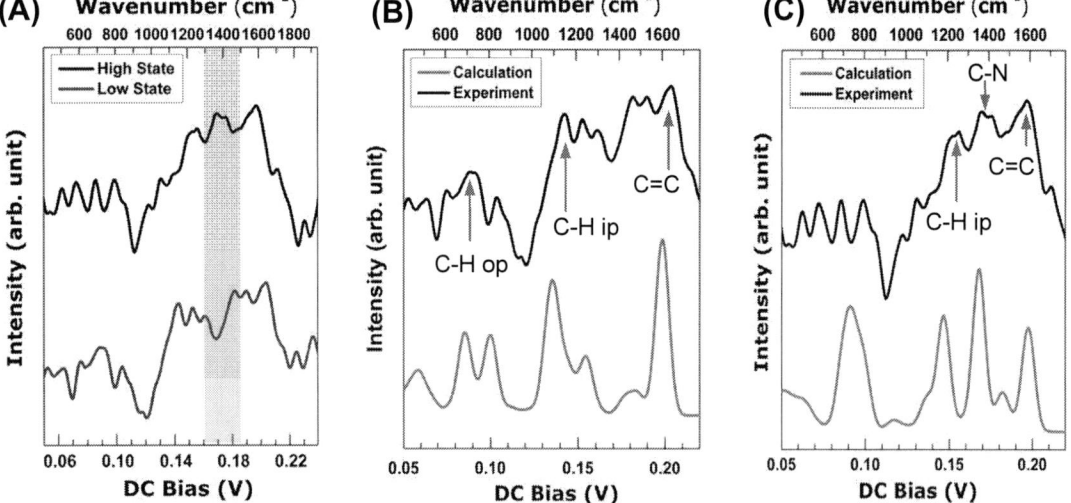

Figure 3: Inelastic Electron Tunneling spectra measured at 4.2 K with 500 Hz, 8 mV excitation voltage **(A)** Peak intensity change observed 170 mV(C-N) in high state, which suggested electron delocalization around –N atom. Observed IETS peaks in high and low state suggest switching behavior attributed to inherent molecular features. Theory-experiment comparisons in high **(B)** and low**(C)** state showed a good agreement.

142

Electric-Field Dependence of Junction Temperature in GaN HEMTs

Vivek Mehrotra, Karim Boutros, and Berinder Brar

Teledyne Scientific Company, 1049 Camino Dos Rios, Thousand Oaks, CA 91360

vmehrotra@teledyne.com

Accurate junction temperatures and thermal models for GaN HEMTs are critical in predicting their reliability in high-power applications. Raman spectroscopy has been used to measure junction temperature [1,2], providing a spatial resolution of 0.5-1 μm, but this technique requires direct access to the semiconductor. This restriction prevents temperature measurement in the vicinity of T-gates, field plates, or air-bridges where the edge of the gate is masked by the overlying metal. Raman spectroscopy on cross-sectioned devices has been attempted, but suffers from poor yield of working devices. More significantly, previous reports relating junction temperature to dissipated power ignore an important effect: That the junction temperature depends not only on the total power dissipated in the device, but also on the bias-dependent volumetric region within the device where the power is dissipated. The device may be biased in the linear regime, creating a uniform power dissipation condition between source and drain, or in the saturation regime, creating a spatially non-uniform power dissipation condition with most of the power being dissipated at the drain-end of the gate. These two cases could be selected to dissipate the same total power, yet are found to produce significantly different junction temperatures. Here we report junction temperature and thermal spreading in GaN HEMTs using a transient measurement technique that utilizes the thermionic Schottky gate characteristics to obtain temperatures.

Our technique first calibrates the gate-to-source forward *I-V* characteristics of the device under test as a function of temperature by mounting on a chuck at a known temperature and under vacuum to eliminate heat loss effects. Fig. 1 shows the forward voltage characteristics of the Schottky gate diode used to provide a direct measurement of temperature in the gate region of the device. The diode stability and robustness was ensured via a 100 hour soak test at ~200°C, and via 20 power cycles. After calibration, the device is pulsed to a pre-determined bias point for 7 seconds and then turned off. After ~100 ms, 1 μA of forward current is forced into the gate and the gate voltage is measured. The shift in gate voltage from the calibration curve is used to determine the junction temperature. Clearly the device has cooled slightly, and this technique does not provide the peak junction temperature. Nonetheless, this simple measurement allows us to measure a number of important thermal properties of the device and develop an accurate thermal model.

Both transient and steady state temperature measurements have been made. The electric field distribution in the channel has a significant effect on the measured temperature rise at fixed dissipated power or energy as shown in Fig. 2. Power dissipation levels of 0.55 W and 0.25 W were used, derived from the recorded drain current and drain-source voltage traces during temperature measurements. At each power dissipation level, we have measured temperature under two different bias conditions: one in the linear regime (circles) with a nominally uniform electric field and one in the saturation regime (triangles), where electric field is highest at the drain–end of the gate. The key result of this paper is that for the *same* dissipated power, significantly different junction temperatures are measured. In the saturation regime, the junction temperature is higher, simply because the power dissipation is localized at the drain-end of the gate. At 0.55 W of instantaneous power dissipation, corresponding to a dissipated energy of 3.865 Joules in the channel, we measured ~45% higher temperature rise. Purely field-dependent memory effects on the gate current, like charging, are confirmed to be negligible by measuring the gate characteristics after pulsing to a high-voltage low-current bias condition, resulting in minimal temperature rise due to little power dissipation.

An extension of this technique allows the temperature of nearby devices (sensors) to be continuously monitored, providing a direct measurement of the chip-scale thermal properties of the material surrounding the device. The surrounding material can vary significantly due to layer design, metal patterns, or process flow, namely ion-implantation versus mesa etching for isolation. Fig. 3 illustrates the typical temperature rise recorded at a sensor device for two different values of the drain-to-source voltage and at ~100 mA of drain current at the heater device. The pulse width of the heating pulse was varied from 1-5 seconds. Thermal spreading measurements indicate a thermal resistance of ~25°C/W per mm between the two sensors located at ~450 μm and ~225 μm from the heater device. Fig. 4 shows the thermal spreading characteristics via temperature measurements on two sensor devices.

This work was supported by ONR under contract N00014-05-C-0120. We thank P. Maki and H. Dietrich for their support.

1. N.Q. Zhang, V. Mehrotra, S. Chandrasekaran, B. Moran, L. Shen, U. Mishra, E. Etzkom, and D. Clarke, "Large Area GaN HEMT Power Devices for Power Electronic Applications: Switching and Temperature Characteristics", *Power Electronics Specialist Conference (PESC'03)*, Vol. 1, June 2003 p. 233-237; J. He, V. Mehrotra and M. C. Shaw, "Ultra-High Resolution Temperature Measurement and Thermal Management of RF Power Devices using Heat Pipes", 11th *International Symposium on Power Semiconductor Devices and ICs Proceedings* (ISPSD) 1999 p. 145.
2. M. Kuball et al, "Measurement of Temperature in Active High-Power AlGaN/GaN HFETs Using Raman Spectroscopy", IEEE Electron Device Letters, Vol. 23, No. 1, Jan. 2002 p. 7-9.

978-1-4244-1101-6/07/$25.00 ©2007 IEEE

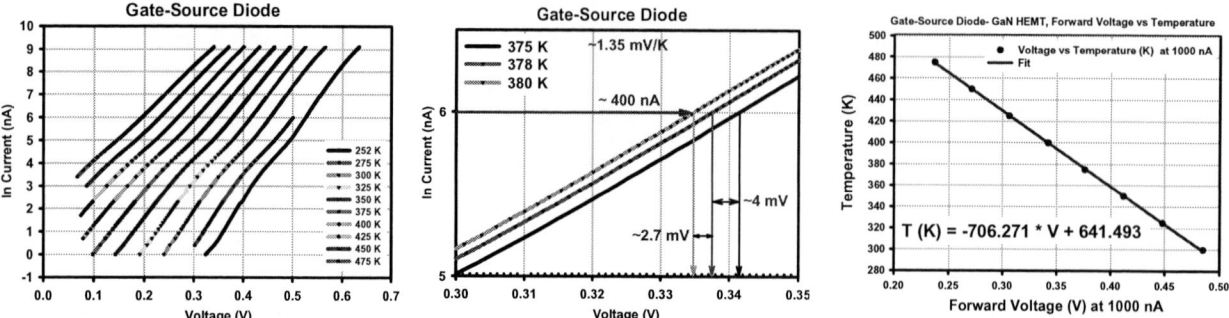

Fig. 1. Gate-source Schottky diode calibration characteristics of a typical GaN HEMT device as a function of temperature (left) provides a temperature resolution of <1 °C (middle). A gate-source probe current of 1 μA is used in subsequent measurements, which shows a linear dependence of forward diode voltage on temperature (right).

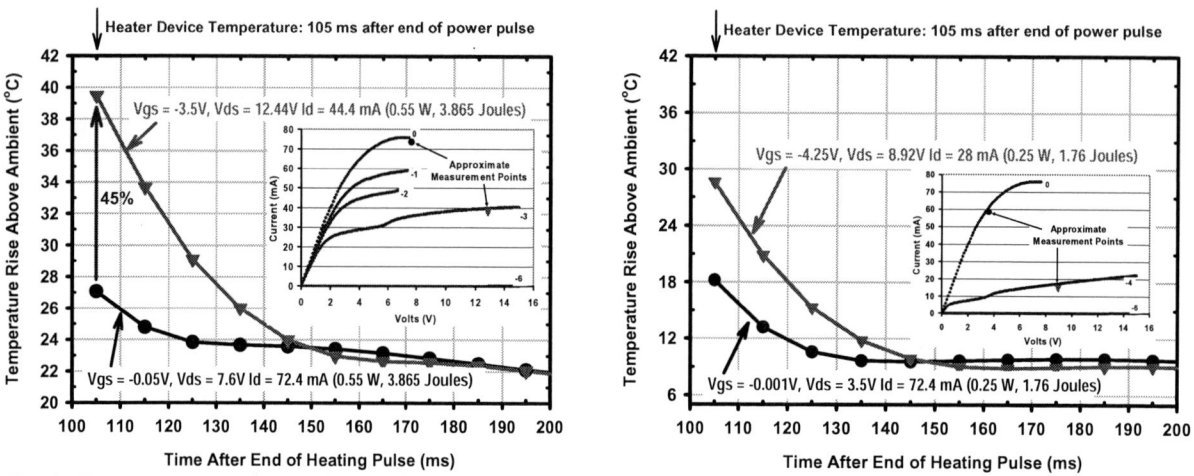

Fig. 2. Temperature-rise above ambient at the gate of the heater device under two different electric field concentrations, as noted in the annotations. The two electric field conditions were chosen to enforce similar power (~0.55 Watts) and energy (~3.865 Joules) dissipation on exactly the same heater device (left). A similar effect was observed under a lower total power dissipation condition (right).

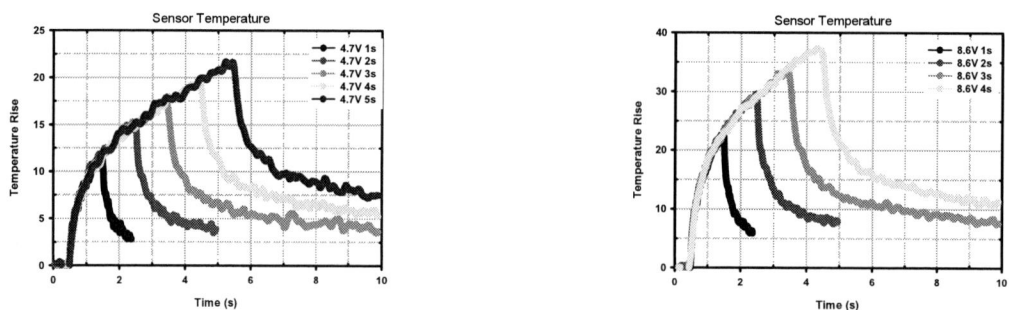

Fig. 3. Temperature-rise above ambient at the gate of a sensor HEMT located ~425 microns from the heater HEMT. The drain-to-source pulse width at the heater device is varied from 1-5 seconds for two different values of the drain-to-source voltage (4.7V and 8.6V), while the heater drain current is held constant at ~100 mA.

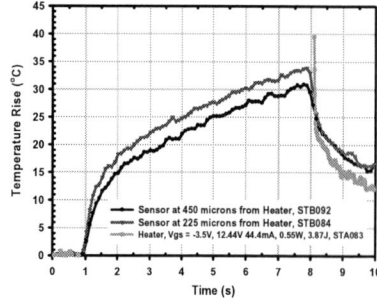

Fig. 4. Temperature-rise above ambient at the gate of the heater device and two sensor devices located at ~450 μm and ~225 μm away, indicating a thermal spreading resistance of ~25 °C/W-mm between the sensor devices.

Session IV.A (DeBartolo Hall Room 102)

High Speed and Terahertz Devices

Tuesday AM, June 19th, 2006

Session Organizer: David Chow, HRL Laboratories and Patrick Fay, University of Notre Dame
Session Chair: Jeff LaRoche, Raytheon RF Components

8:20 AM IV.A-1 Invited Paper
Advanced InP and GaAs HEMT MMIC technologies for MMW commercial products
M. Barsky, M. Biedenbender, X. Mei, P.-H. Liu, and R. Lai, Northrop Grumman Space Technology, Redondo Beach, California, USA

9:00 AM IV.A-2 Student Paper
Ultra-Low Resistance Ohmic Contacts to InGaAs/InP
U. Singisetti[1], A. M. Crook[1], E. Lind[1], J. D. Zimmerman[1], M. A. Wistey[1], A. C. Gossard[1], M. J. W. Rodwell[1], and S. R. Bank[2], [1]ECE and Materials Departments, University of California, Santa Barbara, California, USA and [2]ECE Department, University of Texas at Austin, Austin, Texas, USA

9:20 AM IV.A-3 Student Paper
Sb-heterostructure Millimeter-Wave Detectors with Improved Noise Performance
N. Su[1], Z. Zhang[1], H. P. Moyer[2], R. D. Rajavel[2] J. N. Schulman[2] and P. Fay[1], [1]Department of Electrical Engineering, University of Notre Dame, Notre Dame, Indiana, USA and [2]HRL Laboratories LLC, Malibu, California, USA

9:40 AM IV.A-4 Student Paper
Delta-Doped Si/SiGe Zero-Bias Backward Diodes for Micro-Wave Detection
S.-Y. Park[1], R. Yu[2], S.-Y. Chung[1], P. R. Berger[1,2], P. E. Thompson[3], and P. Fay[4] [1]The Ohio State University, Department of Electrical and Computer Engineering, Columbus, Ohio, USA , [2]The Ohio State University, Department of Physics, Columbus, Ohio, USA, [3]Naval Research Laboratory, Washington, District of Columbia, USA, and [4]The University of Notre Dame, Department of Electrical Engineering, Notre Dame, Indiana, USA

10:00 AM Break

10:20 AM IV.A-5 Invited Paper
Room-Temperature Terahertz Oscillators Using Resonant Tunneling Diodes
M. Asada, Tokyo Institute of Technology, CREST-Japan Science and Technology Agency, Tokyo, JAPAN

11:00 AM IV.A-6
Novel Plasmon-Resonant Terahertz-Wave EmitterUsing a Double-Decked HEMT Structure
T. Suemitsu[1], Y. M. Meziani[1], Y. Hosono[1], M. Hanabe[1], T. Otsuj[1], and E. Sano[2], [1]Research Institute of Electrical Communication, Tohoku University, Aoba, Sendai, JAPAN and [2]Research Center for Integrated Quantum Electronics, Hokkaido University, Sapporo, JAPAN

11:20 AM IV.A-7
THz front-side illuminated quantum well photodetector
M. Patrashin and I. Hosako, National Institute of Information and Communications Technology, Tokyo, JAPAN

11:40 AM IV.A-8

146

Advanced InP and GaAs HEMT MMIC technologies for MMW commercial products

Mike Barsky, Mike Biedenbender, Xiaobing Mei, Po-Hsin Liu, Richard Lai
Northrop Grumman Space Technology, One Space Park, Redondo Beach, CA, 90277, USA

Abstract

NGST is developing advanced high frequency HEMT device and MMIC technologies to address imminent applications at MMW frequencies above 80 GHz through 300 GHz. The improved device transport characteristics, high transconductance, and gain at very high frequencies will benefit next generation communications, radar, imaging and radiometer systems. In this paper, we status the development and production of 0.1 um GaAs HEMT, 0.1 um InP HEMT and sub 0.1 um InP HEMT technologies for high frequency mmW circuits.

Advanced communication, radar, passive imaging and remote sensing instruments are extending frequency and bandwidth requirements to 80 GHz and beyond. These systems will require large arrays and power sources for resolution, range and beam steering. It will be critical therefore not only to achieve high frequency performance, but also high yield, high volume production capability. This paper covers NGST's advanced GaAs and InP HEMT technologies that are producing or developing commercial product to cover the burgeoning high frequency applications.

NGST has developed production GaAs and InP HEMT MMIC processes both on 75 mm and 100 mm substrates and is further pushing to shorter gate lengths and enhanced epitaxy profiles. The epitaxial growth of 100 mm GaAs and InP HEMT wafers are produced on multi-wafer MBE systems. Gate lengths of 0.1 and 0.15 um are currently in production for the InP and GaAs HEMT wafers, while 0.07 um and smaller gate lengths are in development. GaAs wafers are thinned to either 100 or 50 um thickness, while InP wafers are thinned to either 75 or 50 um thickness. All wafers employ a via process through the substrate. A summary of process details and device performance capabilities are given in Figure 2.

The GaAs and InP HEMT MMIC processes have been used to produce a family of high frequency LNAs, power amplifiers and other functions with performance to 300 GHz and potentially beyond to higher frequencies. Several products, especially at W-band have been transitioned to production to achieve higher quantities and lower cost, while newer and higher frequency products are being improved for greater robustness and higher yield.

For greater than a decade, NGST has been developing high performance 0.1 um GaAs HEMT W-band power and low noise amplifiers. More than a thousand 40 dB gain LNAs were used to demonstrate the first passive MMW camera at W-band for all-weather landing. Similar products are currently produced for W-band imaging commercial and military applications. 250 mW and higher W-band power amplifier MMICs have also been fielded. NGST has

recently demonstrated a waveguide combined SSA module with 4.8W output power at 94 GHz.

0.1 um InP HEMT LNAs have been demonstrated and fielded with 3 dB NF from 80-100 GHz for the past 10 years[1]. 112-118 GHz 0.1 um InP HEMT LNA MMIC front-end receiver modules have been flown in space. 400 mW output power, 19% PAE 0.15 um InP HEMT MMIC power amplifiers have also been demonstrated. The operating frequency was increased to 250 GHz through shorter 70 nm gate InP HEMTs[2]. State-of-art 15 dB gain, 5 dB noise figure 160-190 GHz LNA MMIC. Two cascaded MMICs in a G-band LNA module achieved 30 dB gain and 6 dB NF referenced to the waveguide flanges from 160-190 GHz. Nine such LNA modules were built and demonstrated ~ 6 dB NF with only 0.5 dB variation. A power amplifier with 2 cascaded MMICs with 8 total stages, 35 dB small signal gain and 25 mW peak output power referenced to the waveguide flange at 184 GHz and greater than 20 mW output power from 176-190 GHz has been demonstrated.

The reproducibility and production readiness of the 0.1 and 0.07 um InP HEMT process has resulted in state-of-art high frequency amplifier developments through collaborations with external organization. CSIRO utilized NGST's 0.1 um InP HEMT process to achieve state-of-art low noise amplifier designs at 90 and 190 GHz[3]. Jet Propulsion Laboratory utilized NGST's 0.07 um InP HEMT process to achieve a low noise amplifier with 9 dB gain at 240 GHz[4].

[1] R. Lai et al "Production InP MMICs for Low Cost, High Performance Applications" 2005 International InP and Related Materials Conf. Digest

[2] R. Lai et al "InP HEMT Amplifier Development for G-band(140-220GHz)Applications", 2000 International Electron Devices Meeting Digest, pp.175-177

[3] J. Archer et al, "An InP MMIC Amplifier for 180-205 GHz", IEEE Microwave and Wireless Components Lett., vol. 11, pp. 4-6, January 2001.

[4] D. Dawson et al "Beyond G-Band: A 235 GHz InP MMIC Amplifier," IEEE Microwave and Wireless Components Letters, vol. 15, pp 874 – 876, 2005.

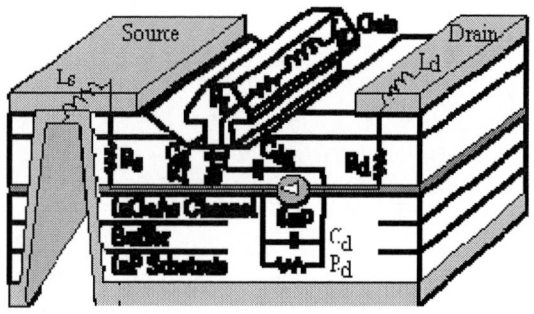

Figure 1. Cross section of HEMT with device equivalent circuit parameter elements

Parameters/ Technology	Units	0.1 GaAs HEMT	0.1 um InP HEMT	0.15 um InP HEMT	0.07 um InP HEMT
Ohmic contact	ohm-mm	0.1	0.1	0.15	0.08
Gm	mS/mm	600	1000	800	1400
fT	GHz	120	190	140	250
fmax (est.)	GHz	250	400	300	500
Max Vds	V	4	1.5	3	1.5
BVgd	V	8	3	5	3
Resistors	ohm/sq.	100	100	100	100
Capacitors	pF/mm^2	300	300	300	300
Interconnect		2 layer with airbridge	2 layer with airbridge	2 layer with airbridge	2 layer with airbridge
Backside		100 and 50 um with vias	75 and 50 um with vias	75 and 50 um with vias	75 and 50 um with vias
Highest frequency		118 GHz MMIC PA	200 GHz LNA	100 GHz MMIC PA	250 GHz LNA
EBL capacity		150 wafers/week per machine	150 wafers/week per machine	150 wafers/week per machine	100 wafers/week per machine
Production Status		production	production	production/ development	development

Figure 2. Table of high frequency HEMT technologies and relevant parameters

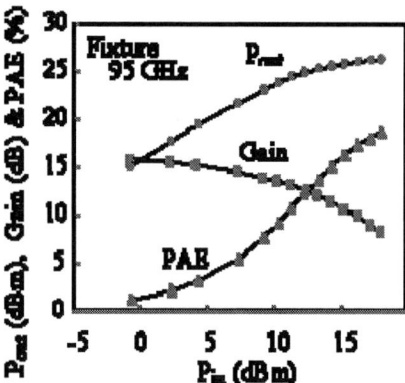

Figure 3. Passive MMW image through fog for a 1040 pixel camera based on front-end 0.1 um 40 dB GaAs HEMT LNA receivers

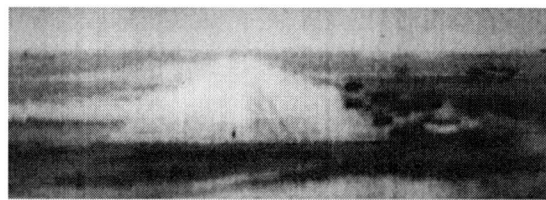

Figure 4. Output power, Gain and PAE for a 2-stage 0.15 um InP HEMT power MMIC at 95 GHz. This amplifier was biased a 3V drain bias.

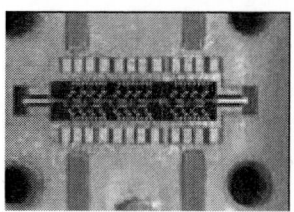

Figure 5. Millimeter wave 160-190 GHz cavity.

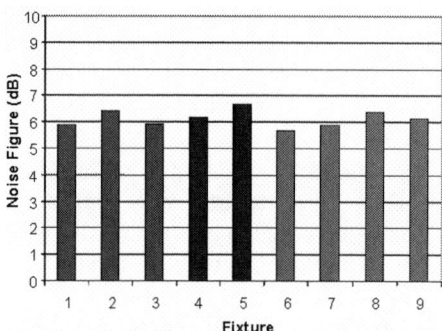

Figure 6. Noise figure measurement at 180 GHz for 9 fixtured 30 dB gain modules

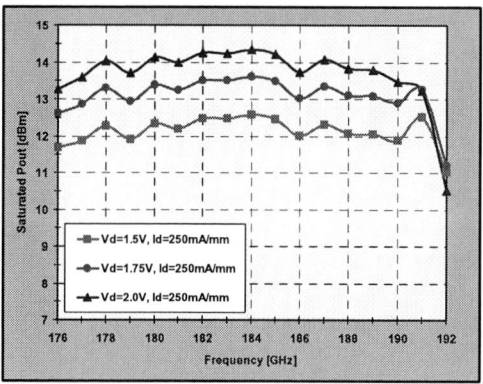

Figure 7. 176-190 GHz SSPA module based on 2 cascaded 70 nm InP HEMT MMICs with 8 total stages with greater than 20 mW output power

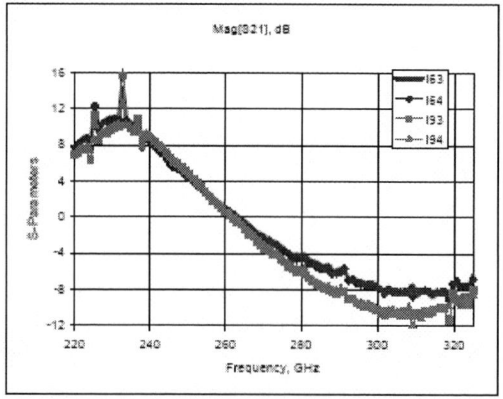

Figure 8. 220-240 GHz 3-stage 70 nm InP HEMT MMIC LNA designed by Jet Propulsion Laboratory

148

Ultra-Low Resistance Ohmic Contacts to InGaAs/InP

U. Singisetti, A. M. Crook, E. Lind, J. D. Zimmerman, M. A. Wistey, A. C. Gossard, and M. J. W. Rodwell

ECE and Materials Departments, University of California, Santa Barbara, CA 93106

S. R. Bank, *ECE Department, University of Texas, Austin, TX, 78712*

Phone: (805) 893-3273, Fax: (805) 893-3262, Email:uttam@ece.ucsb.edu

Very low resistance metal-semiconductor contacts are fundamental to the continued scaling of transistors towards THz bandwidths. The specific emitter contact restivity of heterojunction bipolar transistors (HBTs) and the specific source contact resistivity of III-V field-effect transistors must both decrease in proportion to the inverse square of transistor bandwidth. For both III-V HBTs and FETs, $\sim 1 \cdot 10^{-8}$ Ω -cm^2 contact resistivity is required for transistors having simultaneous 1.5 THz f_t and f_{max} [1]. Emitter access resistance also plays a particularly strong role in determining gate delay in bipolar digital ICs [1]. We report extremely low resistance, non-alloyed ohmic contacts to In$_{0.53}$Ga$_{0.47}$As with specific contact resistivity (ρ_c) $< 1 \cdot 10^{-8}$ Ω -cm^2. We show that these contacts can be formed either *in situ* by metal deposition in the MBE chamber or *ex situ* with better surface cleaning techniques.

We studied two types of *in situ* contacts, contacts in which the metal is deposited in the MBE system without breaking vacuum. This ensures that no oxide is formed at the metal-semiconductor interface. The first approach was to form epitaxial ErAs/ In$_{0.53}$Ga$_{0.47}$As semimetal-semiconductor contacts. ErAs can be grown epitaxially on InGaAs. The ErAs/InGaAs interface shows a continuous As sublattice without any broken bonds [2]. The Fermi level of ErAs is estimated to be ~ 100 meV above the conduction band of InAs [3]. In the second approach, *in situ* electron beam deposited Molybdenum (Mo) contacts to InAs/ In$_{0.53}$Ga$_{0.47}$As were studied. The Mo Fermi level will pin above the conduction band of InAs [4]. Additionally, we studied *ex situ* TiW/In$_{0.53}$Ga$_{0.47}$As contacts. In these, the wafer is grown by MBE and then removed from vacuum. The InGaAs surface was then oxidized by a UV ozone treatment, and the oxide was etched off in concentrated NH$_4$OH before depositing TiW contact metal on the wafer.

All the samples were grown by solid-source MBE on (100) semi-insulating InP substrate. The layer structures are shown in Fig. 1. For the *in situ* contact samples, 40 nm of Mo was deposited in an electron beam deposition system connected to MBE under ultra high vacuum. The Mo was deposited as a cap layer onto the ErAs to prevent its oxidation. The channel doping and thickness were chosen such that 1-D current-flow condition is satisfied in the transmission line measurement (TLM). The active carrier concentrations were confirmed by Hall measurements. The samples were then processed into TLM structures for contact resistance measurement using standard photolithography. The Mo and TiW layers were dry etched in SF$_6$/Ar with Ni as etch mask and the structures were isolated by wet etching the semiconductor (Fig. 2). The TLM pad spacing ranged from 0.6 μm to 25 μm. The pad spacings were verified by scanning electron microscopy imaging. The TLM geometry was designed such that at the smallest spacing the contact resistance is at least 50 % of the total measured resistance. All the contacts show linear Ohmic I-V behavior as formed. Fig. 3 plots the measured resistance vs. the pad spacing in the TLM structures. A four point probe method was employed in the measurement to minimize parasitic resistances. From the slope and intercept of the line, the sheet and contact resistances were calculated. The specific contact resistivities for the *in situ* ErAs/InAs, *in situ* Mo/InAs and *ex situ* TiW/ In$_{0.53}$Ga$_{0.47}$As were $1.5 \cdot 10^{-8}$ Ω -cm^2, $5 \cdot 10^{-9}$ Ω -cm^2 and $7 \cdot 10^{-9}$ Ω -cm^2 respectively. All the samples showed ~ 15-18 Ω sheet resistance. The transfer lengths for *in situ* ErAs/InAs, *in situ* Mo/InAs, and *ex situ* InGaAs contacts were 300 nm, 175 nm, and 190 nm, respectively.

Thermal stability studies on the contacts were carried out by annealing the contacts under N$_2$ flow for 1 minute at various temperatures. Fig. 4 plots the specific contact resistivity as a function of temperature. For *in situ* Mo/InAs, and *ex situ* TiW/InGaAs contacts the specific contact resistivity remains $<1 \cdot 10^{-8}$ Ω -cm^2 even after annealing at 500 C. The ErAs contact resistance increases with annealing temperature. This could be due to reaction of ErAs with Mo, or lateral oxidation of ErAs. SIMS depth analysis on the contacts is shown in Fig 5. The refractory metals TiW and Mo are very effective as diffusion barrier to Au and Ti.

We show for the first time that extremely low *in situ* metal contacts can be formed to (In,Ga)As on InP. We also show that it is possible to form ultra low resistance *ex situ* contacts by improved surface treatment. However, unlike *in situ* contacts, *ex situ* contacts are very sensitive to surface preparation. Similar contact resistivities to InAs on GaAs were reported by Nittono *et al.* [5]. To our knowledge this is the first time such low metal-semiconductor contacts have been demonstrated in In$_{0.53}$Ga$_{0.47}$As system that does not deteriorate at least upto 500 C. These thermally stable, extremely low resistance, Ohmic contacts are an enabling technology for THz bandwidth InGaAs/InP HBTs, mm wave InGaAs HEMT technologies, and the evolving III-V MOSFET technologies.

This work was supported by ONR under the ULTRA LOW RESISTANCE program and Swedish Research Council

[1] M. J. W. Rodwell *et al.*, *Int. Journal of High Speed Electronics and Systems.* **11**,159 (2001).

[2] D. O. Klenov *et al.*, *Appl. Phys. Lett.* **86**, 241901 (2005).

[3] S.R.Bank *et al.*, *Proceedings of 2006 North American Molecular Beam Epitaxy Conference*, Durham, NC, September 2006

[4] S.Bhargava *et al*, *Applied Physics Letters*, **70**, 759 (1997).

[5] T. Nittono et al., *Jap. Journal of Applied Physics.* **27**, 1718 (1988).

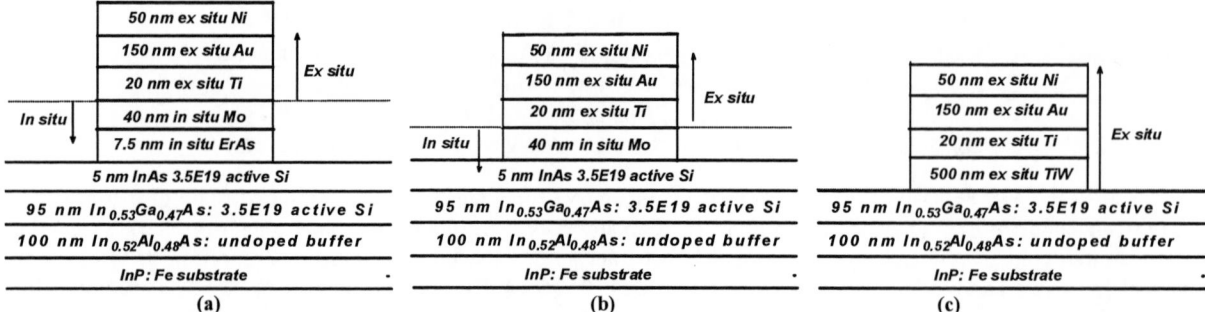

Fig. 1 : Cross-section schematic of the metal-semiconductor contact layer structure grown by solid source MBE, (a) *in situ* ErAs/InAs contact, (b) *in situ* Mo/InAs contact and (c) *ex situ* TiW/InGaAs contact. The Mo was deposited in an electron beam deposition system connected to MBE under ultra high vacuum. TiW is sputtered *ex situ* for the TiW/InGaAs contact.

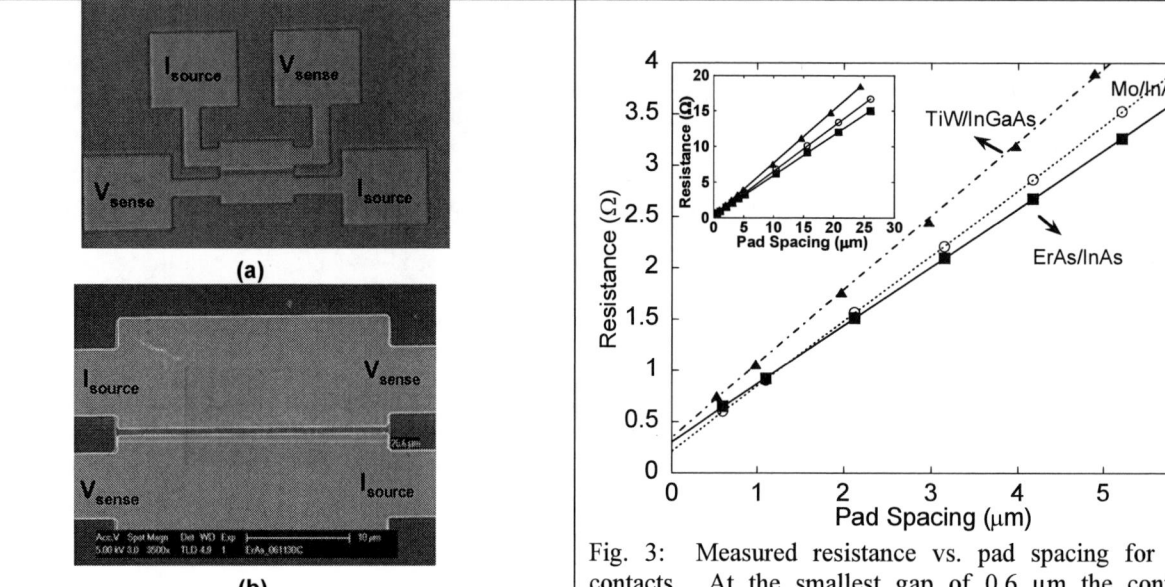

Fig. 2: (a) Optical micrograph and (b) SEM image of the 600 nm gap TLM. A thick interconnect metal (Ti/Au/Ni) was used to minimize metal resistance.

Fig. 3: Measured resistance vs. pad spacing for the contacts. At the smallest gap of 0.6 μm the contact resistance term is at least 50 % of the total measured resistance. The inset plots measured resistance vs. pad spacing ranging from 0.6 μm to 26 μm.

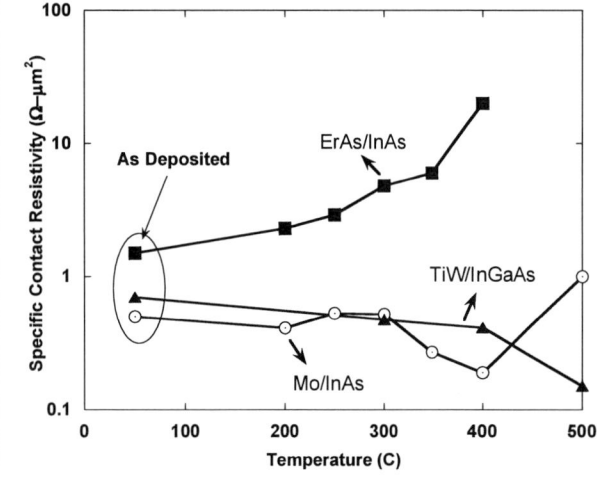

Fig 3: Specific contact resistivity as a function of annealing temperature.

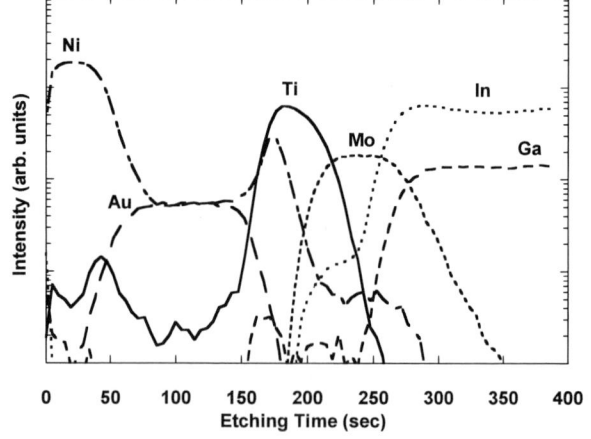

Fig 4: SIMS depth profile of Mo/InAs contact annealed at 400 C.

150

Sb-heterostructure Millimeter-Wave Detectors with Improved Noise Performance

N. Su, Z. Zhang, H. P. Moyer[*], R. D. Rajavel[*], J. N. Schulman[*] and P. Fay

Department of Electrical Engineering, University of Notre Dame,
Notre Dame, IN 46556 Tel: (574)-631-5693, Email: pfay@nd.edu
[*]*HRL Laboratories LLC, 3011 Malibu Canyon Rd, Malibu, CA 90265*

Zero-biased Sb-heterostructure backward diodes are of great interest as a technology for millimeter-wave detection and imaging. Devices have been demonstrated with high sensitivity at W-band and above [1-3], and the devices are inherently insensitive to temperature [4]. For passive imaging applications, detector noise is a critical issue. While reducing the junction resistance, R_j, through the use of thin (<15 Å) barriers leads to a reduced thermal noise floor [5-6], excess low-frequency noise can limit detector performance. Detailed characterization of the low-frequency noise performance of mm-wave Sb-heterostructure detectors and the implications for devices and systems are reported.

The heterostructure and computed energy band diagram of the detectors are shown in Fig. 1. A 10 Å thick AlSb tunneling barrier was used to achieve low R_j, and a 150 Å undoped $Al_{0.1}Ga_{0.9}Sb$ layer was used to optimize the device nonlinearity [4].

Figure 2 shows the DC current-voltage characteristics of a typical 1×1 μm^2 area device with a zero-bias curvature, $\gamma_{IV} = \left(\frac{\partial^2 I}{\partial V^2}\right) / \left(\frac{\partial I}{\partial V}\right)$, of 33 V^{-1} and R_j of 1340 Ω. A junction capacitance C_j of 7.8 fF and series resistance R_s of 32.5 Ω were extracted from on-wafer s-parameters, and a cut-off frequency $f_c=1/(2\pi R_s C_j)$ of 626 GHz at zero bias is obtained. An average RF voltage sensitivity of 3220 V/W from 1 to 50 GHz was measured (Fig. 3). The temperature dependence of the zero-bias γ_{IV} and R_j was characterized (Fig. 4). A modest increase in γ_{IV} and decrease in R_j is observed as temperature is lowered, consistent with shifts in carrier occupation statistics and material bandgaps with temperature [7].

The low-frequency noise as a function of temperature and bias was characterized. While the detectors exhibit only thermal noise at zero bias, the application of either external dc bias or the generation of a self-bias from incident RF power induces low-frequency noise. The low-frequency noise spectrum was measured as a function of applied bias (Fig. 5(a)); the noise equivalent power, NEP, for a matched detector was projected from the spectra for a video bandwidth, Δf_{if}, of 30 Hz (Fig. 5(b)). At zero bias, thermal noise from R_j results in a room-temperature NEP of 0.37 pW. For biases above ~0.5 mV, 1/f noise increases the NEP. Current noise spectra, S_I, for a 2×2 μm^2 device were measured vs. temperature and fit to $S_I = K_f I^m / f^{\gamma_{1/f}}$ (Fig. 6). Based on these spectra, the noise equivalent temperature difference (NEΔT) was projected (Fig. 7) for detectors at zero and 1 mV bias for an RF bandwidth of 30 GHz (30% fractional bandwidth at W-band [5]) and a lossless matching network. A room-temperature NEΔT of 0.88 K is obtained. The NEΔT for detectors biased at 1 mV shows a stronger temperature dependence than the zero-bias case because the 1/f noise power decreases by two orders of magnitude from 300 K to 10 K (Fig. 6(b)), while thermal noise for zero-bias decreases by 30x over this same temperature range. The temperature-dependent noise also shows generation-recombination noise signatures at higher biases (Fig. 8), and several traps with activation energies of 100 meV and below have been observed. Thus the noise analysis provides both engineering insight for detector optimization as well as physical insight into the deep levels and defects within the device.

[1] J. N. Schulman et al., *IEEE Microwave and Wireless Components Lett.*, vol. 14, no. 7, 2004.
[2] P. Fay et al., *IEEE Electron Device Lett.*, vol. 23, no. 10, pp. 585-587, 2002.
[3] R. G. Meyers et al., *IEEE Electron Device Lett.*, vol. 25, no.1, pp. 4-6, 2004.
[4] J. N. Schulman et al., *Electronics Lett.*, vol. 38, no. 2, pp. 94-95, 2002.
[5] H. P. Moyer et al., *IEEE MTT* THIE-02 p.1459, 2006.
[6] J.J. Lynch et al., *IEEE CSIC*, p. 215, 2006.
[7] N. Su et al., *IEEE Electron Device Lett.* vol. 28, 2007.

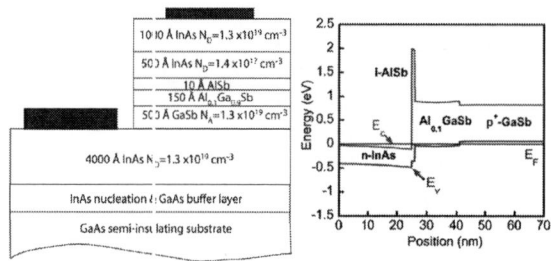

Figure 1: Heterostructure and calculated energy band diagram for the Sb-based detector.

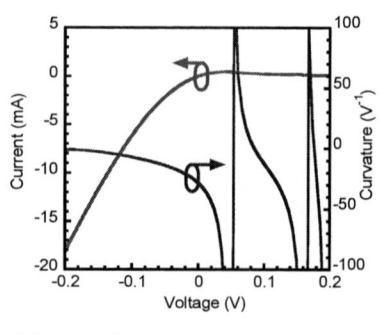

Figure 2: Measured I-V characteristics and curvature (γ_{IV}) for a 1×1 μm² device at room temperature.

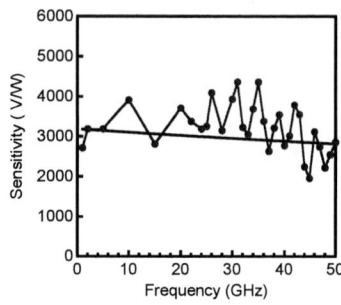

Figure 3: On-wafer measured (•) and modeled (–) RF sensitivity.

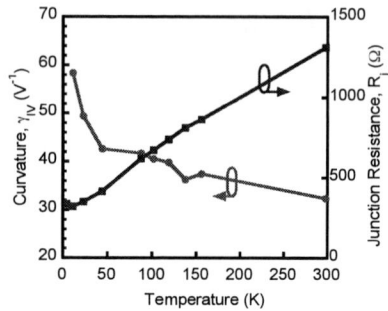

Figure 4: Measured dc curvature (γ_{IV}) and junction resistance (R_j) vs. temperature.

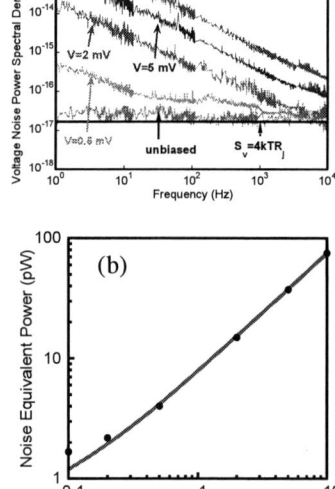

Figure 5: (a) Bias-dependent room-temperature noise spectra; (b) Projected room-temperature NEP vs. bias.

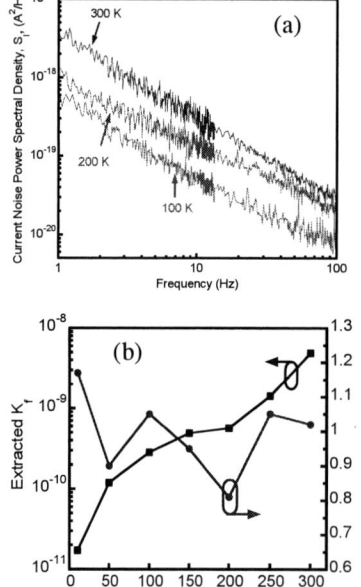

Figure 6: (a) Typical current noise spectra at temperatures of 300 K, 200 K, and 100 K at 20 mV bias; (b) Extracted K_f and $\gamma_{1/f}$ vs. temperature.

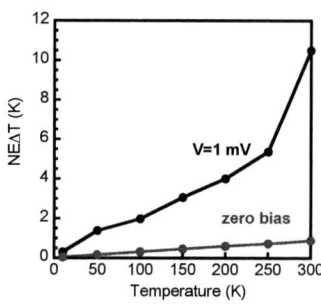

Figure 7: NEΔT as a function of temperature for detectors biased at zero and 1 mV.

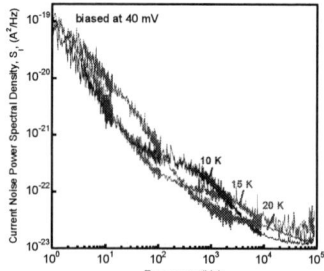

Figure 8: Measured current noise power spectral density as a function of temperature.

152

Delta-Doped Si/SiGe Zero-Bias Backward Diodes for Micro-Wave Detection

Si-Young Park[a], Ronghua Yu[b], Sung-Yong Chung[a], Paul R. Berger[a,b], Phillip E. Thompson[c]
and Patrick Fay[d]

[a]The Ohio State University, Department of Electrical and Computer Engineering, Columbus, OH 43210 USA
[b]The Ohio State University, Department of Physics, Columbus, OH 43210 USA
[c]Naval Research Laboratory, Code 6812, Washington, DC 20375 USA
[d]The University of Notre Dame, Department of Electrical Engineering, Notre Dame, IN 46556 USA

Silicon-based backward diodes incorporating δ-doped active regions for direct detection of micro-wave radiation with zero external dc bias have been demonstrated at room temperature and characterized for their sensitivity. A tunneling backward detector enables the direct detection of micro-wave radiation with zero-bias, therefore no extra bias control circuits are required, resulting in a simplified system and pixel complexity causing a significant reduction of the total cost of fabrication and assembly [1]. Most backward-diode detectors utilize III-V compound based heterojuctions, which are not readily compatible with mainstream silicon technology. Recently, the first high sensitivity Si-based backward diodes, grown by low-temperature molecular beam epitaxial (LT-MBE) growth, were demonstrated by Jin *et al.* [2], who reported a record Si-based curvature coefficient of 31 V^{-1} at zero bias, which is almost double that of a commercial discrete Ge backward diode at room temperature [3].

We report the first directly-measured sensitivity performance of a zero-bias Si-based heterojunciton backward detector, which is readily compatible with mainstream silicon technology. A measured sensitivity of 2376 V/W driven from a 50-Ω source at zero-bias has been obtained. A cutoff frequency of 1.8 GHz was extracted with a series resistance of 290 Ω and a junction capacitance of 0.307 pF using a small signal model established to fit the measured S-parameters for a 5 μm diameter mesa device.

The structure shown in Fig. 1 was modified to significantly reduce the junction resistance (R_j) for better RF performance without a significant reduction in the zero-bias curvature coefficient (γ), which is directly proportional to the RF sensitivity ($\beta_V = 2 Z_s \gamma$). Figure 2 shows the room temperature I-V characteristic of a 5 μm diameter Si/SiGe backward diode illustrating a strong nonlinearity with a curvature coefficient ($\gamma=(d^2I/d^2V)/(dI/dV)$) of 23.2 V^{-1}, and a junction resistance ($R_j= 1/(dI/dV)$) of 687 kΩ for 5μm diameter diode at zero bias. Si/SiGe backward diodes were characterized using on-wafer bias-dependent S-parameter measurements from 1 to 35 GHz in 0.34 GHz steps. De-embedding of the device from pad parasitic was based on S-parameter measurement of on-wafer short- and open-pad test structures. The bias voltage ranged from 0.1 to − 0.2 V in 0.015 V steps. The measured data was fitted to a small-signal equivalent circuit model using Advanced Design System (ADS) software in order to obtain the intrinsic device parameters including series resistance (R_s), junction capacitance (C_j) and junction resistance (R_j). The cutoff frequency, $f_c = 1/(2\pi R_s C_j)$, at zero bias for a 5 μm diameter mesa diode is estimated to be 1.8 GHz based on an extracted series resistance of 290 Ω and junction capacitance of 0.307 pF. This is limited by the relatively large size of the device and capacitive coupling to the conductive substrate.

Measured RF sensitivity of Si/SiGe backward diodes, defined as the DC voltage output developed across the detector divided by the available RF power incident on the detector, was determined on-wafer from 0.01 to 20 GHz as a function of input power level. Figure 3 shows the measured sensitivity and detector voltage characteristics of 5 μm mesa diameter diodes as a function of the input power at an RF frequency of 1 GHz. The highest sensitivity, observed in this study, occurred at 0.01 GHz and ranged up to 2376 V/W with no external bias. Table I. presents the sensitivity measurement as a function of frequency for a 5 μm diameter mesa detector driven by a 50- Ω source.

1. Fay, P., Schulman, J.N., Thomas, S., Chow, D.H., Boegeman, Y.K., and Holabird, K.S.: "High-Perormance Anitmonide-Based Heterostructure Backward Diodes for Millimeter-Wave Detection," *IEEE Electron Device Lett.*, 2002, **23**, pp.585-587.
2. Jin, N., Yu, R., Chung, S-Y., Berger, P. R., Thompson, P.E., and Fay, P.: "High Sensitivity Si-based Backward Diodes for Zero-Biased Square-Law Detection and the Effect of Post-Growth Annealing on Performance," *IEEE Electron Device Lett.*, 2005, **26**, pp.575-578.
3. Burrus, C. A.: "Backward Diodes for Low-Level Millimeter-Wave Detection," *IEEE Trans. Microw. Theory Tech.*, 1963, **11**, pp.357-362.

5 nm undoped Si
P δ-layer
100 nm n$^+$ Si
P δ-layer
2 nm undoped Si
4 nm undoped Si$_{0.6}$Ge$_{0.4}$
B δ-layer
1 nm p$^+$ Si$_{0.6}$Ge$_{0.4}$
80 nm p$^+$ Si
2 nm Si buffer
Si substrate

Fig. 1. A schematic diagram of the Si/SiGe backward diode structure grown by LT-MBE.

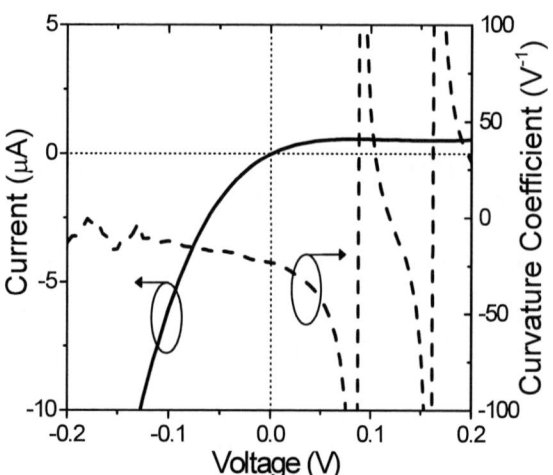

Fig. 2. The room temperature I-V characteristic and the curvature coefficient (γ) as a function of dc bias voltage for a 5 μm diameter Si/SiGe backward diode.

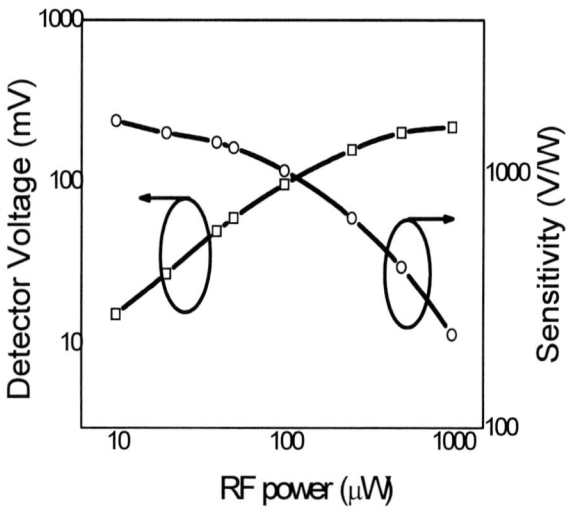

Fig. 3. The measured sensitivity and detector voltage characteristics of a 5 μm diameter mesa as a function of input power at a RF frequency of 1 GHz

Frequency (GHz)	Sensitivity (V/W)
0.01	2376
0.1	2344
0.2	2113
0.5	2022
1	1655
2	1386
5	628
10	269
20	182

Table I. The sensitivity measurement as a function of frequency from 0.01 to 20 GHz for a 5 μm diameter mesa detector

.

Room-Temperature Terahertz Oscillators Using Resonant Tunneling Diodes

M. Asada

Tokyo Institute of Technology, CREST-Japan Science and Technology Agency

2-12-1-S9-3 Ookayama, Meguro-Ku, Tokyo 152-8552, Japan.   E-mail: asada@pe.titech.ac.jp

The terahertz (THz) frequency range is receiving considerable attention because of many possible applications. Compact and coherent solid-state light sources are therefore important key components. Quantum cascade lasers oscillating in the THz range have recently been reported[1]-[3] in the optical device side. Resonant tunneling diodes (RTDs) have been considered as one of the candidates for THz oscillators at room temperature[4]-[6] in the electron device side. In this paper, we report on our recent results of RTD oscillators integrated with slot antennas. Fundamental oscillation up to 0.6THz with 8 μW and harmonic oscillation up to 1.02 THz with 0.6 μW were observed in our device structure up to now. Theoretical analysis shows that fundamental oscillation up to 2.4 THz and 70 μW at 1THz are expected by structure optimization. Voltage-controlled oscillation and mutual injection locking for power combining in array configuration were also observed.

The structure of the fabricated RTD oscillator is shown in Fig. 1. At both ends of the slot antenna, reflectors with metal-SiO_2(100nm)-metal for high frequency are formed. The peak current density of the RTD was typically 300-400 kA/cm^2, and the peak-to-valley current ratio was 2-3.

Fundamental oscillation at 0.6 THz with the output power of 8 μW was obtained for 20-μm-long antenna. In the harmonic mode, the third harmonic oscillation at 1.02 THz with 0.6 μW was obtained for 50-μm-long antenna, as shown in Fig. 2. To our knowledge, this is the highest frequency of single electronic oscillator at room temperature to date. The bias-dependence of the harmonic output was well explained by theoretical analysis[6]. To investigate possibility of fundamental oscillation over 1 THz, we theoretically estimated oscillation frequency, taking into account all the parasitic elements, electron transit time of RTD, and three-dimensional electromagnetic simulation of antenna[7]. The antenna-length dependence of frequency agreed well with theoretical analysis. From this result, fundamental oscillation up to 2.4 THz and 70 μW at 1 THz are expected with RTDs having optimized collector spacer thickness and stub-shaped reflectors at the antenna edges, as shown in Fig. 3. The increase of frequency from about 300 to 400 GHz has been observed by increasing spacer layer thickness. The offset-fed antenna structure is also effective for high power and high frequency[9], a preliminary result of which is shown in Fig. 4.

Frequency tunability of THz sources is a desired property for many applications. We found that the oscillation frequency of RTD varied with bias voltage[8]. Fig. 5 shows frequency change of fundamental and harmonic oscillations. Possible mechanism is the change in transit time with bias voltage, which results in capacitance change. For applications which need high output power, power combining with array configuration is an effective method. Mutual injection locking, which is useful for power combining, was observed between RTD oscillators[10], as shown in Fig. 6, where individual oscillations of two oscillators at 324GHz and 340GHz were locked to the same frequency 330GHz in simultaneous operation. High output power of RTD oscillator is expected by extending this configuration.

[1] R. Köhler, A. Tredicucci, F. Beltram, H. E. Beere, E. H. Linfeld, A. G. Davies, D. A. Ritchie, R. C. Iotti and F. Rossi, Nature, **417** (2002) 156.
[2] B. S. Williams, S. Kumar, Q. Hu, and J. L. Reno, Opt. Exp. **13** (2005) 33315.
[3] C. Walther, G. Scalari, J. Faist, H. Beere, and D. Ritchie, Appl. Phys. Lett. **89** (2006) 231121.
[4] E. R. Brown, J. R. Sönderström, C. D. Parker, L. J. Mahoney, K. M. Molvar and T. C. McGill, Appl. Phys. Lett. **58** (1991) 2291.
[5] M. Reddy, S. C. Martin, A. C. Molnar, R. E. Muller, R. P. Smith, P. H. Siegel, M. J. Mondry, M. J. W. Rodwell, H. Kroemer and S. J. Allen, IEEE Electron Device Lett. **58** (1997) 218.
[6] N.Orihashi, S. Suzuki, and M. Asada, Appl. Phys. Lett., **87** (2005) 233501.
[7] N.Orihashi, S.Hattori and M.Asada, Jpn. J. Appl. Phys. **44** (2005) 7809
[8] N. Orihashi, S. Suzuki, and M. Asada, Electron. Lett. **41** (2005) 872.
[9] S. Suzuki and M. Asada, Jpn. J. Appl. Phys. **46** (2007) 119.
[10]S. Suzuki, N. Orihashi and M. Asada, Jpn. J. Appl. Phys. **44** (2005) L1439.

978-1-4244-1101-6/07/$25.00 ©2007 IEEE

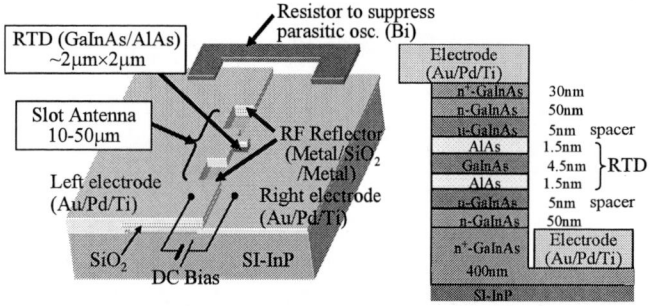

Fig.1 RTD oscillator integrated with slot antenna (left), and layer structure of RTD (right).

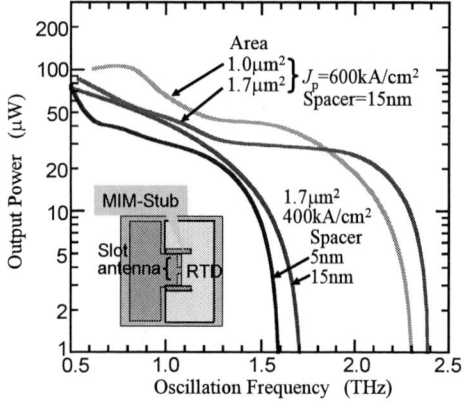

Fig.3 Calculated output power as a function of oscillation frequency for improved structures.

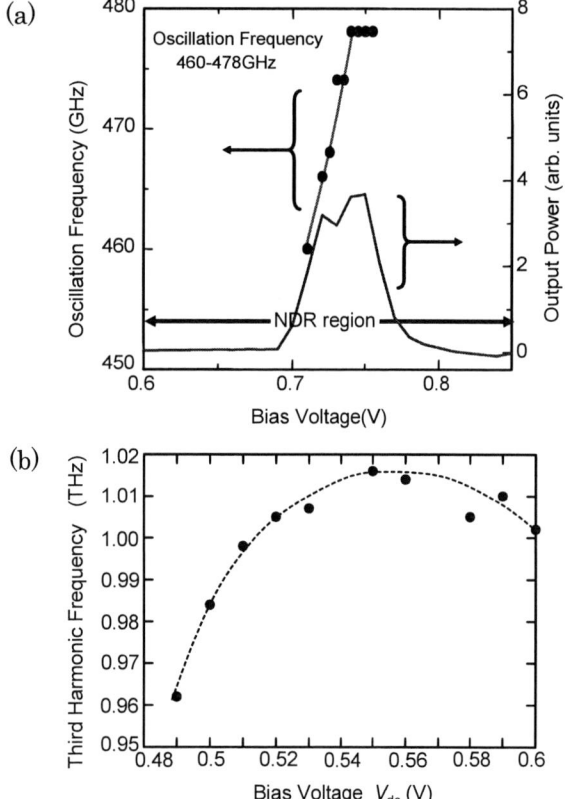

Fig.5 Frequency change with bias voltage for (a) fundamental and (b) harmonic oscillations.

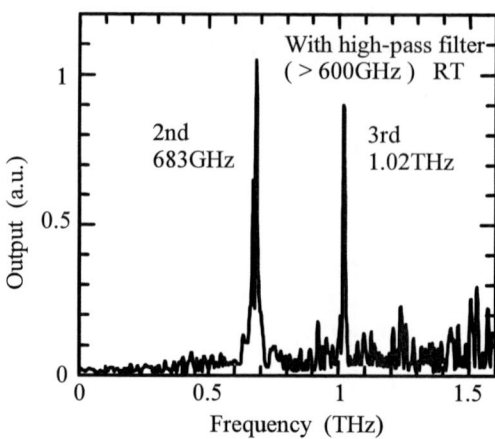

Fig.2 Measured harmonic oscillation >1THz.

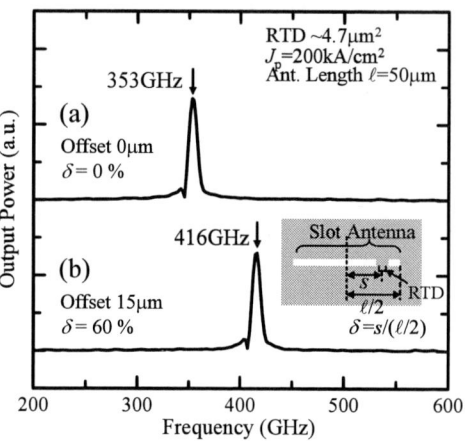

Fig.4 Observation of frequency increase due to offset-fed slot antenna.

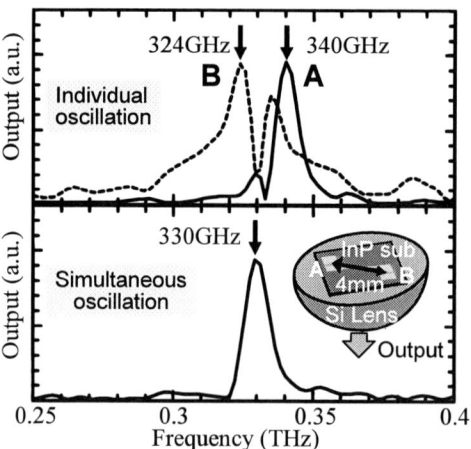

Fig.6 Observation of mutual injection locking between two oscillators. Spectra of individual (upper) and simultaneous (lower) oscillations are shown.

156

Novel Plasmon-Resonant Terahertz-Wave Emitter Using a Double-Decked HEMT Structure

T. Suemitsu, Y. M. Meziani, Y. Hosono, M. Hanabe, T. Otsuji, and E. Sano[†]

Research Institute of Electrical Communication, Tohoku University, 2-1-1 Katahira, Aoba, Sendai 980-8577, Japan
†) Research Center for Integrated Quantum Electronics, Hokkaido University, N13 W8, Sapporo 060-8628, Japan
Phone: +81-22-217-6106, e-mail: sue@riec.tohoku.ac.jp

A new plasmon-resonant THz-wave emitter is fabricated and characterized. The heterostructure of the device consists of double-decked high electron mobility transistor (HEMT) and the upper-deck HEMT works as a grating antenna to convert the non-radiative plasmonic wave in the lower-deck HEMT channel to radiative THz electromagnetic wave. This conversion can be done more efficiently than a metal grating antenna. The experimental observed clear evidence of the THz-wave emission from the double-decked HEMT device.

Exploring the THz frequency range has been drawing increasing attention to realize compact, light-weight, solid-state THz devices for security, sensing, and communication. Two-dimensional (2D) plasma oscillation in semiconductors has been studied as a way to generate, detect, and multiply electromagnetic waves in the THz frequency range [1,2]. In our previous work, the grating dual-gate HEMT structure was employed to produce the 2D plasmon [3]: By applying appropriate biases to the dual gate, 2D electron gas (2DEG) is periodically localized in the channel layer which works as the 2D plasmon cavity to enhance the plasmon instability, resulting in the observation of THz-wave emission at room temperature. In this work, in order to produce the periodically-localized 2DEG, the double-decked HEMT structure is employed. The upper deck channel is then periodically etched as shown in Fig. 1 to form the uncapped region where the 2DEG concentration becomes lower than the capped region without any external gate bias. The HEMT structure consists of the InGaP/InGaAs/GaAs heterostructure with a selective doping in the InGaP layer. For the source/drain ohmic contacts, AuGe/Ni was lifted off and annealed after the upper-deck HEMT was selectively etched. The intrinsic device area has a geometry of 30x75 μm, where the grating pattern is replicated on the upper-deck HEMT layer. The grating consists of 80-nm lines and 350-nm lines aligned alternately with a spacing of 100 nm (Fig. 2).

The plasmon itself is non-radiative. Hence, in order to emit THz wave, an antenna is necessary to transform this longitudinal wave to a radiative transverse wave. The grating gate in the previous work does work as an antenna. However, theoretical study revealed that a low-conductive gate electrode, in which the electron concentration should be comparable to that of 2DEG channel, is preferred in the THz frequency range to enhance the efficiency of field emission [4]. In this case the metal gate has a disadvantage that there are much more electrons in it than the semiconductor channel. In the double-decked HEMT structure in this work, on the other hand, the upper deck channel serves as the grating antenna and its structure is exactly the same as the lower channel. Therefore more intensity in the emitted THz wave is expected.

The experimental setup is shown in Fig. 3. The device is illuminated by a 1.5-μm CW laser from the backside. Electromagnetic wave emitted from the device is detected with a Si bolometer cooled down to 4.2 K through a filter permitting a frequency range between 0.5 to 3 THz. A lock-in technique is used for the measurement. The output of the bolometer is shown in Fig. 4 as a function of the drain bias voltage (V_{ds}). The V_{ds} increases to the knee voltage from which the transistor is operated in the saturation region. The bolometer signal starts increasing at around 6 V and two clear peaks are observed at 8 and 11 V. These features were observed with good reproducibility as shown in Fig. 4. For comparison, Fig. 5 shows the same measurement carried out on the device with the metal grating dual-gate HEMT structure device used in the previous work that confirmed the THz-wave emission by the photo response and electro-optic sampling measurements [3]. Note that the V_{ds} range is different from those of Fig. 4 because the double-decked HEMTs in this work suffer from large parasitic source and drain resistance. Nevertheless the double-decked device exhibits more drastic change in the bolometer signal with increasing V_{ds}. This result supports the idea of low-conductive gate stack to enhance the THz radiation efficiency [4], and therefore indicates that the proposed double-decked HEMT structure is a promising candidate to realize solid-state THz-wave emitters with high power and large efficiency.

In conclusion, strong THz-wave emission is observed from the newly proposed plasmon-resonant THz-wave emitter with a double-decked HEMT structure. Further improvement in the intensity and efficiency of the THz emission will be expected by optimizing device parameters.

This work was partially funded by the Ministry of Internal Affairs and Communications and the Japan Society of the Promotion of Science through the programs of SCOPE and Grant-in-Aid for Scientific Research (Category S), respectively. A part of this work has been carried out at the Laboratory for Nanoelectronics and Spintronics, Research Institute of Electrical Communication in Tohoku University.

---

1. M. Dyakonov and M. Shur, Phys. Rev. Lett., vol. 71, pp. 2465-2468, 1993.
2. V. Ryzhii and M. Shur, Jpn. J. Appl. Phys., vol. 41, pp. L922-L924, 2002.
3. T. Otsuji, Y. M. Meziani, M. Hanabe, T. Ishibashi, T. Uno, E. Sano, Appl. Phys. Lett., vol. 89, pp. 263502, 2006.
4. T. Otsuji, M. Hanabe, T. Nishimura, E. Sano, Optics Express, vol. 14, pp. 4815-4825, 2006.

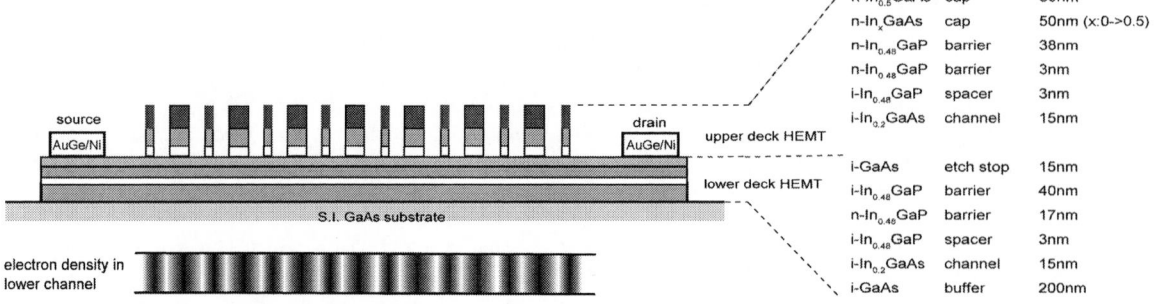

Fig. 1: Cross sectional view of novel plasmon-resonant THz-wave emitter with double-decked HEMT structure. 2D plasmon is generated in the lower-deck HEMT channel and the upper-deck HEMT works as a grating antenna.

Fig. 2: Top view microphotograph of fabricated double-decked HEMT plasmon-resonant THz-wave emitter (top) and cross sectional scanning electron microscope image of grating upper-deck HEMT (bottom).

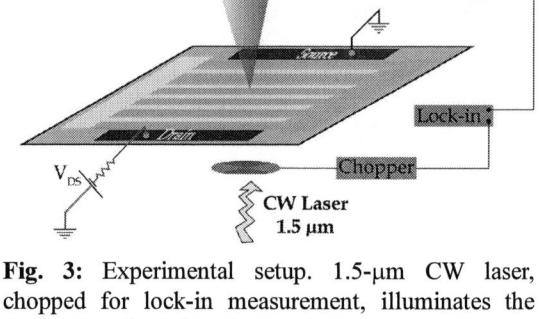

Fig. 3: Experimental setup. 1.5-μm CW laser, chopped for lock-in measurement, illuminates the device from backside. Radiated electromagnetic wave is detected with Si bolometer sensing a frequency range between 0.5 to 3 THz. Measured at room temperature.

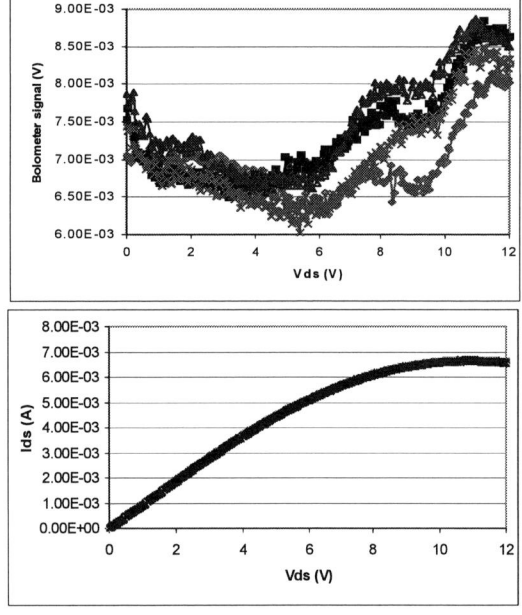

Fig. 4: Bolometer output (top) and drain current (I_{ds}) (bottom) of double-decked HEMT THz-wave emitter as a function of drain voltage (V_{ds}). Measurement took place four times. Bolometer signal increases rapidly at $V_{ds} > 6$ V, which is a clear evidence of the THz-wave emission.

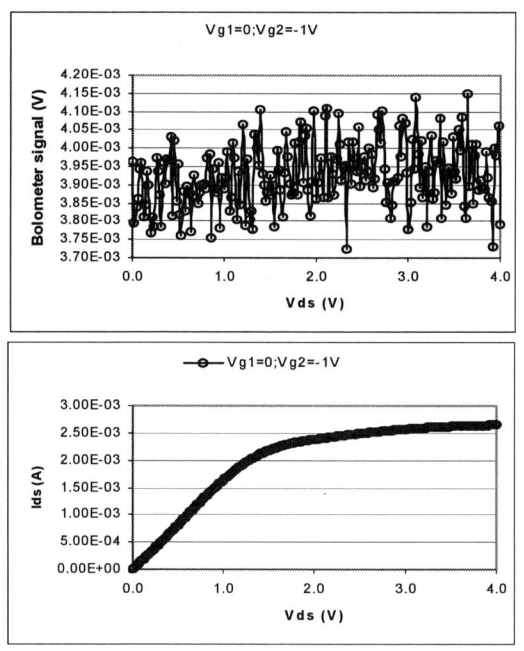

Fig. 5: Same measurement for metal grating dual-gate THz-wave emitter reported previously [3]. The gate biases ($V_{g1} = 0$ V and $V_{g2} = -1$ V) set at point that the highest emission signal was observed. Although THz emission was confirmed also on this device, the bolometer signal is much weaker than that shown in Fig. 4.

158

THz front-side illuminated quantum well photodetector

Mikhail Patrashin, Iwao Hosako

National Institute of Information and Communications Technology, Koganei Tokyo 184-0015, Japan
phone: +81 42 327 6939, fax: +81 42 327 6941, email: mikhail@nict.go.jp

Compound semiconductors, such as GaAs/AlGaAs, provide reliable and well-established material systems for designing quantum-well photodetectors in a targeted spectral range. These were primarily used for near- and middle-infrared devices until recently. The processing technology is fully mature and large-format infrared arrays with up to 1024×1024 pixels have been demonstrated [1]. The optimal design for a QW photodetector with a minimal level of dark current corresponds to a situation where the first excited state in the quantum well is aligned with the top of the barrier (bound to a quasi-bound configuration). If we apply a similar methodology to the range between 1–8 THz, the parameters of the QW structure will correspond to low aluminum fractions of a few percent (1–5%) and QW widths between 10–30 nm [2, 3].

We have designed a front-side illuminated GaAs/AlGaAs detector with a targeted peak frequency of 3 THz (100 μm) in an effort to extend the successful implementation of infrared QW arrays to the terahertz range. A simple multilayer structure has 18-nm GaAs QWs sandwiched between AlGaAs barriers with an Al alloy fraction of 2% (see Fig.1). The effect of different barrier widths and doping concentrations on the expected dark current and spectral response of the structure were numerically simulated, and several samples with various in-well doping concentrations (5×10^{16}–2×10^{17} cm^{-3}) and barrier widths (60 and 80 nm) were grown by MBE on semi-insulating GaAs substrates. X-ray diffraction, Scanning Electron Microscopy/Energy Dispersive Spectrometer (SEM/EDS), and Photoluminescence (PL) measurements were used to verify the composition, period, and energy-level structure of the samples. Only minor deviations from the designed parameters were observed.

The samples were processed into single-element square-shaped mesas with top and bottom ohmic contacts of different sizes using standard photolithography, wet etching, and thermal deposition. A simple grating coupler was implemented on top of the mesas to allow front-side illumination. Despite the low Al content (2%), we obtained consistent results with MBE growth and could control the level of dark current and the impedance of fabricated devices within design-specific requirements. Results of measurements on 1mm × 1mm mesas with 50μm period grating coupler are shown in Fig.2, 3, 4.

The level of dark current in optimized samples was a few μA/cm^2 (Fig.1, 2) and the detector observed response close to the designed detection wavelength (Fig.3). The responsivity of the detector was measured with a calibrated blackbody source. A responsivity of 13 mA/W at an electric bias of 40 mV and an operating temperature of 4 K was obtained by comparing current-voltage characteristics under different photon flux conditions. We believe that optimizing the in-well doping concentration further and improving the design of the grating coupler can produce even better performance. We are currently evaluating the viability of this type of detector for practical applications with a prototype 32-element array.

[1] S. D. Gunapala et al, Proc. SPIE Int. Soc. Opt. Eng. 5783, 789 (2005).
[2] H. C. Liu, C. Y. Song, A. J. SpringThorpe, and J. C. Cao, Appl. Phys. Lett. 84, 4068 (2004).
[3] H. Luo, H. C. Liu, C. Y. Song, and Z. R. Wasilewski, Appl. Phys. Lett. 86, 231103 (2005).

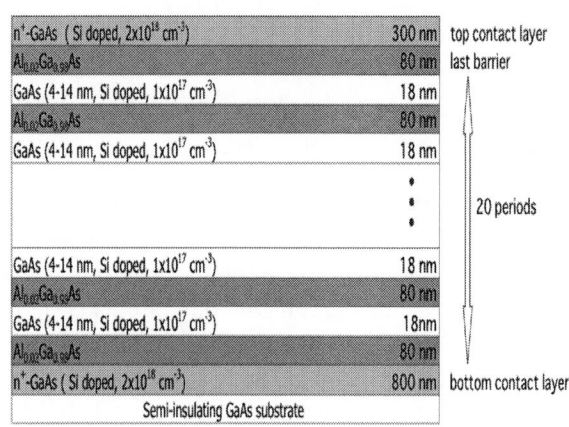

Fig. 1. Schematic layout of MBE grown THz quantum well photodetector structure.

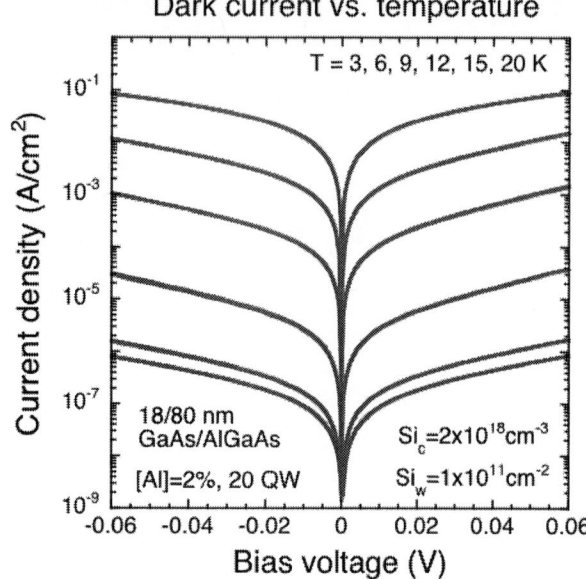

Fig. 2. Bias and temperature dependence of dark current.

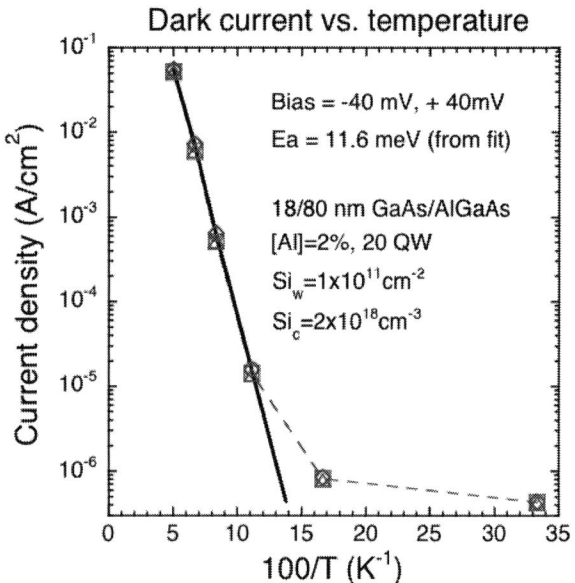

Fig. 3. Temperature dependence of dark current at 40 mV bias voltage. Below 8K the current flattens out at around 1μA/cm².

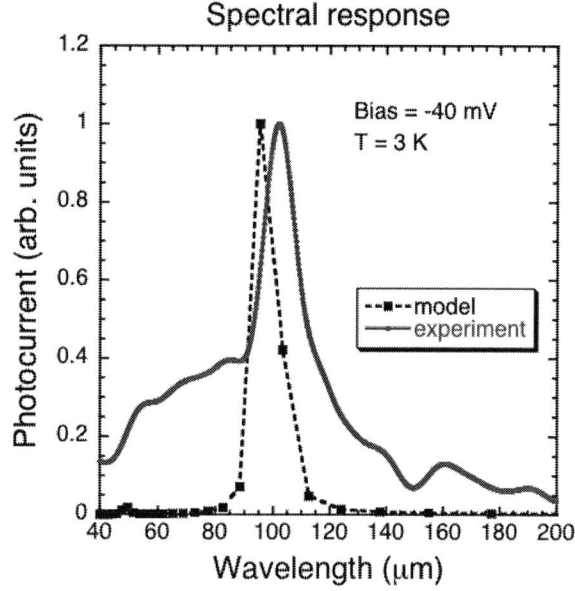

Fig. 4. Normalized spectrum of photoresponse compared with the result of numerical simulations. Detector oserved response close to the designed detection wavelength.

160

Session IV.B (DeBartolo Hall Room 141)

Semiconducting Nanowire Devices

Tuesday AM, June 19th, 2007

Session Organizer: Prabhat Agarwal, NXP Semiconductors
Session Chairman: Jessica Thomas, Nature Nanotechnology

8:20 AM IV.B-1 Invited Paper
Towards vertical III-V nanowire devices on silicon
E. Bakkers[1], M. Borgström[1], W. van den Einden[1], M. van Weert[1], E. Minot[1], F. Kelkensberg[1], M. van Kouwen[1], J. van Dam[1], L. Kouwenhoven[1], V. Zwiller[1], A. Helman[2], O. Wunnicke[2], and M. Verheijen[2], [1]Philips Research Laboratories, Eindhoven, THE NETHERLANDS and [2]Kavli Institute of Nanoscience, Delft, THE NETHERLANDS

9:00 AM IV.B-2
Control of Threshold Voltage in 80 nm Gate Length InAs Vertical Nanowire WIGFETs
T. Löwgren[1], J. Ohlsson[1], L. Samuelson[2], and L.-E. Wernersson[2], [1]QuMat Technologies AB, Lund, SWEDEN and [2],Solid State Physics / Nanometer Consortium, Lund University, Lund, SWEDEN

9:20 AM IV.B-3 Student Paper
Gallium Nitride Nanowire Devices – Fabrication, Characterization, and Simulation
A. Motayed, A. V. Davydov, M. He, and S. N. Mohammad, National Institute of Standards and Technology, Material Science and Engineering Laboratory, Gaithersburg, Maryland, USA

9:40 AM IV.B-4 Student Paper
High performance In2O3 nanowire transistors using organic gate nanodielectrics
S. Ju[1], G. Lu[3], P.-C. Chen[2], A. Facchetti[3], C. Zhou[2], T. J. Marks[3], and D. B. Janes[1], [1]School of Electrical and Computer Engineering and Birck Nanotechnology Center, Purdue University, West Lafayette, Indiana, USA, [2]Dept. of Electrical Engineering, University of Southern California, Los Angeles, California, USA, and [3]Dept. of Chemistry and the Materials Research Center, Northwestern University, Evanston, Illinois, USA

10:00 AM Break

10:20 AM IV.B-5
Impact ionization FETs based on silicon nanowires
M. T. Björk, O. Hayden, J. Knoch, H. Riel, H. Schmid, and W. Riess, IBM Research GmbH, Zurich Research Laboratory, Rueschlikon, Switzerland

10:40 AM IV.B-6 Student Paper
A Low Power, Highly Scalable, Vertical Double Gate MOSFET Using Novel Processes
H. Cho[1], P. Kapur[1], P. Kalavade[2], and K. C. Saraswat[1], [1]Department of Electrical Engineering, Stanford University, CIS, Stanford, California, USA and [2]Intel Corporation, Santa Clara, California, USA

11:00 AM IV.B-7
An n-FET with a Si nanowire channel and doped epitaxially-thickened source and drain regions
G. M. Cohen, P. M. Solomon, S. E. Laux, J. O. Chu, M. J. Rooks, and W. Haensch, IBM T. J. Watson Research Center, Yorktown Heights, New York, USA

11:20 AM IV.B-8 Student Paper
Reduction of Acoustic Phonon Limited Electron Mobility due to Phonon Confinement in Silicon Nanowire MOSFETs
J. Hattoria[1,3], S. Uno[1,3], N. Mori[2], and K. Nakazato[1,3], [1]Department of Electrical Engineering and Computer Science, Nagoya University, Nagoya, JAPAN, [2]Department of Electronic Engineering, Osaka University, Osaka, Japan, and [3]SORST JST

161

11:40 AM IV.B-9 Student Paper
THz probe for nanowire FETs: simulation of few-electron fingerprints
K. M. Indlekofer[1], R. Németh[1], and J. Knoch[2], [1]Institute for Bio and Nanosystems, CNI, Research Center, Jülich, GERMANY and [2]IBM Research GmbH, Zurich Research Laboratory, Rueschlikon, SWITZERLAND

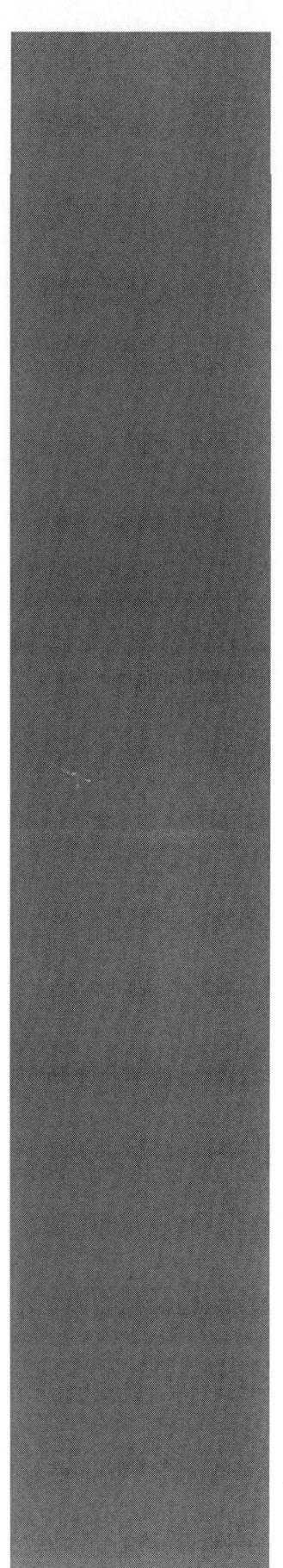

Towards vertical III-V nanowire devices on silicon

Erik P.A.M. Bakkers, Magnus T. Borgström, Wim van den Einden, Maarten van Weert, Ethan D. Minot,
Freek Kelkensberg, Maarten van Kouwen, Jorden A. van Dam, Leo P. Kouwenhoven and Valery Zwiller
Ana Helman, Olaf Wunnicke, Marcel A. Verheijen

Philips Research Laboratories, High Tech Campus 4, 5656 AE, Eindhoven, The Netherlands
Kavli Institute of Nanoscience, Lorentzweg1, TU Delft, Delft, The Netherlands

Higher operation speeds in silicon devices have been achieved by reducing the device dimensions. To make substantial progress, however, effort has been put in the investigation of semiconductor materials that intrinsically have higher mobilities, such as GaAs, InP, and InAs. Nowadays, silicon is the standard for the electronics industry. An approach would be to combine the best parts of these different technologies; i.e. the monolithic integration of the III-V semiconductors into the mature silicon technology. There are some clear advantages for both worlds. However, fundamental issues such as lattice and thermal expansion mismatch and the formation of antiphase domains have prevented the industrial epitaxial integration of III-V with group IV semiconductors. These problems could be avoided by reducing the contact area of the III-V crystals and by making vertical devices. Due to the small dimension the strain could be accommodated at the nanowire surface. In addition, since per crystallite there will only be one nucleation site, we will not suffer from antiphase or twin boundaries. [1]

We have grown III-V nanowires by metal organic vapor phase epitaxy (MOVPE) using the vapor-liquid-solid (VLS)[2] growth mechanism. It is shown that GaP, GaAs, InP and InAs can be grown on silicon with a lattice mismatch ranging between 0.4% (GaP) and 11.6% (InAs). This epitaxial growth on silicon facilitates the addition of new functionalities, such as high-frequency and opto-electronic devices into existing silicon technology. We have developed a process to fabricate vertical two (diodes) and three terminal (transistor) devices (see Figure 1a). Electrical characteristics of these preliminary devices will be discussed (Figure1b).

For appropriate operation of electronic devices quantitative control of impurity doping levels is required. However, for nanowires grown via the VLS mechanism at relatively low temperatures, it is not known whether impurity atoms interact with the metal particle and how they are incorporated into the semiconductor crystal. A possible way to obtain control of doping levels in nanowires is to synthesize core/shell structures (see Figure 1c), where the dopant atoms are in the shell that donates carriers to the undoped core, i.e., to remotely dope nanowires. To demonstrate remote doping in III-V nanowires, we have chosen to remotely dope InAs nanowires by a p-doped InP shell.[3] It is challenging to p-dope InAs nanowires because of surface Fermi level pinning around 0.1 eV above the conduction band edge for p- and n-type doping. We demonstrate that shielding with a p-doped InP shell compensates for the built-in potential and donates free holes to the InAs core (Figure 1d). Moreover, the off-current in field-effect devices can be reduced up to 6 orders of magnitude. The effect of shielding critically depends on the thickness of the InP capping layer and the dopant concentration in the shell.

In addition, I will discuss the fabrication and characterization of reproducible axial InP nanowire LED devices (Figures 1e-h), and show that an active InAsP quantum dot region can be incorporated into these devices.[4] The axial geometry allows for controllable injection of electrons and holes into the precisely defined active region, with the additional advantage of high light-extraction efficiency since the optically active region is not embedded in a high refractive index material. We have investigated the operation of these nanoLEDs with a consistent series of experiments at room temperature and at 10 K, demonstrating the potential of this system for single photon applications.

References

1) E. P. A. M Bakkers, J. A Van Dam, S De Franceschi, L. P Kouwenhoven, M Kaiser, M.Verheijen, H. Wondergem, P Van der Sluis, *Nature Materials* 2004, *3*, 769.

2) R. S. Wagner, Ellis, *Appl. Phys. Lett.* 1964, *4*, 89.

3) H.-Y. Li, O. Wunnicke, M. T. Borgstrom, W. G. G. Immink, M. H. M. van Weert, M. A. Verheijen, and E. P. A. M. Bakkers, *Nano Lett.* 2007, ASAP

4) E. D. Minot, F. Kelkensberg, M. van Kouwen, J. A. van Dam, L. P. Kouwenhoven, V. Zwiller, M. T. Borgstrom, O. Wunnicke, M. A. Verheijen, and E. P. A. M. Bakkers, *Nano Lett.* 2007, 7, 367

978-1-4244-1101-6/07/$25.00 ©2007 IEEE

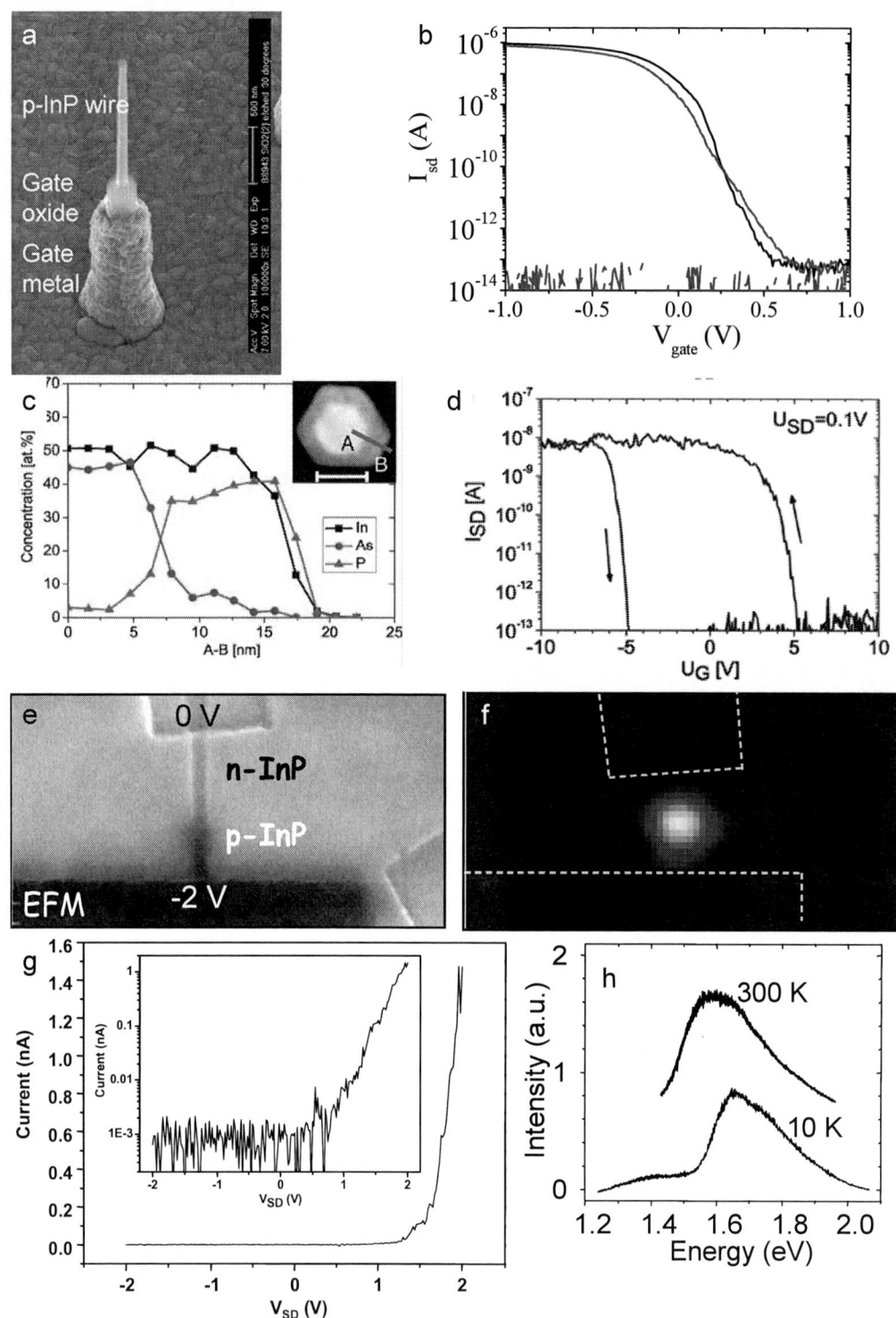

Figure 1 a. Scanning electron micrograph of a part of a vertical transistor based on a p-type InP wire. **b** transfer characteristic of a decice as in a). c Energy-dispersive X-ray scan accros a cross sectional InAs/InP core/shell Nanowire. Inset: cross sectional dark-field TEM image of a radial InAs/InP Heterostructure **d** transfer characteristic of a contacted InAs/InP nanowire showing the p-type behaviour. **e** Electric field microscopy image of a contacted InP wire with an axial p-n junction in reverse bias. The contrast shows that the potential drops in the middle of the wire. **f** Electroluminescence from the same device in forward bias. **g** I-V cureve of the device in e). **h** Electroluminescence spectra of the device at room temperature and 10 K.

Control of Threshold Voltage in 80 nm Gate Length InAs Vertical Nanowire WIGFETs

Truls Löwgren[1], Jonas Ohlsson[1], Lars Samuelson[2], and Lars-Erik Wernersson[2]

[1] *QuMat Technologies AB, Stora Fiskaregatan 13E, S-22224 Lund, Sweden*
[2] *Solid State Physics / Nanometer Consortium, Lund University, Box 118, S-22100 Sweden*
Email: Lars-Erik.Wernersson@ftf.lth.se, Phone: +46 46 2227678

Nanowire transistors are currently being evaluated as an emerging device technology. While numerous investigations have studied the performance of lateral transistors based on single nanowires, there are only a few studies of the fabrication and performance of vertical transistors with wrap-gates, so called WIGFETs. Based on our previous study of 800 nm gate length transistors [1], we have now scaled the technology to sub 100 nm gate length and study the lateral scaling of the nanowires in the transistor structure. It is demonstrated that the scaling provides a mean to adjust the threshold voltage of the transistor.

InAs (n-type, $2-5 \times 10^{17}$ cm^{-3}) nanowires grown by Chemical Beam Epitaxy have been used in this study, since the InAs provides good n-type transport properties, a low ohmic contact resistance, and a useful Fermi-level pinning. In the transistors, 11x11 matrices of 3-μm-long nanowires that were covered with 40 nm SiN_x were used. A Ti/Au gate was then formed in a direct evaporation process [2] and patterned by optical lithography. Notably, in this process the gate length is directly controlled by the thickness of the deposited film. After planarization, Ti/Au drain contacts were formed. Nanowire matrices with two different diameters, 55 and 70 nm, were included on the same sample and they were simultaneously processed to get a direct comparison of performances.

The transistors were evaluated using the substrate as a common source. Both types of transistors showed good transistor characteristics with drive currents reaching 1 mA. The 55 nm diameter transistor showed the lower output conductance, since the transistor with 70 nm diameter nanowires had a punch through at $V_{sd}>1V$ and $V_g\sim1V$, related to insufficient potential control in the body of the wire. Notably, the drive current was not substantially reduced as the diameter was scaled. The transfer characteristics were evaluated at $V_{sd}=0.5V$ for a number of transistors and a maximum hysteresis of $\Delta V=0.1$ V was observed. The transistors showed peak transconductances of about 0.30 mS at $V_g=-0.8$ V and $V_{sd}=0.5$ V with higher values at increased V_{sd}. To evaluate the threshold voltage, the current was plotted as a function of the gate bias, since in the cylindrical geometry used and considering the short gate length, a deviation from the conventional Sqrt(I_d) vs V_g may be expected. The deduced values for the threshold voltages group for the two different diameters used with $V_t=-1.6$ V and $V_t=-1.25$ V for the 70 and 55 nm nanowire diameter, respectively. A minimum spread, $\Delta V_t=0.2$ V, was observed for the transistors with the larger diameter. Finally, we also studied the subthreshold characteristics at $V_{sd}=0.5$ V and observed subthreshold slopes of S=1.4 V/decade and S=1.0 V/decade for 70 and 55 nm nanowire diameter, respectively.

Based on the data, we conclude that sub 100 gate length nanowire transistors with good transistor characteristics may be fabricated in the vertical geometry. It is also shown that lateral scaling of the nanowire diameter may be used to adjust the threshold voltage, as expected in this vertical MISFET configuration. Further improvement in the transistor performance (for instance, transconductance and subthreshold slope) may be achieved with shorter wires that have a lower series resistance, and by adequate scaling of the thickness and dielectric constant of the gate dielectric layer.

[1] T. Bryllert et.al., DRC 2005 pp. 157, and T. Bryllert et. al., IEEE EDL, 27, 323, 2006

[2] L.-E. Wernersson et al, Jap. Jour. Appl. Phys., 46, pp. , 2007

Acknowledgments: This work was supported by QuNano, the Swedish Foundation for Strategic Research, the Swedish Research Council, the EU program NODE 015783, and Knut and Alice Wallenberg Foundation.

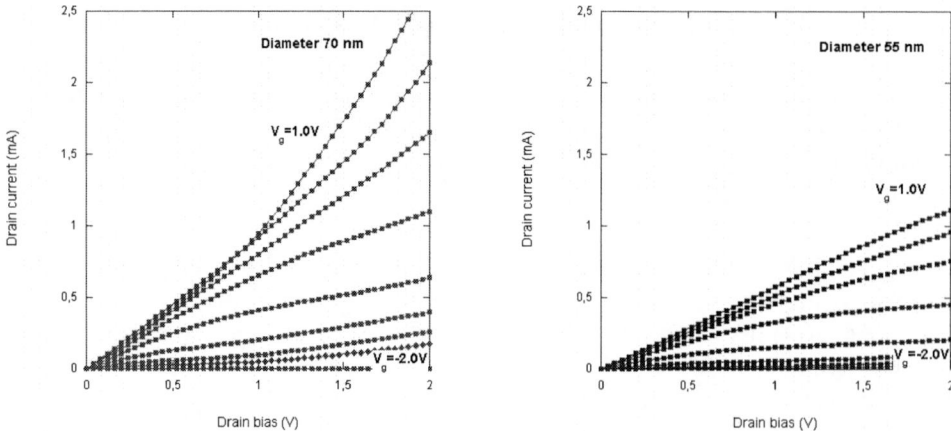

Figure 1. Measured I-V characteristics for 80 nm gate length WIGFETs with diameters of 70 nm (left) and 55 nm (right). The transistor consists of 11x11 nanowires and the data is take in steps of ΔV_g=430 mV.

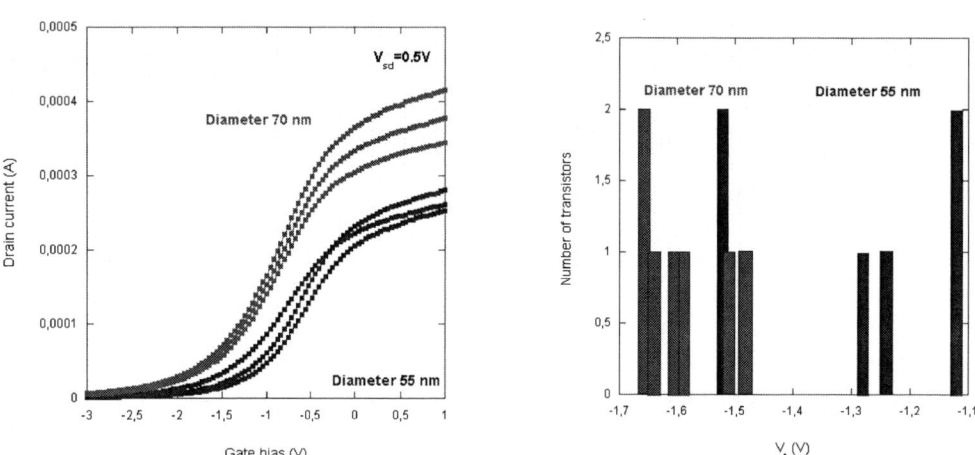

Figure 2. Measured transfer characteristics for a number of different 80 nm gate length WIGFETs with varying diameter (left) and deduced threshold voltages (right).

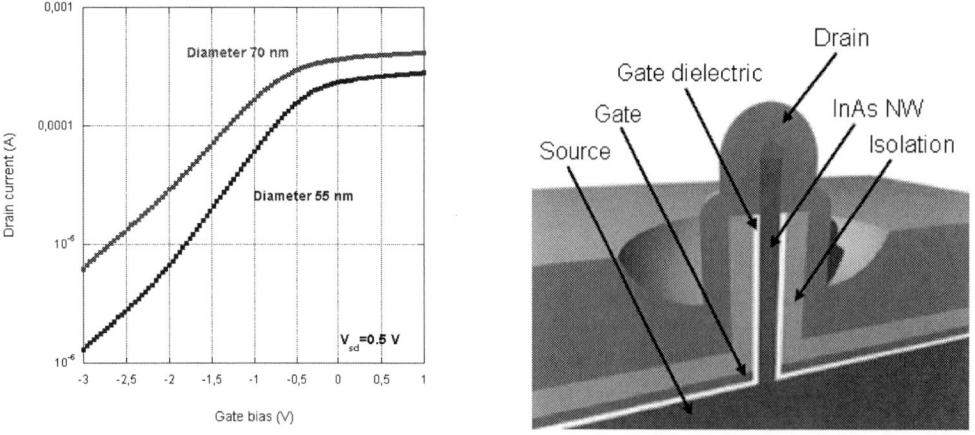

Figure 3 Measured subthreshold characteristics (left) and image of device structure (right).

Gallium Nitride Nanowire Devices – Fabrication, Characterization, and Simulation

Abhishek Motayed,* Albert V. Davydov, Dr. Maoqi He, S. N. Mohammad
*National Institute of Standards and Technology, Material Science and Engineering Laboratory, Gaithersburg, MD
Phone: 301-975-4916, fax: 301-975-4553

Developing integrated devices and systems from nanowires of different material systems would yield more efficient sensors, light emitters, detectors, and ultrahigh density data storage systems. Gallium nitride nanowires offer a great deal of potential as they poses unique material properties like wide direct bandgaps, radiation hardness, mechanical and chemical stability. Although, few results of individual GaN nanowire devices has been reported so far [1]-[2], but most of these often involve fabrication processes unsuitable for large scale nanosystems development. To realize reliable nanowire devices, understanding of the charge transport mechanisms and ability to model the devices and predict various device properties are crucial.

We have demonstrated nanoscale GaN devices made from GaN nanowires utilizing only conventional batch microfabrication techniques. Nanowires were dielectrophoreticaly aligned to achieve large number of reliable devices [3]. These GaN nanowires were grown by direct reaction of Ga and NH_3 and had diameters ranging from 100 nm to 250 nm and lengths up to 200 µm. We have successfully realized different device geometries essential for studying fundamental transport properties in these nanowires. GaN nanowire Field effect transistors (FETs), light emitting diodes (LEDs), metal semiconductor FETs (MESFETs), and four terminal structures (4T) have been realized using this technique. Nanowire FETs with Si substrate as a backgate showed field effect electron mobility values in excess of 300 cm^2 V^{-1} s^{-1} for 200 nm diameter nanowire [4]. Mobility measurements on different diameter nanowires revealed a scaling effect. The reduction of mobility in GaN nanowire FETs with decreasing diameter of the nanowire is attributed to the surface scattering. Using electron beam backscattered diffraction we have also revealed the effect of grain boundary scattering in these nanowire devices.

Effect of different device configurations like bottom gate, omega gate, and cylindrical gate structures on the device performance of these nanowire FETs have been studied. This study revealed that better gate geometries for GaN FETs can result in channel inversion hence enhancement mode in otherwise depletion mode devices. Circular shaped gate geometry resulted in improvement of transconductance per unit gate length by an order of magnitude over omega shaped top gate on the same nanowire. We have modeled the channel conductivity as a function of gate bias using two dimensional self consistent Poisson's equation solver. This has provided us with an insight into the nature of depletion behavior of the nanowire transistors. It is revealed from these simulations that non uniform depletion region in larger diameter nanowire prevent them from completely depleting.

GaN nanowire LEDs have been realized using the same Dielectrophoretic alignment technique on a p- GaN epitaxial layer. The resulting p-n homojunction exhibited 365 nm electroluminescence with a full width half maximum of 25 nm at 300 K. Temperature dependent resistivity measurement from 300 K down to 77 K on 4T structures revealed that the impurity scattering is the most dominant mechanism in transport in these nanowires.

This study encompasses various aspect of nanowire device realization from assembly of nanowire, developing and optimizing fabrication techniques, understanding effectiveness of different device configurations. Present study also revealed fundamental nature of transport through these nanowire devices.

[1] X. Duan et al., *Nature*, vol. 294, pp 241, (2003).
[2] J.-R. Kim et al., *Appl. Phys. Lett*, vol.. 80, pp 3548 (2002).
[3] A. Motayed et al, *J. Appl. Phys.*, vol 100, pp 114310 (2006).
[4] A. Motayed et al, *Appl. Phys. Lett*, vol.. 90, pp 043104 (2007).

978-1-4244-1101-6/07/$25.00 ©2007 IEEE

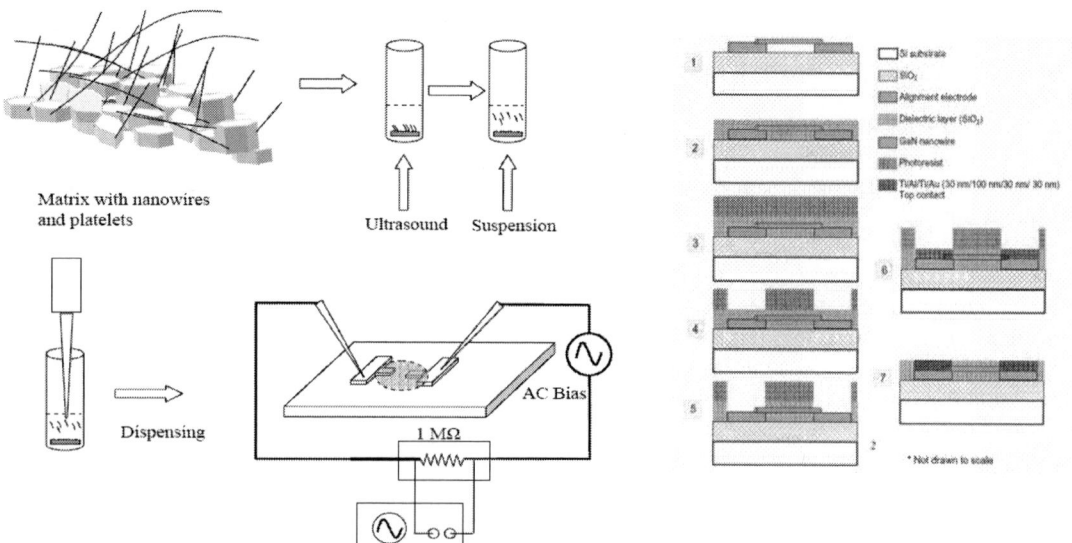

Matrix with nanowires and platelets

Ultrasound   Suspension

Dispensing   AC Bias   1 MΩ

Fig. 1 Dielectrophoretic alignment of GaN nanowires   Fig. 2 Fabrication process scheme

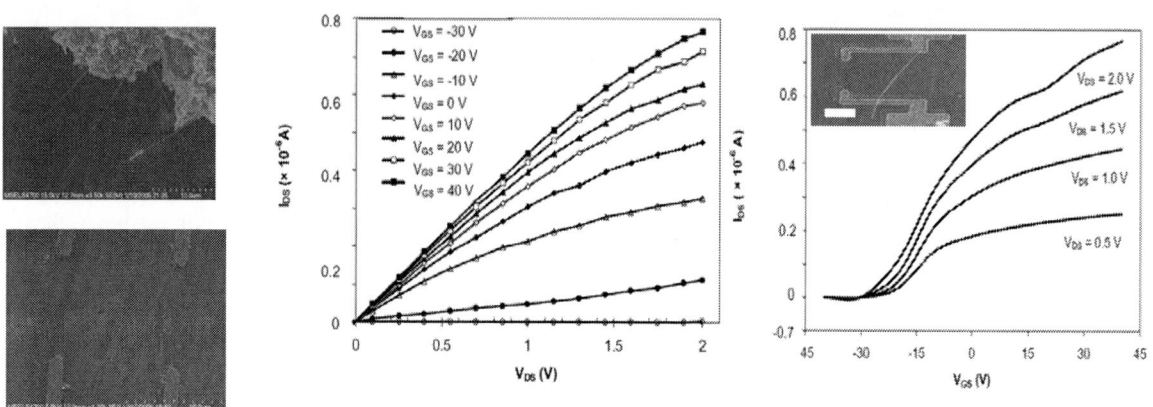

Fig. 3 (a) GaN nanowires aligned between metal pads (b) measured GaN nanowire transistor I_{ds} vs V_{ds} curve (c) I_{ds} vs V_{gs} transconductance curve (inset shows the complete GaN nanowire device)

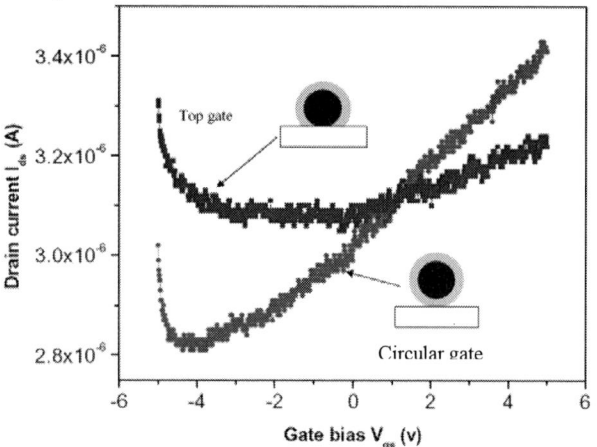

Fig. 4 (a) Transconductance curve of the same nanowire device with two different gate configurations

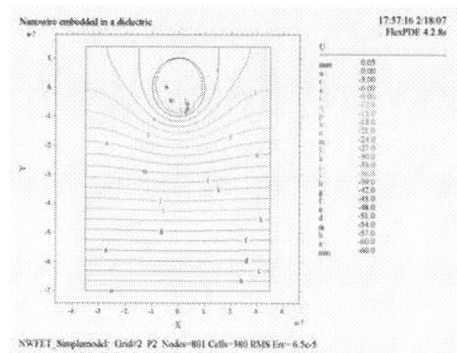

Fig. 4 (b) 2D Simulation showing depletion region inside a nanowire with the application of a gate bias.

Fig. 5 Electroluminescence from a GaN nanowire LED at room temperature at different bias voltages (inset shows the optical image of nanowire under bias emitting.

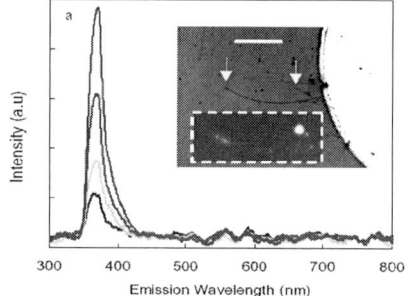

168

High performance In$_2$O$_3$ nanowire transistors using organic gate nanodielectrics

Sanghyun Ju[1], Gang Lu[3], Po-Chiang Chen[2], Antonio Facchetti[3], Chongwu Zhou[2], Tobin J. Marks[3], and David B. Janes[1]

[1]School of Electrical and Computer Engineering and Birck Nanotechnology Center, Purdue University, West Lafayette, IN 47907
[2]Dept. of Electrical Engineering, University of Southern California, Los Angeles, CA 90089
[3]Dept. of Chemistry and the Materials Research Center, Northwestern University, Evanston, IL 60208
Email: janes@ecn.purdue.edu; Phone: (765)494-9263; Fax: (765)494-0811

Recently, there have been several reports of nanowire transistors (NWTs) which attempt to reach high performance and reliable transistor characteristics to replace poly-silicon thin-film transistors (poly-Si TFTs) and amorphous-silicon (α-Si) TFTs.[1-4] NWTs have several advantages compared with poly-Si TFTs and α-Si TFTs, in terms of high mobility, optical transparency and mechanical flexibility. NWTs would also allow a significant reduction in the area of the drive transistor, allowing larger aperture ratios for the light emitting areas for a given pixel spacing. However, the evaluation of reported device performance metrics reveals that there is much room for further improvement of NWT characteristics, so that driving and switching transistors using nanowires can be assembled into useful electronic circuits.

In this study, we report high performance nanowire transistors using individual In$_2$O$_3$ nanowires as channels, a multilayer self-assembled organic nano-dielectric (SAND) as the gate insulator (thickness ~15 nm, capacitance ~180 nF/cm^2, and leakage current density ~1×10^{-6} A/cm^2 up to 2 V). The NWTs use an individually addressable indium zinc oxide (IZO) bottom-gate and Al source/drain electrodes. (Fig.1) The diameter and length of the In$_2$O$_3$ nanowires are 20 nm and 1.6 μm, respectively. Following ozone-treatment to improve and optimize performance, the devices exhibit n-type transistor characteristics (Fig. 2a) with a subthreshold slope (S) of 0.2 V/dec, a current on-off ratio (I_{on}/I_{off}) of 10^6, a threshold voltage (V_T) of 0.0 V, and transconductance (g_m) and channel conductance (g_d) as illustrated in Fig. 3. The g_m at $V_d = 0.5$ V peaks at ~5.87 μS, and the g_d is proportional to gate voltage. The drain current versus drain-source voltage (I_{ds}-V_{ds}) characteristics of a representative SAND-based In$_2$O$_3$ NWT are shown in Fig. 2b, exhibiting high I_{on} ~12 μA for the single In$_2$O$_3$ nanowire at $V_{ds} = 1$ V, $V_{gs} = 1.5$ V, respectively. This current level would be sufficient to drive a 71 x 213 μm pixel at 300 cd/m^2 in current-generation electroluminescent technologies. The field-effect mobility (μ_{eff}) which is extracted from the g_m and g_d of the NWTs, along with an estimated gate-to-channel capacitance, is also plotted versus gate bias in Fig. 2a. The value of μ_{eff} varies from ~1447 cm^2/V-sec to ~300 cm^2/V-sec over the reported gate bias range. The peak value is much higher than recently reported results for In$_2$O$_3$ nanowires (electron mobility of 279 cm^2/V-sec and 98.1 cm^2/V-sec, effective mobility 6.93 cm^2/V-sec)[5-7] and ideal single-crystal In$_2$O$_3$ bulk mobility (~160 cm^2/V-sec).[8] It is expected that both the single crystal nature of the nanowires and quasi-one-dimensional effects, which inhibit low-angle scattering, contribute to the relatively high μ_{eff}. The SAND dielectric has previously been found to be suitable for realizing relatively high performance in other oxide nanowires.[4] As a result, we achieve significantly enhanced device performance for SAND-based NWTs, which are better than other In$_2$O$_3$ nanowire transistors and comparable with poly-Si TFTs and α-Si TFTs, in terms of S and μ_{eff}. The results indicate that SAND-based In$_2$O$_3$ NWTs may support the requirements of a high μ_{eff} and a steep S for fast switching and high-speed logic

[1] J. Xiang *et al*, *Nature*, **441**, 489-493 (2006).
[2] Y. Cui *et al*, *Nano Letters*, **3**, 149-152 (2003).
[3] W. Kim *et al*, *Appl. Phy. Lett.*, **87**, 173101 (2005).
[4] S. Ju *et al*, *5*, 2181-2186 *Nanolett.*, (2005).

[5] B. Lei *et al*, *Appl. Phys. A*, **79**, 439-442 (2004).
[6] D. Zhang *et al*, *Appl. Phy. Lett.* **82**, 112 (2003).
[7] N. Pho *et al*, *Nano Lett.* **4**, 651-657 (2004).
[8] R. L. Weiher *et al*, *J. Appl. Phys.* **33**, 2834 (1962).

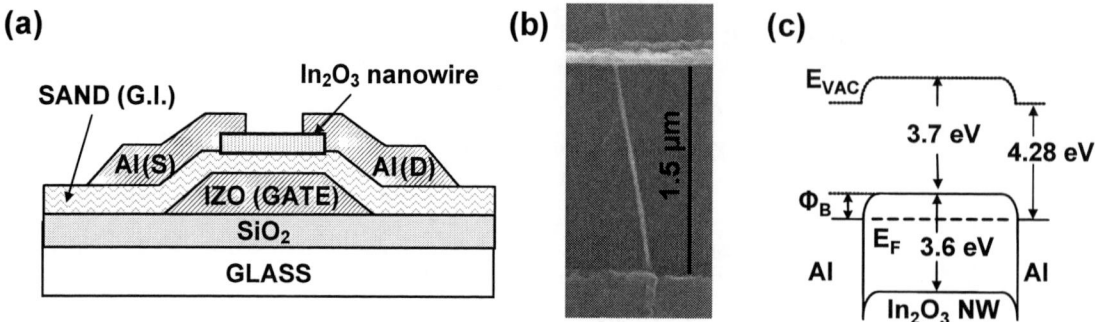

Figure 1. SAND-based In_2O_3 NWTs. (a) Cross-sectional view of the device structure. (b) Top-view FE-SEM images of the device region. Scale bar, 1.5 µm. (c) Source/nanowire/drain cross-sectional band diagram at $V_{gs} = 0$ V.

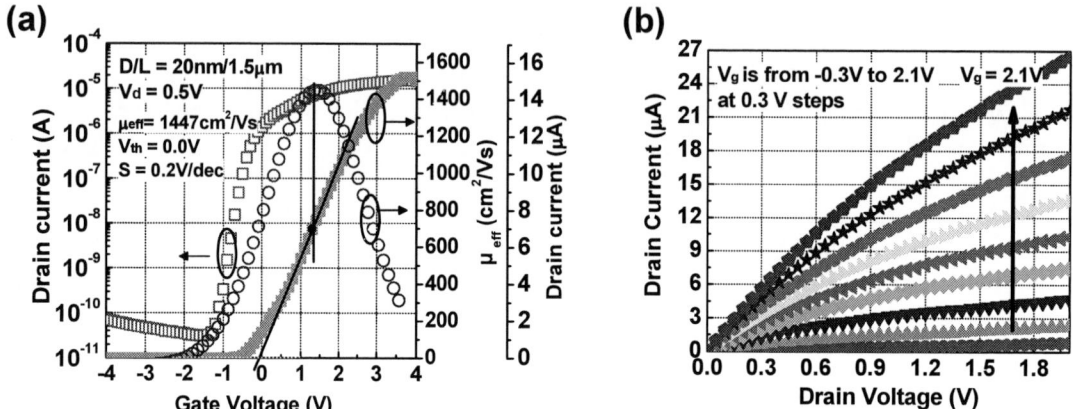

Figure 2 Characteristics of SAND-based In_2O_3 NWTs. (**a**) Drain current versus gate-source voltage (I_{ds}-V_{gs}) characteristics of In_2O_3 NWT at $V_d = 0.5$ V. Green, red and blue data points corresponding to linear-scale I_{ds}-V_{gs}, log-scale I_{ds}-V_{gs} and field effect mobility. **b,** The Drain current versus drain-source voltage (I_{ds}-V_{ds}) characteristic of In_2O_3 NWT.

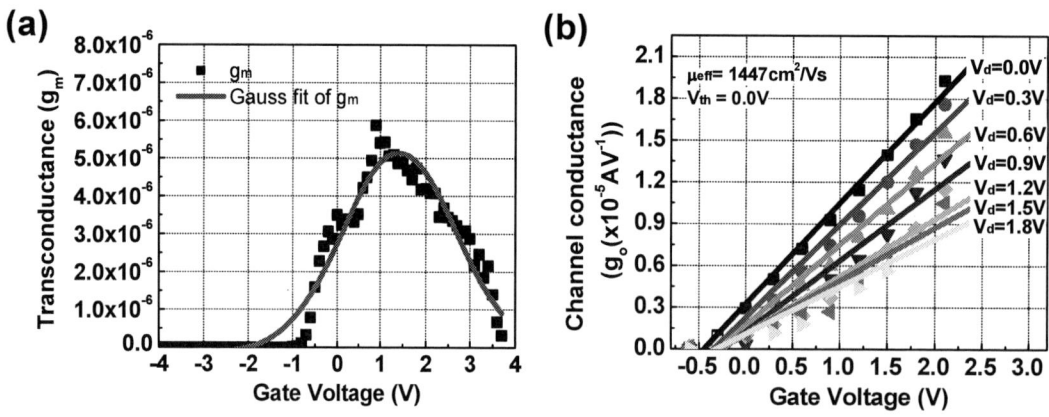

Figure 3 SAND-based In_2O_3 NWTs. (a) Transconductance (g_m) at 5 V_g. (b) Channel conductance (g_d) from 0.0 V_d to 1.8 V_d.

170

Impact ionization FETs based on silicon nanowires

M. T. Björk, O. Hayden, J. Knoch, H. Riel, H. Schmid, and W. Riess

IBM Research GmbH, Zurich Research Laboratory, Saeumerstrasse 4,
CH-8803 Rueschlikon, Switzerland
Email: bjm@zurich.ibm.com

One of the fundamental limits in the scaling of today's MOSFET technology is the room-temperature (RT) limit of 60 mV/decade of the inverse sub-threshold slope. This limit causes an exponentially increasing leakage current as supply, and threshold voltages are scaled down. As a result, the standby power of highly integrated circuits dramatically increases. However, new types of devices based on band-to-band tunneling [1] or impact ionization [2] have recently been demonstrated that circumvent the 60 mV/decade limit thereby offering low leakage currents. Here, we demonstrate vertical integration of a single surround-gated silicon nanowire field-effect transistor (NW FET) with an inverse sub-threshold slope as low as 6 mV/decade at RT that spans four orders of magnitude in current.

Undoped SiNWs were grown via the vapor-liquid-solid growth method in a chemical vapor deposition chamber on n-type Si wafers. One advantage of a vertical approach is that the ungated region between bottom contact and channel is defined by the gate dielectric thickness and not by lithography (see Fig. 1) and is therefore accurately and reproducibly controlled. Conventional lithography and metallization was used to define gates and top contacts to individual NW FETs (details of the fabrication process can be found in Ref. [3]).

Fig. 2 shows the output characteristics of a device biased both in reverse and forward with the top Schottky contact on ground. Due to impact ionization the ON current in reverse bias is roughly two orders of magnitude larger than in forward bias. The transfer characteristic for the same device is displayed in Fig. 3 for bias voltages below 3.5 V. Inverse sub-threshold slopes below 60 mV/decade are obtained in accumulation (hole transport). The large shift of the curves as a function of V_{DS} is due to the strong coupling of the bottom contact to the channel potential. In Fig. 4 the transfer characteristic is again shown, however with the bottom contact (wafer) grounded. In this case, V_{th} changes only slightly with V_{DS} and swing values as low as 6 mV and 5 mV per decade are measured in accumulation and in inversion respectively (Fig 5). The mechanism behind the steep slopes is impact ionization in the ungated regions (close to the wafer for accumulation and at the top of the gate in inversion) as schematically shown in Fig. 6. A disadvantage of IMOS devices is that they are prone to degradation due to hot carrier injection into the gate oxide. Fig. 7 shows that in the present case of a vertical device V_{th} remains constant after several gate sweeps due to the low operating voltages compared to conventional IMOS FETs. These could be lowered even more if considering lower bandgap materials for the channel. The hysteresis observed in Fig. 7 is mainly due to the use of a low temperature gate oxide containing a rather high number of interface states.

In conclusion, vertical impact ionization NW FETs have been demonstrated that allow for very steep sub-threshold swings, low operating voltages and reproducible characteristics.

[1] J. Appenzeller, et al., Phys. Rev. Lett. **93**, 196805 (2004).
[2] K. Gopalakrishnan, et al., IEDM Tech. Dig., 289 (2002).
[3] V. Schmidt et al., Small **2**, 85 (2006).

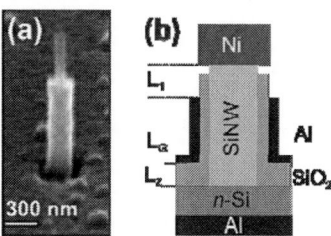

Fig. 1 (a) SEM image of a SiNW with defined Al gate wrapped around the circumference. (b) Schematic of the NW FET architecture showing the critical lengths L_1, L_G, and L_z

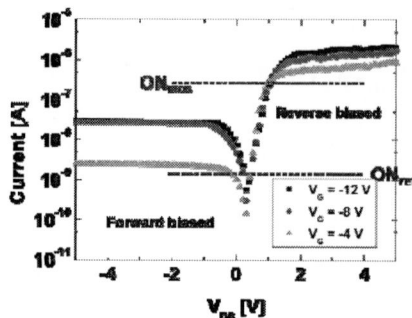

Fig. 2 Output characteristics for an IMOS FET with top contact on ground under forward and reverse biasing conditions. The difference in ON current between the two modes of operation is almost two orders of magnitude.

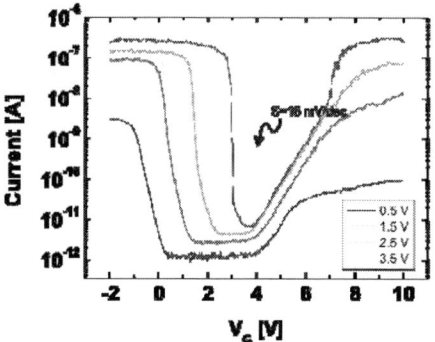

Fig. 3 Transfer characteristics with top contact grounded for different V_{DS}. The inverse subthreshold slope is 16 mV/decade for V_{DS}=3.5 V. The shift of the curves is due to the strong coupling of V_{DS} to the channel

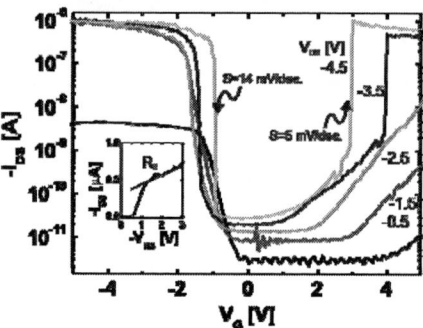

Fig. 4 Transfer characteristics of the same device with the bottom contact grounded. No shift as a function of V_{DS} is observed and SS down to 5 mV/decade is observed both in accumulation and inversion.

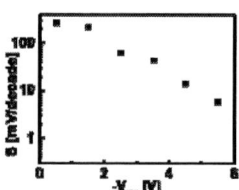

Fig.5 SS as a function of V_{DS} in accumulation.

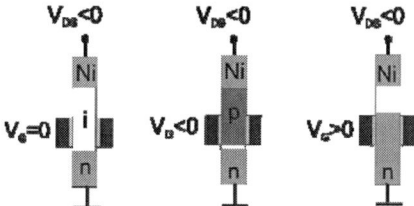

Fig. 6 Schematics of the working principle of the NW IMOS for different biasing conditions.

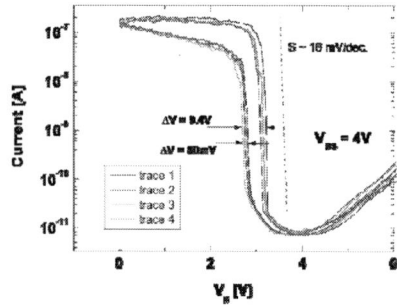

Fig. 7 Repeated measurements show that hysteresis is present due to the gate dielectric but the threshold voltage does not shift.

A Low Power, Highly Scalable, Vertical Double Gate MOSFET Using Novel Processes

Hoon Cho, Pawan Kapur, Pranav Kalavade[*] and Krishna C. Saraswat

Department of Electrical Engineering, Stanford University, CIS, Stanford, CA 94305 USA

*Intel Corporation, Santa Clara, CA 95052 USA

Tell: +1-650-725-3610, Fax: +1-650-723-4659, Email: chozone@stanford.edu

A Sub-45nm body thickness, vertical channel, double gate MOSFETs (VDGFET) is fabricated on bulk-silicon substrate. The process, in principle, is scalable down to sub-5nm body thicknesses. It is realized using very coarse lithography (~1um resolution), does not require CMP, and is capable of being integrated with a planar CMOS flow. It relies on the following key, novel, unit processes: 1) A spacer process obtained using a new set of spacer materials, with demonstrated mask thickness down to 5nm, 2) a self-aligned process for achieving a thicker bottom/corner field isolation oxide, thus, minimizing leakage and parasitic capacitances, 3) a novel drain contact process including etch stop and implant, which overcomes the problem of contacting very thin pillars. Electrical results show excellent short channel effect (SCE) immunity including very low DIBL and GIDL effects. In addition, a close to ideal measured subthreshold slope (64mV/decade) results in a very high I_{ON} to I_{OFF} ratio.

Double gate MOSFETs have been researched aggressively in the recent years due to their scalability advantage [1,2]. However we present, a vertical channel (S/D) (current flow perpendicular to the substrate), symmetric double gate NMOSFET (VDGFET) using several innovated unit processes. Each unit process possesses the versatility to be independently used in other applications. Best implant conditions and their order was determined using extensive TSUPREM4[TM] simulations. Drain was implanted separately for better channel length control. Excellent electrical properties including Id-Vd, Id-Vg and Ig-Vg are observed. In addition, the impact of 1) body thickness (Tsi), 2) drain contact area, and 3) nitride layer (etch-stop/strain-inducer) on channel current enhancement is also studied.

We target a low body doping of 3e15cm^{-3} to obtain a fully depleted operation for sub-50nm body thickness. Despite this, we end-up starting with a 1e17cm^{-3} P-type (100) bulk silicon substrate because of severe boron segregation in thin bodies. In Fig. 1, steps a) through c) highlight the spacer process flow. Silicon-nitride (Si$_3$N$_4$) spacer, formed around poly-Si block, is used as the hard mask [3], differentiating it from spacer processes used before. The process can manufacture spacer-define masks down to 5nm thickness (smallest demonstrated to-date) with very high aspect ratio (Fig. 2). The Si-fin etch, using the spacer mask, was catered to yield a flaired-out base for a low source resistance. The source tilt implant had a low temperature oxide (LTO) sidewall protecting the channel and the nitride spacer protecting the drain. This was followed by a novel self-aligned (to the fin) process to obtain a thicker bottom/corner oxide field isolation for low leakage (Fig. 3). The process exploited a curvature-dependent differential oxidation rate of N+ doped poly-Si [4]. A faster oxidation at the top/sides results in a complete consumption of poly-Si, while, a slower oxidation rate leaves a thin poly-Si at the bottom and corners. This acts as a selective hard mask for the underlying field LTO etch (Fig 4). A high reproducibility renders this process superior to other techniques for getting thick field oxide. A subsequent gate stack deposition/definition and active area isolation (step f, Fig. 1) was followed by a deposition/blanket etch of a nitride (Si$_3$N$_4$) etch stop layer for drain contact etch.

Vertical devices require a robust field isolation and corner leakage reduction strategy. Fig. 5 demonstrates a significant reduction in leakage current of a MOS capacitor using our self-aligned, thicker bottom/corner oxide, process (subsequently also used in the transistor process). Fig. 6 shows the final VDGFET structure (250nm channel length, 45nm Tsi, and thick bottom/corner oxide). A complete process simulation (TSUPREM4[TM]) including the entire thermal budget, shows a channel doping concentration of 3e15cm^{-3} (Fig. 7). Fig. 8 and 9 show the experimentally measured Id-Vd and Id-Vg curves. Despite a large oxide thickness, excellent subtreshold slope (64mV/decade) and extremely low DIBL is observed. Fig. 10 shows the measured Id-Vg curves for three different body thickness (Tsi=50nm, 80nm, and 2.8μm). As expected, a larger Tsi increases subthreshold slope and DIBL. In addition, a low gate-induced drain leakage (GIDL) is seen for both Tsi=50nm and 80nm. Fig. 11 shows that as the drain contact area is increased with controlled etch, a significant increase in I_{dsat} is observed, pointing to a large mitigation of series resistance. An inverted curvature at low Vds (Fig. 9 and 11) is a result of a less than perfect ohmic contact because of P+ Si diffusion from the Al/Si metal [5,6]. A proper silicide and a larger drain contact (Fig. 12 D) can easily address this problem. Finally, the VDGFET with (Fig. 12, D) and without nitride etch stop (Fig. 12, E and F) are compared to quantify its impact on channel current. Fig. 12, E and F, because of their larger size, did not require nitride etch stop. All devices in Fig. 12 have similar drain contact size (2.65μm^2). Device D (thin body with nitride) exhibits a larger I_{dsat} because of tensile stress (S_{zz}) enhanced mobility on a (110) channel [7].

In conclusion, we have demonstrated a vertical double gate MOSFET which is capable of being scaled down to sub-5nm body even with very coarse lithography resolution. It is enabled by several novel unit processes and is capable of being integrated with planar CMOS flow. We show excellent I-V characteristics, especially the short channel immunity, leading to a low power and highly scalable option for next generation.

[1] Wong et al., IEDM p.427 (1997); [2] http://public.itrs.net; [3] Cho et al., Trans. Nanotech., v.5, p.554 (2006); [4] H. Cho et al., ECS v.3 p403 (2006); [5] J. Moers et al., DRC p.191 (2001); [6] M. Mori, Trans. Elec. Dev. v.30, p. 81 (1983); [7] S. Ito et al., IEDM p. 10.7.1 (2000)

978-1-4244-1101-6/07/$25.00 ©2007 IEEE

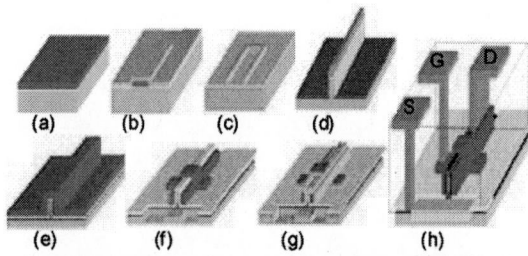

(a) P-type substrate(1e17cm⁻³), dry pad oxidation, poly-Si block deposition
(b) Poly block etch(Mask 1), few nm thick(~Tsi) nitride deposition
(c) Blanket nitride spacer etch, remove poly block by TMAH etch
(d) Silicon pillar(fin) etch, 40nm LTO deposition and blanket etch, Source implant
(e) Remove all nitride and pad oxide, anneal, **self align process** explanted in Fig. 3
(f) Gate poly etch(Mask 2), Isolation trench etch(Mask 3), dry oxidation
(g) **Nitride etch stop** deposition and blanket etch for drain contact, Drain implant
(h) PSG deposition, RTA, contact etch(Mask 4), metal deposition and etch(Mask 5)
*No CMP for the entire process

Fig. 1 Process flow for the vertical double gate transistor, (a)~(c) spacer process using poly-Si, (e) novel self align process

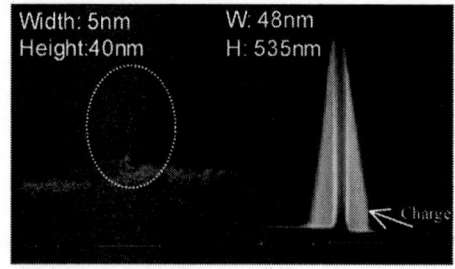

1) Si pillar(fin), 2) LTO depo.
3) N+ doped poly-Si depo.
4) Oxidation until top poly-Si is fully consumed
5) HF dip, remained poly-Si acts as a hard mask
6) Sacrificial oxidation
7) HF dip, gate oxidation
8) N+ Gate poly deposition

Fig. 2 SEM pictures showing very thin(5nm) and high aspect ratio nitride spacers made by our spacer process

Fig. 3 The self-align process flow for thick bottom/corner oxide using the differential oxidation of poly-Si

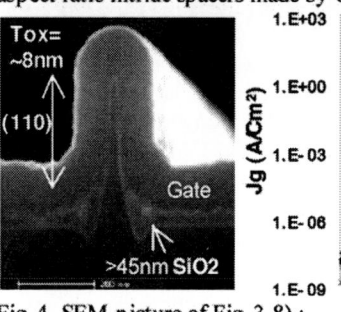

Fig. 4 SEM picture of Fig. 3-8) : thick bottom/corner oxide preventing leakage and parasitic capacitance

Fig. 5 Substrate injection I-V characteristics of vertical MOScaps

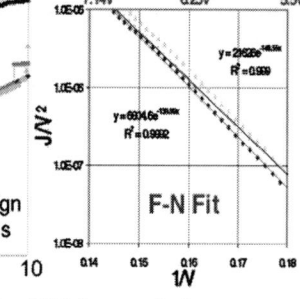

Fig. 6 SEM picture of the vertical double gate FET

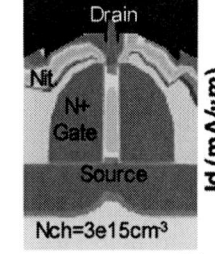
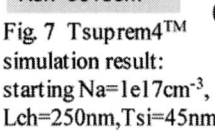

Fig. 7 Tsuprem4™ simulation result: starting Na=1e17cm⁻³, Lch=250nm,Tsi=45nm

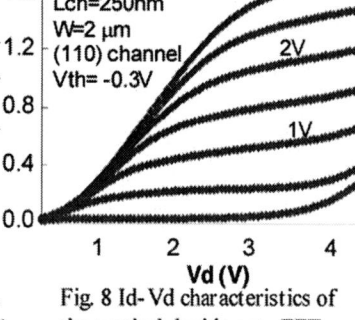

Fig. 8 Id-Vd characteristics of the vertical double gate FET

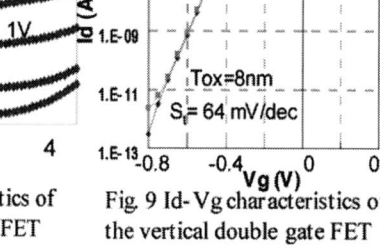

Fig. 9 Id-Vg characteristics of the vertical double gate FET

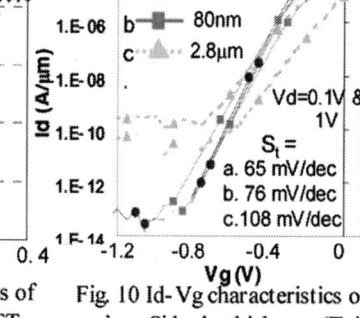

Fig. 10 Id-Vg characteristics of various Si body thickness (Tsi)

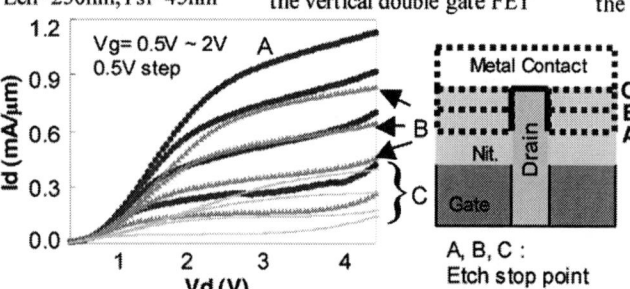

Fig. 11 Id-Vd characteristics of three different contact resistances: A has around three times larger drain contact area than C

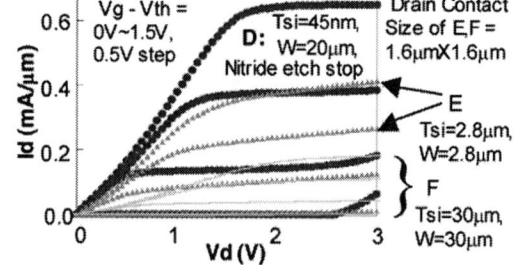

Fig. 12 Id-Vd characteristics of vertical DG FETs with similar drain contact size : E and F have no nitride etch stop

174

An n-FET with a Si nanowire channel and doped epitaxially-thickened source and drain regions

G.M. Cohen, P.M. Solomon, and S.E. Laux, J.O. Chu, M.J. Rooks, and W. Haensch

IBM T. J. Watson Research Center, Yorktown Heights, NY 10598 USA

Nanowire FETs are typically fabricated with Schottky source and drain contacts [1-2]. More recently doped source and drain contacts obtained by modulated in-situ doping [3] or by masked ion implantation [4] were also achieved. In this report we demonstrate greatly improved on-resistance for grown Si nanowire FETs by using epitaxial doped source and drain and, for the first time, estimate a gate-voltage dependent mobility using combined electrical I_d-V_g measurements and 2-D Poisson-Schrödinger simulations.

The main process steps of the device fabrication are shown in Figure 1(a-d). The silicon nanowires were epitaxially grown by the vapor-liquid-solid VLS method in a UHV-CVD chamber, with silane as the silicon precursor and gold as the catalyst. The gold catalyst was etched off and the nanowires were dispensed on a heavily doped host silicon substrate capped by a thin dielectric stack of Si_3N_4 and SiO_2. The host substrate was later used for back-gating the nanowire FETs. A dielectric film of silicon oxide and silicon nitride was blanket deposited over the nanowires, and patterned with contact holes to expose the segments of the nanowire where the source and drain regions were fabricated. The source and drain regions were formed by adding silicon to the exposed nanowire body and implanting it with P. A 1000°C/5 sec rapid thermal anneal was used to activate the implanted dopants. Chemical mechanical polishing (CMP) was applied to electrically isolate the source and drain regions by removing the excess silicon over the silicon nitride. Self-aligned nickel silicide was used to form contacts to the source and drain.

The devices were measured following silicide formation. The wiring resistance from the probe pads to the device terminals was 5.3kΩ. Each device that was measured was verified to have a single nanowire channel. The I_d-V_g characteristics of a nanowire n-FET are shown in Fig. 2(a). The device shows unipolar transport, in contrast with a control device made with Schottky contacts (inset) which exhibits ambipolar transport and a much lower on-current. Figure 2(b) shows I_d vs. V_{ds} for the same device. The subthreshold slope of this device is 150 mV/dec. The large subthreshold slope is not a short-channel effect but is probably due to interface states introduced by the lack of a thermal oxide and the use the Si_3N_4/SiO_2 stack for a gate dielectric.

An estimate of the external resistance R_{ext} was obtained by plotting $R_{on}=R_{ch}+R_{ext}$ as a function of $1/(V_g-V_t)$ as shown in Fig. 3. A linear extrapolation gives R_{ext} of 24kΩ. Subtracting the wiring resistance, and multiplying by the contact area A, gives a contact resistivity of $R \cdot A = 2.4 \cdot 10^{-7}$ Ω·cm². This is much lower than has been reported so far [2].

Since the device capacitance is too small to measure, an *estimate* of the device mobility was obtained using the following procedure. The low drain bias I_{ds}-V_g curve ($V_{ds}=0.1V$) was used to obtain the linearly extrapolated threshold voltage, V_{on}, from which the threshold voltage is obtained as $V_t=V_{on}-V_{ds}/2$ [5]. The charge per unit length, $Q_i(V_g)$, was calculated using a 2-D Poisson-Schrodinger solver [6]. Examples of the calculated charge distribution for low and high gate voltages are shown in Fig. 4. For each device the voltage axis of the calculated charge, $Q_i(V_g)$, is translated to match the measured device threshold voltage V_t. The gate-voltage dependent mobility is obtained by the ratio of the conductance, $g_d=I_{ds}/V_{ds}$, to the charge per unit length, $\mu(V_g)=g_dL/Q_i$, where L is the channel length [5].

To compare the estimated mobility with the universal mobility curve [7], the effective normal electric field was adjusted to account for a cylindrical nanowire channel geometry using the following procedure: Along the nanowire circumference a local effective field was calculated as $F_e=F_s-(1-\eta)\rho_s/\varepsilon$ (see Table 1), where F_s is the local normal field, inside the nanowire, at the surface, as calculated by the Poisson-Schrödinger solver, ρ_s is the integrated mobile charge in a sector of the surface where F_s was calculated, divided by the arc length of the sector, η is the inversion charge factor [7], and ε is the silicon permittivity. The weighted average effective normal field is then calculated as $F=\int F_e\rho_s dl / \int \rho_s dl$. The (110) oriented nanowires have both (100) and (111) facets which correspond [7] to η of 1/2 or 1/3, respectively. In this work we used η=1/3. The estimated electron mobility for two nanowire diameters is plotted in Fig. 5. We note that the procedure used to estimate the mobility does not include interface charge. As a result the total mobile charge may be over estimated, which would yield a lower estimated mobility value.

In conclusion we demonstrated that doped epitaxial contacts to a Si nanowire improve R_{on} and eliminate ambipolar conduction. Our n-FET mobility was shown to be comparable to standard n-channel Si mobility.

[1] Y. Wu et al., Nature, **430**, 61, (2004). [2] W. M. Weber et al., Nano Lett., **6**(12), 2660, (2006). [3] Y. Wang et al., Device. Res. Conf. digest, 175, (2006). [4] O. Hayden et al., Small, **3**(2), 230, (2007). [5] Y. Taur and T.H. Ning, Fundamental of Modern VLSI Devices, 138, (1998). [6] S.E. Laux et al., J. Appl. Phys., **95**(10), 5545, (2004). [7] S. Takagi et al., Trans. Electron Dev, **41**(12), 2363, (1994).

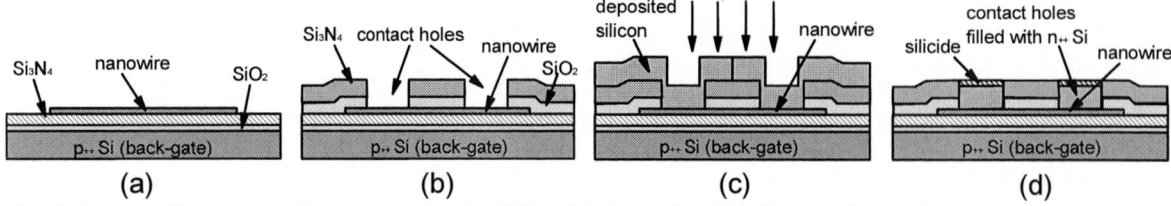

Fig. 1: Process flow used to fabricate nanowire FETs. (a) Nanowires are dispensed on a doped substrate (used as a back-gate) (b) A Si_3N_4 SiO_2 is blanket deposited and patterned to define the source and drain (S/D) regions. (c) Silicon is added to the nanowire in the S/D regions and ion implantation is used to dope the S/D regions. (d) CMP isolates the S/D regions, and a self-aligned silicide is applied to form the contacts to the device.

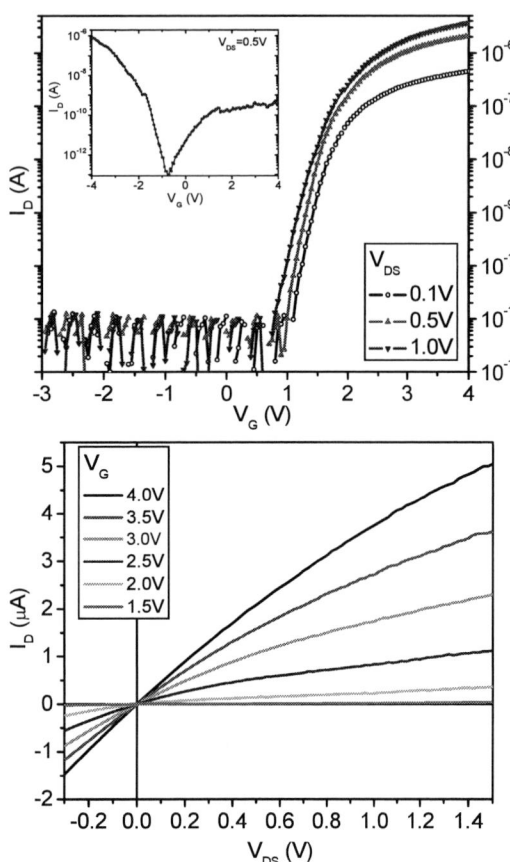

Fig. 2: (a) I_d-V_g characteristics of a device with a 0.95 μm long channel (as defined by the source/drain spacing), a 40 nm diameter nanowire channel and doped source/drain. The inset shows a control device made with a similar nanowire but with Schottky (nickel) contacts which exhibits ambipolar characteristics.
(b) I_d-V_{ds} characteristics of the device.

Table 1:
$$\begin{cases} (1) \quad F_e = (\rho_b + \eta\rho_s)/\varepsilon \\ (2) \quad F_s = (\rho_b + \rho_s)/\varepsilon \end{cases}$$
$$\rightarrow F_e = F_s - (1-\eta)\rho_s/\varepsilon$$
Equation (1) is the effective field definition as in [7].
Equation (2) is the surface field due to impurities and mobile charge.

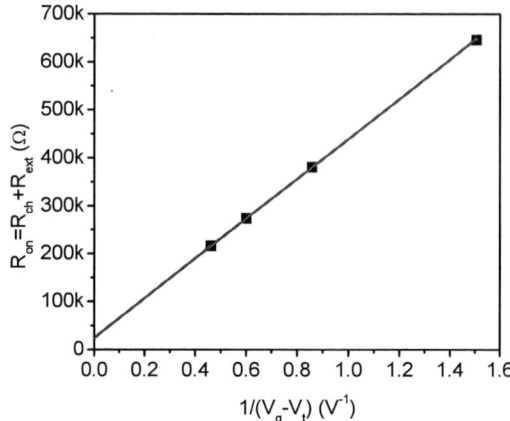

Fig. 3: Extrapolation of R_{ext}.

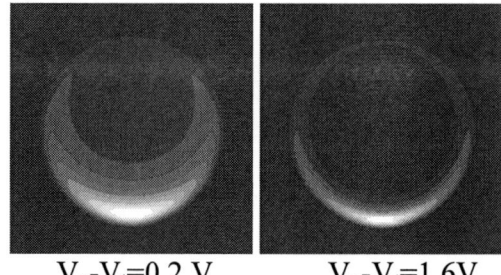

V_g-V_t=0.2 V V_g-V_t=1.6V

Fig. 4: Calculated electron density in a nanowire. A 50×50 nm area is shown. The normalized peak concentration is $8.3\cdot10^{17}$ cm^{-3} (left) and $1.4\cdot10^{19}$ cm^{-3} (right).

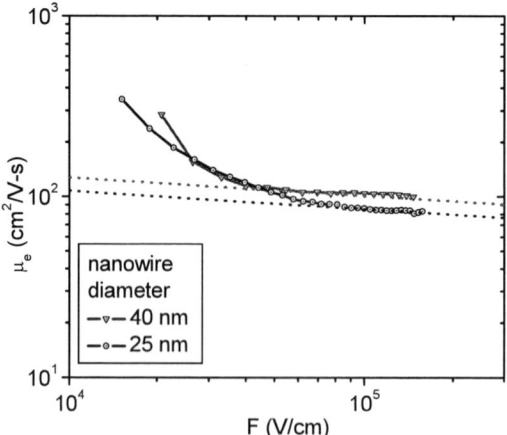

Fig. 5: Estimated electron mobility as a function of the effective field. The upturn at low field could be due to uncertainty in V_t.

Reduction of Acoustic Phonon Limited Electron Mobility due to Phonon Confinement in Silicon Nanowire MOSFETs

Junichi Hattori[a,c], Shigeyasu Uno[a,c], Nobuya Mori[b] and Kazuo Nakazato[a,c]

[a]Department of Electrical Engineering and Computer Science, Nagoya University
Furo-cho, Chikusa-ku, Nagoya 464-8603, Japan
Phone: +81-52-789-2794   Fax: +81-52-789-3139   E-mail: j_hattor@nuee.nagoya-u.ac.jp
[b]Department of Electronic Engineering, Osaka University
2-1 Yamada-oka, Suita, Osaka 565-0871, Japan
[c]SORST JST

1. Introduction

Recently, silicon nanowire (SiNW) MOSFETs, illustrated in Fig. 1, emerged as one of solutions for reducing short channel effects [1]. In predicting device performance, correct modeling of electron mobility is needed. For SiNW MOSFETs, however, impact of confined phonons on electron mobility has yet to be revealed. In this work, acoustic phonons in a SiNW are analyzed in detail, and their impact on electron transport is examined theoretically.

2. Acoustic Deformation Potential Scattering in SiNW with Bulk-like Phonons

Figure 2 shows our model of a SiNW. For simplicity, the SiNW with radius a is assumed to be free-standing. Using relaxation time approximation and isotropic scattering assumption, the acoustic phonon deformation potential (ADP) scattering rate for an initial electron state (m, n, k) reads

$$\frac{1}{\tau_{mn}(k)} = \sum_{m',n'} \frac{D_{aco}^2 k_B T_L}{2\hbar v_l^2 \rho} \times \int I_{mn \to m'n'} \delta \left[E_{m'n'}(k') - E_{mn}(k) \right] dk', \quad (1)$$

where m and m' are electron azimuthal quantum number before and after scattering, n and n' are electron radial quantum number, k and k' are electron wave vector along the z-axis, and $I_{mn \to m'n'}$ is form factor. The other notations have the conventional meanings. Using bulk-like phonons as the phonons in the SiNW, the form factor is given by

$$I_{mn \to m'n'} = \iint |\varphi_{m'n'}(r,\theta)|^2 |\varphi_{mn}(r,\theta)|^2 r\,dr\,d\theta, \quad (2)$$

where φ_{mn} and $\varphi_{m'n'}$ are electron wave functions in confinement plane before and after a scattering event. We use electron wave function in the infinite potential well for simplicity. Figure 3 shows the electron scattering rate in a SiNW with radius 5nm from calculation using Eqs. (1) and (2) plotted as a function of total electron energy. The scattering rate diverges when the total electron energy is equivalent to any electron subband energy. We can calculate electron mobility for each subband using the scattering rate given by Eq. (1), and obtain total electron mobility as wighted average using electron density in each subband as weighting factor. Figure 4 shows electron mobility calculated with bulk-like phonons.

3. Confined Phonon in SiNW

Acoustic phonon normal modes in a free-standing SiNW can be obtained by solving Navier's equation with the free surface boundary condition. Normal modes are categorized into three submodes, torsional, dilatational and flexural. Figure 5 shows the dispersion relations of each mode, and Fig. 6 shows the bird's-eye views of phonon vibrations of each mode at the SiNW surface. The phonon waves of torsional and dilatational modes are axial symmetric. For torsional modes, the phonon waves have only azimuthal component, therefore the SiNW is twisting. In dilatational modes, the phonon waves have only radial and axial components in contrast to the torsional modes. As a result, the cross-section of the SiNW repeats dilatations and compressions. The phonon waves of flexural modes are not axial symmetric and have all three components. The phonon azimuthal quantum number $|m_p|$ classifies these modes into finer modes such as flexural modes of order $|m_p|$. Figure 5(c) and Fig. 6(c) correspond to the lowest order flexural modes ($|m_p| = 1$).

4. ADP Scattering in SiNW with Confined Phonons

In calculating the scattering rate according to Eq. (1) using confined phonons, form factor is given by

$$I_{mn \to m'n'} = \frac{L_z \rho}{v_l^2} \sum_{q_l} \omega_{q_l} \left| \langle n' | A J_{m_p}(q_l r) e^{j m_p \theta} | n \rangle \right|^2, \quad (3)$$

where q_l is phonon wave vector along the radial direction, A is normalization factor of phonon wave function, and ω_{q_l} is phonon frequency. When $m = m'$, m_p becomes zero and the only dilatational modes contribute to the scatterings, otherwise m_p becomes $m' - m$ and flexural modes corresponding to $|m_p|$ contribute to the scatterings. Meanwhile torsional modes do not contribute to the scattering events at all. Figure 7 shows the ratio of form factor calculated using confined phonons with respect to that using bulk-like phonons. Note that the form factor with confined phonons is equal to that with bulk-like phonons at large $q_z a$. On the other hand, the form factor with confined phonons is larger than that with bulk phonons at small $q_z a$. This causes the increase of the scattering rate through Eq. (1). For this reason, electron mobility calculated using confined phonons might be smaller than that using bulk-like phonons. In our presentation, further results on the scattering rate and mobility would be shown. Then we discuss the differences from those of bulk-like phonons and the mechanism that causes them.

5. Conclusion

Confined phonons and their impact on electron transport in a free-standing SiNW have been investigated. The form factor with confined phonons was found to be larger than that with bulk-like phonons. This increase would make electron mobility with confined phonons smaller than that with bulk-like phonons.

References

[1] K. H. Yeo, et al., IEDM Tech. Dig., p. 539 (2006).

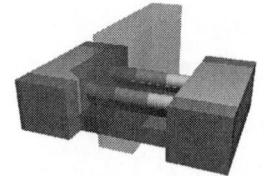

Fig. 1: The schematic diagram of a SiNW MOSFET.

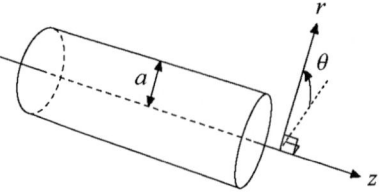

Fig. 2: The SiNW model used in our calculation. The SiNW with radius a is free-standing. The coordinates used in this work are cylindrical as shown in this figure.

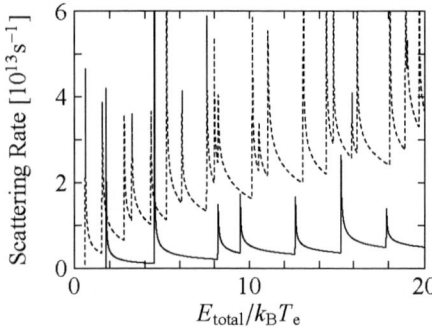

Fig. 3: The scattering rates using bulk-like phonons for 5nm Si nanowire plotted as a function of total electron energy. The solid curve corresponds to the 2-fold electron while the dashed curve to the 4-fold.

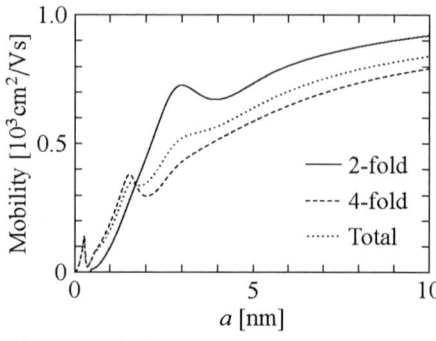

Fig. 4: The ADP limited electron mobility using bulk-like phonons plotted as a function of Si nanowire radius a for 2-fold (solid curve), 4-fold (dashed curve) and average (dotted curve).

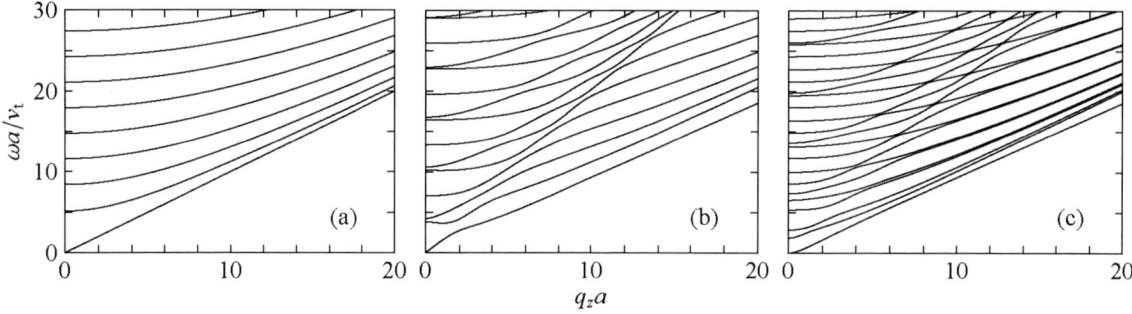

Fig. 5: The dispersion relations for (a) torsional modes, (b) dilatational modes and (c) lowest order flexural modes ($|m_{\mathrm{p}}| = 1$). Each horizontal axis represents phonon wavevector along the z-axis multiplied by a, and each vertical axis represents frequency multiplied by v_{t}/a, where v_{t} is the transverse sound velocilty in Si ($v_{\mathrm{t}} = 4.7 \times 10^3\,\mathrm{m/s}$).

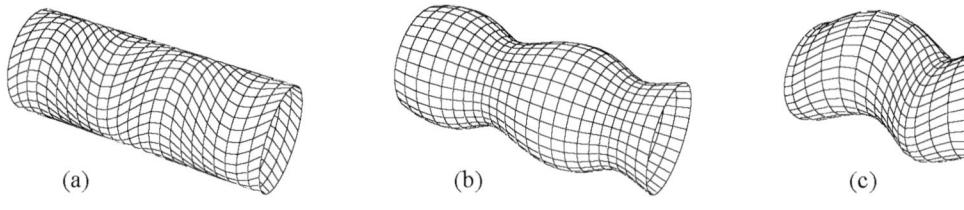

Fig. 6: The bird's-eye views of phonon waves at the Si nanowire surface for (a) torsional modes, (b) dilatational modes and (c) lowest order flexural modes ($|m_{\mathrm{p}}| = 1$).

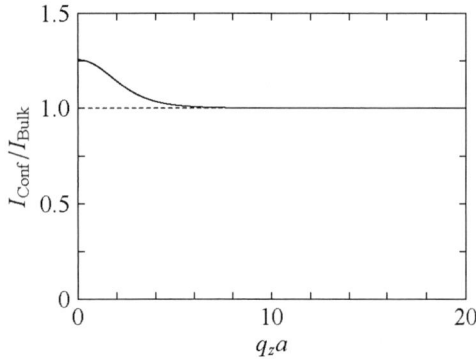

Fig. 7: The form factor of SiNW for intra-lowest-subband scattering ($(m, n) = (m', n') = (0, 1)$) using confined phonons divided by that using bulk-like phonons. Note that the form factor with confined phonons is equal to that with bulk-like phonons at large $q_z a$. On the other hand, the form factor with confined phonons is larger than that with bulk phonons at small $q_z a$. This causes the increase of the scattering rate through Eq. (1). For this reason, electron mobility calculated using confined phonons might be smaller than that using bulk-like phonons.

178

THz probe for nanowire FETs: simulation of few-electron fingerprints

Klaus Michael Indlekofer[1], Radoslav Németh [1], and Joachim Knoch [2]

[1]Institute for Bio and Nanosystems, CNI, Research Center Jülich, D-52425 Jülich, Germany
phone: +49-2461-613580, fax: +49-2461-612940, email: m.indlekofer@fz-juelich.de
[2]IBM Research GmbH, Zurich Research Laboratory, 8803 Rueschlikon, Switzerland

Nanowire-based field effect transistors (FET) represent prototypes for the study of technological as well as physical challenges in future transistor designs. In contrast to conventional devices, future FETs will be strongly influenced by two non-classical effects: quantization of electronic states due to confinement and few-electron Coulomb interaction effects. On the other hand, these effects can be used to obtain valuable information about application-relevant system parameters such as capacitances.

So far, common characterization methods for FETs involve signal frequencies from DC up to 100 GHz. However, typical energy scales of quantized states in realistic nanowire FETs are on the order of a few meV corresponding to the THz frequency range. In this paper, we therefore consider signals in the THz regime to directly probe quantum transitions between few-electron states within a nanowire FET. For the first time, we simulate the intra-band THz response of such devices by means of a many-body quantum approach, taking quantization and Coulomb interaction effects beyond mean-field into account. Combining this spectroscopic approach with a multi-gate design, we obtain spatially resolved information about the electronic spectra inside the FET, far beyond the limitations of standard characterization methods.

Fig.1 shows a schematic sketch of the considered nanowire FET with multiple gate segments. We assume that only one lateral subband needs to be taken into account, corresponding to channel diameters in the sub-50 nm range for InAs. A realistic quantum simulation of such a nanodevice requires the consideration of a sufficiently large number $Z \approx$ 50-500 of single-particle states, resulting in a many-body problem which scales exponentially in Z and thus grows beyond any computational limit. In this context, we have recently introduced a multi-configurational approach (MCSCG, [1,2]) which employs a reduced adaptive basis for the simulation of Coulomb blockade phenomena in nanowire FETs. For the simulation of THz response, however, the number of considered basis states has yet to be increased. As a solution, we have therefore extended the MCSCG by a novel "bucket-brigade" algorithm [3] which enables us to select a few thousand relevant many-body configurations. The following simulation results are based on this algorithm in combination with first-order Kubo theory [4].

For the following idealized example, the outer ends of the channel are fully covered by gate electrodes (gates 1 and 5 in Fig.1). Applying a sufficiently large negative voltage, barriers arise and we thus can assume an almost isolated channel system. For simplicity, we consider a thermodynamical equilibrium state in the low temperature limit in order to obtain a well-resolved occupation of quantum states. The input THz excitation is applied to gate 1 (see Fig.1), whereas the intermediate gate fingers (gates 2,3,4 in Fig.1) serve as spatially resolved probes. (A future experimental realization might employ alternating screen and signal gates, combined with an optimized THz layout.) In our example, we consider two different InAs-based nanowire MOSFETs with SiO$_2$ shell. In both cases we keep the channel length $L = 200$ nm, the total oxide capacitance $C_{ox} = 63$ aF, and the total electron number $N = 4$ constant, whereas the channel diameters are chosen as $d = 4$ nm and $d = 20$ nm with corresponding oxide thicknesses of $d_{ox} = 2$ nm and $d_{ox} = 10$ nm, respectively. Fig.2 shows the simulated charge density along the channel axis. For the 20 nm device (Fig.2b), the single-particle confinement energy dominates, resulting in a charge distribution which resembles the shape of a non-interacting system. In contrast, for the 4 nm case (Fig.2a) which exhibits a larger Coulomb interaction but shorter screening length, one can clearly identify a charge density wave, indicating the onset of a Wigner-like regime with separated electrons owing to the dominating Coulomb repulsion. Finally, Fig.3 shows the simulated spatially resolved THz response spectra for the two cases with a resolution of 40 intermediate probe gates. One has to note that the maxima in this *transition* spectrum are located at those positions where the charge *oscillates* the most, and thus need not coincide with the charge density maxima. Fundamental resonances can be found at 0.85 THz and 1.8 THz for $d = 4$ nm and $d = 20$ nm, respectively. The higher value for the 20 nm case stems from a narrower electron distribution (see Fig.2b). Most noticeably, comparing the qualitative form of the two spectra, signatures of the Wigner-like regime for the 4 nm case (Fig.3a) can be identified via a spatial peak multiplication. The obtained THz spectra thus can be considered as "fingerprints" of the concrete electronic configuration of the channel.

In summary, we have considered a THz probe for a spatially resolved analysis of electronic spectra in nanowire-based transistors employing a multi-segment gate design. For the first time, we have simulated the THz response of few-electron quantum states in nanowire FETs by use of a recently developed numerical many-body technique. The discussed example of an InAs-based device demonstrates that signatures of Wigner-like charge density waves can be identified by use of this method, which lies beyond the scope of standard characterization methods for FETs. Such detailed information about the electronic states within the channel will prove useful for the design and optimization of future one-dimensional nanoelectronic devices.

[1] K. M. Indlekofer et al., Phys. Rev. B **72**, 125308 (2005) [2] K. M. Indlekofer et al., Phys. Rev. B **74**, 113310 (2006)
[3] K. M. Indlekofer and R. Németh, cond-mat/0609540 [4] R. Kubo, J. Phys. Soc. Jpn. **12**, 570 (1957)

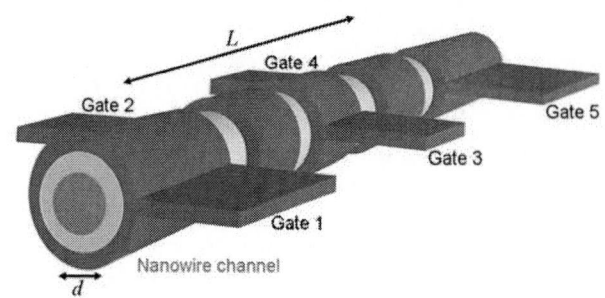

Fig.1: Schematic view of an idealized nanowire MOSFET in multi-gate configuration.
Gates 1 and 5 are negatively biased and thus provide barriers. Gates 2,3,4 act as probes.
In the simulations, the inter-gate gaps are assumed to be negligible.
Scaling of d for constant C_{ox} and L implies an oxide thickness $d_{ox} \sim d$.

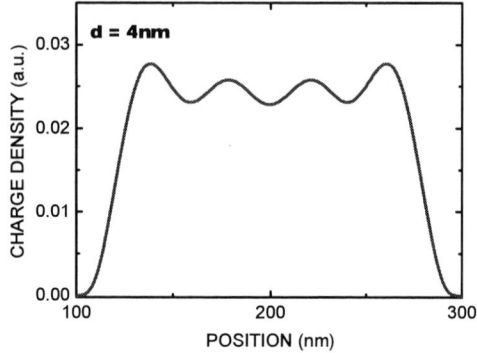

Fig. 2a: Simulated electron charge density for an InAs nanowire FET with $L = 200$ nm and $d = \mathbf{4\ nm}$. The number of electrons is $N = 4$.

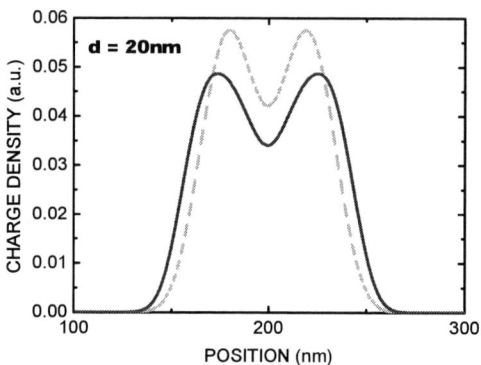

Fig. 2b: Simulated electron charge density for an InAs nanowire FET with $L = 200$ nm and $d = \mathbf{20\ nm}$. The number of electrons is $N = 4$.
For comparison, the dashed curve shows the case without Coulomb interaction between electrons.

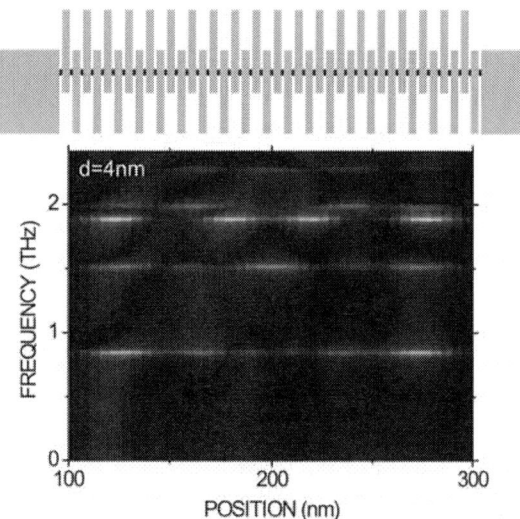

Fig. 3a: Top-view of the nanowire FET (40+2 gates) and the simulated THz transition spectrum for an InAs nanowire with $L = 200$ nm and $d = \mathbf{4\ nm}$. White corresponds to a strong response signal. The number of electrons is $N = 4$.

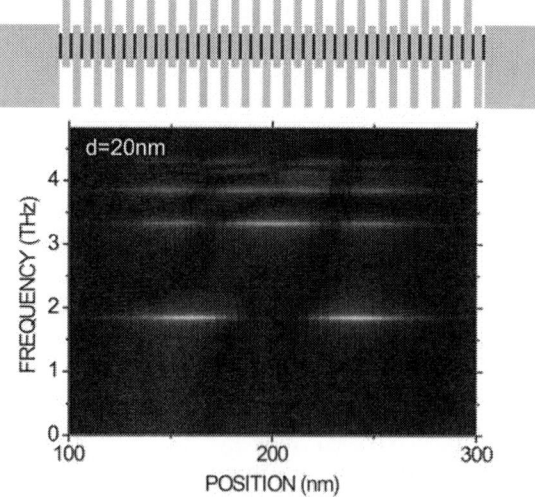

Fig. 3b: Top-view of the nanowire FET (40+2 gates) and the simulated THz transition spectrum for an InAs nanowire with $L = 200$ nm and $d = \mathbf{20\ nm}$. White corresponds to a strong response signal. The number of electrons is $N = 4$.

Session V.A (DeBartolo Hall Room 102)

Photonic Devices

Tuesday PM, June 20th, 2007

Session Organizers: Mike Larson, Agility Communications, Inc.
Session Chairman: TBA

1:30 PM V.A-1 Invited Paper
The first commercial large-scale InP photonic integrated circuits: current status and performance
S. Hurtt, A. G. Dentai, J. L. Pleumeekers, A. Mathur, R. Muthiah, C. Joyner, R. P. Schneider, R. Nagarajan, F. A. Kish, and D. F. Welch, Infinera Corporation, Sunnyvale, California, USA

2:10 PM V.A-2 Student Paper
A hybrid silicon evanescent photodetector
H. Park[1], A. W. Fang[1], R. Jones[2], O. Cohen[3], O. Raday[3], M. N. Sysak[1], M. J. Paniccia[2], and J. E. Bowers[1], [1]University of California Santa Barbara, ECE Department, Santa Barbara, California, USA, [2]Intel Corporation, Santa Clara, California, USA and [3]Intel Corporation, Jerusalem, ISRAEL

2:30 PM V.A-3
High-Frequency Performance of a High-Power Traveling Wave Photodetector with Parallel Optical Feed
A. Beling[1], H.-G. Bach[2], G. G. Mekonnen[2], R. Kunkel[2], D. Schmidt[2] and J. C. Campbell[1], [1]Department of Electrical and Computer Engineering, University of Virginia, Charlottesville, Virginia, USA, and [2]Fraunhofer Institute for Telecommunications, Heinrich-Hertz-Institut, Berlin, GERMANY

2:50 PM V.A-4
Single Ultra Violet Photon Detection with 4H-SiC Avalanche Photodiodes
X. Bai, D. Mcintosh, H. Liu, J. Campbell, Electrical and Computer Engineering, University of Virginia, Charlottesville, Virginia, USA

3:10 PM Break

3:30 PM V.A-5 Invited Paper
The challenges and opportunities for dilute nitride antimonides in photonic devices
Jim Harris , Stanford University, Stanford, California, USA

4:10 PM V.A-6
1.65 μm buffer-free GaSb/AlGaSb quantum-well diode lasers grown on a GaAs substrate operating at room temperature
M. Mehta, G. Balakrishnan, A. Jallapali, M. N. Kutty, L. R. Dawson, and D. L. Huffaker, Center for High Technology Materials, University of New Mexico, Albuquerque, New Mexico, USA

4:30 PM V.A-7
High Temperature CW Operation of Interband Cascade Lasers at ? ˜ 4.0 μm
C. S. Kim, M. Kim, W. W. Bewley, C. L. Canedy, D. C. Larrabee, J. A. Nolde, J. R. Lindle, I. Vurgaftman, and J. R. Meyer, Naval Research Laboratory, Washington District of Columbia, USA

4:50 PM V.A-8 Student Paper
Strain-Compensated AlAs-InGaAs Quantum-Cascade Lasers with Emission Wavelength 3–5 μm
M.P. Semtsiv[1], S. Dressler[1], M. Wienold[1], I. Bayrakli[1], M. Ziegler[2], K. Kennedy[3], R. Hogg[3], and W.T. Masselink[1], [1]Humboldt University, Berlin, Germany, [2]Max-Born-Institut, Berlin, Germany, and [3]University of Sheffield, Sheffield, UK

The first commercial large-scale InP photonic integrated circuits: current status and performance

Sheila Hurtt, Andrew G. Dentai, Jacco L. Pleumeekers, Atul Mathur, Ranjani Muthiah, Charles Joyner, Richard P. Schneider, Radha Nagarajan, Fred A. Kish, David F. Welch

Infinera has commercialized the first digital transport network (DTN) systems as a highly reliable, small footprint system to the telecommunications carriers. The key technology in the DTN is optical-electrical-optical (O-E-O) digital signal processing. O-E-O processing allows the carriers to install systems at more locations, which translates into more entry points into the network. The photonic integrated circuit, or PIC, enables cost-effective O-E-O processing. While research had been ongoing for several decades, Infinera systems are the first to incorporate large-scale photonic integration. Infinera's commercially deployed PICs operate at 100Gb/s on 10 channels (10Gb/s per channel). Each transmit PIC channel consists of a tunable DFB, modulator, power monitor and variable optical attenuator (VOA). The PIC has been deployed worldwide, demonstrating an excellent reliability track-record. The challenges met in commercializing this chip, including the tradeoff between performance, reliability, and a robust processing methodology, will be discussed as well the latest development results, wherein we have demonstrated transmitter chips integrating over 240 elements with a aggregate capacity of 1.6 Tb/s.

A hybrid silicon evanescent photodetector

Hyundai Park[1], Alexander W. Fang[1], Richard Jones[2], Oded Cohen[3], Omri Raday[3], Matthew N. Sysak[1], Mario J. Paniccia[2], & John E. Bowers[1]

[1]University of California Santa Barbara, ECE Department, Santa Barbara, CA 93106, USA
[2]Intel Corporation, 2200 Mission College Blvd, SC-12-326, Santa Clara, CA 95054, USA
[3]Intel Corporation, SBI Park Har Hotzvim, Jerusalem, 91031, Israel
Phone: (805) 893-4235, Fax: (805) 893-7990, Email: hdpark@engr.ucsb.edu

Significant research effort in silicon photonics has focused on realizing individual optical components that are suitable for photonic integrated circuits, including active devices such as lasers, modulators, and photodetectors as well as passive waveguide devices. Photodetectors are one of the important components that convert optical signals into the electrical domain for further signal processing and data manipulation. Germanium waveguide photodetectors (WPD) have been demonstrated using selective growth on a silicon-on-insulator platform [1], and a SiGe WPD has been investigated to reduce the lattice mismatch experienced by Ge photodetectors [2] in the wavelength regime of 1.3 µm and 1.5 µm. The work presented here is a silicon evanescent waveguide photodetector utilizing AlGaInAs quantum wells as an absorbing region, covering a wavelength range up to 1600 nm with a quantum efficiency of 90 %. The materials and processing are compatible with the hybrid silicon photonic integrated circuit (PIC) technology platform which has already demonstrated lasers [3] and optical amplifiers [4].

The device is comprised of AlGaInAs quantum wells bonded to a silicon waveguide as shown in Fig. 1. As light propagates through the hybrid waveguide, it is absorbed in the III-V region, generating electron hole pairs which are swept away as shown with the three arrows in Fig 1a. The input to the photodetector is a passive silicon waveguide. At the junction of the hybrid waveguide and the passive silicon waveguide, the III-V region is tilted by 7° to reduce the reflection at the waveguide transition. The final III-V absorbing region length in the hybrid photodetector is 400 µm with a silicon waveguide width of 2 µm. An SEM image of the final fabricated hybrid photodetector is shown in Fig. 1b.

Figure 2 shows the measured TE spectral response with different biases. TE responsivity at 1550 nm is 0.31 to 0.32 A/W, and is roughly constant over a range of reverse bias conditions from 0.5V to 3V. At a reverse bias of 3 V the internal quantum efficiency is ~ 90 % at 1550 nm using the measured -5.5 dB fiber coupling loss. The TE material absorption coefficient is estimated to be 1594 cm^{-1} at zero bias by measuring output power from a silicon output waveguide and using the calculated confinement factors. The edge of the spectral response is red-shifted with a higher reverse bias since the applied electric field increases the absorption at longer wavelength. Figure 3 shows the photocurrent output as a function of the coupled input power at 1550 nm with different reverse biases. The 1-dB saturation input power is 1.8 mW and 8.8 mW for 0 V and 1 V reverse bias, respectively. No output current saturation is observed beyond a reverse bias of 4 V for the available 14 mW of fiber coupled power. The dark current is typically 50 nA to 200 nA with a reverse bias range of 1V to 4 V, and breakdown occurs when the reverse bias exceeds 16 V. The dark current increases exponentially as reverse bias is increased and it indicates that the dark current is likely dominated by band-to-band tunneling. The 11 ohm series resistance is due to the thin InP n-layer and the contact resistances. The measured device capacitance with different reverse biases is shown in Fig. 4. The capacitance is 7.5 pF under zero bias and decreases down to 5.3 pF as the reverse bias increases. This large capacitance is mainly due to the large III-V mesa size (12 µm x 400 µm). The capacitance of the III-V mesa is calculated to be 3.8 pF with zero bias ignoring the air fringe capacitance. Moreover, two p-probe pads contribute an additional capacitance of 2.95 pF from a 450 nm thick SiNx layer (ε=7.5) sandwiched between the p-probe pad and the n-layers. The capacitance of the mesa and p-pad capacitance can be minimized by reducing the width and length of the III-V mesa, and changing the SiNx insulation layer to a several micron thick benzocyclobutene (BCB, ε =2.6) layer respectively. The bandwidth of the photodetector was measured to be 470 MHz at a reverse bias of 4 V with a 50 Ω termination. This agrees with an RC limited bandwidth of 480 MHz calculated from the measured series resistance and capacitance. Higher bandwidth can be achieved by minimizing the capacitance of the device and incorporating traveling wave electrode designs [5]. Figure 5 and 6 show the calculated quantum efficiency and band width of silicon evanescent traveling wave photodetectors. The bandwidth calculation includes the effects of the velocity mismatch between optical and electrical signal, impedance mismatch to 50 ohm transmission line, microwave loss and carrier transit time in the undoped absorber layers. The calculation shows more than 10 GHz bandwidth is achievable with more than 50 % quantum efficiency.

We thank Mike Haney, Jag Shah and Wayne Chang for supporting this research through DARPA contracts, W911NF-04-9-0001 and W911NF-05-1-0175.

[1]   J. F. Liu, et al., Appl. Phys. Lett, 87, 011110 (2005).
[2]   J. F. Liu, Proc. Group IV Photonics Conference, ThA2, (2006).

[3]   A. Fang et al., Optics Express15, 2315-2322 (2007).
[4]   H. Park et al., IEEE Photonics Technology Letter 19, 230-232, (2007).
[5]   K. S. Giboney, et. al., IEEE Journal of Selected Topics Quantum Electronics 2, 622-628, (1996).

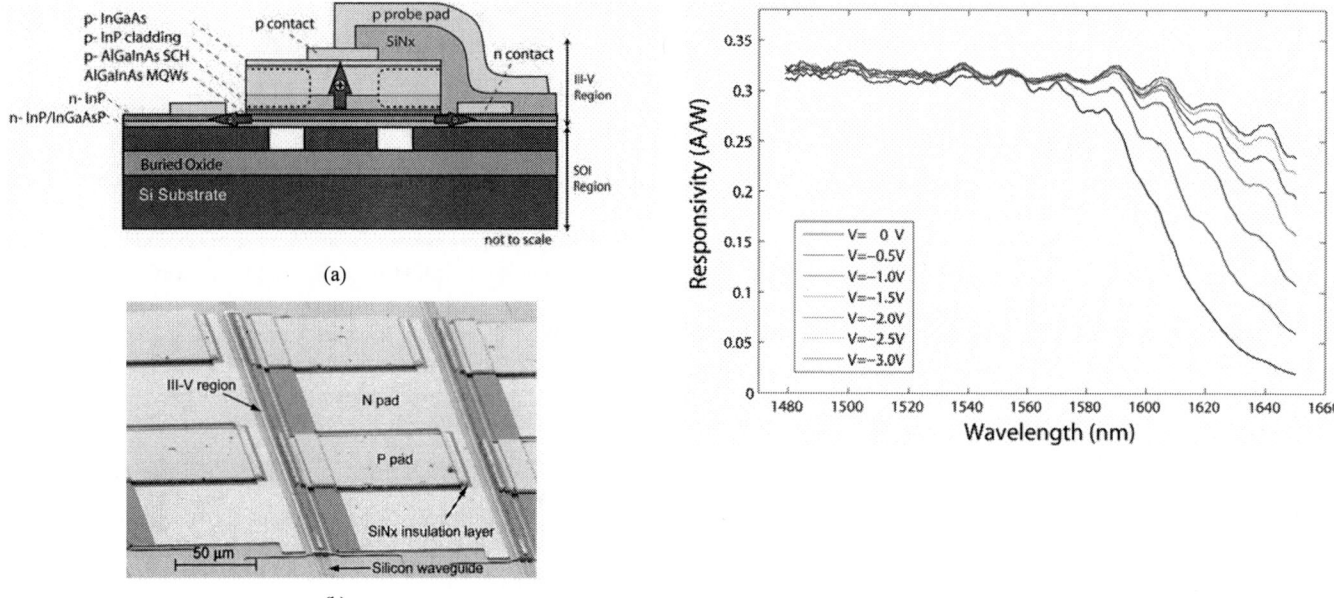

(a)

(b)

Fig. 1. (a) Device cross section (b) A SEM image of a fabricated silicon evanescent photodetector.

Fig. 2. Spectral response for TE polarization.

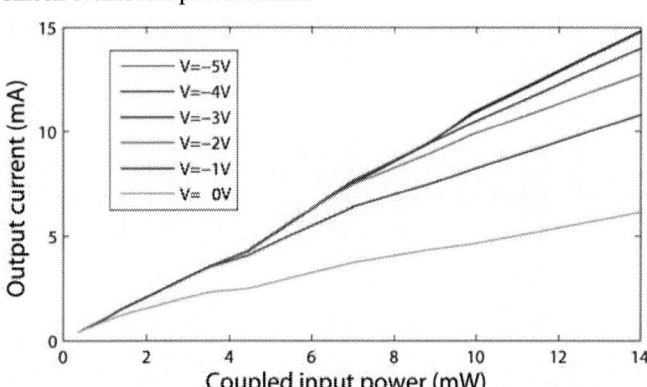

Fig. 3. Saturation characteristics with different biases.

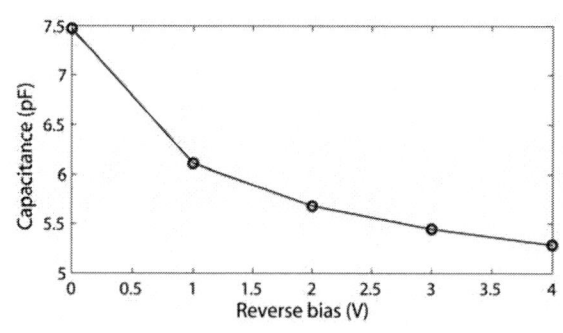

Fig. 4. Dependence of capacitance on reverse bias.

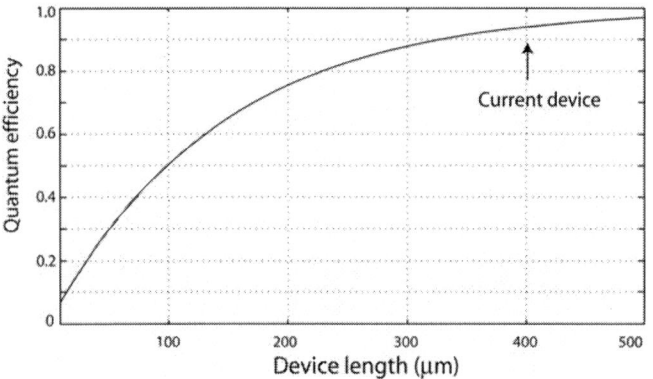

Fig. 5. Calculated quantum efficiency

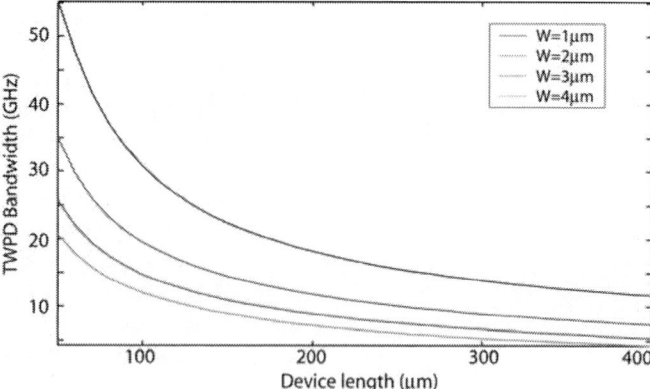

Fig. 6. Calculated bandwidth of traveling waveguide photodetectors. Each color represents a different mesa width.

High-Frequency Performance of a High-Power Traveling Wave Photodetector with Parallel Optical Feed

A. Beling[1], H.-G. Bach[2], G. G. Mekonnen[2], R. Kunkel[2], D. Schmidt[2] and J. C. Campbell[1]

[1]Department of Electrical and Computer Engineering, University of Virginia, 351 McCormick Road, Charlottesville, VA 22904, USA, Tel.: +1 434 243 2147, Fax: +1 434 924 8818, E-mail: ab3pj@virginia.edu
[2]Fraunhofer Institute for Telecommunications, Heinrich-Hertz-Institut, Einsteinufer 37, D-10587 Berlin, Germany

Fast photodetectors with large saturation photocurrent are key components in high-bit-rate fiber networks and photonic microwave applications incorporating optical pre-amplification. For high-speed operation, the photodetector has to be designed for low capacitance and small carrier transit times. These considerations lead to a reduced size of the photodiode (PD), which, however, results in less high-power handling capability and a lower saturation photocurrent. One way to overcome this trade-off between speed and saturation photocurrent is to distribute symmetrically the optical signal to several photodiodes and combine their photocurrents by means of a transmission line [1,2]. Now, due to the uniform optical power distribution, the unsaturated output photocurrent scales directly with the number of photodiodes. By embedding the discrete PDs within a transmission line, a traveling wave photodetector (TWPD) can be formed of which characteristic impedance can be matched to that of the external microwave circuit and a phase match between the propagating optical and electrical signals can be achieved [3]. Since the frequency response is not limited by the overall RC time constant the bandwidth of the single photodiode can be retained, to a large extent, within the Bragg limit.

The studied traveling wave photodetector chip comprises a mode field converter for effective fiber-chip coupling [4] and a 1 x 4 multi-mode interference (MMI) power splitter (width: 43 µm, length: 1038 µm) of which output waveguides feed four p-i-n PDs, each with an active area of 4 x 7 µm^2. The InGaAs/InGaAsP heterostructure PDs with an intrinsic InGaAs absorption layer thickness of 200 nm were optimized to provide high responsivity [5]. A coplanar waveguide (CPW) transmission line connects the PDs in parallel and collects the electrical output signal. By choosing a CPW with an impedance of 85 Ω and a spacing between adjacent PDs of d = 90 µm, an impedance match of the TWPD to the 50 Ω-environment as well as a phase match were calculated. The electrical Bragg frequency was determined to be >200 GHz, which is sufficiently high to provide a smooth frequency response up to more than 100 GHz. In order to eliminate electrical reflections at the input of the transmission line a matching resistor (R_{50} in Figs. 1 and 2) was integrated at the expense of half of the radio frequency (RF) photocurrent being lost. Details on the epitaxial layer stack and the fabrication process can be found in [5]. Using a cleaved fiber for the input, a responsivity of 0.24 A/W with a polarization dependent loss of only 0.2 dB was measured at 1.55 µm wavelength. In order to determine the characteristic impedance and the electrical phase velocity the four S-parameters were measured using a network analyzer and contacting both ends of the CPW within a TWPD without R_{50}. At a bias voltage $V_{bias} = -3.5$ V, an impedance match of >84 % up to 50 GHz is found. The increase in the characteristic impedance with increasing reverse bias can be attributed to the decreasing p-n junction capacitances and thus a decrease in the capacitive loading of the CPW. A similar trend is measured for the electrical phase velocity, which compares well with the optical signal velocity in the singlemode waveguide. Since the electrical and optical paths between adjacent PDs have approximately identical lengths, the phase match is achieved by equalizing both signal velocities. The frequency characteristics of the TWPD were determined with the optical heterodyne technique and reveal a 3 dB bandwidth of 80 GHz. At 170 GHz, the response has dropped by only 10 dB. Circuit simulation suggests that the bandwidth is primarily limited by the series resistances of the photodiodes and transmission line losses. The high-power characteristics of the TWPD were measured at several fixed beat frequencies. A maximum electrical output power of -2.5 dBm is measured at 150 GHz. Compared to a single PD from the same wafer with an active area of 4 x 7 µm^2, this is an improvement of 7 dB available power. An enhancement up to 12 dB can be expected if both devices have identical frequency responses. The 1 dB compression point amounts to 6 mA and 22 mA for the single PD and the TWPD, respectively. At 200 GHz, the available power from the TWPD still amounts to -9 dBm, and even at 400 GHz a power of -32 dBm was detected. Due to the lack of calibration data beyond 170 GHz, the output power can be anticipated to be somewhat higher due to the losses in the experimental setup that were not considered. In conclusion, a phase- and impedance-matched TWPD based on high-speed waveguide-integrated photodiodes is demonstrated. The device exhibits high-power detection capability up to 400 GHz, which recommends the parallel-fed TWPD as an efficient photomixer at sub-THz frequencies.

[1] C. L. Goldsmith et al., *IEEE Trans. Microwave Theory Tech.*, vol. 45 (8), 1997, p. 1342.
[2] S. Murthy et al., *Electronics Letters*, Vol. 38 (2), 2002, p. 78.
[3] Y. Hirota et al., *J. Lightw. Technol.*, Vol. 19 (11), 2001, p. 1751.
[4] A. Umbach et al., *Technical Digest Optical Fiber Communication Conf. (OFC 2000)*, p. 117.
[5] A. Beling et al., *IEEE Photonics Technol. Letters*, Vol. 17 (10), 2005, p. 2152.

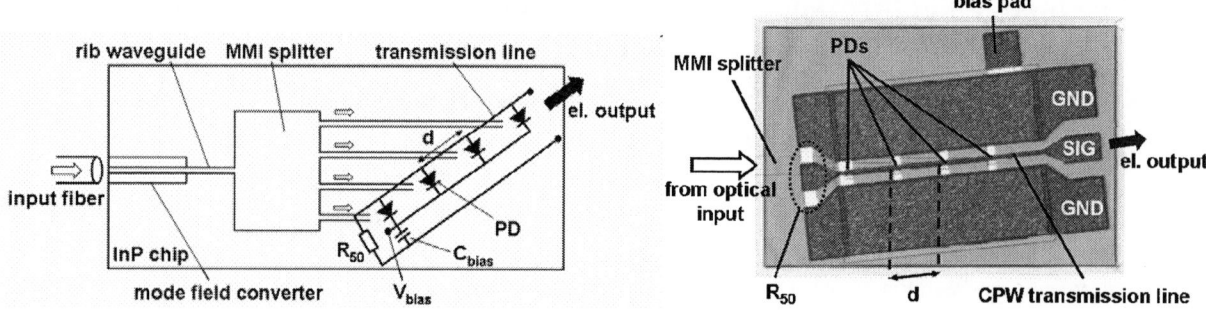

Fig. 1: Schematic view of the TWPD. A metal-insulator-metal capacitor C_{bias} is implemented to decouple the bias voltage from RF ground and to avoid the short-circuit current at the matching resistor R_{50}.

Fig. 2: Micrograph of the fabricated TWPD chip with d = 90 µm. The mode field converter is not shown.

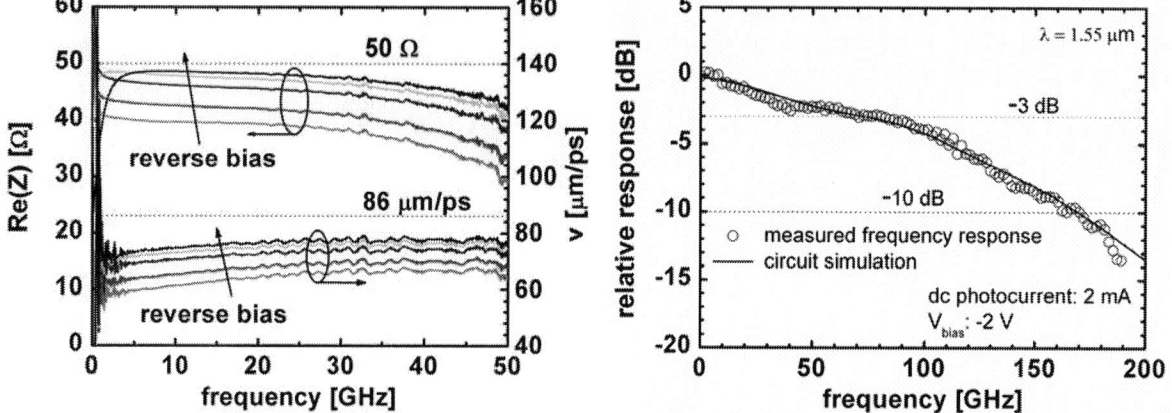

Fig. 3: Real part of the characteristic impedance Re(Z) and electrical phase velocity v of a TWPD without matching resistor. Bias voltages: 0 V, -0.5 V, -1.5 V, -2.5 V and -3.5 V. The signal velocity in the optical waveguide amounts to 86 µm/ps.

Fig. 4: Simulated and measured optoelectronic frequency responses of a TWPD with matching resistor. The influence of the experimental setup was taken into account up to 170 GHz.

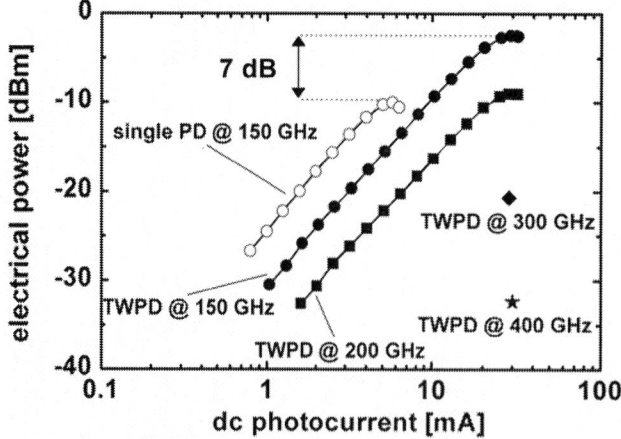

Fig. 5: RF output power over dc photocurrent at V_{bias} = -2 V for a single PD and TWPD, both with integrated R_{50}. In all measurements the dc photocurrent increased linearly with the optical input power.

Single Ultra Violet Photon Detection with 4H-SiC Avalanche Photodiodes

Xiaogang Bai, Dion Mcintosh, Handin Liu, Joe Campbell

Electrical and Computer Engineering, University of Virginia, Charlottesville, VA 22904

Recently, the emergence of several application areas that utilize ultra-violet (UV) signals, e.g. laser-induced fluorescence biological-agent detection, missile detection, unattended ground sensors, and non-line-of-sight covert communications, has peaked interest in UV detectors that can achieve high sensitivity and low noise. At present, photomultiplier tubes (PMTs) are widely used however, their low quantum efficiency, large size, high cost, high bias voltage, and fragility limit practical deployment for many applications. A solid-state replacement for PMTs has been the focus of numerous research efforts. In this paper we report SiC APDs that achieve gain comparable to PMTs while exhibiting higher quantum efficiency and lower dark current. In addition, we report single photon detection at 265 nm.

The 4H SiC APDs have the following surface to substrate structure: 200 nm p+ layer (N_A=1.0×10^{19} cm^{-3}), 200 nm p+ doped layer (N_A=2.4×10^{18} cm^{-3}), 480 nm p$^-$ layer (N_A=2.8×10^{15} cm^{-3}), 2000 nm n+ layer (N_D=3.0×10^{19} cm^{-3}), n-type 4H SiC substrate.[1] Mesa structures (100 μm diameter) were defined by photolithography and reactive ion etching. Then ohmic contacts are defined to both p+ and n+ layer with Ni/Ti/Al/Au metallization. Silicon dioxide was used as passivation. The devices are top illuminated. A typical I-V is shown in Fig. 1.

The light source was a pulsed UV laser at 265 nm. The laser pulse has a repetition rate of 7.3 kHz and pulse width of 400ps. The laser is coupled into a UV fiber and focused on the device through a microscope objective. The incident laser power is extracted from the photo current at unity gain bias, which is normally ~10 pW. The laser can be precisely attenuated using calibrated neutral density filters. Before breakdown, the SiC APD shows gain larger than 10^6 (equal to that of PMTs) and operates in linear mode. Fig. 2 shows the IV for different incident laser powers. At V_{Bias}=145 V, the SiC APD can be used to detect laser with power of 147 fW. With a lock-in amplifier setup, the noise from the dark current is suppressed and the APD can detect laser with power down to single photon per pulse. The data is shown in Fig. 3. At a gain of 1000 the dark current is 5.4×10^{-11} A, from which it follows that the primary unmultiplied dark current is 5.4×10^{-14} A.

The 4H SiC APD were also operated in Geiger mode. In Geiger mode, the APD is biased above breakdown. The current is low until an avalanche process is initiated by either a photo-excited carrier or a thermally excited carrier (dark count). The current increases until it reaches the limit set by the external circuit. A quenching circuit resets the bias voltage below breakdown to a low-current level. Then the circuit rearms the APD above breakdown to start the next detection cycle [2]. We used a gated quenching circuit shown in Fig. 4. The SiC APD was biased at 140 V. A short AC voltage pulse was applied to the device simultaneous to the arrival of the UV photon. The incident laser was attenuated to approximately 0.2 photons per pulse. For single photon detection, the two most important figures of merit are single photon detection efficiency (SPDE) and dark count probability (DCP). SPDE is the probability of detection of a single incident UV photon. DCP is the probability of a false positive or error detection per gate due to thermally excited carriers. Fig. 5 shows the SPDE and DCP vs. excess bias (voltage over breakdown voltage). Fig. 6 shows SPDE vs. DCP. The external quantum efficiency of the device at 265nm is 21%, which is ~ 2x that of PMTs at this wavelength. A SPDE of 13.7% means that approximately two thirds of the absorbed incident photons are detected.

In conclusion, we report 4H SiC APDs for the UV single photon detection. In the linear mode operation, a lock-in technique reveals the device is capable of detecting single photon pulses. In Geiger mode, the device

978-1-4244-1101-6/07/$25.00 ©2007 IEEE

exhibits a SPDE of 13% and DCP of 1.7×10^{-4} at room temperature, respectively.

[1] Xiangyi Guo, A. L. Beck, J. C. Campbell, D. Emerson, and J. Sumakeris, IEEE J. Quantum Electron., 41, no. 10, pp. 1213-1216, 2005.

[2] S. Cova et al., *Appl. Optics*, **35** n. 12, pp.1956-1958 (1996)

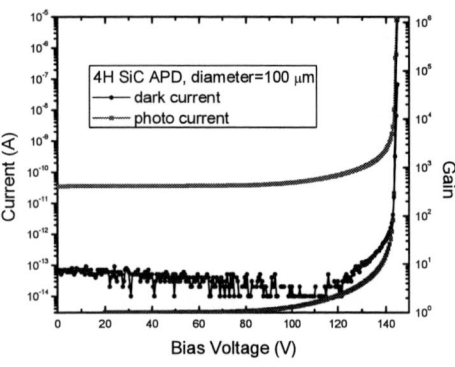

Fig.1 4H SiC APD

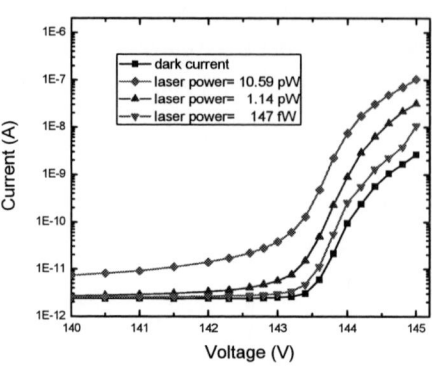

Fig. 2 I-V curve (DC)

Photo-current, dark current and gain vs. reverse bias voltage.   I-V curves under different power levels of UV illumination.

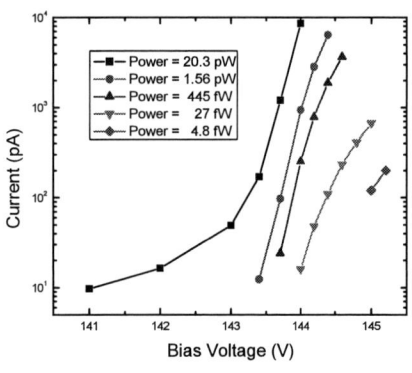

Fig. 3 I-V Curve (lock-in amplifier)

I-V curve measured with a lock-in amplifier. The lowest
power level of 4.8 fW corresponds to 0.88 photon per pulse.

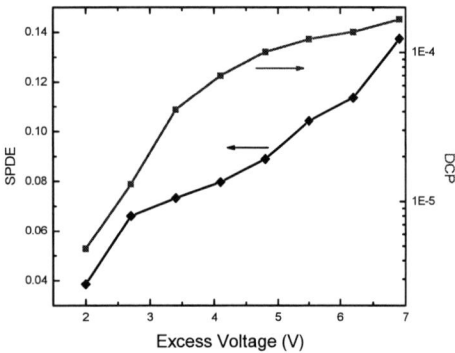

Fig. 4 Gated quenching scheme

The AC voltage pulse and laser arrive at device simultaneously.

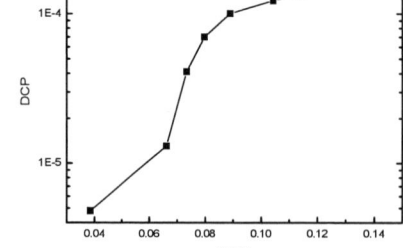

Fig. 5 SPDE, DCP vs. excess voltage

Both SPDE and DCP increase with excess voltage.

Fig. 6 SPDE vs. DCP

SPDE=13.7% with DCP=1.66×10^{-4}.

The challenges and opportunities for dilute nitride antimonides in photonic devices
Jim Harris, Stanford University

Abstract not available at the time of publication.

1.65 μm buffer-free GaSb/AlGaSb quantum-well diode lasers grown on a GaAs substrate operating at room temperature.

M.Mehta, G.Balakrishnan, A. Jallapali, M.N.Kutty, L.R. Dawson, and D.L. Huffaker

Center for High Technology Materials, University of New Mexico, Albuquerque, NM-87106.

huffaker@chtm.unm.edu

We present a 1.65 μm GaSb/AlGaSb quantum-well (QW) laser operating at room temperature, grown buffer-free on a GaAs substrate. The enabling technology in this device is a layer of interfacial misfit dislocations (IMF) that completely relieve the strain between the GaAs substrate and the GaSb epi-layer without the use of thick metamorphic buffers. The IMF array is localized to the GaAs/GaSb interface, allowing current to pass through electrically pumped devices without the detrimental effects of threading dislocations that result in large voltage drops and non-radiative recombination.[1],[2] This growth technology provides an effective way of integrating III-Sb based devices such as IR-lasers, detectors and transistors with a higher quality and more scalable substrate such as GaAs. Our group has in the past demonstrated super-luminescent LEDs emitting at 2 μm and optically pumped VCSELs at 1.65 μm using similar IMF layers on Si(100) substrates.[3]

The epitaxial layer growth is performed completely by molecular beam epitaxy (MBE) and commences with the GaSb IMF formation on an n-type GaAs substrate followed by a 5 nm un-doped GaSb layer. The growth then proceeds directly to the 2.3 μm $Al_{0.45}Ga_{0.55}Sb$ n-type cladding layer, followed by the active region, the 1.5 μm $Al_{0.45}Ga_{0.55}Sb$ p-type cladding layer and a highly doped 50 nm GaSb p-type contact layer. The active region is comprised of six GaSb QWs and seven $Al_{0.35}Ga_{0.65}Sb$ barrier layers embedded between separate 300 nm un-doped $Al_{0.35}Ga_{0.65}Sb$ waveguide layers. The samples are processed such that they form stripe lasers varying in width between 25 μm and 100 μm. The simple fabrication method involves only a Ti/Pt/Au top p-metal evaporation, an inductively-coupled reactive ion etch (ICP-RIE) of the p-cladding layer to just above the un-doped waveguide, lapping of the GaAs substrate to ~150 μm, and a Ge/Au/Ni/Au bottom side n-metal evaporation. The samples are then cleaved to bar lengths of 560 μm.

The lasers operate in the pulsed mode up to room temperature. Under operating conditions of a 500 ns pulse and 0.5 % duty cycle, the devices show 15 mW peak output power at -10 °C at a drive current of 1 A. Under the same drive conditions, the devices show 1 mW peak output power at 20 °C. The lasing wavelength is ~1650 nm, one of the longest reported wavelengths for a device grown on GaAs [4]. The threshold current density rises from 2 kA/cm² to 3 kA/cm² between -10 °C and 20 °C, resulting in a room temperature T_0 value of 83 K. This value is higher than those previously reported for GaSb QW lasers, possibly due to a smaller reduction in internal efficiency versus temperature resulting from the transition from a GaSb to a GaAs substrate [5].

The current-voltage (I-V) characteristics indicate a diode turn-on of 2.5 V, which far exceeds the ~ 1 V built-in potential of the laser diode. Possible contributions to the excess voltage include a predicted 0.7 V potential drop at the GaAs/GaSb IMF interface, and a contact potential resulting from an un-annealed Ge/Au/Ni/Au contact to the n-type GaAs substrate. We expect to improve the IMF voltage using delta doping at the interface. The high threshold current values likely arise due to the relatively low material gain in unstrained GaSb QWs. The material gain will be increased in future devices by adding In and strain to the QWs..

[1] M. Mehta et. al., *Appl. Phys. Lett.*, Vol. **89**, pp. 211110 (2006)
[2] S. Huang et. al., *Appl. Phys. Lett.*, Vol. **88**, pp. 131911 (2006)
[3] G. balakrishnan et. al., *IEEE Journal of Selected Topics in Quantum Electronics*, Vol. **12**, pp. 1636 (2006)
[4] S.R. Bank et. al., *Elec. Lett.*, Vol. **42**, pp. 156 (2006)
[5] Y. Ohmori et. al., *Jap. Journ. of Appl. Phys.*, Vol. **24**, pp. L657 (1985)

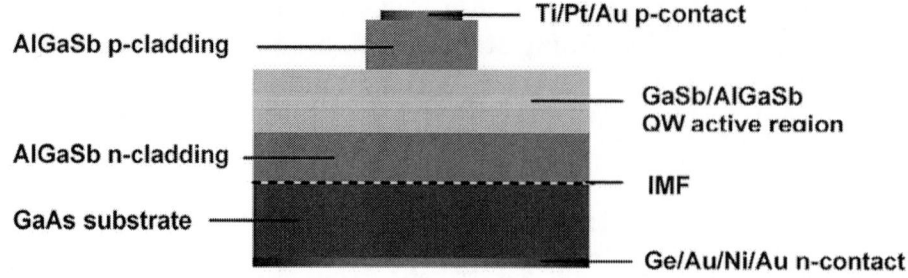

Fig 1: GaSb/AlGaSb laser schematic indicating position of IMF between contacts.

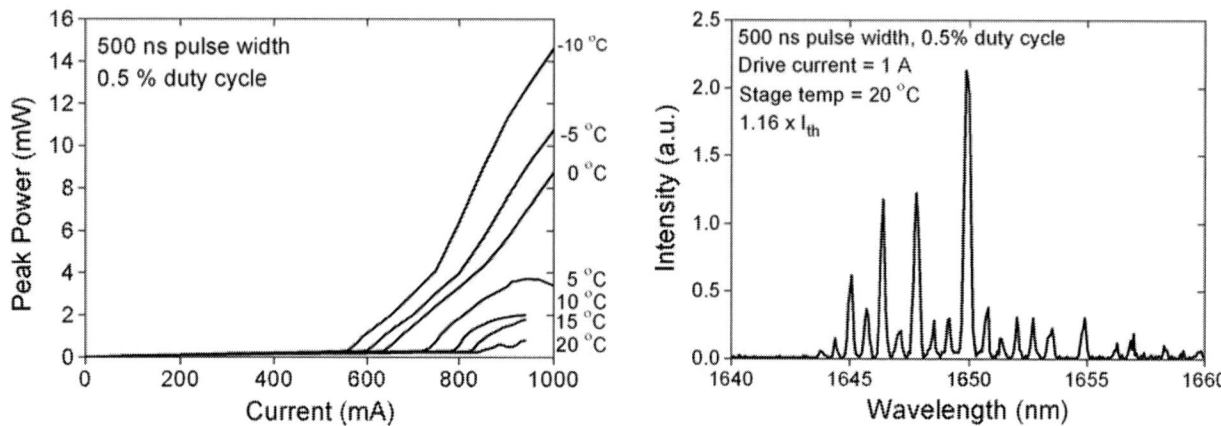

Fig 2: Light-current (L-I) characteristics of GaSb/AlGaSb laser over temperature range covering -10 °C to 20 °C.

Fig 3: Spectral characteristics of GaSb/AlGaSb indicating lasing wavelength of 1650 nm at room temperature.

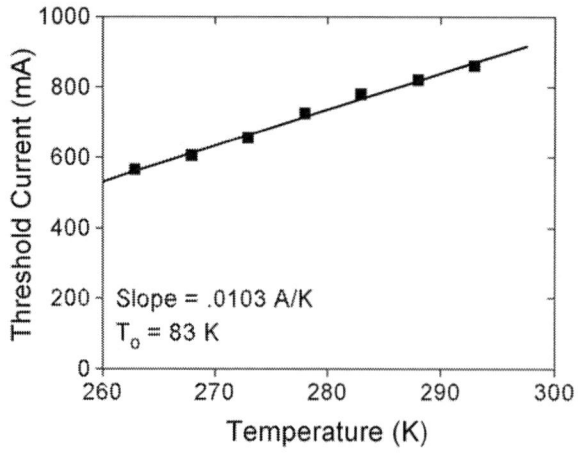

Fig 4: Characteristic temperature (T_o) analysis indicating a T_o value of 83 K.

High Temperature CW Operation of Interband Cascade Lasers at $\lambda \approx 4.0$ µm

Chul Soo Kim, Mijin Kim, William W. Bewley, Chadwick L. Canedy, <u>Diane C. Larrabee</u>, Jill A. Nolde, J. Ryan Lindle, Igor Vurgaftman, Jerry R. Meyer

Naval Research Laboratory, Washington DC

The mid-infrared spectral band encompasses a variety of important applications, including chemical sensing, free-space optical communication, and IR countermeasures. Currently, however, most of the wavelength range between 3 and 4 µm is not accessible with semiconductor sources operating in continuous mode at room temperature. We believe that the most promising structure for bridging this gap is the interband cascade laser (ICL).[1]

Here we report on the design, growth, processing, and characterization of 3.6-4.1 µm ICLs that exhibit the highest cw operating temperatures, highest wallplug efficiencies, and lowest threshold current densities of any interband diodes emitting at $\lambda \geq 3.2$ µm. The structures were grown by molecular beam epitaxy on n-GaSb substrates. The active regions contain 10 type-II "W" interband gain sections separated by InAs/AlSb superlattice injectors and GaSb tunneling layers. The top and bottom optical cladding layers, consisting of Si-doped InAs/AlSb superlattices, are 2.3 and 4.0 µm thick, respectively. Narrow ridges were formed by contact lithography and wet etching through the active region. The ridges were coated with 200 nm of sputtered Si, followed by 400 nm of sputtered SiO_2. A 5-µm-wide contact window was formed on top of each ridge by contact lithography and metallization with Ag/Ti/Pt/Au. The samples were thinned to 170 µm, contacted on the back with Ag/Cr/Sn/Pt/Au, and then annealed at 300 °C for one minute. The top metal was electroplated with 3.5-4.5 µm of gold for improved thermal transport. After cleaving, high reflectivity and/or anti-reflectivity facet coatings were deposited and the bars were mounted epitaxial-side-up on a copper block. Currently, we are also developing an epi-down mounting process for improved heat-sinking.

Several of the devices broke performance records. For example, an 11 µm × 0.5 mm ICL with HR/AR facet coatings reached a maximum wallplug efficiency of 27% at 78 K, where the wavelength was 3.6 µm. Another 13 µm × 4 mm, HR/uncoated ridge lased cw to 269 K (λ = 4.05 µm). The threshold current density for a 150 µm × 2 mm, uncoated/uncoated wide stripe was only 4.8 A/cm^2 at 78 K (cw) and 1.15 kA/cm^2 at 300 K (pulsed). Output powers and beam qualities were also favorable, with an 11 µm × 3 mm, HR/AR coated device producing 200 mW at 78 K, into a beam that was narrower than three times the diffraction limit. A sister wafer produced ICLs operating at λ = 3.8 − 4.3 µm (for T = 78-320 K) which similarly performed better than any previous interband diodes in that wavelength range. Devices from that wafer yielded T_{max}^{cw} = 243 K, and for 78 K operation had a maximum wallplug efficiency of 18% and threshold current density as low as 6 A/cm^2.

---

[1] R. Q. Yang, *Superlatt. Microstruct.* **17**, 77 (1995).

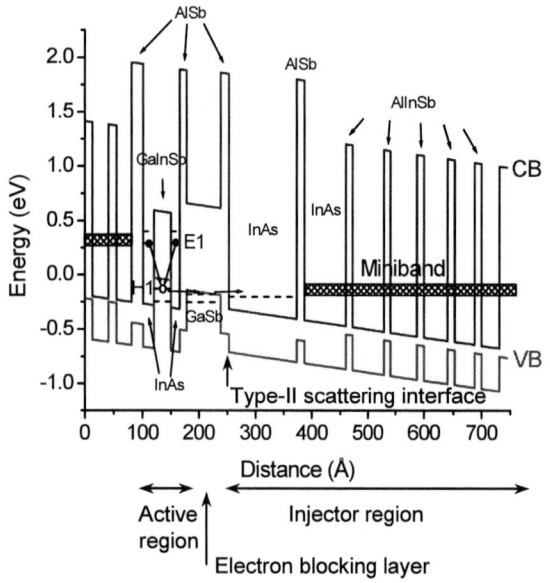

Fig. 1: Schematic of the conduction and valence band profiles in the interband cascade laser active region.

Fig. 2: Wavelength *vs.* temperature for pulsed excitation.

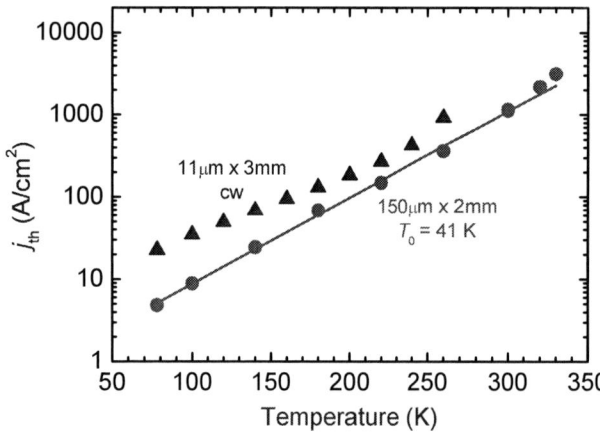

Fig. 3: Threshold current density *vs.* temperature for pulsed (150 μm stripe) and cw (11 μm ridge) operation.

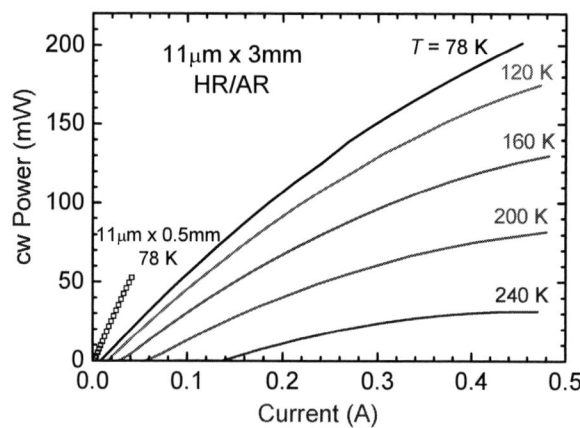

Fig. 4: CW *L-I* characteristics for the long (3 mm) and short (0.5 mm) narrow ridges ($w = 11$ μm). The maximum output powers shown are not limited by thermal rollover.

Strain-Compensated AlAs-InGaAs Quantum-Cascade Lasers with Emission Wavelength 3–5 μm

M.P. Semtsiv[1], S. Dressler[1], M. Wienold[1], I. Bayrakli[1], M. Ziegler[2], K. Kennedy[3], R. Hogg[3], and <u>W.T. Masselink</u>[1]

1) Humboldt University, Berlin, Germany, 2) Max-Born-Institut, Berlin, Germany
3) University of Sheffield, Sheffield, UK
Email: masselink@physik.hu-berlin.de

The development of quantum-cascade lasers (QCLs) operating in the 3–5 μm spectral region is driven by a number of applications including gas sensing for both environmental and medical uses, communication, and military countermeasures. QCLs emitting at wavelengths shorter than 4 μm have been especially challenging, requiring a very large conduction band discontinuity, a small electron effective mass, but also a relatively mature materials system. We have developed QCLs emitting in this short wavelength part of the spectrum based on the use of strain compensation with very high levels of strain in the individual layers; barriers based on AlAs, wells on $In_{0.73}Ga_{0.27}As$, and the entire structure on average lattice-matched to InP [1]. For more flexibility to control both strain and conduction band potential, "composite barriers" are used, composed of AlAs and $Al_{0.5}In_{0.5}As$. For wavelengths near 3 μm, InAs added to the otherwise $In_{0.73}Ga_{0.27}As$ well. A remaining obstacle to very short wavelength emission is the presence of the Γ and X valleys within the well material that can limit the photon energy to the energy difference between these valleys and the lower laser state. By locating the upper laser state in a well based on a material with a larger band gap than the material where the lower laser state is located, however, leakage into the Γ and X valleys can also be avoided. Combining these design components, we have produced QCLs emitting at wavelengths covering the entire range down to 3.05 μm [2–4].

By inhibiting non-radiative transitions out of the upper laser state, we have obtained differential quantum efficiencies as high as 45% per cascade at 8 K below 4 μm. For these lasers, T_0 is as high as 160 K, total pulsed emission power as high as 12 W, and room-temperature power 1.4 W. QCLs emitting at 3.6 μm emit cw at 77 K and in pulsed mode up to room temperature with non-optimized heat sinking. Threshold currents as low as 200 mA (600 A/cm^2) are realized at cryogenic temperatures and $T_0 = 140$ K. By locating the upper laser state in an $In_{0.55}Al_{0.45}As$/AlAs quantum well and the lower laser state in an $In_{0.55}Al_{0.45}As$/ $In_{0.73}Ga_{0.27}As$/ InAs composite quantum well, lasers emitting at 3.3 and 3.05 μm were realized. At 3.3 μm, $J_{th} = 2.5$ kA/cm^2 at 70 K and 10 kA/cm^2 at 200 K for 10 μm × 4.5 mm laser stripes; P_{max} was about 300 mW/facet. QCLs emitting at 3.05 μm were also realized with a similar design; laser emission was maintained up to 145 K. The energetic position of the Γ and X valleys in the $In_{0.73}Ga_{0.27}As$ is probably lower than the upper laser state; this result demonstrates that population inversion in such a case is possible, but the rapid increase in J_{th} above 150 K also indicates the limitations imposed by the indirect valleys.

QCLs were also fabricated with narrow 5-μm stripes resulting in an approximately circular beam profile. These QCLs emitted an average power of 60 mW using thermoelectric cooling with a beam quality of $M^2 \approx 1.6$ in the lateral direction and approximately diffraction limited in the vertical direction.

[1] M.P. Semtsiv et al, *Appl. Phys. Lett.* **85**, 1478–1480 (2004).
[2] M.P. Semtsiv et al, *Appl. Phys. Lett.* **89**, 211124 (2006).
[3] M.P. Semtsiv et al, *IEEE J. Quantum Electron.* **43**, 42, (2007).
[4] M.P. Semtsiv et al, *Appl. Phys. Lett.* **90**, 051111 (2007).

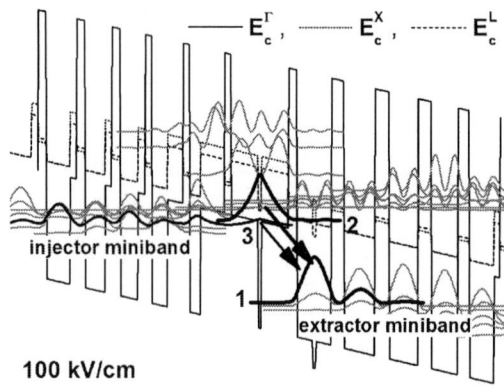

Fig. 1 – One cascade period of the active region of a QCL emitting at 3.3 μm.

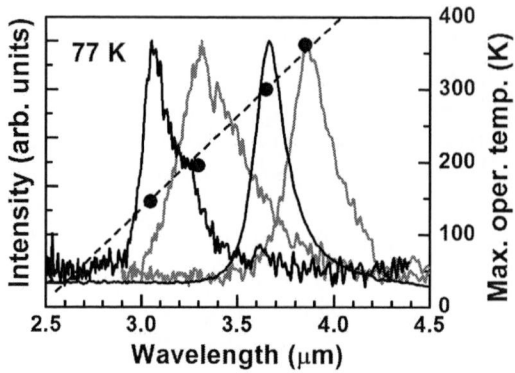

Fig. 2 – Sub-threshold emission for 4 QCLs similar in design to Fig. 1, also indicating the maximum lasing temperature.

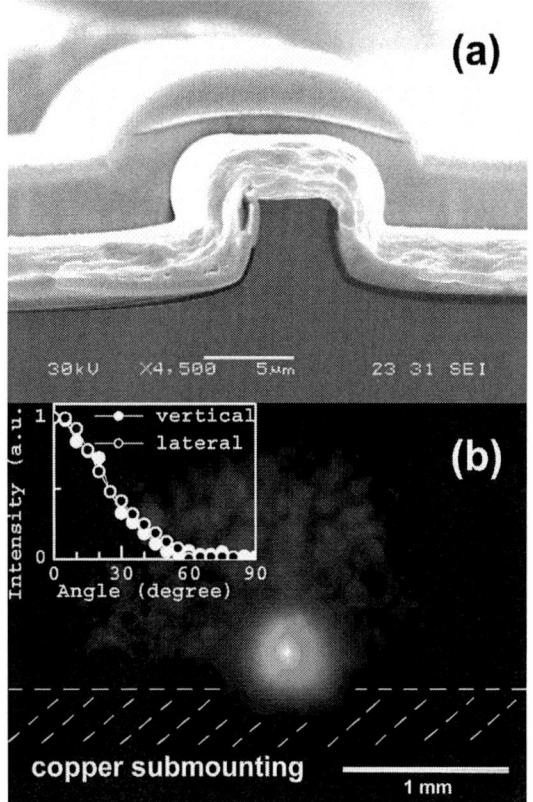

Fig. 3 – SEM of cleaved facet of a laser with 5-μm stripe width emitting at 3.9 μm. The IR camera image shows the circular beam profile.

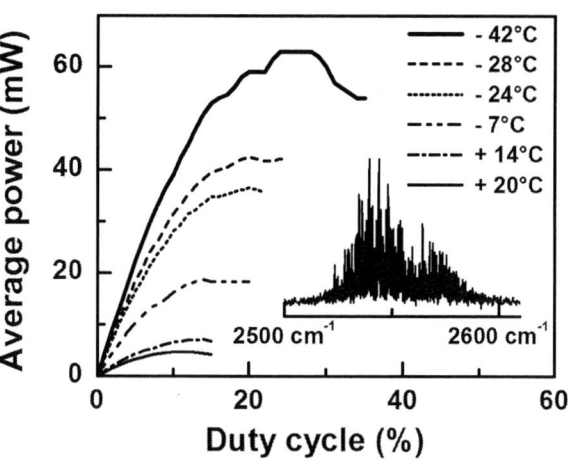

Fig. 4 – Total average emission power from a 5-μm × 3-mm QCL emitting at 3.9 μm for a variety of TE temperatures.

Strain-Compensated AlAs-InGaAs Quantum-Cascade Lasers with Emission Wavelength 3–5 μm

M.P. Semtsiv[1], S. Dressler[1], M. Wienold[1], I. Bayrakli[1], M. Ziegler[2], K. Kennedy[3], R. Hogg[3], and <u>W.T. Masselink</u>[1]

1) Humboldt University, Berlin, Germany, 2) Max-Born-Institut, Berlin, Germany
3) University of Sheffield, Sheffield, UK
Email: masselink@physik.hu-berlin.de

The development of quantum-cascade lasers (QCLs) operating in the 3–5 μm spectral region is driven by a number of applications including gas sensing for both environmental and medical uses, communication, and military countermeasures. QCLs emitting at wavelengths shorter than 4 μm have been especially challenging, requiring a very large conduction band discontinuity, a small electron effective mass, but also a relatively mature materials system. We have developed QCLs emitting in this short wavelength part of the spectrum based on the use of strain compensation with very high levels of strain in the individual layers; barriers based on AlAs, wells on $In_{0.73}Ga_{0.27}As$, and the entire structure on average lattice-matched to InP [1]. For more flexibility to control both strain and conduction band potential, "composite barriers" are used, composed of AlAs and $Al_{0.5}In_{0.5}As$. For wavelengths near 3 μm, InAs added to the otherwise $In_{0.73}Ga_{0.27}As$ well. A remaining obstacle to very short wavelength emission is the presence of the Γ and X valleys within the well material that can limit the photon energy to the energy difference between these valleys and the lower laser state. By locating the upper laser state in a well based on a material with a larger band gap than the material where the lower laser state is located, however, leakage into the Γ and X valleys can also be avoided. Combining these design components, we have produced QCLs emitting at wavelengths covering the entire range down to 3.05 μm [2–4].

By inhibiting non-radiative transitions out of the upper laser state, we have obtained differential quantum efficiencies as high as 45% per cascade at 8 K below 4 μm. For these lasers, T_0 is as high as 160 K, total pulsed emission power as high as 12 W, and room-temperature power 1.4 W. QCLs emitting at 3.6 μm emit cw at 77 K and in pulsed mode up to room temperature with non-optimized heat sinking. Threshold currents as low as 200 mA (600 A/cm^2) are realized at cryogenic temperatures and $T_0 = 140$ K. By locating the upper laser state in an $In_{0.55}Al_{0.45}As$/AlAs quantum well and the lower laser state in an $In_{0.55}Al_{0.45}As$/ $In_{0.73}Ga_{0.27}As$/ InAs composite quantum well, lasers emitting at 3.3 and 3.05 μm were realized. At 3.3 μm, $J_{th} = 2.5$ kA/cm^2 at 70 K and 10 kA/cm^2 at 200 K for 10 μm × 4.5 mm laser stripes; P_{max} was about 300 mW/facet. QCLs emitting at 3.05 μm were also realized with a similar design; laser emission was maintained up to 145 K. The energetic position of the Γ and X valleys in the $In_{0.73}Ga_{0.27}As$ is probably lower than the upper laser state; this result demonstrates that population inversion in such a case is possible, but the rapid increase in J_{th} above 150 K also indicates the limitations imposed by the indirect valleys.

QCLs were also fabricated with narrow 5-μm stripes resulting in an approximately circular beam profile. These QCLs emitted an average power of 60 mW using thermoelectric cooling with a beam quality of $M^2 \approx 1.6$ in the lateral direction and approximately diffraction limited in the vertical direction.

[1] M.P. Semtsiv et al, *Appl. Phys. Lett.* **85**, 1478–1480 (2004).
[2] M.P. Semtsiv et al, *Appl. Phys. Lett.* **89**, 211124 (2006).
[3] M.P. Semtsiv et al, *IEEE J. Quantum Electron.* **43**, 42, (2007).
[4] M.P. Semtsiv et al, *Appl. Phys. Lett.* **90**, 051111 (2007).

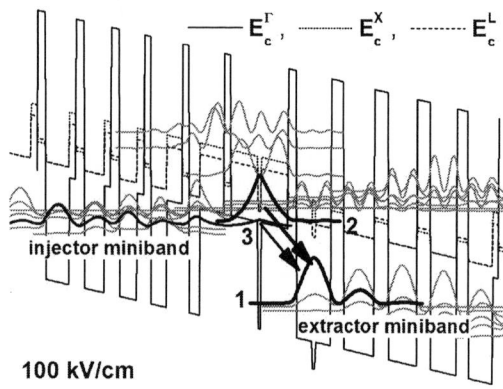

Fig. 1 – One cascade period of the active region of a QCL emitting at 3.3 μm.

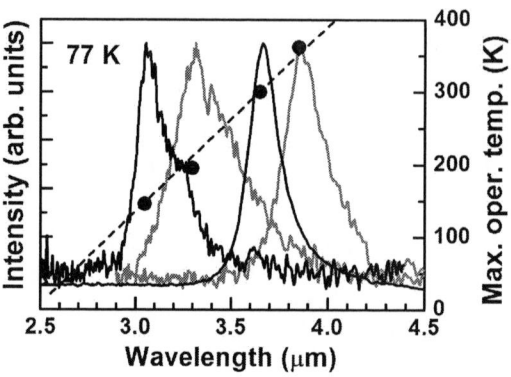

Fig. 2 – Sub-threshold emission for 4 QCLs similar in design to Fig. 1, also indicating the maximum lasing temperature.

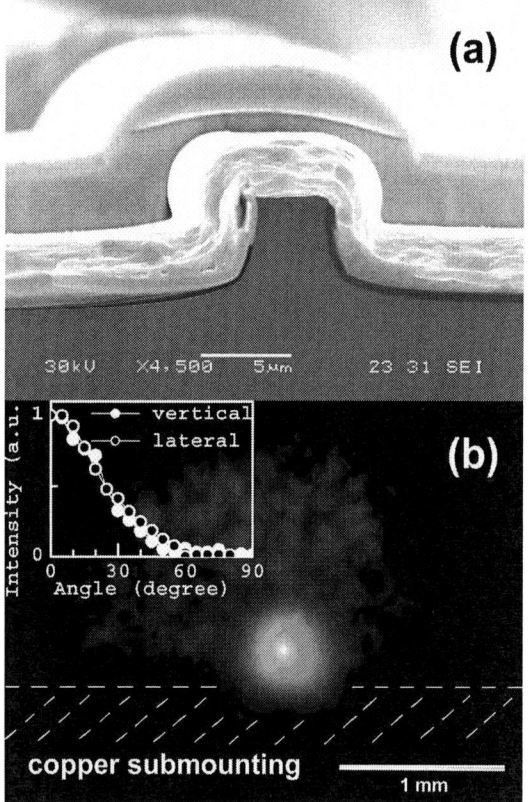

Fig. 3 – SEM of cleaved facet of a laser with 5-μm stripe width emitting at 3.9 μm. The IR camera image shows the circular beam profile.

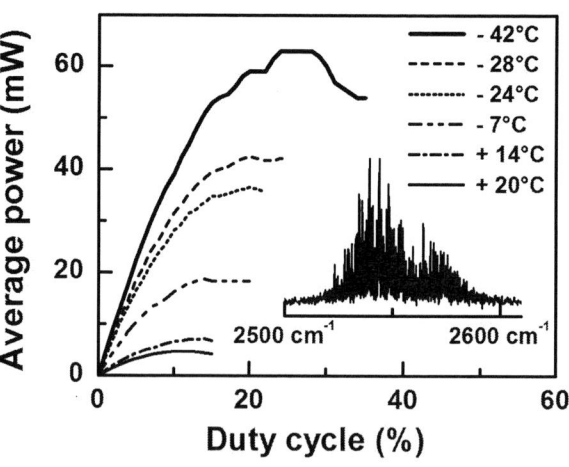

Fig. 4 – Total average emission power from a 5-μm × 3-mm QCL emitting at 3.9 μm for a variety of TE temperatures.

Session V.B (DeBartolo Hall Room 141)

III-V CMOS

Tuesday PM, June 20th, 2007

Session Organizer: Judy Hoyt, Massachusetts Institute of Technology and Steve Koester, IBM
Session Chair: Tomas Palacios, Massachusetts Institute of Technology

1:30 PM V.B-1 Invited Paper
InGaAs CMOS: a "Beyond-the-Roadmap" Logic Technology?
J. A. del Alamo and D. H. Kim, Massachusetts Institute of Technology, Cambridge, Massachusetts, USA

2:10 PM V.B-2
InGaAs and GaAs/InGaAs Channel Enhancement Mode n-MOSFETs With HfO2 Gate Oxide and a-Si Interface Passivation Layer
S. Oktyabrsky[1,] S. Koveshnikov[2], V. Tokranov[1], M.I Yakimov[1], R. Kambhampati[1], H. Bakhru[1], F. Zhu[2], J. Lee[3], and W. Tsaib[2], [1]College of Nanoscale Science and Engineering, University at Albany-SUNY, New York, USA, [2]Intel Corporation, Santa Clara, California, USA, and [3]The University of Texas at Austin, Department of Electrical and Computer Engineering, Austin, Texas, USA

2:30 PM V.B-3
Enhancement Mode n-MOSFET with High-? Dielectric On GaAs Substrate
K. Rajagopalan[2], P. Zurcher[2], J. Abrokwah[2], R. Droopad[2], D. A. J. Moran[1], R. J. W. Hill[1], X. Li[1], H. Zhou[1], D. McIntyre[1], S. Thoms[1], I.G. Thayne[1] and M. Passlack[2], [1]Department of Electronics & Electrical Engineering, University of Glasgow. Glasgow, UK and [2]Freescale Semiconductor Inc., Tempe, Arizona, USA

2:50 PM V.B-4
High-performance submicron inversion-type enhancement-mode InGaAs MOSFET with maximum drain current of 360 mA/mm and transconductance of 130 mS/mm
Y. Xuan, Y. Q. Wu, H. C. Lin, T. Shen and P. D. Ye, School of Electrical and Computer Engineering, Purdue University, West Lafayette, Indiana, USA

3:10 PM Break

3:30 PM V.B-5
Enhancement-mode In0.70Ga0.30As-channel MOSFETs with ALD Al2O3
Y. Sun, E. W. Kiewra, J. P. De Souza, S. J. Koester, K. E. Fogel, D. K. Sadana, IBM Thomas J. Watson Research Center, Yorktown Heights, New York, USA

3:50 PM V.B-6 Student Paper
Performance of Sub-micron Gate Length InAlP Native Oxide GaAs-channel MOSFETs
J. Zhang, T. H. Kosel, D. C. Hall, and P. Fay, Dept. of Electrical Engineering, University of Notre Dame, Notre Dame, Indiana, USA

202

InGaAs CMOS: a "Beyond-the-Roadmap" Logic Technology?

J. A. del Alamo and D. H. Kim

Massachusetts Institute of Technology (MIT), Cambridge, MA 02139, U.S.A, E-mail: alamo@mit.edu

It is a great time to be a semiconductor device technologist. While a matter of considerable debate, the semiconductor device technology fueling the microelectronics revolution appears to be reaching the end of the road. There are severe doubts that it will make economic sense for Silicon CMOS as we know it, to scale beyond the 22 nm node. The virtuous cycle that propelled CMOS to deliver the exponential improvements in density, power and speed that underpins Moore's law is quickly unraveling. For the first time in 40 years since CMOS was invented, smaller is no longer better.

In the eclectic landscape of alternatives to Si CMOS for beyond-the-roadmap logic applications, III-V compound semiconductors shine bright. With room temperature bulk electron mobilities that span from about 7,000 cm^2/V-s for GaAs to about 70,000 cm^2/V-s for InSb, these materials promise improvements in electron velocity that can no longer be obtained out of Silicon.

This is not the first time that III-V compound semiconductors have been discussed as an alternative to Si CMOS for logic applications. In the 80's, vigorous research was conducted on MESFET-, HEMT-, and HBT-based III-V technologies for logic. While some of these technologies got inserted in a few high performance applications, none of these approaches ever posed any serious challenge to Si CMOS.

What has changed? First, as things look today, III-V's will most likely not compete against Si. The goal is to continue beyond the point where Si can reach. In addition, when compared with many of the alternatives, III-V FETs look positively attractive. In some of the III-V's, these are real devices with impressive performance, reasonably well understood physics, acceptable reliability and quite well established manufacturing technology.

Still, a future III-V CMOS logic technology faces daunting challenges. First, one has to demonstrate high quality epitaxial growth of III-V heterostructures on very large Si wafers with very thin buffer layers (for thermal reasons, but also for manufacturing cost reasons). In addition, III-V enhancement-mode FETs are needed with self-aligned device architecture for minimum footprint and parasitics. In order to meet the density requirements of the 15 nm node, exceptional electrostatic integrity will be essential. This will demand the development of MIS-type FETs with a thin and reliable high-K dielectric. This, in turn, will require tackling what is probably the hardest problem of all, avoiding Fermi level pinning and obtaining a high quality passivated III-V surface that approaches the quality of the SiO2/Si system. A future III-V CMOS logic technology might also require 3D-type device structures, such as Fin-FET or triple gate, something never quite demonstrated with III-V's. Finally, one will have to contend with the poor hole mobility. Here the hope is that, similar to Si and Ge where enhancements as high as 5X have been demonstrated, the introduction of strain will allow us to engineer the valence band so as to boost the hole mobility.

Among the III-Vs, InGaAs with compositions closely lattice matched to InP, looks rather unique. With an electron mobility in excess of 9,000 cm^2/V-s at room temperature and a rather mature processing technology, InGaAs-based transistors on InP (HBTs and HEMTs) have held the frequency record for the higest f_T transistors for nearly 20 years in a row. InGaAs HEMT and HBT technologies have been used to demonstrate >100 GHz and >100 Gb/s communications IC's with SSI-level complexity. InGaAs IC technology is also space qualified. This means that sufficient reliability levels, adequate for very expensive satellite systems were repairs are not possible, have been obtained. Among III-V's, InGaAs perhaps represents the best balance between performance and maturity.

This talk will discuss the general issues associated with III-V CMOS that are enunciated above. It will also describe the authors' research activities in the area of InGaAs HEMTs for logic. In particular, we will summarize the findings of a recent scaling study of InGaAs HEMTs down to 60 nm in gate length [1]. In this work, we fabricated HEMTs with a 70% InAs composition in the channel and with varying gate lengths and InAlAs barrier thicknesses (from 11 to 3 nm). This study resulted in devices that have substantially more current drive than state-of-the-art 65 nm CMOS at a voltage of 0.5 V. It also showed that InGaAs HEMTs scale according to a simple electrostatics law similar to fully-depleted SOI MOSFETs. Our research reveals that HEMTs are excellent test vehicles to study topics of great relevance to future III-V MISFETs, such as self-aligned device architectures, scaling limit of planar devices, impact of strain on transport physics, and the consequences of a low density of states on current drive of deeply scaled devices..

[1] D.-H. Kim *et al.*, IEDM 2006.

Acknowledgements: This work has been sponsored by Intel and MARCO-MSD. The heterostructures are supplied by MBE Technology. The InGaAs HEMTs have been fabricated at the facilities of the Microsystems Technology Laboratories, NanosStructures Laboratory and Scanning Electron Beam Lithography Facility at MIT.

978-1-4244-1101-6/07/$25.00 ©2007 IEEE

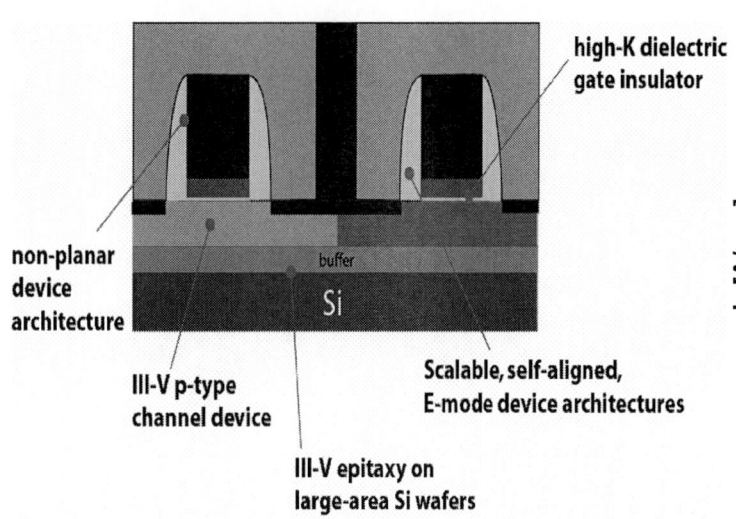

Fig. 1 The "grand challenges" for III-V CMOS.

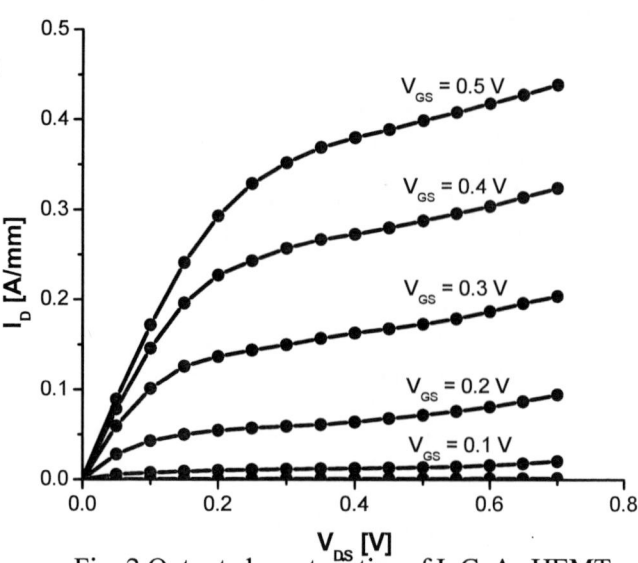

Fig. 2 Output characteristics of InGaAs HEMT with L_g = 60 nm and t_{ins} = 3 nm.

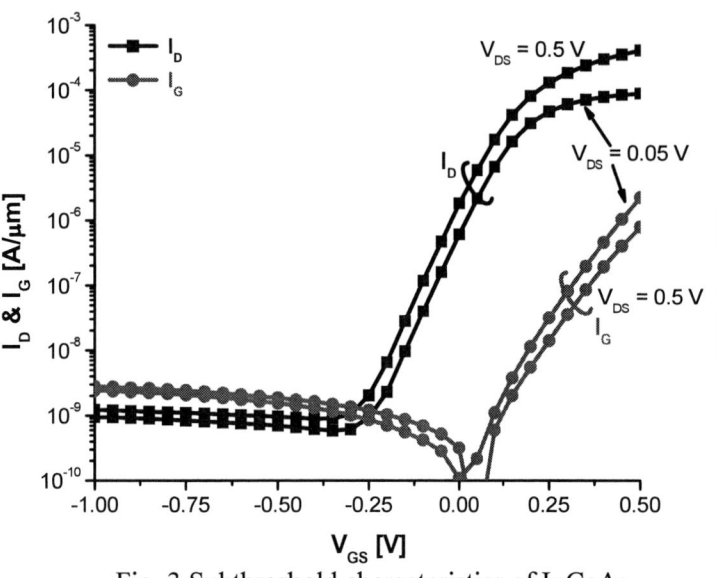

Fig. 3 Subthreshold characteristics of InGaAs HEMT with L_g = 60 nm and t_{ins} = 3 nm.

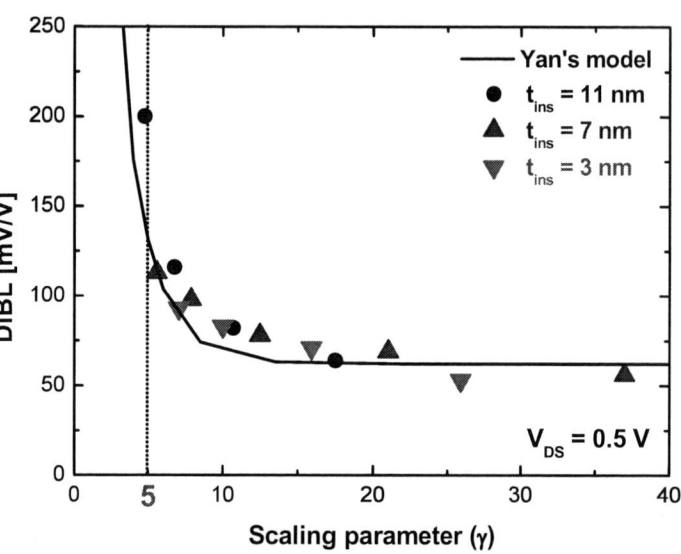

Fig. 4 DIBL vs. scaling parameter for InGaAs HEMTs [1].

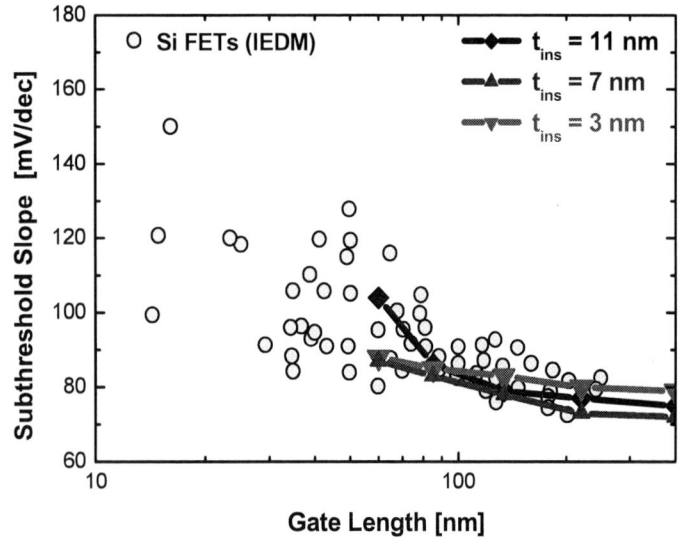

Fig. 5 Subthreshold slope vs. L_g for InGaAs HEMT & CMOS.

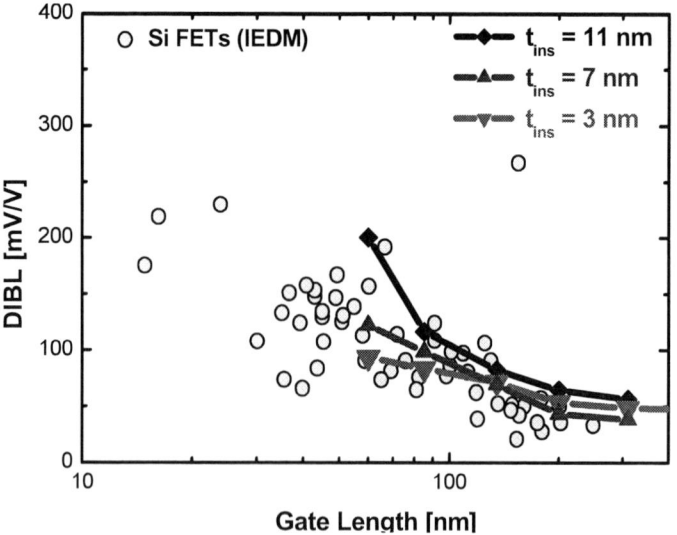

Fig. 6 DIBL vs. L_g for InGaAs HEMTs & CMOS.

InGaAs and GaAs/InGaAs Channel Enhancement Mode n-MOSFETs With HfO$_2$ Gate Oxide and *a*-Si Interface Passivation Layer

Serge Oktyabrsky,[a] Sergei Koveshnikov,[b] Vadim Tokranov,[a] Michael Yakimov,[a] Rama Kambhampati,[a] Hassaram Bakhru,[a] Feng Zhu,[c] Jack Lee,[c] and Wilman Tsai[b]

(a) College of Nanoscale Science and Engineering, University at Albany-SUNY, NY 12203 phone:518 437 8688, fax 518 437 8687, email: soktyabrsky@uamail.albany.edu

(b) Intel Corporation, Santa Clara, CA 95052

(c) The University of Texas at Austin, Department of Electrical and Computer Engineering
Austin TX 78712

We demonstrate for the first time self-aligned, gate-first, enhancement mode n-MOSFETs with InGaAs and GaAs/InGaAs channels, Atomic Layer Deposition HfO$_2$ gate oxide and TaN gate. Good control of the drain current was achieved due to effective passivation of the III-V-high-k interface with ultra-thin MBE *in-situ* grown *a*-Si layer. High transconductance and electron channel mobility along with negligible shift of the threshold voltage and small gate bias hysteresis of the drain current are demonstrated.

As CMOS scaling is continued, new channel materials including III-V semiconductors would be needed for better transistor performance [1]. Successful operation of new generation devices will also require integration of III-V materials with high-k dielectrics. The key challenge for III-V-based MOSFETs is the lack of high-quality, thermodynamically stable insulators that prevent Fermi level pinning at III-V-gate oxide interface. In this work, we demonstrate for the first time enhancement mode n-MOSFETs with surface InGaAs and buried GaAs/InGaAs channels and HfO$_2$ gate oxide. An ultra-thin MBE *in-situ* grown *a*-Si passivation layer is a key element preventing Fermi level pinning at the III-V-oxide interface during *ex-situ* high-k deposition [2].

Self-aligned, gate-first n-MOSFETs with ALD HfO$_2$ gate oxide were fabricated on semi-insulating GaAs wafers (Fig. 1). A carbon doped AlGaAs layer (2×10^{17} cm^{-3}) was grown to provide modulation p-type doping of the 12 nm thick In$_{0.2}$Ga$_{0.8}$As channel. For buried channel devices, a 3 nm thick undoped GaAs layer was deposited on top of the InGaAs channel. An amorphous Si layer with the thickness of 1.5 nm was deposited in the MBE chamber at ~100°C. *Ex-situ* deposition of 4.0 nm thick HfO$_2$ was carried out about 48 hours later using the Atomic Layer Deposition (ALD) technique. Source and drain regions were implanted with 50 keV Si$^+$ ions with a dose of 1×10^{14} cm^{-2} followed by a post implantation anneal at 740°C for 10 sec in N$_2$.

Fig. 2 demonstrates characteristics of a MOSFET with a surface InGaAs channel. The transistor is normally off with threshold voltage, V$_t$ = 0.2V. Good control of the drain current by the gate voltage is demonstrated while the gate current remains low, thus providing strong evidence of a non-pinned Fermi level at the III-V-high-k interface. Owing to small charges in the ALD HfO$_2$, a negligible shift of V$_t$ and small hysteresis of the drain current under the gate bias are observed.

Characteristics of a MOSFET with a buried GaAs/InGaAs channel are shown in Fig. 3. High stability of the threshold voltage at 0.65 V and small hysteresis were also obtained on these MOSFETs. Transconductance of 0.88 mS was measured for 5 μm channel length, corresponding to intrinsic transonductance scaled to 1 μm channel of 49 mS/mm. By measuring gate-channel capacitance and channel conductance in the linear regime the effective channel electron mobility of 450 cm^2/V-s was obtained.

Thus, to the best of our knowledge, this is the first demonstration of enhancement mode MOSFET with InGaAs channel and *ex-situ* deposited high-k gate dielectric. Effective passivation of the III-V surface with an MBE *in-situ* grown *a*-Si layer and high thermal stability of the III-V-HfO$_2$ interface enables integration of this passivation technique into MOSFET fabrication. The major trade-off of this passivation method for *ex-situ* high-k deposition is a notable Si layer thickness of ~1.5 nm needed to prevent III-V surface oxidation. By using our recently developed *in-situ* HfO$_2$ deposition the Si layer thickness can be drastically reduced (Fig. 4) thus providing further scaling of the equivalent oxide thickness.

[1] R. Chau, S. Datta, A. Majumdar, " Opportunities and challenges of III-V nanoelectronics for future high-speed, low-power logic applications, "IEEE Compound Semiconductor Integrated Circuit Symposium November 2005.

[2] S. Koveshnikov, W. Tsai, I. Ok and J. C. Lee, V. Torkanov, M. Yakimov, and S. Oktyabrsky, "Metal-oxide-semiconductor capacitors on GaAs with high-k gate oxide and amorphous silicon interface passivation layer," *Appl. Phys. Lett.,* vol. 88, pp. 022106-8, 2006.

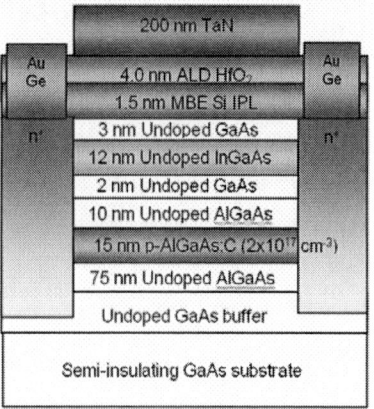

Fig. 1. MOSFET structure with buried GaAs/ InGaAs channel.

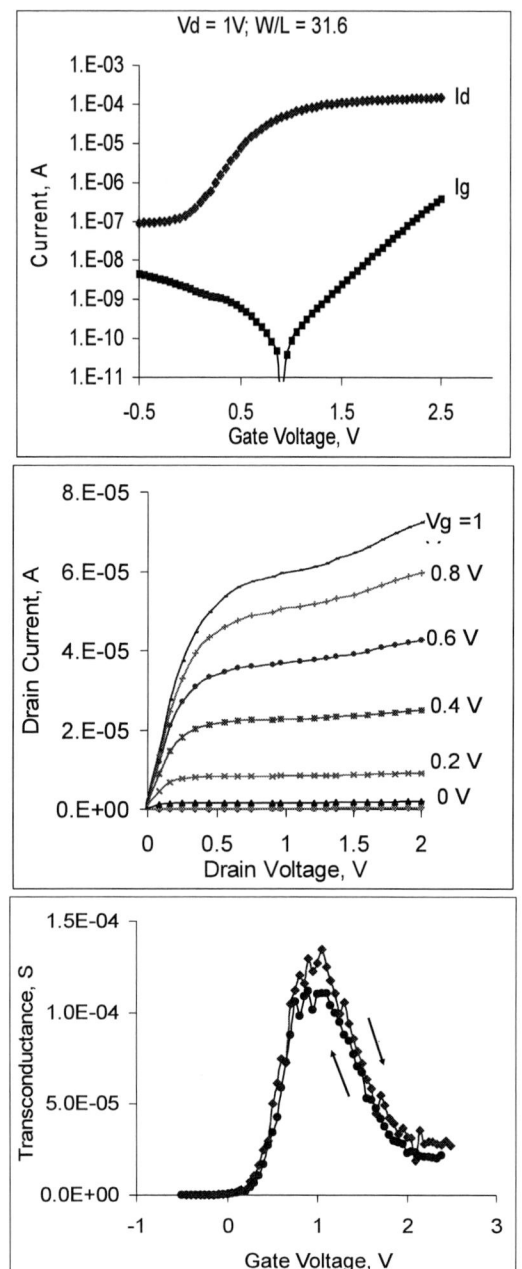

Fig. 2. I_d-V_g, I_g-V_g (top), I_d-V_g (center) and hysteresis of g_m-V_g (bottom) for surface InGaAs channel MOSFET.

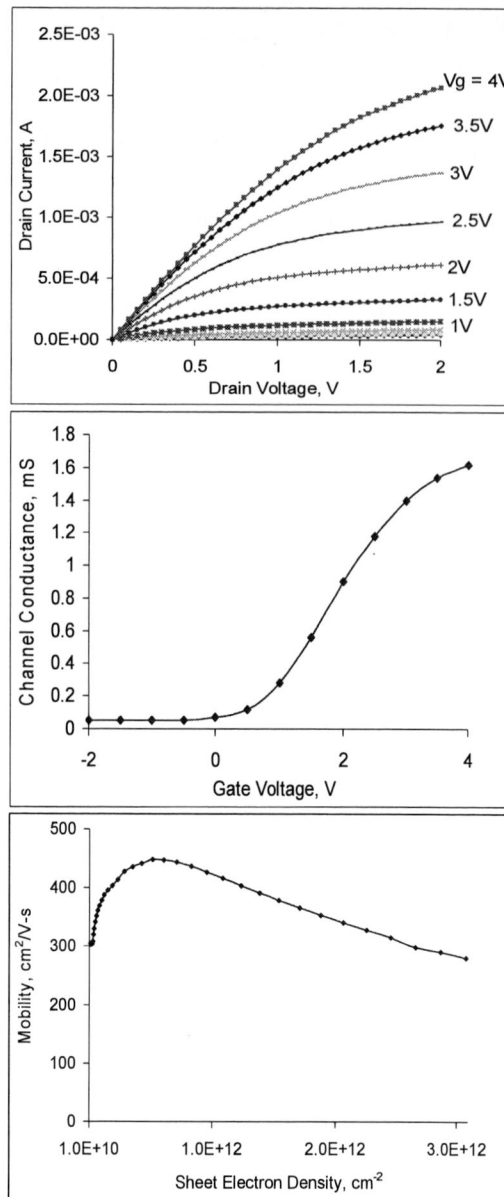

Fig. 3. I_d-V_d (top), g_d vs. V_g (center), and μ_{eff} vs. channel sheet density (bottom) for buried GaAs/InGaAs channel.

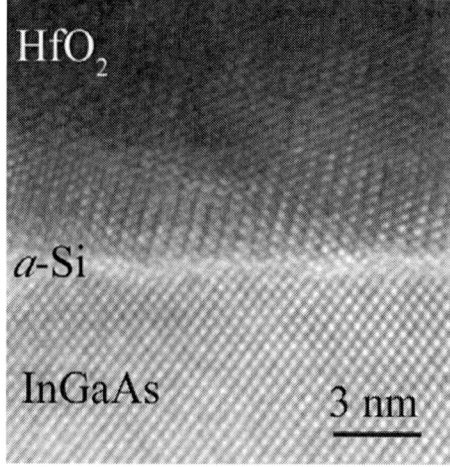

Fig. 4. Cross-sectional TEM micrograph of InGaAs with HfO_2 and 0.2 nm Si.

Enhancement Mode n-MOSFET with High-κ Dielectric On GaAs Substrate

K. Rajagopalan, P. Zurcher, J. Abrokwah, R. Droopad, D. A. J. Moran*, R. J. W. Hill*, X. Li*, H. Zhou*, D. McIntyre*, S. Thoms*, I.G. Thayne* and M. Passlack

Freescale Semiconductor Inc. 2100, E Elliott Road, Tempe, AZ-85284.

* Department of Electronics & Electrical Engineering, University of Glasgow. Glasgow, UK.

Challenges faced by Silicon CMOS technology, as device scaling reaches physical limits, have prompted the proposal of solutions based on compound semiconductor channels in the recent ITRS Roadmap for Semiconductors 2006. The key drawback of compound semiconductors like GaAs has been the lack of a suitable surface passivating dielectric that could yield a good quality electrical interface between the oxide and the semiconductor.

Here we report MOS heterostructures grown by molecular beam epitaxy on III-V substrates, employing a high-κ dielectric stack ($\kappa \approx 20$) comprised of gallium oxide and gadolinium gallium oxide. Mobilities exceeding 12,000 and 6,000 cm^2/Vs, for sheet carrier concentration n_s of about 2.5×10^{12} cm^{-2} were measured on MOSFET structures on InP and GaAs substrates, respectively. These structures were designed for enhancement mode operation and include a 10 nm thick strained In$_x$Ga$_{1-x}$As channel layer with In mole fraction x of 0.3 and 0.75 on GaAs and InP substrates, respectively. Fig. 1 shows the heterostructure on GaAs substrate and its corresponding TEM image.

n-MOSFETs with a gate length of 1 μm and a source-drain spacing of 3 μm were fabricated on the heterostructure shown in Fig. 1. Fig. 2 shows the I-V characteristics of the 1 μm x 100 μm device. The drain current at 2V gate bias was measured to be in excess of 400 mA/mm. A threshold voltage of 0.27 V, transconductance of 428 mS/mm and on-resistance of 2 ohm-mm were also measured. The output conductance was measured to be 10 mS/mm. Fig. 3 shows a comparison of the transconductance measured on our fabricated devices as compared to g_m values reported for similar devices, over a period of thirty years.

*Corresp. author: karthik.rajagopalan@freescale.com, Phone: +1-480-413-7032, Fax: +1-480-413-4453

Figures

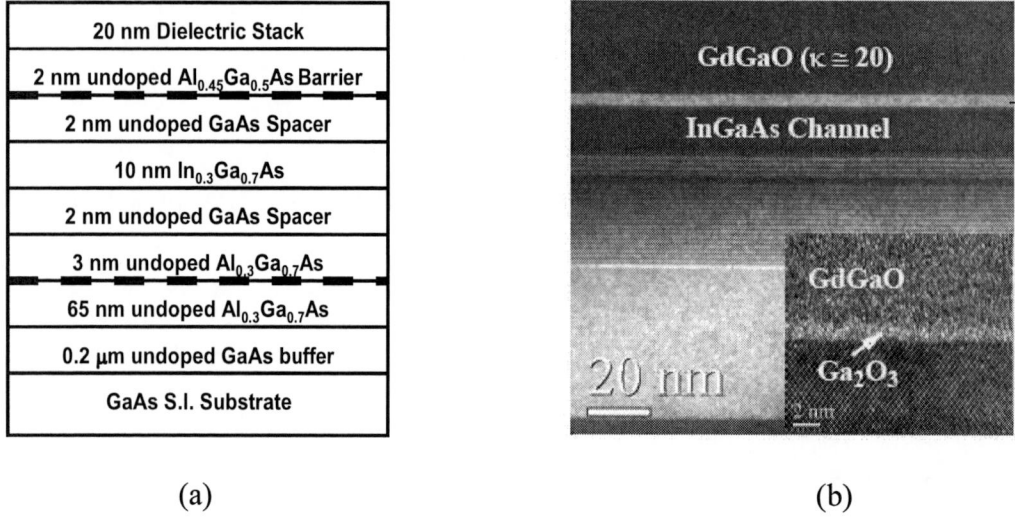

(a)   (b)

Fig. 1. (a) n-MOSFET layer structure on GaAs substrate (b) TEM image of n-MOSFET layer structure on GaAs Substrate

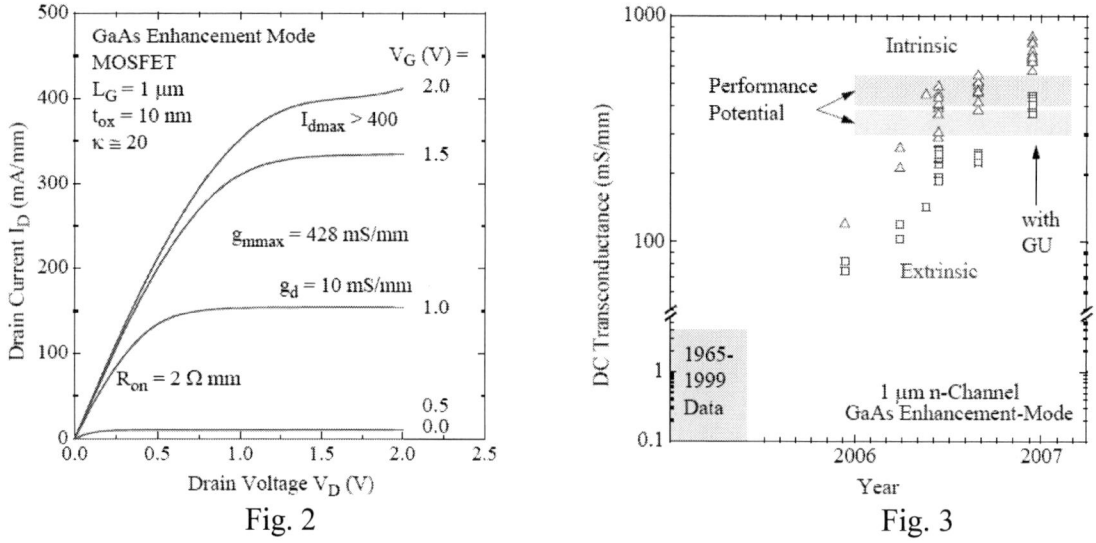

Fig. 2   Fig. 3

Fig. 2. ID-VD of 1 μm x 100 μm n-MOSFET on GaAs substrate

Fig. 3. Comparison of g_m from this work with prior data. The range of expected intrinsic and extrinsic performance potential of our 1 μm MOSFETs is also shown.

High-performance submicron inversion-type enhancement-mode InGaAs MOSFET with maximum drain current of 360 mA/mm and transconductance of 130 mS/mm

Y. Xuan, Y. Q. Wu, H. C. Lin, T. Shen and P. D. Ye[*]

School of Electrical and Computer Engineering, Purdue University, West Lafayette, IN 47906
* Tel: 765-494-7611, Fax: 765-494-0676, E-mail: yep@purdue.edu

After implementation of ALD high-k gate dielectrics and metal gates in high-volume manufacturing for upcoming CMOS integrated circuits by Intel and IBM, the hope is growing that the ALD high-k dielectrics developed for Si may also be applicable to compound semiconductors. Although *in-situ* MBE grown $Ga_2O_3(Gd_2O_3)$ shows promising results as a good gate dielectric on III-V compound semiconductors,[1-2] the current research is mainly focused on *ex-situ* ALD or PVD Al_2O_3 and HfO_2 due to its potential manufacturability. But so far the reported *inversion-type* E-mode GaAs MOSFETs, which has the scalability for future 22 nm technology node or beyond, suffer from low drain currents.[3-4]

In this paper, we report, for the first time, submicron inversion-type E-mode n-channel MOSFETs on $In_{0.53}Ga_{0.47}As$ using ALD Al_2O_3 as high-k gate dielectric with more than 360 mA/mm maximum drain current and 130 mS/mm transconductance. The device performance has a significant leap with *3000 times increase* of the maximum drain current, compared to our previous results on $In_{0.2}Ga_{0.8}As$ MOSFET.[3] We ascribe this improvement to the fact that $In_{0.53}Ga_{0.47}As$ is the more forgiving material with respect to Fermi level pinning and has a narrower bandgap easier to be inverted. Fig. 1 and Table 1 show the schematic cross section of the device structure and the device fabrication flow. A well-behaved I-V characteristic of a 0.5 μm-gate-length inversion-type $In_{0.53}Ga_{0.47}As$ NMOSFET is demonstrated in Fig. 2 with maximum drain current of 367 mA/mm. From measured channel resistance R_{Ch} vs. designed mask gate length L_{Mask} in Fig. 3, the source-drain contact resistance R_{SD} and the difference between L_{Mask} and effective gate length (ΔL) are 15 Ω and <0.05 μm, respectively. After subtracting the R_{SD} of 15Ω, the resulting intrinsic maximum drain current and transconductance for 0.5 μm device are 425 mA/mm and 145 mS/mm, respectively. By the conventional linear region extrapolation method or second derivative method, the extrinsic threshold voltage is determined around 0.25 V. From drain current vs. $1/L_{mask}$ under V_{gs}=5V and V_{ds}=2V in Fig. 4, the drain current or transconductance is linearly inversely proportional to L_{mask} as expected and start to saturate at L_{mask}=0.75 μm. Fig. 5 shows C-V characteristics of the Al_2O_3 (8nm)/$In_{0.53}Ga_{0.47}As$ (p-type) MOS structure with multiple frequency measurement at room temperature in dark. Clear transitions from accumulation to depletion for HF C-V and the inversion features for LF C-V below 1KHz are observed. The mid-gap D_{it} is estimated to be around 1.4×10^{12} /cm^2-eV by HF - LF method. The "split-CV" and I_d-V_d method is used for obtaining the effective mobility μ_{eff}. As shown in Fig. 7, the extracted Al_2O_3/$In_{0.53}Ga_{0.47}As$ μ_{eff} has a peak value of 1100 cm^2/Vs, twice higher than Si universal mobility at normal electric field of 0.5 MV/cm. Although the effective mobility for $In_{0.53}Ga_{0.47}As$ is expected to be further improved in a near term by further optimizing the process, the reported μ_{eff} here in Fig. 7 is one of the best among the reported values at oxide/III-V interface with reliable extraction methods.

In summary, we have demonstrated high-performance inversion-type E-mode $In_{0.53}Ga_{0.47}As$ MOSFETs using ALD high-k gate dielectrics. These results suggest $In_{0.53}Ga_{0.47}As$ could be an ideal channel material which has higher electron effective mobility and wide enough bandgap for low drain voltage ultimate CMOS applications.

[1] F. Ren et al., *IEEE Electron Device Letters* **19**, 309 (1998).
[2] K. Rajagopalan et al., *IEEE Electron Device Letters* **27**, 959 (2006).
[3] Y. Xuan et al., *Applied Physics Letters* **88**, 263518 (2006).
[4] S. Oktyabrsky et al., *Materials Science and Engineering* B **135**, 272 (2006).

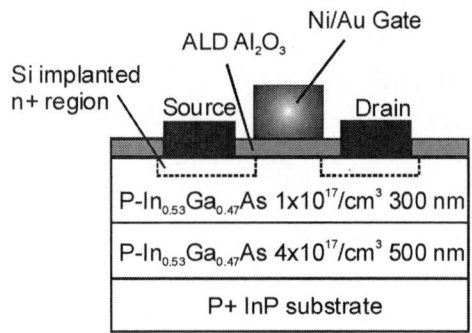

Fig 1. Schematic cross section of $In_{0.53}Ga_{0.47}As/InP$ n-MOSFET with ALD Al_2O_3 gate dielectric.

Tabel 1. Gate-last process of ALD Al_2O_3/InGaAs MOSFET

1) ALD Al_2O_3 30nm growth on MBE $In_{0.53}Ga_{0.47}As$/InP
2) S/D patterning and Si implantation (30KeV, 1E14 & 80KeV, 1E14)
3) S/D activation by RTA (650-850°C 10s in N_2)
4) Remove 30nm Al_2O_3 by BOE solution
5) $(NH_4)_2S$ surface preparation
6) ALD Al_2O_3 8nm re-growth and PDA at 600°C in N_2
7) S/D contact formation and Au/Ge/Ni ohmic metal evaporation and 400°C metallization
8) Gate patterning and Ni/Au evaporation

Table 1. Device fabrication flow of $In_{0.53}Ga_{0.47}As$/InP n-MOSFET.

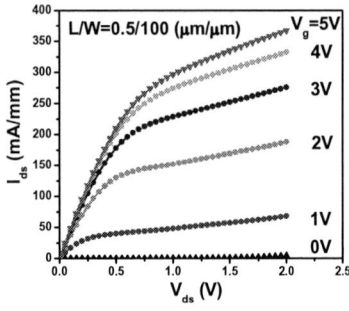

Fig 2. Drain current versus drain volatge at different gate biases for a 0.5μm-gate-length MOSFET with 8nm ALD Al_2O_3 gate dielectric.

Fig 3. Channel resistance versus different mask gate length at different gate biases. R_{SD}=15 Ω and ΔL≤0.05μm.

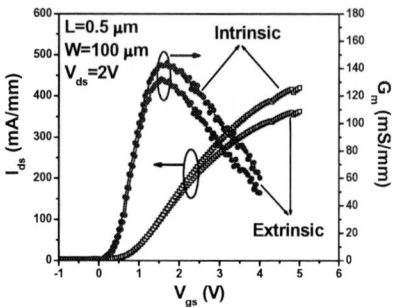

Fig 4. Extrinsic and intrinsic drain current and trans-conductance versus gate bias.

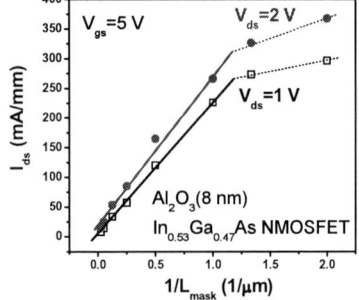

Fig 5. Drain current versus gate length (1/L_{mask}) at different drain voltages.

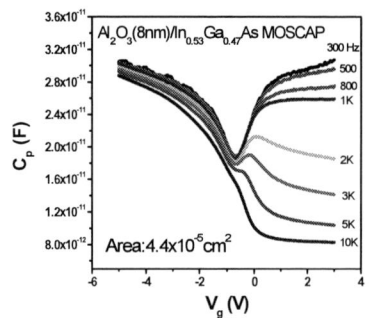

Fig 6. *C-V* characteristics of an Al_2O_3 (8nm) / $In_{0.53}Ga_{0.47}As$ MOS structure at multiple frequencies.

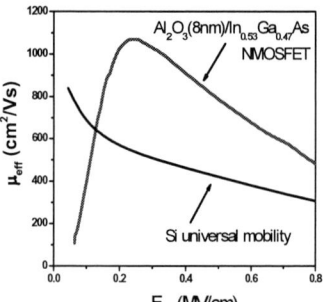

Fig 7. $In_{0.53}Ga_{0.47}As$ n-MOSFET effective mobility versus effective electric normal field.

Enhancement-mode In$_{0.70}$Ga$_{0.30}$As-channel MOSFETs with ALD Al$_2$O$_3$

Yanning Sun, E. W. Kiewra, J. P. De Souza, S. J. Koester, K. E. Fogel, D. K. Sadana

IBM Thomas J. Watson Research Center, Yorktown Heights, NY 10598, USA

phone: (914) 945-3083, fax: (914) 945-2141, email: yansun@us.ibm.com

Introduction: Compound III-V semiconductors are receiving renewed attention for use as channel materials for advanced VLSI logic applications due to their high electron mobility, as well as the increasing difficulty of achieving enhanced performance by conventional scaling. InGaAs-channel HEMTs have produced g$_m$ values over 2 S/mm [1], f$_T$ of 562 GHz [2], and compared favorably in terms of power-delay product [3]. Still, for VLSI logic applications, InGaAs-channel FETs will ultimately need to incorporate high-κ insulating gate dielectrics in order to meet ITRS leakage requirements. Previous work on InGaAs-channel MOSFETs has mainly focused on surface-channel device geometries [4], but these devices would require the formation of an extremely-high-quality semiconductor-dielectric interface in order to preserve low D_{it} near the surface-layer conduction-band edge. For this reason, a buried-channel MOSFET design may be preferable, as shown in Fig. 1, to relax the requirements for low D_{it} near the surface-semiconductor conduction-band edge, as well as improve the carrier mobility.

In our previous work, we demonstrated the operation of depletion-mode In$_{0.7}$Ga$_{0.3}$As buried-channel MOSFETs with HfO$_2$ gate dielectrics [5]. However, the structure used in that work was not optimal due to its large suface-to-channel distance. In this work, we report enhancement-mode buried-channel MOSFETs, with largely reduced surface-to-channel distance, and greatly improved gate dielectric deposition. These devices display good output and pinch-off characteristics, low leakage, and high I$_{on}$/I$_{off}$ of ~10^5. The effective channel mobility extracted from these devices is the highest reported value for enhancement-mode III-V MOSFETs.

Layer Structure Design and Device Fabrication: In this work, an In$_{0.7}$Ga$_{0.3}$As/In$_{0.52}$Al$_{0.48}$As quantum well layer structure was used as shown in Fig. 2. The layer structure was grown on an InP substrate, and consisted of a 300 nm In$_{0.52}$Al$_{0.48}$As buffer layer, a 10 nm In$_{0.70}$Ga$_{0.30}$As strained quantum well, a 8 nm In$_{0.52}$Al$_{0.48}$As top barrier layer, and n$^+$-InGaAs cap layer. A Si δ-doped layer with doping density of 5×10^{11} cm^{-2} is 3 nm above the strained quantum well. Long-channel ringFET devices were made by patterning the heterostructure with optical lithography followed by selectively etching the InGaAs cap layer to form a gate recess area. After 6 nm ALD Al$_2$O$_3$ deposition and annealing, Ohmic contacts were then formed. The gate regions were then defined by optical lithography and Al evaporation and lift off. A cross-sectional diagram and SEM micrograph of the final long-channel MOSFETs are shown in Figs. 2 and 3.

Enhancement-mode Long-Channel MOSFETs: The DC output characteristics of a typical buried In$_{0.7}$Ga$_{0.3}$As-channel MOSFET with L_g = 10 μm are shown in Fig. 4. The devices show good saturation and pinch off characteristics with external series resistance (R_{ext}) of 5.2 Ω · mm. The gate leakage characteristic of the same device is shown in Fig. 5, and the results are compared with HEMTs from our previous work [5] and the recent work from Kim et al. [3]. The gate leakage current density of the MOSFET is more than six orders of magnitude lower than that of both HEMT devices. The sub-threshold characteristics and corresponding transconductance are shown in Fig. 6 and Fig. 7. The devices operate in enhancement mode and have a threshold voltage of +0.1 V, as determined by linear extrapolation from the peak trans-conductance at V_{ds} = 50 mV. The drain current on-off ratio is ~10^5, and the devices have a subthreshold slope of 150 mV/decade. The extrinsic transconductance, g_{mext}, has a peak value of 60 mS/mm at V_{gs} = +0.9 V and V_{ds} = +2 V, with corresponding intrinsic transconductance (g_{mi}) of 71 mS/mm. The capacitance-voltage results for the MOSFETs are shown in Fig. 7. The effective-oxide thickness (EOT) extracted from the $C_g - V_g$ characteristics at strong forward bias (Vg = +2 V) is 4.8 $\pm$ 0.3 nm, a value that agrees fairly well with expectations for the given device dimensions. The effective drift mobility and the sheet density were calculated from the C_g vs. V_g (200 kHz) and linear $I_d - V_{gs}$ characteristics (V_{ds} = 50 mV). The resulting mobility vs. sheet density plot is shown in Fig. 9. A peak mobility of ~1280 cm^2/Vs (~1350 cm^2/Vs after correcting for R_{ext}) was determined at a carrier density of 1.1 x 10^{12} cm^{-2}. The extracted effective channel mobility is the highest reported value for enhancement-mode III-V MOSFETs and is also much higher than that of the Si MOSFET with high-κ gate dielectrics. However, the effective channel mobility is much smaller than the reported Hall-effect mobility of In$_{0.7}$Ga$_{0.3}$As [2], due to high D_{it} at the dielectric/InAlAs interface. Further improvements should be possible through optimization of the interface properties.

In conclusion, we have demonstrated long-channel enhancement-mode In$_{0.7}$Ga$_{0.3}$As MOSFETs with ALD Al$_2$O$_3$. The devices show good output and pinch-off characteristics, have I$_{on}$/I$_{on}$ of 10^5, six orders of magnitude lower gate leakage than Schottky-gated devices, and a peak effective mobility of 1280 cm^2/Vs. These results are promising for realizing scalable InGaAs-channel MOSFETs suitable for VLSI applications.

978-1-4244-1101-6/07/$25.00 ©2007 IEEE

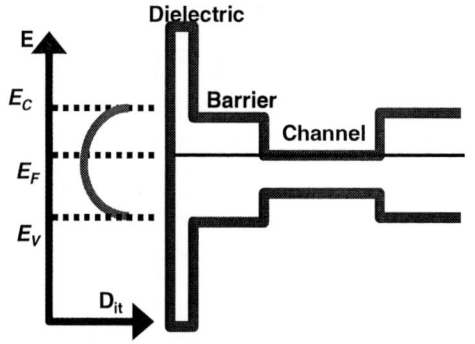

Fig. 1. Schematic band diagram of a buried-channel MOSFET

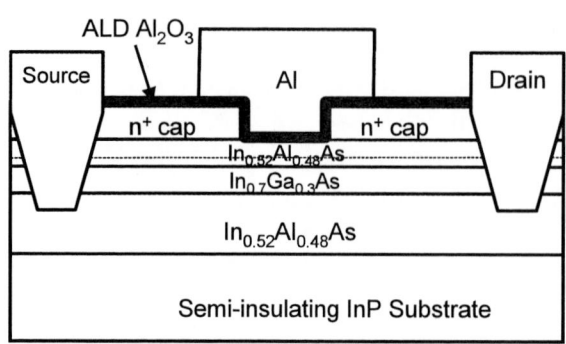

Fig. 2. Schematic cross-sectional diagram of an enhancement-mode InGaAs-channel MOSFET

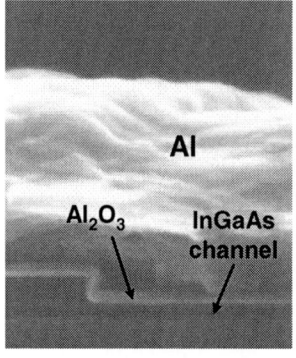

Fig. 3. XSEM of an enhancement-mode InGaAs-channel MOSFET

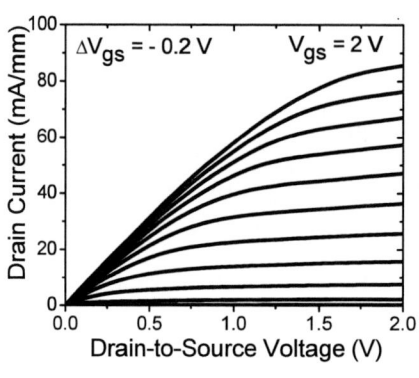

Fig. 4. Output characteristics of a long-channel MOSFET with L_g = 10 μm

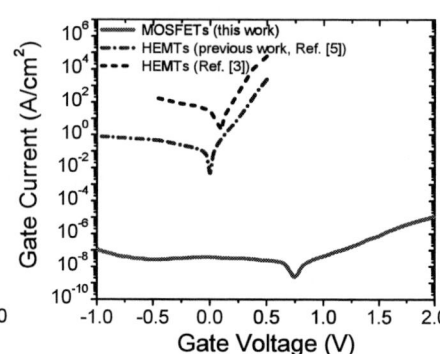

Fig. 5. Gate leakage characteristics of a long-channel MOSFET, compared with previous HEMTs

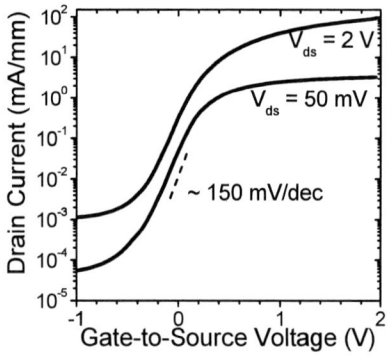

Fig. 6. Subthreshold characteristics of a long-channel MOSFET with L_g = 10 μm

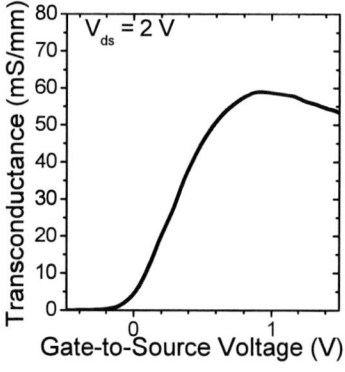

Fig. 7. Transconductance of a long-channel MOSFET with L_g = 10 μm

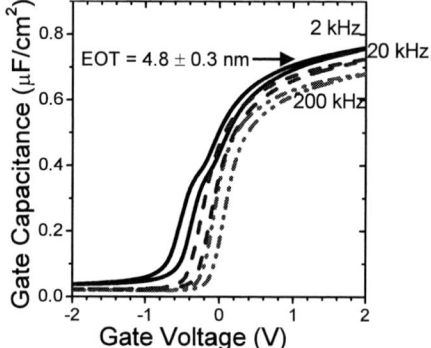

Fig. 8. Multi-frequency capacitance vs. voltage characteristics of a buried-channel MOSFET

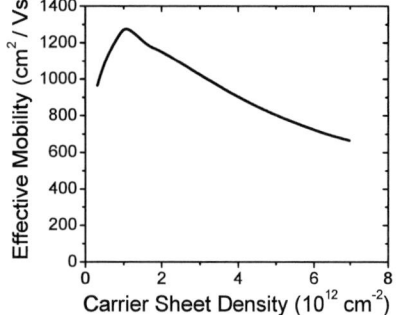

Fig. 9. Effective mobility vs. carrier sheet density of a buried-channel MOSFET using 200 kHz C-V data

Acknowledgement

The authors would like to thank Ghavam Shahidi for management support and technical guidance, and Robert Sandstrom for fabrication support.

References

[1] D. Xu et al, IEEE Elect. Dev. Lett. 20, 206 (1999).
[2] Y. Yamashita et al, IEEE Elec. Dev. Lett. 23, 573 (2002).
[3] D.H. Kim et al, IEDM Tech. Dig., 787 (2005).
[4] F. Ren et al, IEEE Elec. Dev. Let., 19, 309 (1998)
[5] Yanning Sun et al, DRC, 49 (2006)

Performance of Sub-micron Gate Length InAlP Native Oxide GaAs-channel MOSFETs

J. Zhang, T. H. Kosel, D. C. Hall, and P. Fay

[1]Dept. of Electrical Engineering, University of Notre Dame, Notre Dame, IN 46556
Tel: (574) 631-5693, Email: pfay@nd.edu

GaAs based MOSFETs have attracted significant interest as a potential technology for both digital and RF applications. Among many candidate gate insulating materials (e.g. [1-4]), the native oxide of InAlP offers a low leakage current [4] and modest interface state density [5] while at the same time being simple to fabricate and offering a path to low-cost devices. Both enhancement-mode [6] and depletion-mode MOSFETs [7-8] with gate lengths ≥ 1 μm have been demonstrated with InAlP native oxide gate dielectrics. We present here the first sub-micron gate length InAlP native oxide GaAs-channel devices, with record microwave performance.

GaAs-channel MOSFETs are fabricated on heterostructures grown by metal-organic chemical vapor deposition (MOCVD). The device structure includes a 100 nm doped GaAs channel, an InGaP spacer and oxidation stop layer, an n$^+$ doped InAlP layer, and an n$^+$ GaAs cap (Fig. 1(a)). Mesa-isolated devices were fabricated using mix-and-match optical/electron-beam lithography. Optical lithography is used to define the mesa and source/drain contacts; electron-beam lithography is used to define the region for gate oxidation and the gate metallization. The fabrication processing is similar to a conventional HEMT, except that for the formation of the gate oxide a window around the gate area is exposed and the GaAs cap is selectively removed. A 15 minute, 440 °C wet thermal oxidation is performed, resulting in a 5 nm thick InAlP oxide in the gate region. The gate lithography is then performed and T-gates are formed by liftoff. A schematic cross-section of the finished device is shown in Fig. 1(b). Hall effect measurement of the as-grown heterostructure shows a channel carrier concentration of 6.8×10^{12} cm^{-2} and mobility of 4000 cm^2/Vs; after oxidation, a mobility of 3800 cm^2/Vs is measured.

Capacitance-voltage characteristics (1 MHz) are measured on a typical 1×150 μm^2 transistor (Fig. 2), demonstrating clear channel modulation. The common-source current-voltage characteristics for a 0.25 μm gate length device show clear pinch-off (Fig. 3). A peak extrinsic transconductance of 71.4 mS/mm at a gate voltage of −1.8 V is measured (Fig. 4). On-wafer s-parameter measurements from 1 to 40 GHz are used to characterize the microwave performance. The current gain h_{21} and maximum available gain are plotted in Fig. 5. An f_t of 28 GHz and an f_{max} of 50 GHz are obtained. To the authors' knowledge, this is the first report of sub-micron native-oxide based GaAs-channel MOSFETs; the f_t reported represents a 65% improvement over the best previously reported devices [8].

Transistor performance vs. gate length was also evaluated; the extrinsic transconductance and output conductance vs. gate length show the onset of short-channel effects for $L_g < 1$ μm (Fig. 6), arising from a lack of channel confinement at the channel/buffer interface. Maximum voltage gains (g_m/g_{ds}) of 17.4 are obtained at long gate lengths, decreasing to 9.7 for $L_g = 0.25$ μm. The microwave performance vs. L_g is shown in Fig. 7.

Sub-micron gate length GaAs-channel MOSFETs with native InAlP oxide gate dielectric are demonstrated with record microwave performance, indicating InAlP native oxide is a promising gate dielectric for high speed GaAs-based MOSFETs.

[1] Y.C.Wang et al., IEEE Electron Device Lett., 20, p. 457 1999.
[2] P. D. Ye et al., Appl. Phys. Lett., vol. 84, p. 434, 2004.
[3] M. Passlack et al., IEEE Electron Device Lett., 26, p. 713, 2005.
[4] Y. Cao et al., Appl. Phys. Lett. 86, 062105, 2005.
[5] X. Li et al., J. Appl. Phys., 95, p. 4209, 2004.
[6] X. Li et al., IEEE Electron Dev. Lett., 25, p. 772, 2004.
[7] Y. Cao et al., IEEE Electron Dev. Lett., 27, p. 317, 2006.
[8] Y. Cao et al., Proc. Compound Semiconductor Integrated Circuit Symp., p. 43, 2006.

(a)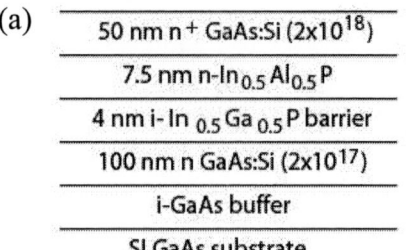

50 nm n+ GaAs:Si (2×10^{18})
7.5 nm n-$In_{0.5}Al_{0.5}P$
4 nm i-$In_{0.5}Ga_{0.5}P$ barrier
100 nm n GaAs:Si (2×10^{17})
i-GaAs buffer
SI GaAs substrate

(b)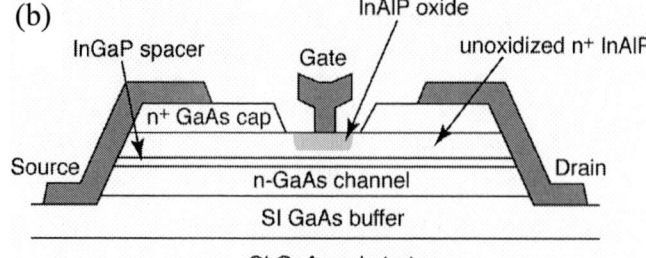

Fig. 1. (a) Schematic diagram of the heterostructure; (b) cross-sectional diagram of fabricated device.

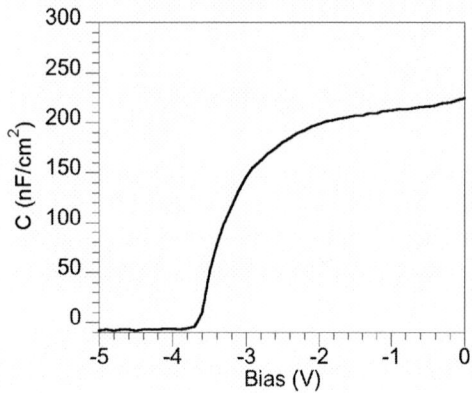

Fig. 2. Capacitance-voltage characteristic (1 MHz) of a typical 1x150 μm MOSFET.

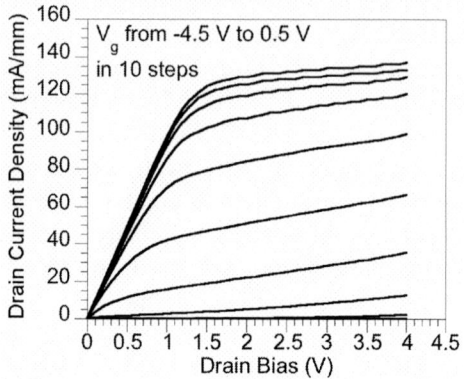

Fig. 3. Common-source current-voltage characteristics for a 0.25 μm gate length MOSFET.

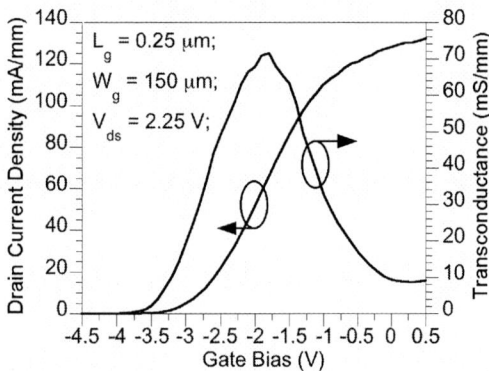

Fig. 4. Drain current and transconductance as a function of gate voltage.

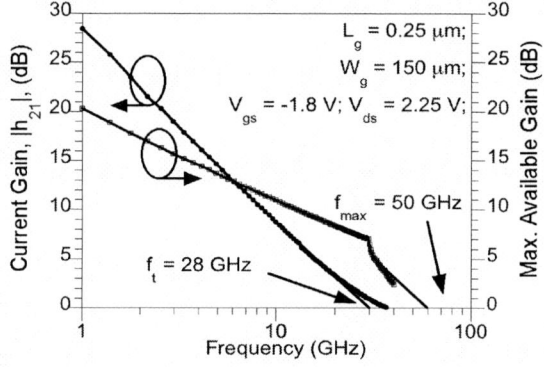

Fig. 5. Microwave performance of a typical MOSFET.

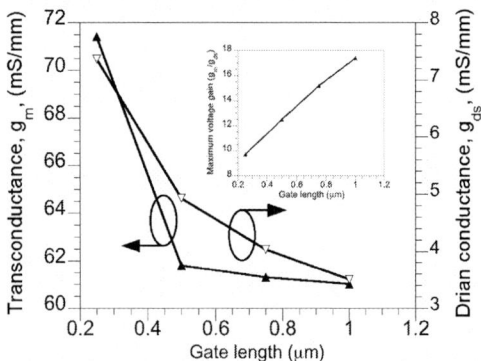

Fig. 6. Transconductance and drain conductance vs. gate length.

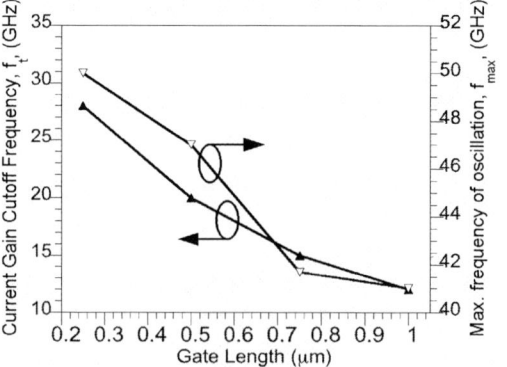

Fig. 7. f_t and f_{max} vs. gate length.

214

Session V.C (DeBartolo Hall Room 155)

Memory Devices

Tuesday PM, June 20th,2007

Session Organizer: Pranav Kalavade, Intel
Session Chair: Andrei Mihnea, Micron

1:30 PM V.C-1 Student Paper
Hybrid ALD-SiN/Si-nanocrystals/ALD-SiN FinFET device with large P/E window for MLC NAND Flash memory application
J.-D. Choe[1], S.-H. Lee[2], Y. J. Ahn[2], D. Jang[2], Y.-B. Yoon[2], J. J. Lee[2], I. Chung[1], K. Park[2] and D. Park[2], [1]School of Information and Communication Engineering, Sungkyunkwan University, Kyungki-Do, KOREA and [2]Technology Development Team 2, Samsung Electronics Co., Yongin-City, Kyungki-Do, KOREA

1:50 PM V.C-2 Student Paper
Ge/Si hetero-nanocrystalnonvolatile floating gate memory
B. Li, Y. Zhu and J. Liu, Quantum Structures Laboratory, Department of Electrical Engineering, University of California, Riverside, California, USA

2:10 PM V.C-3 Student Paper
Memory Effects in Metal-Oxide-Semiconductor Capacitors Incorporating Dispensed Highly Mono-disperse One-Nanometer Silicon Nanoparticles
O. M. Nayfeh[1], D. A. Antoniadis[1], K. Mantey[2] and M. H. Nayfeh[2], [1]Microsystems Technology Laboratories, Massachusetts Institute of Technology, Cambridge, Massachusetts and [2]Department of Physics, University of Illinois at Urbana-Champaign, Urbana, Illinois, USA

2:30 PM V.C-4 Student Paper
Modeling of Multi-layer Nanocrystal Memory
T.-H. Hou, C. Lee, and E. C. Kan, School of Electrical and Computer Engineering, Cornell University, Ithaca, New York, USA

2:50 PM V.C-5

3:10 PM Break

3:30 PM V.C-6 Invited Paper
Phase-change Memory
C. Lam, IBM Qimonda Macronix PCRAM Joint Project, IBM T.J. Watson Research Center, Yorktown Heights, New York, USA

4:10 PM V.C-7
Novel Cross Point Switch based on Zn1-xCdxS memory devices for FPGA
K. Abe[1], Z. Wang[2], S. Fujita[1], T. H. Lee[2] and Y. Nishi[2], [1]Frontier Research Laboratory, Corporate R&D Center, Toshiba Corporation, JAPAN, and [2]Center for Integrated Systems, Stanford University, Stanford, California, USA

Hybrid ALD-SiN/Si-nanocrystals/ALD-SiN FinFET device with large P/E window for MLC NAND Flash memory application

Jeong-Dong Choe[1], Se-Hoon Lee, Young Joon Ahn, Donghoon Jang, Young-Bae Yoon,
Jong Jin Lee, Ilsub Chung[1], Kyucharn Park and Donggun Park

[1]School of Information and Communication Engineering, Sungkyunkwan University, Kyungki-Do, KOREA
Technology Development Team 2, Samsung Electronics Co., Yongin-City, Kyungki-Do, KOREA,
Phone: 82-31-209-9349, Fax: 82-31-209-9861, E-mail: ivo.choe@samsung.com

Nonvolatile memory (NVM) technology, especially NAND Flash technology is going through the fastest evolution among the silicon technologies as manifested in current NAND technology trend. However as the NAND technology approaches 40nm node and beyond, many challenges will appear from the scaling limits [1]. In many efforts to overcome the issues, silicon nanocrystal memory is a promise candidate because it breaks up the floating gate of a nonvolatile memory into discrete electrically isolated charge storage elements and thereby mitigates charge loss through tunnel oxide defects. This permits scaling of tunnel oxide and operating voltages. In spite of these advantages of nanocrystal memory, large program and erase (P/E) window over 8 V for multi-level cell (MLC) operation has not been presented yet [2,3]. In addition to that, many efforts were performed just in metal-oxide-semiconductor (MOS) structures [3-5]. To improve P/E window of real NVM device, we have fabricated hybrid nanocrystal FinFET structures using state-of-the-art technologies such as TaN metal gate, atomic layer deposition (ALD) method, Al_2O_3 blocking dielectrics, post-deposition anneal (PDA) and plasma doping (PLAD) in order to provide the product-wise solutions for the next generation NAND Flash memory. We present results of P/E characteristics of our device with above-mentioned technology and propose the process integration.

For the hybrid nanocrystal NAND Flash with Fin-shaped active structure, the following processes have been performed. Si fin was formed using the conventional shallow trench isolation (STI) process, including the trimming oxidation and field oxide recess to control the fin width and height, respectively. 3 nm- thick SiO_2 were grown by clean and dry oxidation after the active fin formation as a tunnel oxide. FinFETs with different widths (20/40/60 nm) and heights (30nm/60nm) were prepared, including planar devices, for the comparison. For the trap layer engineering, Si nanocrystals were deposited using a cyclic CVD method with SiH_4 as a seed forming gas at about 550℃. Before and after nanocrystal formation, thin ALD SiN layers of 1 nm-thick were deposited, respectively. A cross-sectional high resolution transmission electron microscopy (HRTEM) of the trap layer and process sequences are shown in Fig. 1, and well-formed silicon nanocrystals are observed between the SiN layers. The mean size of nanocrystals and aerial density turned out to be 3 nm and $1 \times 10^{12} cm^{-2}$, respectively. Right after pre-treatment for the blocking oxide, the ALD Al_xO_y are deposited followed by 950℃, 30 sec PDA process. TaN or n^+ poly-Si was then deposited and patterned as a gate electrode. We used PLAD process for the formation of lightly doped drain (LDD) on the fin-shaped active structure.

The P/E characteristics of TaN and n^+ poly-Si gate are compared as shown in Fig. 2 (a) and (b). Owing to the larger work-function of TaN, the P/E speed and the back tunneling phenomenon are improved and P/E memory window is over 8.3 V, which is applicable to the MLC NAND Flash operation. We can also confirm the improved work-function engineering from the P/E threshold voltage - electric field relations as shown in Fig. 4. From our measurement data (Table 1.), best-on-cell current (BOC), worst-on-cell (WOC) current and off-current of planar MOSFET were 0.93 μA, 0.3 μA and 2.4 fA, respectively. For the hybrid nanocrystal FinFET, BOC, WOC and off-current were 1.91 μA, 1.1 μA and 1.3 fA, respectively. The characteristics of cell currents were very improved by using state-of-the-art technologies. Experimental results on the retention characteristics after 200℃, 2-hours-bake (Hot Temperature Stress, HTS) are presented in Fig. 4. From this figure and Fig. 3, we can see that hybrid trap structure with SiN/nanocrystal/SiN has better performance and reliability features than nanocrystal or SiN alone structure. Finally, current results indicate that nanocrystal and nitride trap memory are no longer the separate parts. We propose the hybrid nanocrystal FinFET with larger P/E window which is applicable to next generation Flash memory device.

By introduction of the technologies such as work-function engineering, hybrid ALD-SiN/nanocrystal/ALD-SiN trap engineering, high-k blocking dielectrics, PDA and PLAD processes combined with FinFET structure in order to provide the product-wise solutions for the next generation MLC NAND Flash memory, we could achieved the higher performance and improved device characteristics.

References
[1] Kim K. et al., *IEEE International symposium on VLSI-TSA*, 2005, Tech. Digest, pp. 88-94
[2] Kim S. S. et al., *Applied Physics Letters*, 2006, **88**, p. 223502
[3] Coffin H. et al., *Materials Science and Engineering*, 2005, **B124-125**, pp. 499-503
[4] Ng C. Y. et al., *Thin Solid Films*, 2006, **504**, pp. 25-27
[5] Lu T. Z. et al., *Applied Physics Letters*, 2005, **87**, p. 202110

978-1-4244-1101-6/07/$25.00 ©2007 IEEE

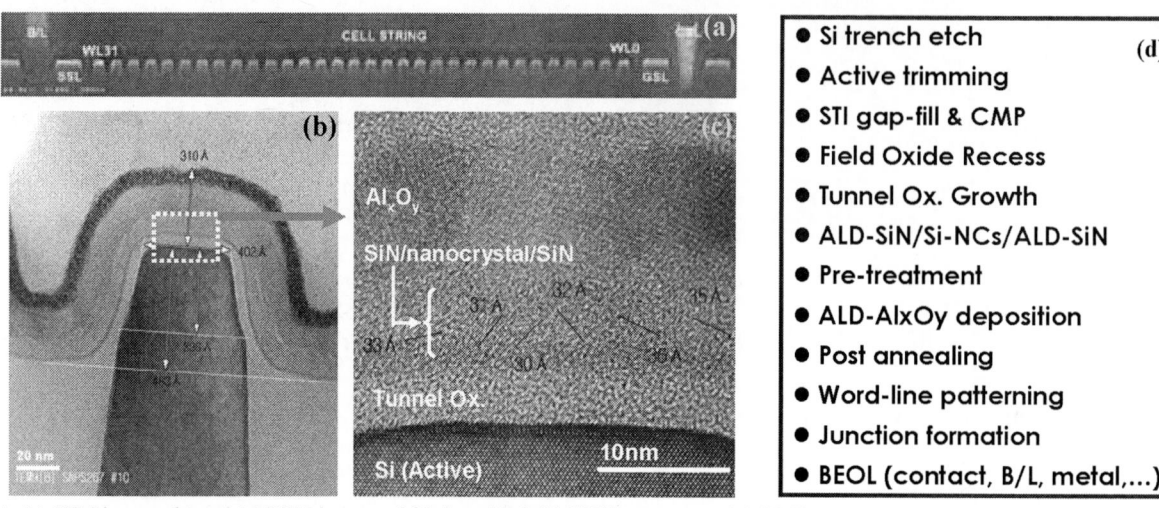

Fig. 1. (a) SEM image. (b) and (c) TEM images of fabricated hybrid SiN/Si-nanocrystals/SiN FinFET NAND strings. (d) Process flow sequence.

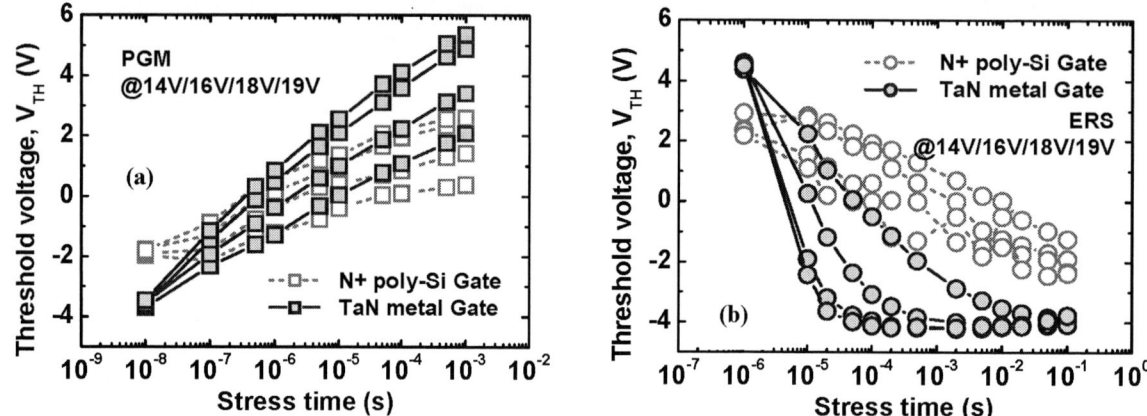

Fig. 2. (a) program and (b)erase characteristics of hybrid FinFET NAND flash devices with TaN metal gate and n$^+$ poly-Si gate.

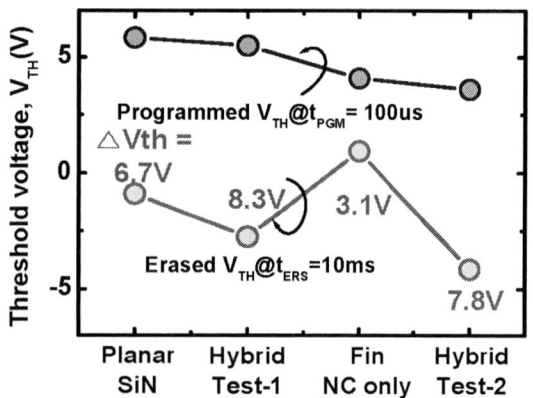

Fig. 3. P/E windows measured at planar SONOS, FinFET NC, and hybrid structure, respectively.

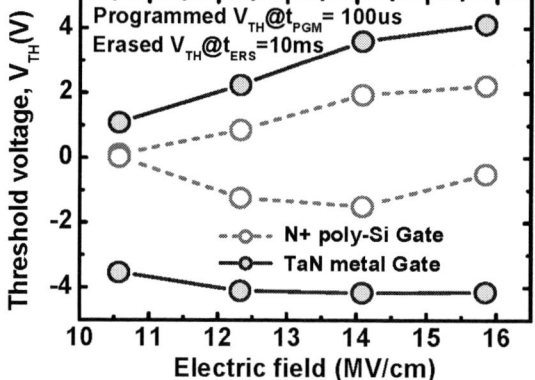

Fig. 4 P/E threshold voltages as a function of the electric field for different gate electrodes.

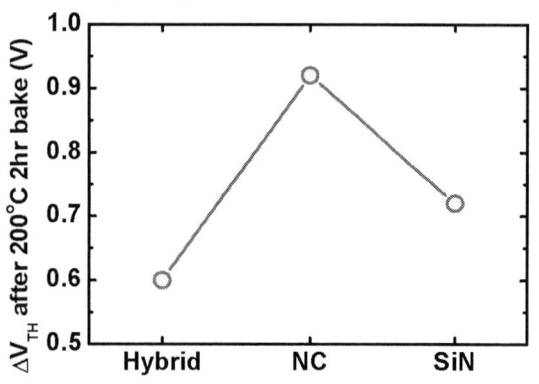

Fig. 5. The HTS characteristics for different trap layers after 200 ℃ bake.

Table 1. Cell currents at planar and Fin structures, respectively.

ITEMS	BOC (μA)	WOC (μA)	Off (fA)
Hybrid planar	0.93	0.3	2.4
Hybrid Fin	1.91	1.1	1.3
Up/down	Up	Up	Down

218

Ge/Si hetero-nanocrystal nonvolatile floating gate memory

Bei Li, Yan Zhu and Jianlin Liu

Quantum Structures Laboratory, Department of Electrical Engineering, University of California, Riverside, CA 92521

Phone: 951-827-6275, Fax: 951-827-2425, email: bli003@ucr.edu

Silicon nanocrystal (NC) [1] has been used as the charge storage in nonvolatile memory devices owing to their advantages of low operation voltage, small size and compatibility to the logic circuit. However a trade-off between the programming speed and retention time remains. In order to improve the retention characteristics and the programming speed simultaneously, Ge/Si hetero-nanocrystal (HNC) was proposed [2-3] to be used as the floating gate. In this presentation, we report the fabrication and characterization of Ge/Si HNC MOS memories.

Figure 1 (a) and (b) show the device structure and energy band diagram of MOS memory with Ge/Si HNC embedded in the SiO_2. Type II band alignment for germanium on silicon provides a quantum well of about 0.47eV deep for holes to store in the germanium side. The retention time prolongs due to the fact that the holes have to be thermally excited to the valance band edge of a Si NC first during the tunneling-back process from the Ge dot to the substrate. Briefly, a thin layer (~5nm) of thermal oxide is grown on n-Si (100) substrate. Si NCs are grown by LPCVD. Ge NCs are selectively grown on Si NCs in the same chamber. Then a control oxide layer is deposited on NCs in another LPCVD furnace. Finally, aluminum electrodes on the back and front side of the sample are deposited and patterned.

Figure 2 shows the atomic force microscopy (AFM) images of Si NCs and Ge/Si HNCs grown on the SiO_2/Si substrate. Comparing these two figures, we find that there is insignificant change for the dot density while the Ge/Si dots height are larger than that of Si dots, which indicates the Ge/Si HNCs are formed on the substrate. The presence of Ge is also proved by x-ray photoemission spectroscopy (XPS) as shown in Figure 3. Ge 2p1/2 and 2p3/2 are clearly shown on the Ge/Si curve.

Figure 4 is the typical CV sweep of Ge/Si MOS capacitor and the inset is CV curve for the reference sample where no dots are embedded between control oxide and tunneling oxide. The flat band voltage shift between forward and backward sweep indicates the charge storage in the NCs. The CV sweep of reference sample shows no clear flat band voltage shift. This result suggests that the memory effect comes from NC, rather than defect or interface states.

Figure 5 shows the programming/erasing characteristics of Ge/Si MOS memory as a function of applied voltage on the gate. In the programming mode, a negative voltage is applied on the gate. As the voltage increases to a point that the holes in the inversion layer begin to charge into the nanocrystals, the flat band voltage shifts. As shown in Figure 5, the shift appears around -10V. During the erasing, a positive voltage is applied on the gate, flat band voltage shifts at smaller bias due to the fact that electrons tunnel into the nanocrystals, which is easier and faster than holes charging.

Figure 6 shows the normalized capacitance during the retention as a function of waiting time after charging at 20V for 2s. It is found that the retention time of Ge/Si HNC memory is significantly longer than that of Si NC MOS memory.

[1] S. Tiwari, F. Rana, H. Hanafi, A. Hartstein, E. F. Crabbe, and K. Chan, "A silicon nanocrystals based memory," Appl. Phys. Lett., vol. 68, pp. 1377-1379, 1996.

[2] H. G. Yang, Y. Shi, L. Pu, J. Wu, R. Zhang, B. Shen, P. Han, S. L. Gu, and Y. D. Zheng, "Nonvolatile memory based on Ge/Si hetero-nanocrystals," Appl. Surf. Sci., vol. 224, pp. 394-398, 2004.

[3] D. T. Zhao, Y. Zhu, R. G. Li and J. L. Liu, "Simulation of a Ge-Si Hetero-Nanocrystal Memory," IEEE Trans. On Nanotechnology, 5, 37 (2006)

978-1-4244-1101-6/07/$25.00 ©2007 IEEE

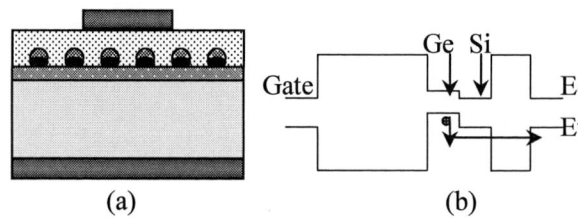

(a)   (b)

Fig.1 (a) Ge/Si HNCs MOS memory device structure, (b) energy band diagram of Ge/Si HNCs MOS structure

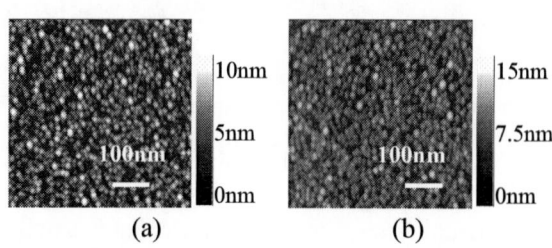

(a)   (b)

Fig.2 AFM images of (a) Si NCs, (b) Ge/Si HNCs

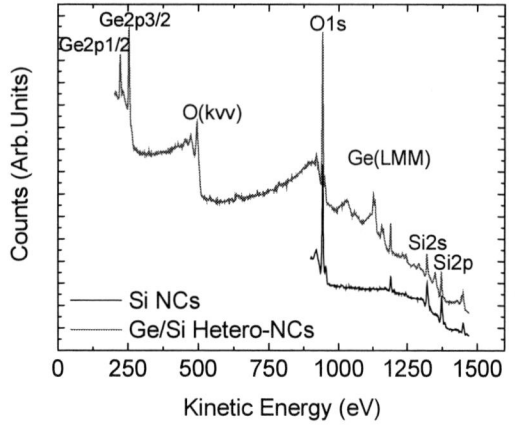

Fig.3 XPS of Si NCs and Ge/Si HNCs. The one for Ge/Si HNCs shows two sharp peaks for Ge 2p1/2 and Ge 2p3/2.

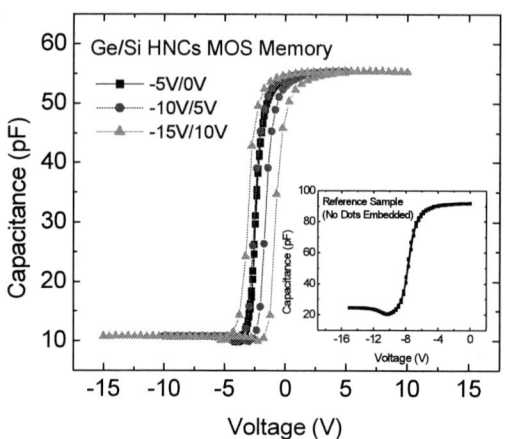

Fig.4 CV sweep of Ge/Si HNCs under -5V~0V, -10V~5V, and -15V~10V. The inset is a CV curve for reference sample where no dots are embedded between control oxide and tunneling oxide.

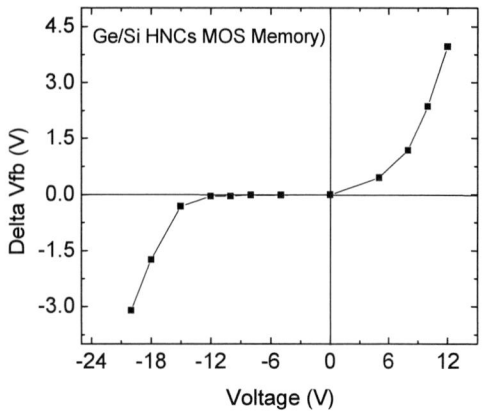

Fig.5 Flat band voltage shift as a function of applied voltage on the gate during writing and erasing of Ge/Si HNCs MOS memory.

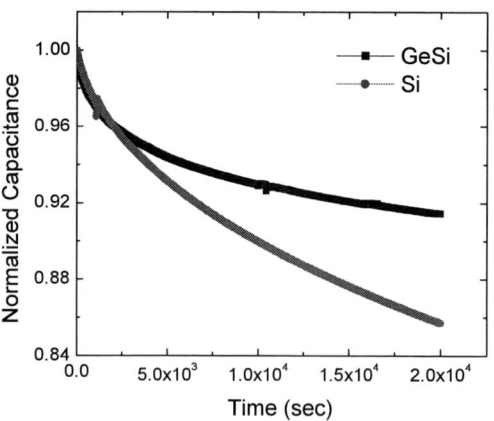

Fig.6 Retention characteristics of Si NCs MOS memory and Ge/Si HNCs MOS memory.

Memory Effects in Metal-Oxide-Semiconductor Capacitors Incorporating Dispensed Highly Mono-disperse One-Nanometer Silicon Nanoparticles

Osama M. Nayfeh* and Dimitri A. Antoniadis
Microsystems Technology Laboratories, Massachusetts Institute of Technology,
60 Vassar St, Cambridge, MA 02139, *onayfeh@mtl.mit.edu, 617-253-0450

Kevin Mantey and Munir H. Nayfeh
Department of Physics, University of Illinois at Urbana-Champaign, 1110 W. Green St, Urbana, IL 61801

MOS capacitors incorporating ex-*situ* produced, colloidal, highly mono-disperse, spherical, 1 nm Si nanoparticles were fabricated and evaluated for potential use as charge storage elements in future non-volatile memory devices. The CV characteristics are well behaved and agree with similarly fabricated zero-nanoparticle control samples and with an ideal simulation. Unlike larger particle systems, the demonstrated memory effect exhibits effectively pure hole storage. The nature of charging i.e. hole-type vs. electron-type may be understood in terms of the novel characteristics of ultra-small Si particles: large energy gap, large charging energy, and consequently a small electron affinity.

The highly mono-disperse, colloidal Si nanoparticles ($Si_{29}H_{24}$), 1 nm in diameter and H-terminated, were prepared using electrochemical dispersion of device quality Si wafers in a mixture of HF/H_2O_2 [1-2]. Si substrates doped with boron at a level of $\sim 10^{15}$ cm^{-3} were used for device fabrication. Standard RCA clean was performed, followed by dry oxidation at 800 °C to grow a $\sim$ 4.2 nm SiO_2 tunnel oxide layer. The Si nanoparticle/isoproponol (IPA) colloid was then dispensed via spin-coating delivery directly over the tunnel oxide. Following the delivery, a $\sim$ 10 nm SiO_2 cap layer was deposited by LPCVD at 400 °C. Gate and backside Al films were then deposited by PVD, and capacitors were defined. A N_2/H_2 anneal at 450 °C for ten minutes completed the fabrication process. Control MOS capacitors were also containing zero-nanoparticles.

The charge storage in the MOS capacitors was analyzed using capacitance-voltage (C-V) measurements at a frequency of 1 MHz as in [3]. **Fig. 1** presents the C-V hysteresis loops for an active device with nanoparticles and a zero-nanoparticle control device. The inset shows a schematic of the device. The hysteresis curve shows that the positive half voltage of the cycle (+7 V to -7 V) involves charging with holes while the negative half (-7 V to +7 V) involves complete erasing and a return to the uncharged state. The negligible hysteresis in the control samples indicates that the effect is nanoparticle related. We estimate by integration of the hysteresis loop, a stored charge density of $\sim 5 \times 10^{11}$/cm^2 for the active device. Also shown in **Fig. 1** is an ideal simulated C-V curve. The similarity in the control, active and simulated curves, in regards to the absence of kinks, humps, or differences in slope point to the fact that the charging in the active device is due to charging of the nanoparticles and not of interface states.

We have tested the charge retention. We measured at room temperature the time-dependence of the stored hole charge loss rate (measured as a voltage shift). **Fig 2** showing 25% charge loss after $\sim$ 100s, and 50% in $\sim$1 hour. Extrapolation of the measured data, indicates that full charge loss should occur in $\sim$1 year. **Fig. 3** shows endurance measurements. We observe only slight decrease in the hysteresis after 10^5 cycles.

The left shift of the programmed C-V with respect to the uncharged curve of the active device in **Fig. 1** with bias of +7 V indicates charging of the particles with holes. With the application of an erase bias of -7 V, the stored holes are removed and the curve shifts back to the right to the uncharged condition. The effectively pure hole-charging characteristic may be understood in terms of properties of the 1 nm particles and their high degree of mono-dispersity. For instance, Quantum Monte Carlo (QMC) and time dependent local density theory (TDLDT) calculations predict that the energy-gap and the charging energy for Si particles in the 1-3 nm regime rise due to the effects of quantum confinement, and consequently the electron affinity decreases, approaching that of the SiO_2 matrix [4-5]. The excitation and emission measurements of [6] for the 1 nm $Si_{29}H_{24}$ particles used in this study yielded 3.5 eV for the energy gap, the difference between the highest unoccupied molecular orbital (HOMO), and the lowest unoccupied molecular orbital (LUMO). In addition, [7] measured $\sim$ 1.8 eV for the charging energy of individual 1 nm silicon particles by UHV-STM tunneling spectroscopy. Because of the increased band-gap and large charging energy, the electron affinity of 1 nm Si nanoparticles is greatly reduced from that of bulk Si and becomes comparable to that of $SiO_2 \sim 1.0$ eV. Consequently, there is no or very little conduction band-offset between the nanoparticle and SiO_2, thus inhibiting electron storage in the conduction band of the particle. However, the well-known asymmetry in the electron and hole barrier heights for the Si/SiO2 system ($\sim$ 3.1 eV for electrons, and $\sim$ 4.7 eV for holes) maintains a valence band-offset on the order of 1.7 eV between the nanoparticle and SiO_2 suitable for hole-storage in the valence band of the particles. **Fig. 4** shows a proposed energy band diagram.

Recently, Spinelli et. al [8] performed theoretical calculations of the retention time as a function of the barrier-offset for a structure with similar geometry (4 nm tunnel oxide). By extrapolating the calculated results, shown in **Fig. 5** to our measured retention time data, we deduce a barrier offset of ~1.5 eV for our system, which is on the order of our analysis. Based on this analysis, we can further increase the retention time of the devices by using a thicker tunnel oxide, or a high-k dielectric with a larger valence band-offset than SiO_2 such as Al_2O_3.

ACKNOLWEDEGEMENT

We acknowledge the MIT SMA Program for support

REFERENCES

1. O. Akcakir, J.Therrien, G. Belomoin, N. Barry, E.Gratton, M. Nayfeh, Applied Physics Letters. 76, 1857 (2000)
2. L. Mitas, J. Therrien, G. Belomoin, and M. Nayfeh, Applied Physics Letters. 78, 1918 (2001)
3. L. W. Teo, W. K. Choi, W. K. Chim, V. Ho, C. M. Moey, M. S. Tay, C. L. Heng, and Y. Lei, D. A. Antoniadis and E. A. Fitzgerald, Applied Physics Letters. 81, 3639 (2002)
4. S. Rao, J. Sutin, R. Clegg, E. Gratton, and M. H. Nayfeh, S. Habbal, A. Tsolakidis and R. M. Martin, Physical Review B. 69, 205319 (2004)
5. D.V. Melnikov and J.R. Chelikowsky, Physical Review B. 69, 113305 (2004)
6. G. Belomoin, J. Therrien, A. Smith, S. Rao, S. Chaieb, M. H. Nayfeh, Applied Physics Letters. 80, 841 (2002)
7. Therrien, J. (2002). *Size Dependence of the Electical Characterisitcs of Silicon Nanoparticles.* PhD thesis, University of Illinois at Urbana-Champaign.
8. C.M. Compagnoni, D. Ielmini, A.S. Spinelli, A.L. Lacaita, C. Previtali, C. Gerardi, Reliability Physics Symposium Proceedings, 2003. 506 (2003)

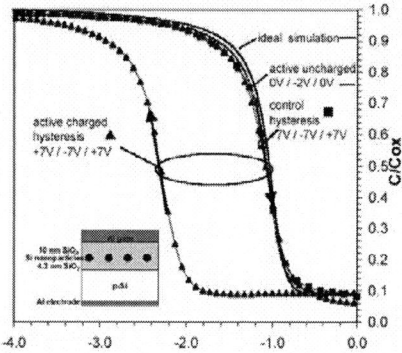

Fig. 1. The C-V hysteresis loops. The control and active devices have negligible hysteresis under uncharged conditions (sweeping forward and backward between Vg = 0 V, and Vg = -2.0 V (light-solid, and labeled). Under charging/discharging conditions (sweeping voltage between Vg=+/- 7.0 V, forward and backward), the control device (squares, labeled) has a hysteresis of only ~0.01 V, however, the active device (triangles, labeled) has a hysteresis of ~1.2 V, corresponding to an estimated stored hole charge density by integration of the C-V of ~4.8x10^{11} cm^{-2}. The C-V curves of both control and active devices are well-behaved and agree with the ideal simulated CV curve (dark solid, labeled). The inset shows a schematic of the MOS capacitor structure.

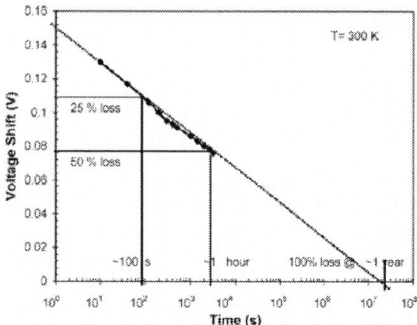

Fig. 2. The charge retention measured at room temperature. 25% of the stored hole charge is lost after ~ 100s, 50% after ~1 hour, and by extrapolation, full charge loss should occur in ~1 year

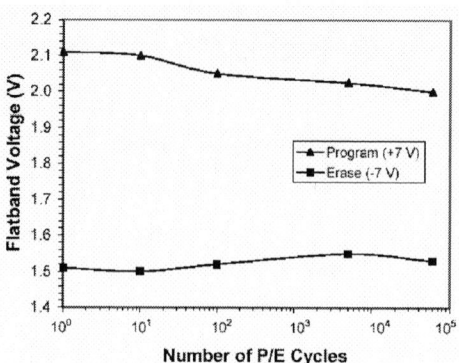

Fig. 3. The endurance characteristics of the MOS device. Programming/Erasing were performed with 35 msec pulses at +/- 7 V. Only slight decrease in the hysteresis is observed after ~10^5 cycles.

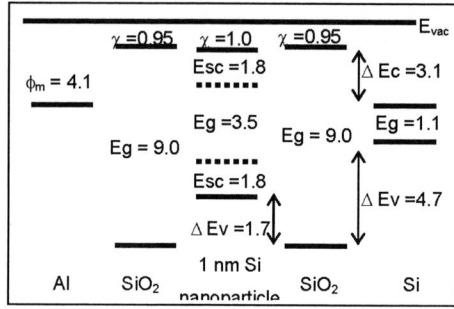

Fig. 4. Proposed energy band diagram of MOS system. Shown are adjustments to the band-gap of the 1 nm particles (~ 3.5 eV), and the charging energy (~1.8 eV). There still remains valence band offset on the order of 1.7 eV for the storage of holes due to the asymmetry in barrier heights for electrons and holes in the Si/SiO2 system.

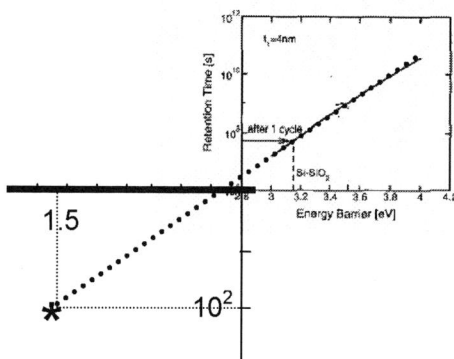

Fig. 5. Extrapolation of calculated retention time in [8] for tox=4 nm. From this extrapolation, we extract a barrier offset of ~1.5 eV for our system, which is on the order of our analysis in fig. 4.

222

Modeling of Multi-layer Nanocrystal Memory

Tuo-Hung Hou, Chungho Lee, and Edwin C. Kan

School of Electrical and Computer Engineering, Cornell University, Ithaca, NY 14853, USA

Phone: 1-607-255-4181, Fax: 1-607-254-3508, Email: th273@cornell.edu

The double-layer (DL) nanocrystal (NC) memory utilizing two layers of NCs for charge storage was shown superior in the ratio of retention to programming/erasing (P/E) time and magnitude of the memory window over the conventional single-layer (SL) NC memory [1-2]. To accurately simulate the P/E and retention characteristics of the SL NC memory, we recently reported a physical model based on the three-dimensional (3D) electrostatics and the one-dimensional (1D) WKB tunneling current calculation [3]. In this paper, we show that the same compact physical model is applicable to the DL NC memory. To further optimize memory performance, design criteria of the size ratio between two layers of NCs and the configuration with multi-layer NCs are explicitly examined.

Figure 1 illustrates the unit cell of the DL NC memory, assuming azimuthal symmetry around the central axis in the z direction. Two layers of spherical metal NCs [2], the lower-layer NC1 with diameter D_{NC1} and the upper-layer NC2 with diameter D_{NC2}, are embedded in a trap-free dielectric, which consists of tunneling oxide, inter-NC oxide, and control oxide with thickness of T_{tunl}, T_{IL}, and T_{conl}, respectively. The NC unit cell diameter C is derived from the NC number density N. The currents through the tunneling oxide I_1, the inter-NC oxide I_2, and the control oxide I_3 depend on the charge states in NC1 and NC2. They are calculated by solving the 3D electrostatic potential profile, the 1D WKB tunneling approximation along the least-action path, and appropriate capture cross sections as fitting parameters extracted from the experimental write characteristics [3]. Notice that the NC Coulomb charging energy E_{CH} for the given geometric configuration is included in the simulation without ambiguity. When D_{NC1} and D_{NC2} are smaller than 5 nm, the energy penalty E_{QM} from the quantum size effect will become prominent and are also accounted for. Furthermore, the flatband shift ΔV_{FB} under the influence of the charges in the discrete NCs is also directly derived from the 3-D electrostatic solution without assuming a coupling ratio [3]. Finally, the P/E and retention dynamics are evaluated with appropriate time evolution of charge states through continuous minimization of current flows, which will however cause some wiggling in ΔV_{FB} during its time evolution.

Figure 2 demonstrates good agreement between simulated and experimental programming transients in SL and DL Au NC memory. Details of the device fabrication were similar to those in [2]. The same simulation parameters were applied in both SL and DL

cases. The programming speed in the DL sample is slightly slower as a result of the addition of inter-NC oxide. However, the maximum memory window is slightly larger because of the larger charge storage capability provided by NC2. In Fig. 3, the DL sample exhibits 10 to 100 times longer retention time than the SL sample, consistent with the previous experimental results [1-2]. It also shows two distinct regimes in the the DL retention. Depending on the initial charge state, in a time scale of seconds, the redistribution of charges through the inter-NC transfer results in increasing or decreasing ΔV_{FB}. The long-term retention is still set by the charge loss to the substrate, which is determined by the band offset between NC and the silicon conduction band edge, independent of the initial charge state. With the same number of charges stored, the NC potentials in the DL sample (see Fig. 4) are consistently lower than that in the SL sample, thanks to the charge distribution in a larger storage volume. Therefore, better retention is achieved as a result of larger band offset.

Similar to the benefits with large-size NC1 [3], enlarging the NC2 size is favorable for memory window and retention time due to its larger charge storage capacity (see Fig. 5). However, it seems an impractical solution for aggressively scaled devices where high NC number density is required. By replacing one layer of oversize NC2, multiple upper layers of NCs with smaller size each can retain similar advantages as shown in Fig. 6. This approach ensures both good scalability and performance simultaneously.

Another design perspective for better retention is to exploit the blockade effect by shrinking the NC1 size. The additional blockade barrier of NC1 can prevent charges in NC2 tunneling back to NC1 and eventually to the substrate. Figure 7 depicts the Coulomb charging energy E_{CH} and the total blockade energy including quantum confinement energy E_{QM} as a function of the NC1 diameter. The improvement on the retention time over the nominal device in Fig. 8 can only be obtained for a NC diameter of 1 nm [4] and the blockade energy up to 1 eV. The sharply decreased blockade energy of NC1 around 2 nm is no longer sufficient to prevent back-tunneling from NC2. For $D_{NC1} < D_{NC2}$, because of the large E_{CH} and E_{QM}, NC1 is ineffective to hold charges. These contribute to the abrupt degradation in retention time with slightly larger NCs.

[1] R. Ohba. et. al., *IEEE Trans. Electron Devices*, vol. 49, p. 1392, 2002. [2] C. Lee et. al., *IEDM Tech. Dig.*, 2003, p. 557. [3] T.-H Hou et. al., *IEEE Trans. Electron Devices*, vol. 53, p. 3095, 2006. [4] R. Ohba. et. al., *IEDM Tech. Dig.*, 2006, p. 959.

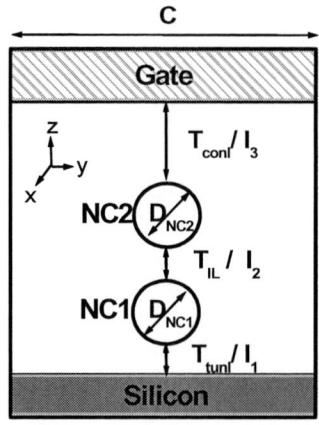

Fig. 1. Schematic of the unit cell of the DL NC memory, assuming azimuthal symmetry around the central axis in the z direction.

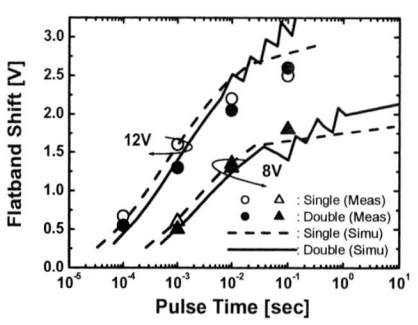

Fig. 2. Measured and calculated flatband voltage shift versus programming pulse time for both SL and DL devices at $V_G =$ 8V and 12V. The same simulation parameters were applied in both SL and DL cases, including a p-type doping of 2×10^{17} cm^{-3}, NC work function $\Phi_{NC} = -5.1$ eV, $T_{tunl} = 2$ nm, $T_{IL} = 2.5$ nm, $T_{conl} = 27$ nm, $D_{NC1} = D_{NC2} = 5$ nm, $N = 4 \times 10^{11}$ cm^{-2}, and the capture cross sections of NC1 and NC2 $\sigma_{NC1} = \sigma_{NC2} = 5.3 \times 10^{-14}$ cm^2 per NC.

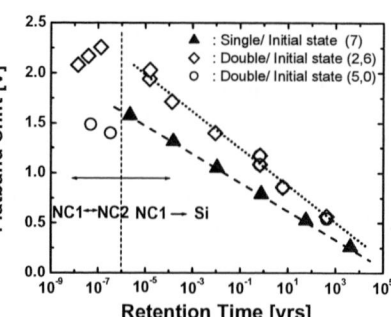

Fig. 3. Calculated room-temperature retention time at $V_G = 0$ V for both SL and DL devices. The DL devices with two different initial states (Q_{NC1}, Q_{NC2}) after programming exhibit similar retention characteristics.

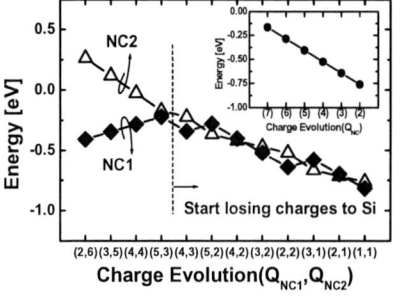

Fig. 4. Evolution of charge states (Q_{NC1}, Q_{NC2}) during retention and the corresponding NC potential energies, with the inset showing a comparison to the SL device.

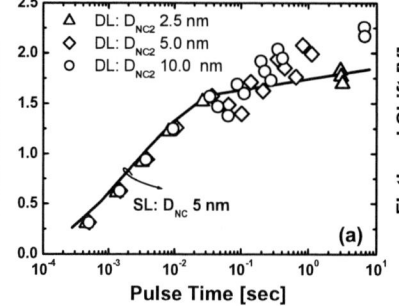

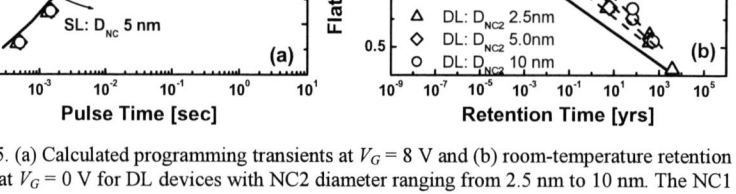

Fig. 5. (a) Calculated programming transients at $V_G = 8$ V and (b) room-temperature retention time at $V_G = 0$ V for DL devices with NC2 diameter ranging from 2.5 nm to 10 nm. The NC1 diameter of 5 nm and the NC number density of 4×10^{11} cm^{-2} are fixed. The results from SL devices with the same configuration are also plotted for comparison.

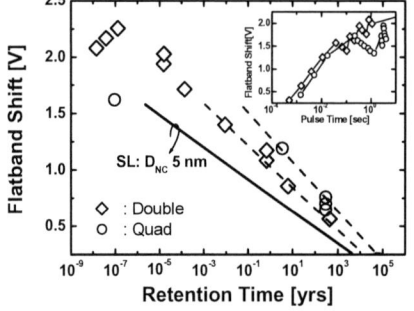

Fig. 6. Calculated room-temperature retention time at $V_G = 0$ V for DL and quad-layer devices. The inter-NC oxide is 2.5 nm, the NC diameter is 5 nm and the NC number density is 4×10^{11} cm^{-2} in all layers. The inset shows the calculated programming transients at $V_G = 8$ V.

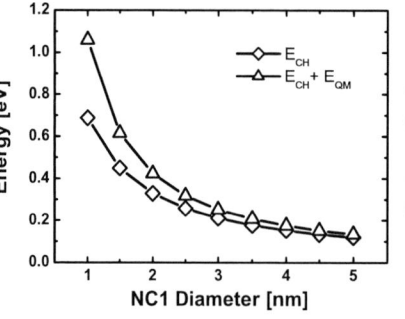

Fig. 7. Calculated Coulomb charging energy E_{CH} and total blockade energy including quantum confinement energy E_{QM} as a function of the NC1 diameter. The NC2 diameter of 5 nm and the NC number density of 4×10^{11} cm^{-2} are fixed.

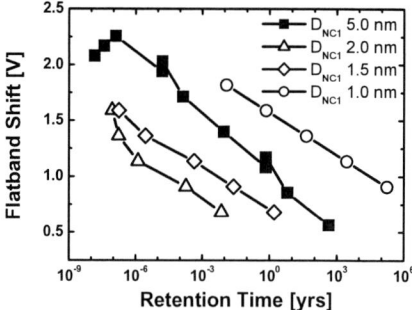

Fig. 8. Calculated room-temperature retention time at $V_G = 0$ V for DL devices with NC1 diameter ranging from 1 nm to 5 nm. The NC2 diameter of 5 nm and the NC number density of 4×10^{11} cm^{-2} are fixed.

Phase-change Memory

Chung Lam

IBM Qimonda Macronix PCRAM Joint Project
IBM T.J. Watson Research Center, 1101 Kitchawan Road, Yorktown Heights, N.Y. 10598, USA
Tel:+1-914-945-3902, email: clam@us.ibm.com

Abstract

This paper reviews the current development status of Phase-Change Memory (PCM), discusses advanced scaling of this technology along with a scaling demonstration. A probable development road map and a prospective view of future possible applications of PCM will also be presented.

Introduction

Since the discovery of reversible switching phenomena in chacolgenide materials first published in 1968 [1], advancements towards realizable applications were slow to come due to the high energy requirement in programming. Thirty years passed before new possibilities for this technology were re-discovered, namely, the invention of the CDROM. Rapid advancement in lithography has brought the critical dimension down to less than 1 µm in the late 1990's and greatly reduces the energy requirement to manageable level for semiconductor applications of PCM. The first major advancement in PCM was published in 2001 [2]. Since then, hundreds of papers and patents were published. The motivations driving this competition and the reasons why vast resources are being devoted to developing PCM have become clear. In comparison to the alternative well established non-volatile memories, PCM is scalable, has lower voltage operation, has lower power consumption, has lower fabrication costs, and has a much faster programming speed which is comparable to Dynamic Random Access Memory (DRAM). These demonstrated benefits are convincing the industry that this technology has a chance to replace NOR Flash, NAND Flash, and even DRAM.

In this paper, we first discuss the current development status of the PCM technology. Next, the most advanced scaling demonstration results are presented. Finally, we discuss possible future applications and the development roadmap for PCM.

Current Status of Phase-change Memory

Patent distribution: Fig. 1 shows the distribution of PCM related patents from 1980 through the end of 2006. The amount of PCM related patents begins to increase exponentially between 2002 and 2006, indicating that PCM is heavily pursued within the semiconductor industry.

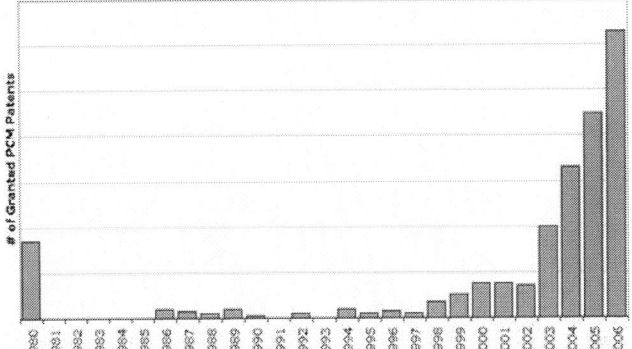

Fig. 1 US granted PCM patents from 1990 to 2006.

Cell Structure: Most of the major players are currently using a "mushroom" structure as their memory cell which is comprised of a bottom heater in contact with the phase-change material. Fig. 2 shows the mushroom cell structures; transistors or diodes are used for select devices in the array. Comparing the option of using a transistor vs. a diode for the select device, a transistor provides a higher on/off ratio, a simpler integration scheme, and a cell size bigger then the diode select device, whereas a diode provides more current per area which leads to a smaller cell size.

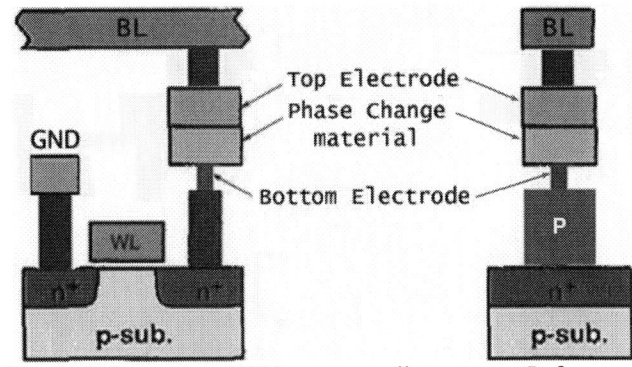

Fig. 2 Mushroom type PCM memory cell structures. Left: transistor select structure. Right: diode select structure.

Material: $Ge_2Sb_2Te_5$ (GST) is the industry standard phase-change material. The material properties are improved by doping the $Ge_2Sb_2Te_5$ with nitrogen. The incorporation of nitrogen into the material inhibits grain growth thereby increasing the retention of the amorphous state (Fig. 3) and also increases the resistivity of the material resulting in a reduction of the crystalline-to-amorphous programming current.

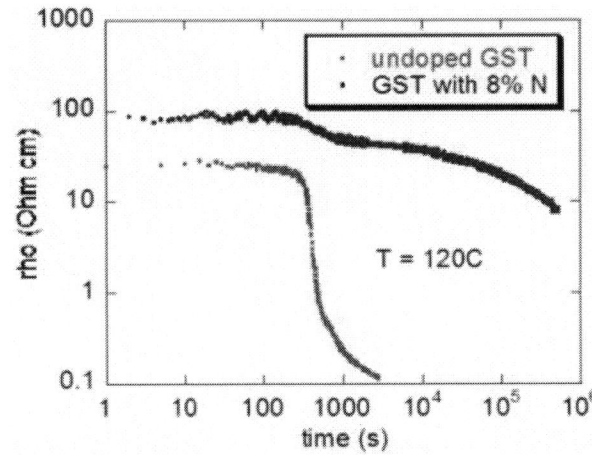

Fig. 3 Resistivity vs. annealing time for undoped and doped GST.

Programming current: The major challenge for PCM technology is to reduce the RESET current of the memory element. The RESET current limits the size of the driving device (either diode or transistor), as the driving device must be large enough to support the programming of the memory element. Many different cell structures have been proposed [3, 4] to reduce the RESET current. The RESET current can be reduced by decreasing the bottom heater dimension and by decreasing the volume of the region of the phase-change material which undergoes switching. Other methods of reducing the RESET current include increasing the resistivity of the bottom heater and incorporating additional dopants into the phase-change material [5]. Fig. 4 shows the RESET current vs. critical area of four different cell structures. Each cell structure exhibits a different dependence on the critical dimension, indicating that the choice of cell structure is critical and strongly influences how the RESET current will scale for the memory element in future technologies.

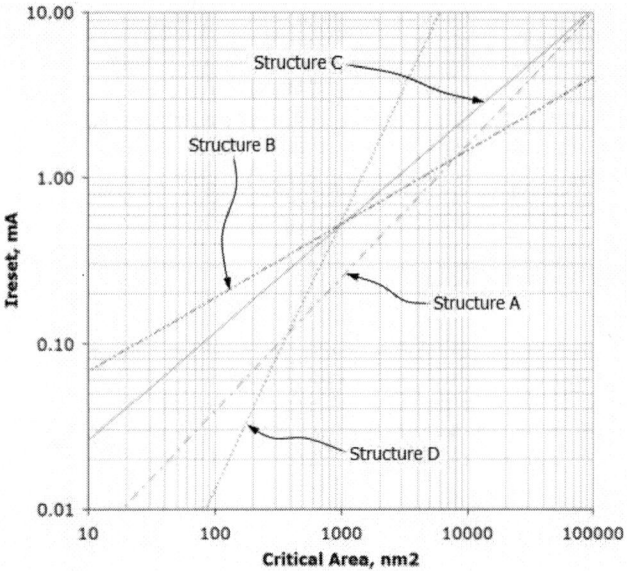

Fig. 4 RESET current vs. critical area for different memory cell structures.

Phase-change Memory Scaling

In Fig. 5, the smallest phase-change device reported to date is depicted. A very thin (3nm) and narrow (20nm) doped GeSb line was fabricated above of a pair of TiN bottom electrodes. The memory element can be switched with RESET currents as low as 100uA and shows SET-RESET cycling up to 10^5 without fail. Further experimental results show that the scaling of this structure is limited by the e-beam lithography and not by the memory structure itself.

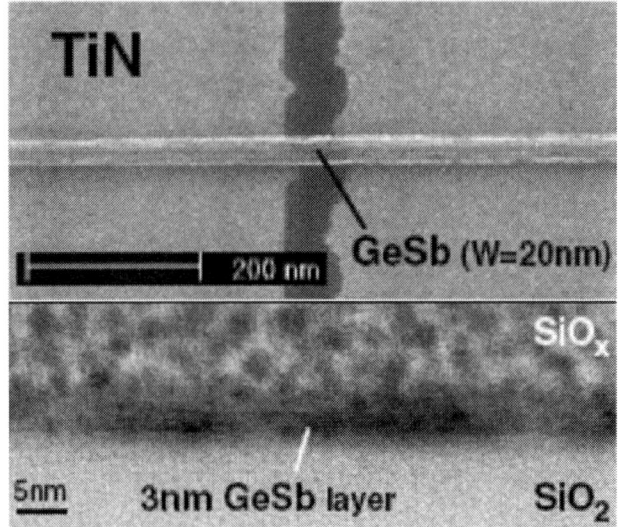

Fig. 5 TEM of a 20nm width, 3nm thick GeSb nano-line formed a" bridge" cell on top of a pair of TiN electrodes.

Fig. 6 shows the RESET current as a function of the phase-change material width for different values of the thickness, with a bottom electrode separation of 50nm. The RESET current can be as small as 80uA. Ultra-thin phase-change material was investigated in order to further understand the scaling limitation of phase-change material. As seen in Fig. 7, GeSb thin film thickness thinner than 1.1nm still shows crystallization after annealing. The fcc phase will disappear but the hcp phase remains. This indicates the GeSb can be scaled down to 100Å and continue to change phase.

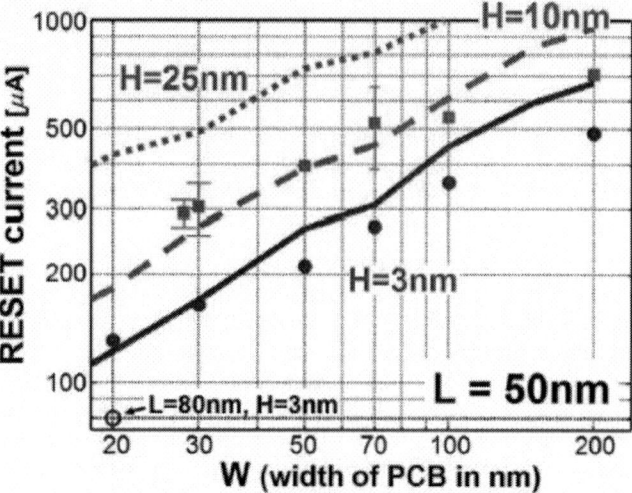

Fig. 6 RESET current vs. line width of "bridge" memory cell.

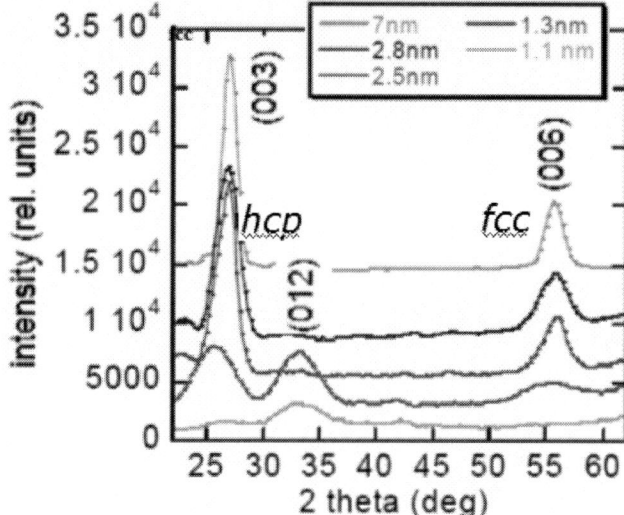

Fig. 7 XRD theta-2theta scans for thin GeSb films after 430°C annealing.

Future of Phase-change Memory

Application: Stand-alone and embedded NOR Flash probably will be the first technology replaced by PCM. For embedded Flash memory, typical cell size is around 15~20F². PCM cell with field effect transistor select device has similar cell size at 90nm node [6]. PCM does not need high voltage devices, requires less complex process, and has much faster programming time. Fig. 8 shows a one-mask only PCM cell which can provide a much lower cost solution for embedded Flash application. The top electrode and phase-change material are deposited after the front-end of the line processes are completed and before the back-end of the line processes begin. The only extra mask is used for patterning the phase-change memory element. Stand-alone NOR Flash faces serious scaling issues beyond 45nm node. Diode select PCM cell [7] has demonstrated cell size smaller than 6F² at 90nm capable of supplying greater than 1.5mA for RESET operation. The diode select PCM cell is also shown to scale beyond 22nm.

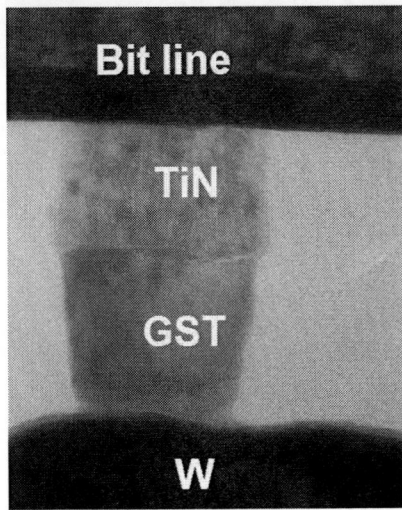

Fig. 8 One mask only needed phase-change memory cell.

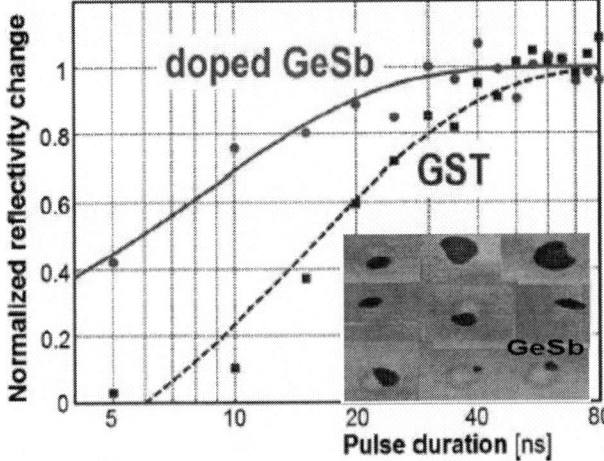

Fig. 9 Normalized reflectivity change as a function of optical pulse duration for doped GeSb. The inset shows the AFM imaging of partially crystallized doped GeSb optical spots.

It will be challenging for PCM to compete with DRAM due to DRAM's fast programming speed and high write endurance requirement. The SET (transition from amorphous to crystalline) performance of the programming operation is the issue here. By changing the phase-change material from GST to GeSb, the phase-change SET speed can be improved from 16ns to 5ns as shown 0.4 reflectivity change line in Fig. 9 in an optical demonstration. Hence, by utilizing a faster phase-change material PCM has a chance to meet the performance requirements of DRAM, however, the 10^{15} cycling requirement is still a large hurdle for PCM to overcome.

NAND Flash has achieved the highest density and smallest memory cell size compared to all other semiconductor memories. Furthermore, the predicted scaling limit for NAND Flash is expected to extend beyond the 25nm node. For PCM to compete with NAND FLASH, the cell size needs to be smaller then $4F^2$, which requires the use of a diode as the driving device. A diode of this cell size in the 25nm node requires the RESET current of the memory element to be less than 150uA. Another "must have" feature in order for PCM to compete with NAND is the capability of multiple bits per cell operation. Fig. 10 shows a 2-bits-per-cell demonstration of a PCM cell. Four different resistance states are programmed with only one-shot program condition for each state. The key for a successful Multiple-Level-Cell (MLC) PCM is the ability to control the resistance distribution of each state.

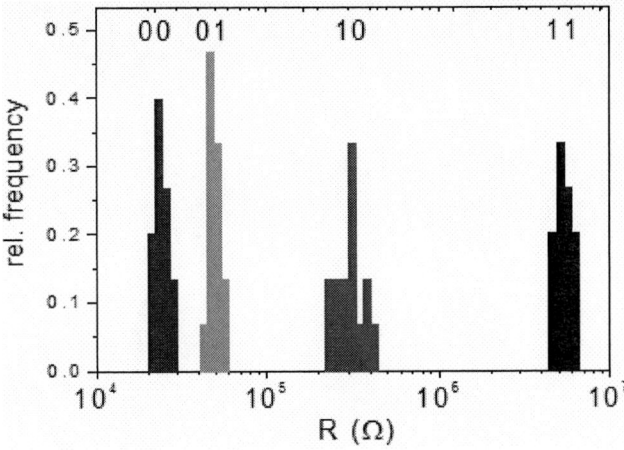

Fig. 10 4 level 2-bits-per-cell demonstration at set state.

Technology development roadmap: Fig. 11 shows the estimated PCM technology roadmap. The diode (BJT) select array will always have a smaller cell size than the FET select array. 40nm node will be achieved around 2011 and will have a cell size of approximately $5F^2$ with a bottom heater diameter of approximately 15nm.

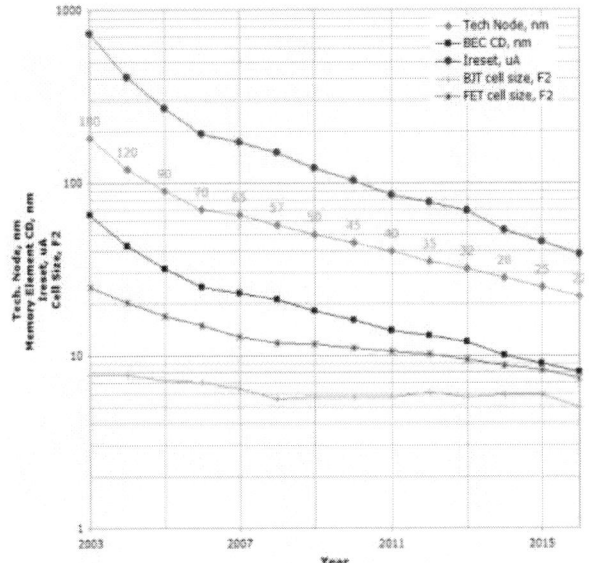

Fig. 11 PCM technology road map.

There are two way to further reduce the memory cell size: one is MLC, and another is to have multiple layers of memory elements stacked on top of one another. Fig. 12 compares the benefits of these two approaches. Its shows the multi-layer's benefits will saturate after 8 layers, whereas the cost of MLC approach continues to decrease with increasing number of bits. These results further emphasized the importance of MLC operation for PCM in the future.

Table 1 shows the key research and development items. The structure of the memory element is crucial. The size of the heater and phase-change memory CD needs to be reduced while maintaining good control to minimize variations; this is necessary to achieve tight RESET and SET resistance distributions. Development of a diode or 3D transistor is needed for increasing the driving current capability, which ultimately limits the cell size for a given memory element. Finally, materials research needs to continue to study the effect of doping on resistively and re-crystallization which in turn affects the retention characteristic.

227

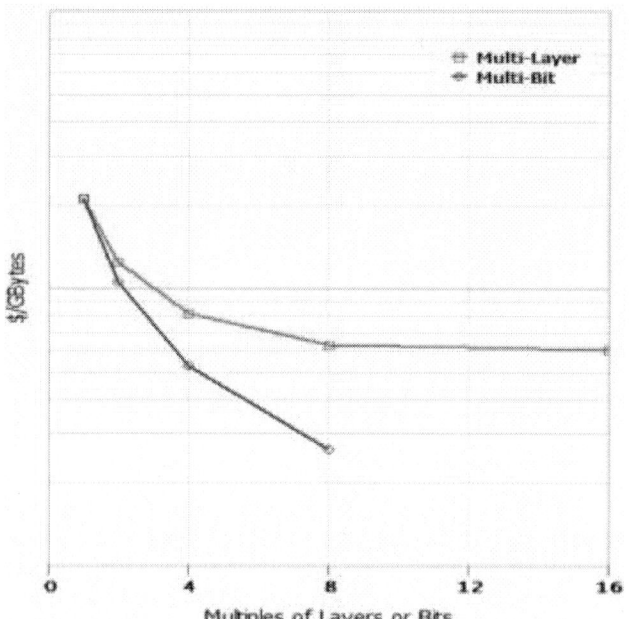

Fig. 12 Giga-bits cost comparison of the MLC approach and the multiple-layer approach for future PCM technologies.

Memory Element Structure			
Contact CD		Phase Change CD	
Isolation/Select Device			
Bipolar		FET	
Bipolar	Diode	Planar	3D
Material Engineering			
Material Search			
Doping		Resistivity modulation	
		Crystallization Temp	
		Melting point	

Table 1 Main PCM future research and development items.

Conclusion

PCM has been receiving a great deal of attention in recent years as it provides a possible way to continue scaling for and to reduce the cost of future semiconductor memory technologies. Scaling of the PCM element down to at least 20nm appears to be feasible without showing serious problems. Decreasing the heater and phase-change element CD, developing high current support diode, developing MLC operation, and developing advanced phase-change materials are the most important ways to make PCM successful.

Acknowledgements

Authors would like to thank the PCRAM joint project team members and the expert processing support from the Microelectronic Research Line at IBM Watson Research Center. Valuable discussions with W. Gallagher, R. Liu, and G. Mueller are gratefully acknowledged

References

[1] S. R. Ovshinsky, "Reversible Electrical Switching Phenomena in Disordered Structures," Phys. Rev. Lett. Vol. 21, 1968, p. 1450.

[2] S. Lai, T. Lowrey, "OUM - A 180 nm nonvolatile memory cell element technology for stand-alone and embedded applications," 36.5, IEDM Tech. Dig., 2001.

[3] Y. C. Chen, et. al., "Ultra-Thin Phase-Change Bridge Memory Device using GeSb," 30.3, IEDM Tech. Dig., 2006.

[4] T. D. Happ, et. al., "Novel One-Mask Self-Heating Pillar Phase-change Memory," Symp. VLSI Tech.. 2006.

[5] W. Czubatyj et al, "Current Reduction in Ovonic Memory Devices," E*PCOS 2006.

[6] S. J. Ahn et al, "Highly Reliable 50nm Contact Cell Technology for 256Mb PRAM," Symp. VLSI Tech. 2005.

[7] J.H. Oh et al, "Full Integration of Highly Manufacturable 512Mb PRAM based on 90nm Technology," IEDM Dig. 2006.

Novel Cross Point Switch based on $Zn_{1-x}Cd_xS$ memory devices for FPGA

Keiko Abe, Zheng Wang*, Shinobu Fujita, Thomas H. Lee* and Yoshio Nishi*

Frontier Research Laboratory, Corporate R&D center, Toshiba Corporation, Japan, *Center for Integrated Systems, Stanford University, CA, USA 94305, Phone: +81-44-549-2455; Fax: +81-44-549-2455; e-mail: keiko2.abe@toshiba.co.jp

[Application of new cross point switch devices to FPGA and related issues] Application area of FPGA (Field Programmable Gate Array) is growing by replacing ASIC. If the large overhead of FPGA of area, delay and power can be reduced, its application area will expand more dramatically. The large overhead is principally attributed to cross point switch (CPS), as shown in Fig. 1, which connects each configurable logic block (CLB), consisting of a pass transistor and an SRAM cell in widely used SRAM-based FPGA. Here CPS has more than $120f^2$ area, where f is the feature size, and power of SRAM is huge, and the delay at the CPS is large since the pass transistor has ON-resistance as high as several $k\Omega$. Recently, it has been reported that the overhead can be reduced drastically using Cu_2S memory devices having superior scalability as CPS (Fig. 1)[1]. The area of CPS idealistically is $4f^2$. There is no power consumption during normal FPGA operation, since CPS device is nonvolatile. Also, ON-resistance (R_{ON}) is around 50 Ω, much lower than that of SRAM-based. However, Cu_2S memory device itself has several issues considering actual usage. One issue is high leakage current due to relatively low OFF-resistance (R_{OFF}), since total power strongly depends on the leakage current, since FPGA contains "millions of" CPSs. It has also issues related to long-term (LT-) reliability. Since the threshold voltage is as low as 0.3V (OFF-to-ON) and 0.1V(ON-to-OFF), much lower than CMOS supply voltage(0.8~1.5V) in FPGA, it easily causes *disturbance* of OFF-state device (Fig. 2.). As for ON-state, though the disturbance of ON-state device does not occur(Fig.2), LT-reliability is difficult due to very short retention of Cu_2S, less than 3 months[1]. In this work, we show that $Zn_{1-x}Cd_xS$ (ZnCdS)- based memory devices are superior CPS having potential for usage in FPGA. We also present *current limitation effect* and *excess-current programming method* to ensure LT-reliability for OFF- and ON-state of $Zn_{1-x}Cd_xS$ -CPS, respectively.

[Advantage of ZnCdS-CPS over CuS-CPS] ZnCdS is a solid electrolyte memory like AgGeSe, AgS and Cu_2S. The switching mechanism involves the formation of nano-scale conductive bridge by segregation after metal ion migration through the film, as shown in Fig.3. We fabricated CPS using two-terminal RF sputtered $Zn_{0.4}Cd_{0.6}S$ with Pt bottom electrode and Ag top electrode (3μm x 3μm) on CMOS (0.25μm process technology) used for FPGA test circuits, as shown in Fig. 4. Its I-V characteristic exhibits a bipolar-type hysteresis curve as shown in Fig. 3. Figure 5 shows R_{off}, which is known to be dominated by leakage current of the film that depends on the device area, is much higher than that of CuS device by at least 2 orders by comparing at the same device area. This preferable feature is attributable to ZnCdS having bandgap larger than that of CuS. Although R_{ON} is around 150 Ω, inferior to that of CuS device, this value can be decreased to 30 Ω, as explained later. As a result, the R_{OFF}/R_{ON} ratio reaches more than 10^7 for the device area less than 0.01μm^2, whereas the ratio of CuS is 10^5.

[LT-reliability of OFF-state-CPS by "current limitation effect" by CMOS] For the *DC* measurement, average OFF-to-ON threshold voltage of $Zn_{0.4}Cd_{0.6}S$ was 0.5 V, higher than that of Cu_2S but still smaller than the CMOS supply voltage. For the *AC* bias disturbance test, higher threshold voltage can be expected. Figures 6 show results of the *AC* bias disturbance test with 1 μsec duration for the OFF-state CPS *without CMOS*. Whereas the ON-state was stable for the 0.5V pulse application, the disturbance was seen for the 1V pulse application, as shown in Fig. 6(a). However, we found that this disturbance cannot occur for the CPS *with CMOS* by *current limitation* effect, as shown in Fig. 6(b). Here the selected width of connected transistor was 1μm in Fig. 4, which can reduce maximum current through the CPS to much lower level (~60 μA) than that required for OFF-to-ON set-programming (~1mA). When the current is reduced to lower level than the threshold current, the CPS state cannot be changed even if the voltage over the threshold level is applied to the CPS, which can assure LT-reliability of OFF-state.

[LT-reliability of ON-state-CPS by "excess-current programming method"] We also found that, when much larger current ("*excess-current*") than the threshold current was applied for set-programming (*OFF-to-ON*), as the excess-current over the threshold current (~1mA) is increased, the threshold current for reset-programming (*ON-to-OFF*) is also increased, as shown in Fig. 7(a).These new findings suggest that the size of conductive bridge formed in ZnCdS is increased by the excess-current programming, as shown in Fig. 7(b). It can be considered that the retention of CPS having large conductive bridge is longer than the general one. Figure 8 shows ZnCdS-based CPS programmed by the excess-current method keep almost the same R_{ON} over 6 months, while the retention of ON-state-$Cu_{0.2}S$ is smaller than 3 months. In addition, we found that the excess-current programming method is effective to reduce the ON-resistance of CPS, as shown in Fig. 5, which is consistent with the model in Fig.7(b). While On-resistance is about 150 Ω by general programming, it reaches 30 Ω, which is lower than that of CuS.

[Conclusion] Novel cross point switch is realized by using ZnCdS which exhibits superior characteristics over Cu_2S-based one as shown in Table 1 with particularly large R_{OFF}/R_{ON} ratio and much longer ON state retention.

[1] S. Kaeriyama et al., *IEEE J. Solid-State Circuits*, Vol. 40, 1, p.p. 168 – 176, Jan. 2005.

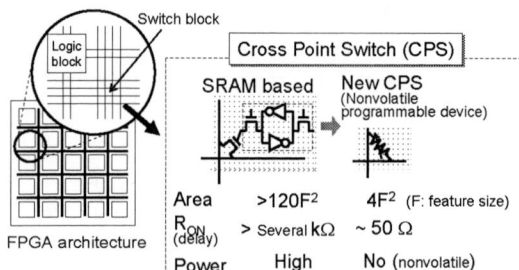

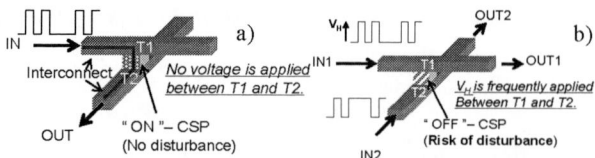

Fig.2 . Signal transfer through two interconnects with ON-state CSP (a) and OFF-state CSP (b). Basically, disturbance for ON-state cannot occur, since no voltage is applied between the two electrodes of the ON-state CPS. However, disturbance for OFF-state can occur, when the CSP has low threshold voltage.

Fig.1 . FPGA architecture and two kinds of cross point switches (CPS). Conventional SRAM-based CPS has large overhead of area, delay and power, where the delay depends on ON-resistance of CPS. Such large overhead can be effectively reduced using new CPS.

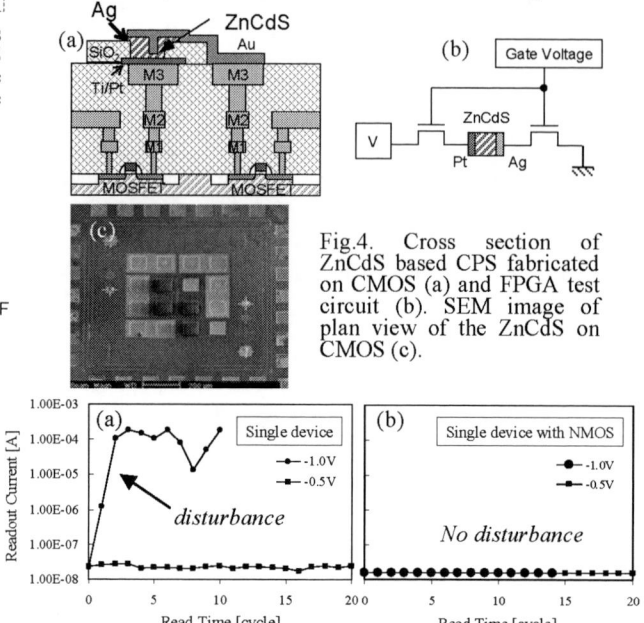

Fig.4. Cross section of ZnCdS based CPS fabricated on CMOS (a) and FPGA test circuit (b). SEM image of plan view of the ZnCdS on CMOS (c).

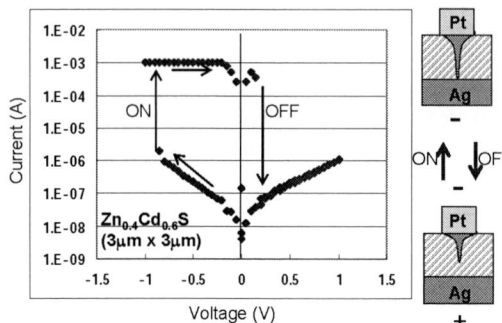

Fig.3 . Typical *I-V* characteristics of ZnCdS memory device and its mechanism.

Fig.6. Results of AC disturbance test (500kHz) for single ZnCdS-CPS and that with CMOS. Due to *current limitation effect* through CPS by CMOS, no disturbance was observed even at 1V.

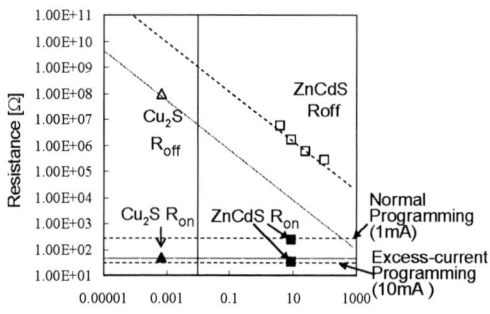

Fig.5. R_{on} and R_{off} of fabricated $Zn_{0.4}Cd_{0.6}S$ device and Cu_2S device.[1]. R_{off} depends on device contact area, however, it is known that R_{on} dose not depend on it, since the size of conductive bridge formed is much smaller than device area of CPS.

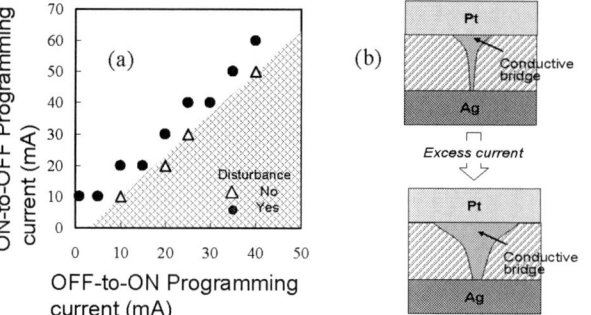

Fig.7. Relationship between OFF-to-ON set-programming current (*excess current*) and that for ON-to-OFF reset-programming (a), and a model of increasing size of conductive bridge by excess current programming (b) .

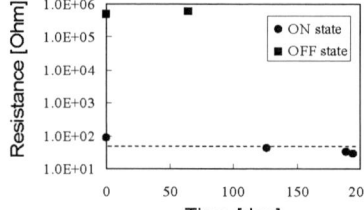

Fig.8. Long term retention of ZnCdS-CPS by the excess programming method. (Reliability of OFF-state device is not a issue, since it is known that its retention is much longer that that of OFF-state for solid electrolyte memory devices.)

Acknowledgement: The authors would thank for valuable discussion with Prof. S. Wong, W. Wang, and S.-W. Kim of Stanford University.

	V_{th} (V) ON-to-OFF @average	V_{th} (V) OFF-to-ON @average	$R_{on}(\Omega)$ @0.01um²	$R_{off}(\Omega)$ @0.01um²	R_{on}/R_{of} Ratio @0.01um²	OFF-state disturbance	ON-sate retention (month)
ZnCdS	0.3	0.5	30	1G	3×10^7	No	>6
CuS	0.1	0.3	50	5M	1×10^5	probable	<3

Table 1. Comparison of CPS characteristics for FPGA application. Even if *current limitation effect* and *excess-current method* might be applicable to Cu_2S, effectiveness is not better than that for $Zn_{0.4}Cd_{0.6}S$, since metal ions easily migrate in Cu_2S and its band gap is small.

McKenna Hall, Rooms 100-104, 112-114, Auditorium

Rump Sessions

Tuesday PM, June 19th, 2007

8:30 PM (McKenna Hall, Rooms 100-104**)**
The Future of Nanowire Devices: Top-down or bottom-up?

Session Organizers:
Prabhat Agarwal, NXP
Steven Koester, IBM
Eric Pop, University of Illinois/Urbana Champaign

Panelists:
Erik Bakkers, Philips Research
Narayan Balasubramanian, IME Singapore
Xiangfeng Duan, NanoSys
Ali Keshavarzi, Intel
Paul Solomon, IBM Research

Over the past years, Nanowires and Nanotubes have been touted as the panacea for future integrated circuits. The components are generally grown and made into circuits in an approach generally known as "bottom-up" assembly. The lure of III/V on silicon co-integration, high-mobility channels, and useful quantum-effects have created much excitement.

However, grown nanowires are strongly limited by their fab-contaminating catalysts and huge device parasitics, amongst other things. Recently, several groups have therefore proposed a "top-down" approach, manufacturing silicon nanowires using etching technology. These devices are automatically fab-compatible, and can have much smaller device parasitics.

8:30 PM (McKenna Hall, Rooms 112-114**)**
Why do we need nano for biosensors?

Organizers:
Sang-Hyun Oh, University of Minnesota
Janos Voros, ETH Zurich

Panelists:
Peter Fromherz, Max Planck Institute
Ashraf Alam, Purdue University
Jeffrey B.-H. Tok, Lawrence Livermore National Laboratory
Yang-Kyu Choi, KAIST
Mike Wiltshire, Imperial College

Optical and/or electrochemical instruments have stood as dominant tools for biosensing. As our understanding of the device physics in nanostructures (e.g. nanowires, CNT, FETs, nanoparicles) and technology to make them in a more controllable manner have advanced, we come to a question. Can these nanoscale devices revolutionize the field of biosensing as they did for IC industry? A number of recent advances suggest that such nanobiosensors might provide unprecedented capability in terms of detection sensitivity, miniaturization, low cost and portability. At the same time, significant challenges still remain before they could become realistic alternatives to existing biosensing methods.

The purpose of this panel session is to provide a forum to address the question of whether or not the potential benefits of such nanoscale devices make adoption of this technology inevitable.

8:30 PM (McKenna Hall, Auditorium)
The THz Bridge

Session Organizer:
Mike Wojtowicz

Panelists:
Mark Rosker, DARPA/MTO
Rich Lai, Northrop Grumman Space Technology
Michael Shur, Rensselaer Polytechnic Institute

Closing the THz gap will enable the next generation of imaging and sensing applications. With recent advances in both electronic and photonic research both technologies are poised to close the THz gap. The session will address the topic of which technology will be able to close and dominate the gap. The applicability and issues related to solid state and photonic operation at the THz frequencies will be discussed.

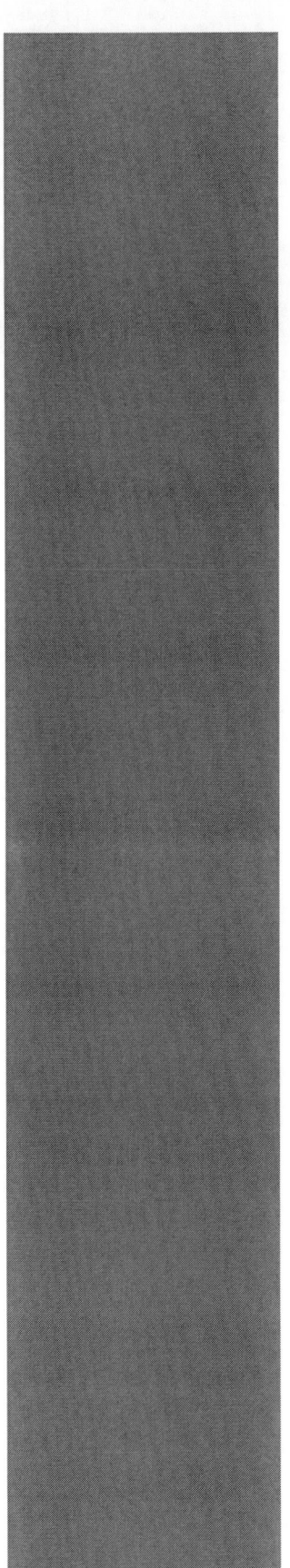

(Leighton Concert Hall - DeBartolo Performing Arts Center)

Plenary Session

Wednesday AM, June 20, 2007

Joint DRC/EMC Plenary Session

8:20 AM Plenary Paper
Low Cost "Plastic" Solar Cells: A Dream or Reality???
Alan J. Heeger, University of California, Santa Barbara

9:20 AM Break

234

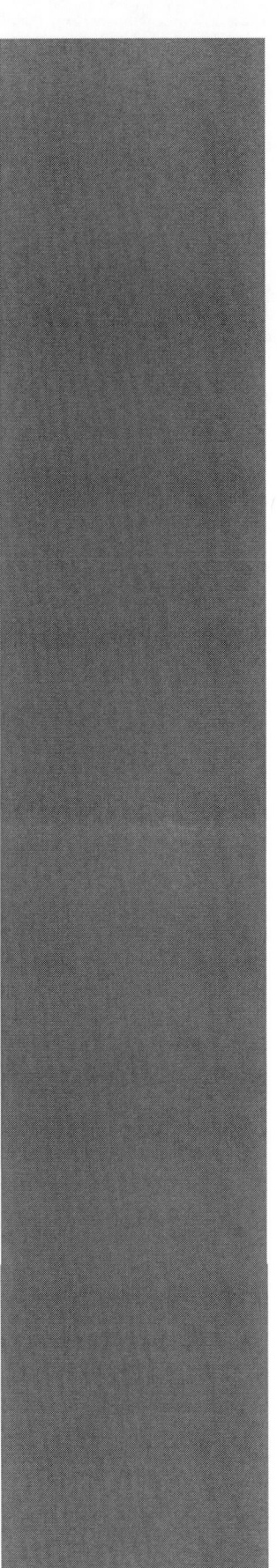

Session VI.A (McKenna Auditorium)

Nanobiotechnology

Wednesday AM, June 20th, 2007

Session Organizer: Sang-Hyun Oh, University of Minnesota
Session Chairman: Yang-Kyu Choi, Korea Advanced Institute of Science and Technology

10:00 AM VI.A-1 Invited Paper
K. Boahen, Stanford University, Bioengineering Dept., Stanford, California, USA

10:40 AM VI.A-2 Invited Paper
Joining Microelectronics and Microionics: Nerve Cells and Brain Tissue on Semiconductor Chips
P. Fromherz, Department of Membrane and Neurophysics, Max Planck Institute for Biochemistry, Munich, GERMANY

11:20 AM VI.A-3
Biolithography: DNA-assisted Manufacturing of Nanodevices for Optical and Electronic Biosensing
J. Vörös, Laboratory of Biosensors and Bioelectronics, Institute for Biomedical Engineering, Zurich, SWITZERLAND

235

The brain and the computer

Kwabena Boahen, Associate Professor
Stanford University, Bioengineering Dept., 318 Campus Drive West, Stanford, CA 94305
boahen@stanford.edu; 650 724 5633

ABSTRACT

Exactly fifty years ago, when he published "The computer and the brain" in 1957, John von Neumann foresaw computer designers benefiting from basing their designs on the brain. Interest in this topic has been renewed by the convergence in properties of transistors and ion-channels (the brain's transistors). At ten nanometers, a transistor's channel becomes so narrow that an electron trapped by a dangling bond at the surface (an unavoidable atomistic defect) blocks electron flow. The current turns off and on at random, as trapping and detrapping occurs stochastically. Such stochastic behavior, which corrupts the deterministic on/off states computers rely on to perform binary arithmetic, is also displayed by ion-channels. At under a nanometer in size, an ion-channel's gate is agitated by thermal forces, opening and closing randomly. I will describe efforts in the neuromorphic engineering community to explore how the brain computes with stochastic devices by emulating an ion-channel's ionic current directly with a transistor's electronic current. While a present-day transistor, at a hundred-nanometers wide, corresponds to a small population of ion-channels, not a single ion-channel, this analog approach provides an extremely efficient method to simulate the brain while at the same time laying the groundwork for building brain-like computers out of next decade's nanotransistors. John von Neumann was prescient in anticipating that computer designers could profit by modeling features of the brain in their designs—even though he did not foresee the remarkable device-level convergence.

978-1-4244-1101-6/07/$25.00 ©2007 IEEE

238

Joining Microelectronics and Microionics:
Nerve Cells and Brain Tissue on Semiconductor Chips

Peter Fromherz

Department of Membrane and Neurophysics, Max Planck Institute for Biochemistry
82152 Martinsried / Munich, Germany
phone: +49 89 8578 2820, fax: +49 89 8578 2822, email: fromherz@biochem.mpg.de

It is a challenging idea to create information processing systems that are hybrids of "thinking" neuronal parts and of computing microelectronic devices. To achieve that goal, we must solve two fundamental problems: On one hand, we have to "understand" the brain, i.e. to know the "trick" how the slow dynamics of nerve cells is able to perform astonishing tasks such as fast pattern recognition and fast control of motor dynamics. On the other hand, we have to interface the neuronal and electronic systems on a microscopic level in both directions in order to take full technological advantage of "thinking" in a hybrid. It is the latter problem that we consider in our work. We assemble simple hybrid devices of nerve cells and semiconductor chips and study the basic physical chemistry of interfacing. By enhancing the complexity of the hybrids step by step, we look how far we come in the implementation of novel functions. Another aspect of the approach is that it may help to solve the first problem, i.e. to understand the brain, by developing novel neurophysiological techniques.

On the side of the neuronal systems, we consider the three levels of (i) molecular ion channels, (ii) individual nerve cells and (iii) brain tissue with neuronal networks. On the side of microelectronics, we study the basis of interfacing in both directions using electrolyte-oxide-semiconductor (EOS) transistors and EOS capacitors of simple silicon chips, and we transfer those methods to more involved chips that are fabricated by CMOS technology.

Crucial for the transmission of electrical signals from chips to cells and from cells to chips are the microscopic tructure and the electrical property of cell-silicon junctions. They are elucidated with luminescent dyes that are attached to the cell membrane. The reflecting surface of silicon gives rise to a change of fluorescence due to interference effects such that the distance of cell membrane and chip can be determined. Alternate voltages applied to silicon induce a spectral shift of fluorescence that is used to determine the electrical resistance of the cell-chip junction.

The mechanism of electrical interfacing is studied with defined molecular channels for Na^+ and K^+ ions in the membrane of cells that are cultured on capacitors and transistors. When voltage ramps are applied to a chip, the capacitive current gives rise to a voltage across the cell membrane that opens the channels. When the channels are open, ionic current flows along the resistance of the cell-chip junction and gives rise to a gate-voltage that modulates the source-drain current of a field-effect transistor.

Two-way interfacing to silicon chips has been implemented with nerve cells from invertebrates (snails) and of mammals (rats). The interfacing of invertebrate neurons is more advanced, because these cells are larger such that the coupling to capacitors and transistors is stronger. By improving the quality of the EOS capacitors (high-k insulators) and of the EOS transistors (low-noise design), significant progress was recently achieved with mammalian neurons.

Two kinds of elementary hybrid circuits have been assembled with two neurons: (i) One neuron is stimulated from a capacitor, its activity is coupled to a second neuron through a synapse and the excitation of the second neuron is recorded with a transistor. (ii) An excited neuron is coupled to a transistor, its signal is shaped on the chip and used to stimulate a second neuron with a capacitor. More complicated networks have been cultured on chips with defined network geometry using controlled outgrowth by chemical and topographical patterns. The yield of the resulting hybrids, however, is still rather low. In a first step towards a recording not only of neuronal excitation, but also of synaptic activity, the release of individual vesicles from chromaffine cells as a model system was detected with transistors.

Neuronal networks from rat brain have been interfaced to silicon chips by culturing brain slices from the hippocampus. The two-way interfacing with EOS capacitors and EOS transistors was implemented through local populations of neurons. It was possible to induce learning effects (LTP, LTD) from the chips.

CMOS chips with 16384 sensor transistors on one squaremillimeter were developed with an inert surface of titaniumdioxide. The mechanism of interfacing between ion channels, nerve cells and brain tissue is the same as with the simple silicon chips. Multi-site interfacing was achieved with individual nerve cells, with small networks of snail neurons and with brain tissue (slices from rat brain, retinae from rabbits). Due to the high resolution in space and time (8 μm, 6 kHz) time-resolved electrical maps of individual nerve cells, of neuronal networks and of brain tissue were obtained. Experiments with more involved CMOS chips for two-way multisite interfacing are in progress.

P. Fromherz, The Neuron-Semiconductor Interface. in: Bioelectronics – from theory to applications. Eds. I. Willner & E. Katz (Wiley-VCH, Weinheim 2005) pp 339-394.

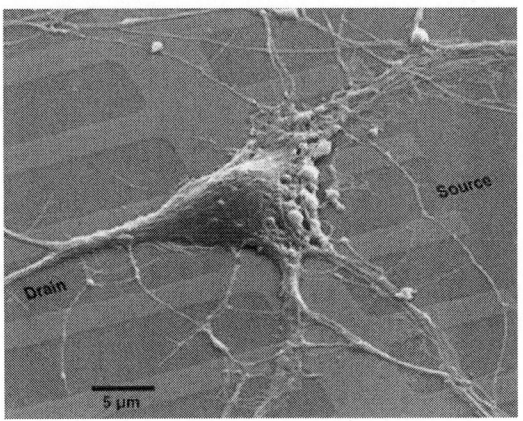 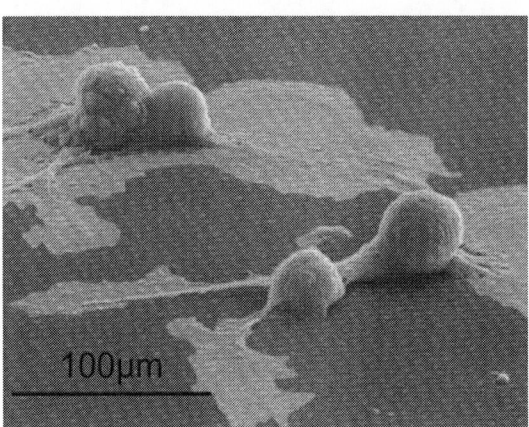

Fig. 1. Nerve cell from rat brain on the SiO_2 surface of a silicon chip with a linear array of low-noise electrolyte-oxide-silicon transistors.

Fig. 2. Nerve cells from pond snail on the TiO_2 surface of a CMOS chip with a multi-transistor array (MTA) of 16384 sensor transistors per mm^2 (pitch 7.8 μm).

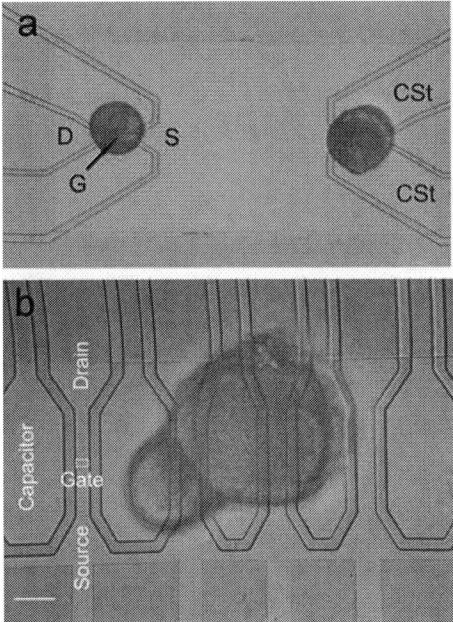

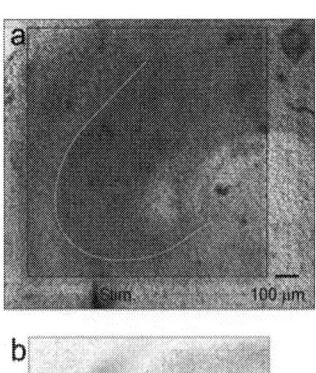

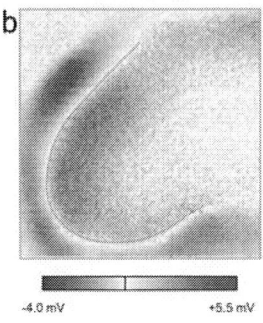

Fig. 3. Elementary hybrid devices. (a) Cellular neuroprosthetics with input neuron (left) on a transistor with source (S), drain (D) and gate (G) and output neuron (right) on two wings of a capacitive stimulator (CSt). Transistor and capacitor are connected by a circuit on the chip. (b) Neuronal memory on chip with input neuron (small, left) on a capacitor and output neuron (large, right) on a transistor. The neurons are coupled by a synapse that is potentiated by capacitive stimulation.

Fig. 4. Electrical mapping of brain slice by multitransistor array. (a) Hippocampus slice on CMOS chip. The square marks the area of the transistor array. The slice is stimulated by a tungsten electrode at the bottom (Stim). The gray line is redrawn as a mark from the activity map shown below. (b) Electrical voltage patternp 5 ms after stimulation. The chip maps the activity of synapses in the area of inward current and of compensating outward current.

241

Biolithography: DNA-assisted Manufacturing of Nanodevices for Optical and Electronic Biosensing

Janos Vörös

Laboratory of Biosensors and Bioelectronics, Institute for Biomedical Engineering, ETH Zurich, Switzerland; voros@ethz.ch; http://www.lbb.ethz.ch

The success of medical devices critically depends on the ability to interact with the biological environment. The bioresponse is often determined by the properties of the biointerface which requires the precise control of this interface on the micron- and nanometer scale. On the other hand a living cell is a beautiful example of a highly sophisticated, controlled system of nature's nanomachines, proteins and other biomolecules, which self-assemble into various supramolecular architectures and interact in a well-defined manner at the nanometer scale.

Recent advancements in nanotechnology have created a variety of top-down techniques that can reach feature sizes of 100 nm or less, thus approaching a size range very relevant to biology. At the same time a number of self-assembly based techniques have been developed and can be used to create artificial nanostructures mimicking biological systems with similar or even superior performance.[1]

The combination of these novel top-down and bottom-up approaches enables us to interact with complex biological systems: tissues, cells, proteins and DNA in an unprecedented manner. In addition, new tools, such as arrays of nanoparticles and nanowires can be created on a large scale with promising applications in electronic and optical biosensing. (See Figure 1.)

Examples for the handling of natural and artificial macromolecular complexes at the nanometer scale will be presented. Proteins, vesicles, macromolecular assemblies, and nanoparticles specifically placed onto predefined artificial patterns can trigger defined functions in cells, reveal the details of cell-surface interactions and allow for the ultimate miniaturization of array-type sensors down to the single molecule level.

Recently, we have put a lot of efforts into achieving not only spatial but also a dynamic control over the properties of biointerfaces. Surfaces that change upon external stimuli provide us with new research tools for studying complex biological problems and for tissue engineering. Highlights for the use of novel, electronically- or photo-active surfaces for applications in biosensing and local drug delivery will also be presented.[2,3] (See Figure 2.)

References

1. J. Vörös, T. Blättler, M. Textor, MRS Bulletin, 30(3):202-206, 2005.
2. C.S. Tang, M. Dusseiller, S. Makohliso, M. Heuschkel, S. Sharma, B. Keller, J. Vörös; Analytical Chemistry, 78:711-717, 2006.
3. B. Städler, C. Huwiler, J. Vörös, H.M. Grandin; IEEE Transactions on Nanobioscience, 5(3):215-219, 2006.

978-1-4244-1101-6/07/$25.00 ©2007 IEEE

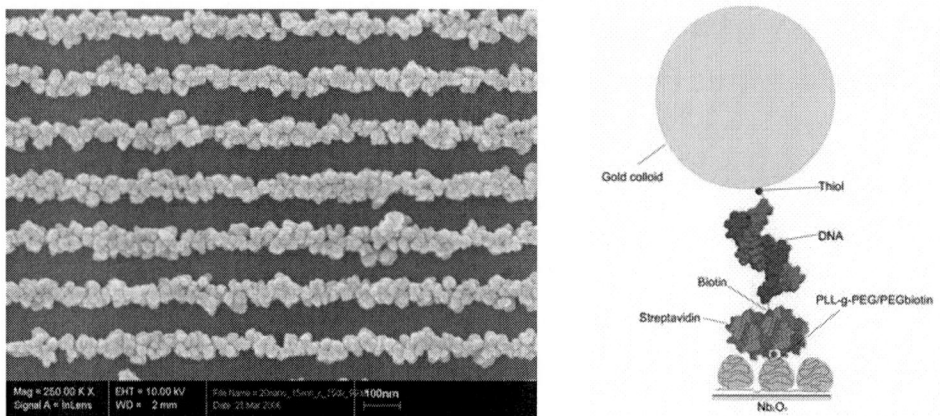

Figure 1: Nanowires can be prepared by self-assembly of DNA-tagged gold colloids. The image on the right depicts the multi-step surface modification process, the SEM image on the left shows the electrically conducting wires after the gold enhancement.

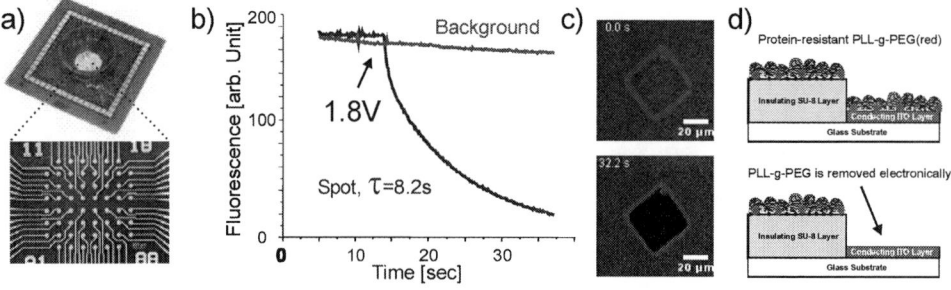

Figure 2: Principles of the use of electronically controllable surfaces for microarray biosensing: a) Commercially available ITO microelectrode array (MEA) from Ayanda Biosystems. (The SEM image shows the 1x1mm² active area of the chip.) b) The loss of fluorescent intensity, due to the electronic removal of the PLL-g-PEG-Alexa633 upon a 1.8V applied potential, follows exponential kinetics with a 8.2s time constant. c) CLSM images of the electrically activated spot of the microelectrode array before and after the removal of the PLL-g-PEG-Alexa633. d) Schematics of the surface architecture: the ca. 5μm thick SU8 film surrounds the spots and insulates the connecting ITO wires.

Session VI!.A (DeBartolo Hall Room 141)

Magnetoelectronics Devices & Optical Resonators

Wednesday PM, June 20th, 2007

Session Organizer: Michael Flatte, University of Iowa and Taiichi Otsuji, Tohoku University
Session Chairs: Craig Pryor, University of Iowa and Tom Jackson, Penn State University

1:30 PM VII.A-1 Invited Paper
Recent Advances in MRAM Technology
J. M. Slaughter, Freescale Semiconductor, Inc., Chandler, Arizona, USA

2:10 PM VII.A-2
Magnetic Sensitivity in Mesoscopic EMR Devices in I-V-I-V Configuration
T. Boone, L. Folks, J. A. Katine, E. Marinero, N. Smith and B. A. Gurney, Hitachi Global Storage Technologies , San Jose Research Center, San Jose, California, USA

2:30 PM VII.A-3
Simulation of Spin Torque Devices with Inelastic Spin flip Scattering
S. Salahuddin and S. Datta, School of Electrical and Computer Engineering and NSF Network for Computational Nanotechnology, Purdue University, West Lafayette, Indiana, USA

2:50 PM VII.A-4

3:10 PM Break

3:30 PM VII.A-5 Invited Paper
Characterization and Application of Large Magnetoresistance in Organic Semiconductors
M. Wohlgenannt, G. Veeraraghavan, Y. Sheng, O. Mermer, and T. D. Nguyen, Department of Physics and Astronomy, The University of Iowa, Iowa City, Iowa, USA

4:10 PM VII.A-6 Invited Paper
Metamaterials - Negative Indices with Positive Benefits?
M. C. K. Wiltshire, Imaging Sciences Department, Imperial College London, London, UK

4:50 PM VII.A-7 Student Paper
High Performance Polycrystalline Diamond Micro Resonators
N. Sepúlveda[1], J. Lu[1], D. M. Aslam[1], and J. P. Sullivan[2] , [1]Electrical and Computer Engineering, Michigan State University, E. Lansing, Michigan, USA and [2]Sandia National Laboratories, Albuquerque, New Mexico, USA

Recent Advances in MRAM Technology

Jon M. Slaughter

Freescale Semiconductor, Inc., 1300 North Alma School Road, Chandler, Arizona 84224
e-mail: jon.slaughter@freescale.com, Tel: (480) 814-2168

Magnetoresistive Random Access Memory (MRAM) technology has the attributes of non-volatility, high-speed operation and unlimited read and write endurance, a combination not found in any other existing memory technology. MRAM is based on magnetic tunnel junction (MTJ) devices integrated with CMOS. The MTJ has a low resistance when the magnetic moment of the free layer is parallel to the fixed layer and a high resistance when the free layer moment is oriented anti-parallel to the fixed layer moment. Figure 1 shows an MRAM cell undergoing a) read and b) write operations with magnetic fields created by adjacent write lines. Unlike most other semiconductor memory technologies, the data is stored as a magnetic state, rather than charge, and sensed by measuring the resistance without disturbing the magnetic state. In this paper we review recent progress, present specifics of Freescale's commercial 4Mbit MRAM device, and discuss the future outlook for MRAM technology, including the extension of Toggle MRAM to meet industrial/automotive requirements, and scaling behavior to the reduced dimensions of advanced technology nodes.

The first MRAM product to market, Freescale's 4Mb MR2A16A, is built on 180 nm CMOS technology with magnetic bit cells of 300 nm minimum dimensions integrated in the upper layers of metal. At these dimensions, both the magnetic switching and magnetoresistive property distributions are governed by a combination of material and patterning variations. One of the keys to controlling these distributions and insuring manufacturability was the invention of the Toggle Write mode. (1) This mode uses a balanced synthetic antiferromagnetic free layer combined with a phased write pulse sequence to achieve robust magnetic switching margin by eliminating the half-select disturb issue found in conventional approaches. Another crucial solution was the ability to deposit and pattern high-quality, high-TMR magnetic tunnel junctions with narrow bit-to-bit resistance variation, low defect density and long-term reliability.

We have performed extensive operational characterization and accelerated stress testing of our extended-range MRAM technology and validated its full functionality and high reliability in these extreme environments. Figure 2 demonstrates that the Toggle switching operating region is maintained at the highest temperatures with more than ample margin. Figure 3 shows dielectric-breakdown and resistance-drift lifetime results that exceed the requirements for both industrial and automotive applications.

MRAM cells can be scaled to dimensions significantly below the 0.18 μm technology used for the first production part. A recent demonstration has shown successful integration of new-generation, high-MR MTJ material with a 90nm CMOS front-end logic process (2). The new material, with MgO tunnel barriers, was used to significantly increase the read signal over standard AlOx-based material. Higher MR enables faster read and helps overcome the increasing resistance variability inherent in scaling interconnects and devices to smaller dimensions. Read and toggle-write operations were demonstrated in 8kb test arrays. The new MgO-based MTJ material also has been evaluated in the MR2A16A 4Mb circuit(3), demonstrating improved separation of high and low states as shown in Figure 4. This material enabled faster access times since it is almost double the MR of a similar structure with an AlOx barrier.

In conclusion, recent progress demonstrates the ability of Toggle MRAM to meet industrial and automotive requirements. New research results on higher-performance materials and devices show the potential for advanced scaling and further performance improvements in both read and write properties.

---

(1) B. N. Engel, J. Åkerman, B. Butcher, R. W. Dave, M. DeHerrera, M. Durlam, G. Grynkewich, J. Janesky, S. V. Pietambaram, N. D. Rizzo, J. M. Slaughter, K. Smith, J. J. Sun, and S. Tehrani, "A 4-Mbit Toggle MRAM Based on a Novel Bit and Switching Method ," IEEE Transactions on Magnetics, vol. 41, pp. 132-136, 2005.

(2) J.M. Slaughter, R.W. Dave, M. Durlam, G. Kerszykowski, K. Smith, K. Nagel, B. Feil, J. Calder, M. DeHerrera, B. Garni, and S. Tehrani, "High Speed Toggle MRAM with MgO-Based Tunnel Junctions," *Electron Devices Meeting, 2005. IEDM Technical Digest. IEEE International* , pp. 893-896, Dec. 5, 2005.

(3) R. W. Dave, G. Steiner, J.M. Slaughter, J.J. Sun, B. Craigo, S. Pietambaram, K. Smith, G. Grynkewich, M. DeHerrera, J. Åkerman, and S. Tehrani, "MgO-based Tunnel Junction Material for High-Speed Toggle MRAM," IEEE Trans. Magn., Vol. 42, pp. 1935 - 1939, August 2006.

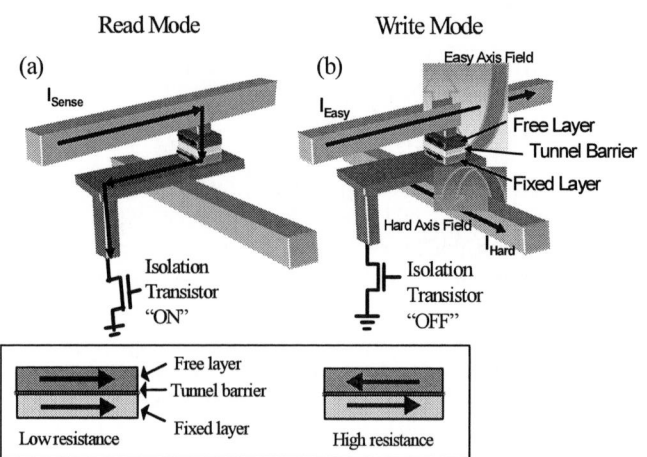

Figure 1. Illustration of a field-switched MRAM cell undergoing a) read and b) write operations. The inset shows the magnetic configuration of the layers in the MTJ for the two resistance states.

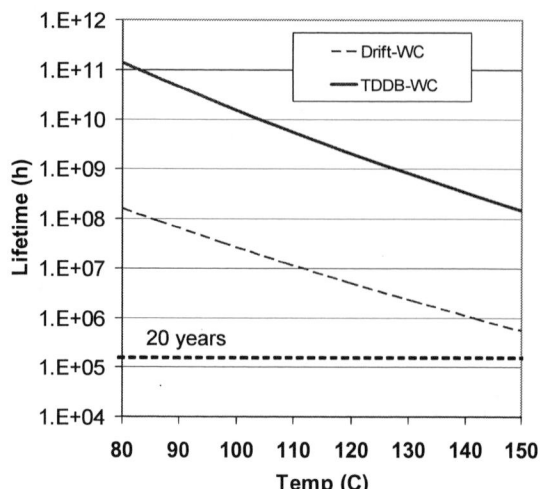

Figure 3. Time Dependent Dielectric Breakdown (TDDB) and Resistance Drift lifetimes, measured from 1 kb test arrays. Both parameters easily exceed the product reliability requirements for both the industrial and automotive environments.

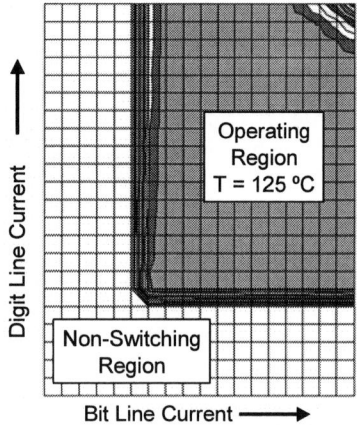

Figure 2 Toggle switching map at elevated temperature (T= 125 ºC) showing large operating region. Saturation of the free layer SAF exchange can start to be observed at over driven currents at the upper right in the figure.

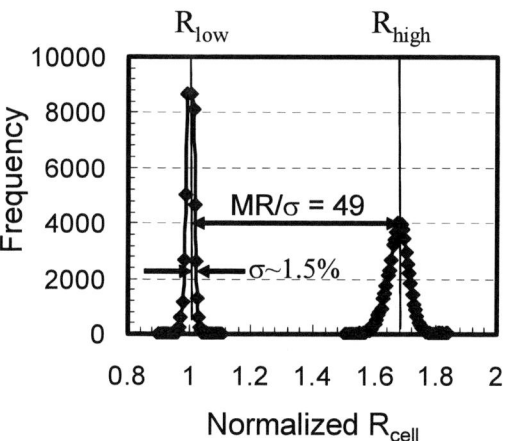

Figure 4. Bit-to-bit resistance distributions measured in 4Mb arrays made with MgO-based MTJ material having a CoFeB free layer.

Magnetic Sensitivity in Mesoscopic EMR Devices in I-V-I-V Configuration

T. Boone, L. Folks, J. A. Katine, E. Marinero, N. Smith and B.A. Gurney

Hitachi Global Storage Technologies , San Jose Research Center, San Jose, CA 95135

Email:thomas.boone@hitachigst.com/ Phone: (408) 717-5919

Magnetic field sensors utilizing the Extraordinary Magnetoresistance effect (EMR) have been proposed for application in future magnetic recording applications [1]. For a decade, Giant Magnetoresistance (GMR) sensor technology has scaled phenomenally well and has resulted in the successful demonstration of recording areal density greater than 230 Gbit/in^2 [2]. However, as critical dimensions decrease below 50 nm, [3] deleterious effects associated with thermal magnetic-noise and spin torque become increasingly difficult to avoid in GMR and related devices. EMR devices are hybrid distributed resistors comprised of a high mobility semiconductor in parallel and in contact with a low resistance metallic shunt. A magnetic field applied perpendicular to the wafer plane modulates the device resistance by selectively steering the current between the semiconductor and the shunt. Although this phenomenon is similar to the Hall Effect, modeling and experiments have suggested that the sensitivity is significantly greater than traditional Hall sensors. Additionally, no ferromagnetic materials are incorporated in EMR precluding vulnerability to magnetic noise sources.

In previous investigations, mesoscopic EMR devices were created with abutted junctions formed by depositing metal directly on exposed edges of quantum wells. These devices were characterized by sensing the voltage in an I-V-V-I four wire resistance measurement. Although these results were intriguing, unconventional fabrication techniques employed, such as edge cleaving and metal depositions through shadow mask, are impractical for most industrial applications [1, 4]. Alternatively, we demonstrated EMR devices with minimum feature sizes less than 100 nm using standard planar processing techniques [5]. Rather than the conventional I-V-V-I configuration, these devices were characterized in a unique I-V-I-V four wire non-local resistance measurement [5]. Finite element modeling predicts that this geometry will produce superior sensitivity to the I-V-V-I configuration [5-7]. Sensitivities of 5 µV/Oe were observed in these devices, which are about three times larger than any previously reported for a non-Corbino disk geometry and comparable to GMR sensors.

Here we present recent characterization results from EMR devices with dimensions approaching that needed for magnetic recording head applications in the I-V-I-V configuration. Minimum feature sizes have been reduced to 50 nm and the magnetically sensitive surface area is designed to be 150 nm x 50 nm. The 2-DEGs used for these devices consisted of AlSb(2 nm)/InAs(12.5 nm)/AlSb(2 nm) quantum wells with room temperature mobility up to 20,000 cm^2/V·s. Devices have been characterized at room temperature in magnetic fields of +/- 450 Oe and applied currents up to 1 mA. Unlike the I-V-V-I geometry, which has a parabolic sensitivity with respect to magnetic field, the behavior in the I-V-I-V configuration is linear. Additionally, the sensitivity appears to be linear with applied current up to current densities well beyond 1×10^7 A/cm^2. Due to robustness of these devices and their remarkable current sinking capability, signal amplitudes greater than 2.5 mV have been routinely observed. To the best of our knowledge, this signal amplitude is more than ten times greater than any previously reported for similarly sized devices. This presentation will include detailed results from various devices sizes and geometries, as well as comparisons with more typical hall sensor geometries fabricated concurrently with the EMR devices and finite element modeling.

[1] S. Solin et al, Appl. Phys. Lett. 80, 4012 (2002).
[2] C. Tsang et al, TMRC 2005, Stanford University, Palo Alto, CA. (Aug 2005).
[3] J. Childress et al, IEEE Trans. Mag, v. 12 n. 10, (Oct. 2006).
[4] T. Zhou et al, Appl. Phys. Lett. 78, 667 (2001).
[5] T. Boone et al, IEEE Trans. Mag, v. 12 n. 10, (Oct. 2006).
[6] J. Moussa et al, J. Appl. Phys. V94, n2 (July 2003).
[7] M. Holz et al, Appl. Phys. Lett. 86 (2005).

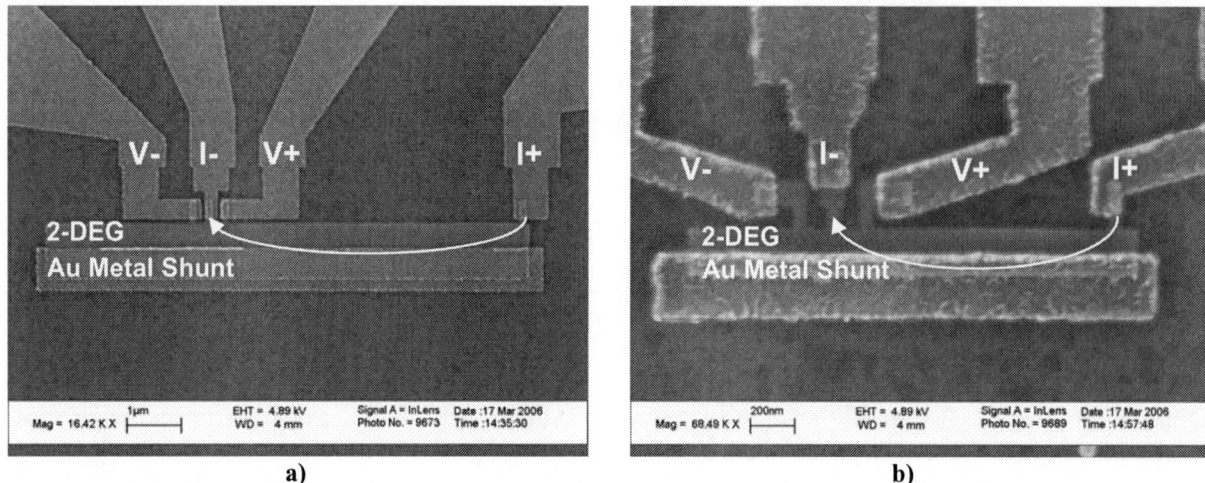

Fig. 1. SEM micrographs of EMR devices developed in 2-DEGs generated by in AlSb (2 nm) / InAs (12.5 nm) / AlSb (2 nm) heterostructures with a) 100 nm and b) 50 nm tab minimum feature sizes. Magnetic sensitivity is determined by measuring the non-local voltage in an I-V-I-V electrode contact configuration. The magnetically sensitive surface area for the device in b) is predicted to be 150 nm x 50 nm, which is approaching the size required for magnetic recording applications.

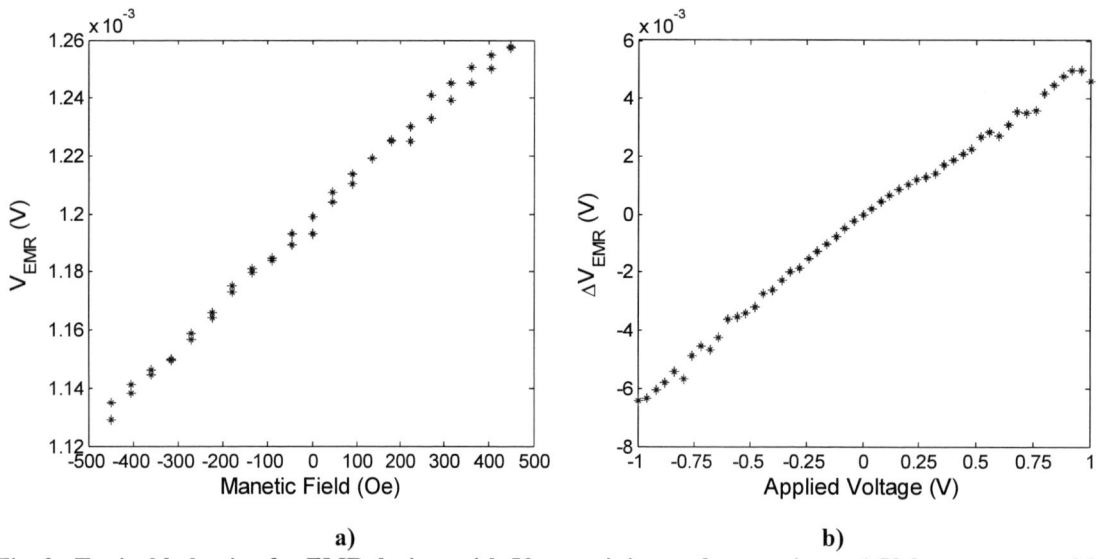

Fig. 2. Typical behavior for EMR devices with 50 nm minimum feature sizes. a) Voltage measured between V-V tabs in the I-V-I-V configuration as a function of magnetic field for an applied current of 10 µA. b) Change in EMR voltage from H= -450 to 450 Oe as a function of applied bias voltage.

Simulation of Spin Torque Devices with Inelastic Spin flip Scattering

Sayeef Salahuddin and Supriyo Datta

School of Electrical and Computer Engineering and NSF Network for Computational Nanotechnology, Purdue University, West Lafayette, IN-47907. (email: ssalahud@purdue.edu, datta@purdue.edu)

In this paper we present simulation results for spin torque devices. Our simulation is based on self consistent solution of Non-Equilibrium Green's Function (NEGF) method, which is used to describe the transport mechanism, and Landau-Lifshitz-Gilbert (LLG) Equation, which accounts for the magnetization dynamics. In addition to purely barrier dependent reflection/transmission, which is normally considered to be the sole contributor to spin torque, we also study the effects of spin flip scattering. From this, we show here, for the first time, that the presence of spin flip scattering may explain experimentally observed values for both (i) tunneling magneto resistance (TMR) and (ii) critical current at the same time, which, otherwise, can not be explained in a coherent manner.

Spin torque devices (1,2) utilize exchange interaction between spin polarized current and ferromagnetic layers to achieve current induced switching, which is important since this process can be scaled down unlike when the magnetization is flipped by an external magnetic field. A typical spin torque device is shown in Fig. 1. The left contact is a heavy ferromagnet which has a much larger coercive field compared to the soft ferromagnet at the right contact that can be flipped by a current. Traditionally, spin torque is understood as the following: due to the magnetization of the contact, the up and down spin components experience different reflection/transmission and the resulting difference in spin angular momentum is absorbed by the soft magnet, since the total angular momentum has to be conserved. However, it has been shown that spin –flip scattering can also contribute to the change in spin angular momentum (3, 4). In this paper, we show that the presence of scattering can give experimentally observed values for both TMR and switching currents at the same time, which, otherwise, can not be explained in a coherent manner.

The simulation has two parts: (i) the electronic transport, which is done using the NEGF method and (ii) the magnetization dynamics governed by the LLG equation [5]. Both these processes have to be solved self consistently (see Fig. 2.). The self consistent method has been described in [6]. To calculate the spin torque, one needs to calculate the spin current as shown in Fig. 3. To find the spin torque at a point k, one needs to find the difference of spin currents to the left and right of point k. For details of how spin currents and torque are calculated, the reader is referred to [7]. What we are concentrating here is the spin flip scattering that adds an additional component to the net difference in spin current given by

$$\left[I_{spin} \right](E) = (e/\hbar)i \left[Trace\{ G \, \Sigma_s^{in} - \Sigma_s^{in} G^\dagger - \Sigma_s G^n + G^n \Sigma_s^\dagger \} \right], \qquad (1)$$

where G is the Green's function and Gn is the electron correlation function and Σ_s^{in} and Σ_s are respectively the inscattering and total scattering functions [7].

Our self consistent simulations give R-I hsyteresis curves shown in Fig. 4. The critical current values are similar to those found in experiments (1,2). We also plot (see Fig. 5) the TMR vs. voltage curve for two cases (i) with spin flip scattering (ii) without it (for two different parameter sets for case (ii)). For comparison, experimental [8] R-I curves are shown in Fig. 6. Note the similarity in TMR decrease with voltage with experiment for the case when spin flip scattering is present. This shows that spin flip scattering should be included in the simulations of spin torque devices.

By taking into account the effects of spin flip scattering we have been able to explain, for the first time, experimentally observed values for both TMR and critical currents in a coherent manner. These results show that it is extremely important to include spin flip scattering in simulation of spin torque devices. In a related note, in a recent paper [9], we have shown that interacting systems like magnets may enable switching while dissipating considerably less energy than the thermodynamic limit. We believe that spin torque based systems can play an important role in designing low power switching circuits with magnets.

(1) Kubota et. al., JJAP, 44, 40, 1237,2005 (2) M.Hosomi et. al, IEDM, 2005 (3) S. Datta, Proc. Int. Sch. Phy.,Enrico Fermi, Italy, p. 1, 2005. (4) S.Salahuddin et. al, PRB, R73,081301 (5) T. Arrott, Ultrathin magnetic Structures, Vol. IV. Pg. 101, 2005. (6) S. Salahuddin and S. Datta, APL, 89,152504.,2006. (7) S. Salahuddin and S. Datta, IEDM, 2006. (8) Fuchs et. al., APL, 86,152509,2005 (9) S. Salahuddin and S. Datta, APL, 90, 0611569, 2007.

978-1-4244-1101-6/07/$25.00 ©2007 IEEE

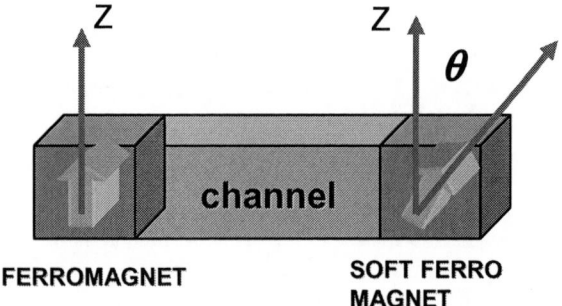

Fig 1. A typical geometry of a spin torque device. The left contact is a heavy ferromagnet which has a much larger coercive field compared to the soft ferromagnet at the right contact that can be flipped by a current.

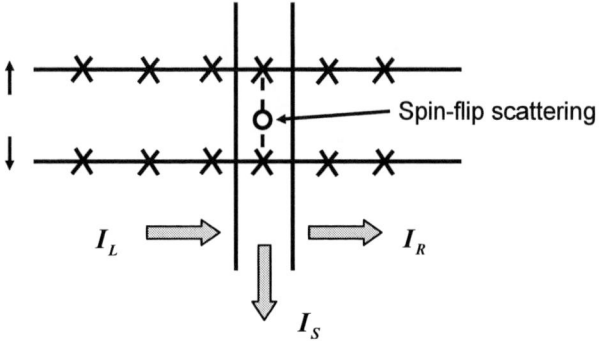

$$I_s = I_{spin-flip\_scattering} + I_{spin\_dependent\_reflection}$$

Fig. 3. The difference of spin currents to the left and right of a point gives the measure of the net transfer of spin angular momentum. Note that this difference arises from two individual components: (i) spin dependent reflection/transmission (ii) spin dependent scattering.

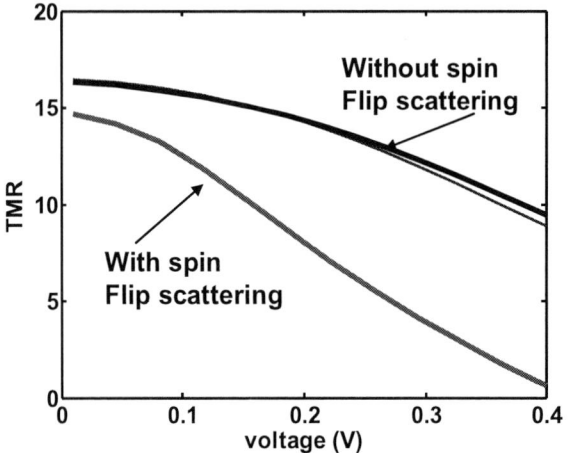

Fig. 5. TMR vs. voltage with and without spin scattering. The curve with scattering shows remarkable agreement with experimental observation as shown in Fig. 6.

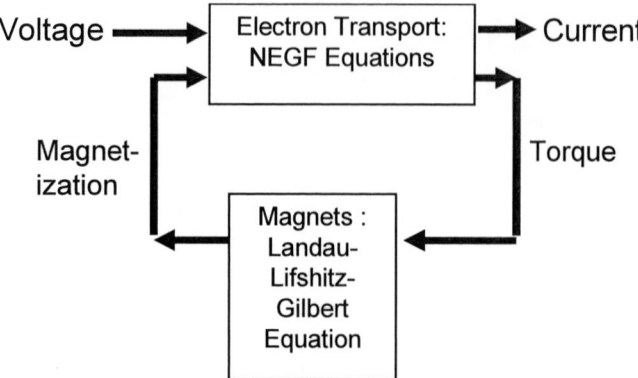

Fig 2. The schematic of the simulation procedure. The conventional NEGF method is extended to take magnetization as input in addition to the voltage and give a resultant torque in addition to the current. This torque is then used as a source term in the Landau-Lifshitz-Gilbert equation that describes the magnetization dynamics. The transport and magnetization dynamics have to be solved self consistently

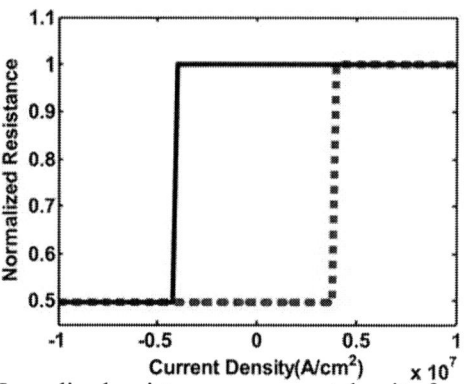

Fig.4. Normalized resistance vs. current density for a device shown in Fig. 1 using a self consistent solution of NEGF and LLG equations. The torque component has contributions from both (i) spin dependent reflection (ii) spin dependent scattering. The critical current density is similar to that seen in several experiments (1, 2).

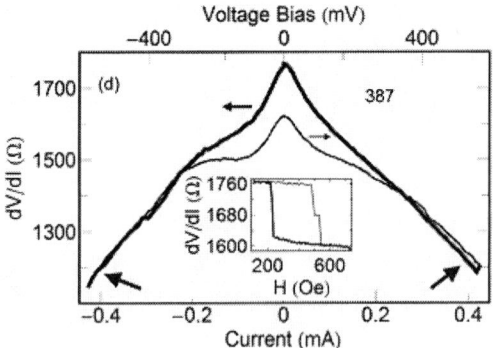

Fig. 6. Experimental resistance vs. voltage for a spin torque device [8] showing sharp decrease of TMR as predicted in the simulation for the case when spin flip scattering is present (see Fig. 6).

Characterization and Application of Large Magnetoresistance in Organic Semiconductors

Markus Wohlgenannt, Govindarajan Veeraraghavan, Yugang Sheng, Omer Mermer, Tho D. Nguyen
Department of Physics and Astronomy, University of Iowa, Iowa City, IA 52242, USA phone: +1 319 353 1974,
fax: +1 319 353 1115, e-mail: markus-wohlgenannt@uiowa.edu

In the short term the most promising application of organic semiconductors appears to be as displays based on organic light emitting diodes (OLEDs) which are widely tipped to replace the current liquid crystal display technology. Yet the future may see a wider range of applications for this technology, with an entirely new generation of ultra low-cost, lightweight and even flexible electronic devices in the offing, which will perform functions traditionally accomplished using much more expensive components based on conventional semiconductor materials. Recent years have seen a surge in interest in magnetoresistive [1] and spintronic [2] properties of organic semiconductors, whereas this field was previously almost exclusively concerned with their electrooptical properties.

We report on the extensive experimental characterization of a recently discovered [1] large and intriguing magnetoresistive effect in OLEDs that reaches up to 10% at room temperature for magnetic fields, B = 10mT. This magnetoresistive effect is therefore amongst the largest of any bulk material. The existence of this effect is highly surprising, since it has generally been believed that large room-temperature magnetoresistive effects can exist only in ferromagnetic devices, whereas our devices are constructed entirely from non-magnetic materials. Our study includes a range of materials that show greatly different chemical structure, mobility, hyperfine and spin-orbit coupling strength. By demonstrating that the effect is critically altered by the presence of strong spin-orbit coupling and that it does not occur in fullerene devices, we prove that the transport in organics sensitively depends on spin-dynamics induced by hyperfine interaction with the hydrogen protons. We discuss a possible relation between organic magnetoresistance and other magnetic field effects in organics that were known long before its discovery [3, 4].

We propose a possible theoretical explanation of the effect: Briefly, the model is based on carrier hopping and assumes that the hopping sites can be either unoccupied, singly occupied or doubly occupied, and that double occupation is only allowed in a singlet configuration. Sites close to the Fermi energy may already be occupied, and hopping onto these sites can therefore occur only in an overall singlet state. The density of potential target sites for hops is therefore restricted by Pauli's principle, and this restriction is partially lifted in the case of mixing of Zeeman states by the hyperfine interaction.

Since the devices we describe can be manufactured cheaply they hold promise for applications where large numbers of magnetoresistive devices are needed, such as magnetic random-access-memory (MRAM); and applications related to organic light-emitting diode displays such as touch screens where the position of a magnetic stylus is detected (patent pending). These interactive displays will have functions similar to existing personal digital assistants (PDAs) and tablet PCs. The envisioned product will consist of a magnetic pen and an array of organic magnetoresistive sensors fabricated directly on the display substrate. The display pixels may actually double as the pen sensors, leading to a significant reduction in fabrication costs of pen-input displays. We will show a video of a simple demonstrator device.

References

[1] T. L. Francis, O. Mermer, G. Veeraraghavan, and M. Wohlgenannt, "Large magnetoresistance at room temperature in semiconducting polymer sandwich devices," *New J. Phys.*, vol. 6, p. 185, 2004.

[2] Z. H. Xiong, D. Wu, Z. V. Vardeny, and J. Shi, "Giant magnetoresistance in organic spin-valves," *Nature*, vol. 427, pp. 821–824, 2004.

[3] K. Schulten and P. Wolynes *J. Chem. Phys.*, vol. 68, p. 3292, 1978.

[4] E. Frankevich, A. Lymarev, I. Sokolik, F. Karasz, S. Blumstengel, R. Baughman, and H. Hoerhold, "Polaron pair generation in poly(phenylene vinylene)s," *Phys. Rev. B*, vol. 46, pp. 9320–9324, 1992.

978-1-4244-1101-6/07/$25.00 ©2007 IEEE

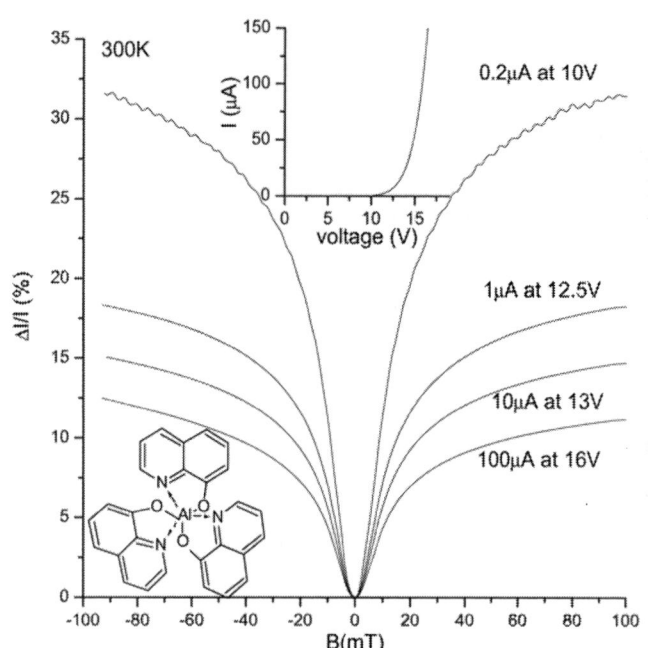

Magnetoconductance, $\Delta I / I$, curves, measured in an Alq_3 OLED at different voltages at room temperature. The inset shows the current-voltage characteristics.

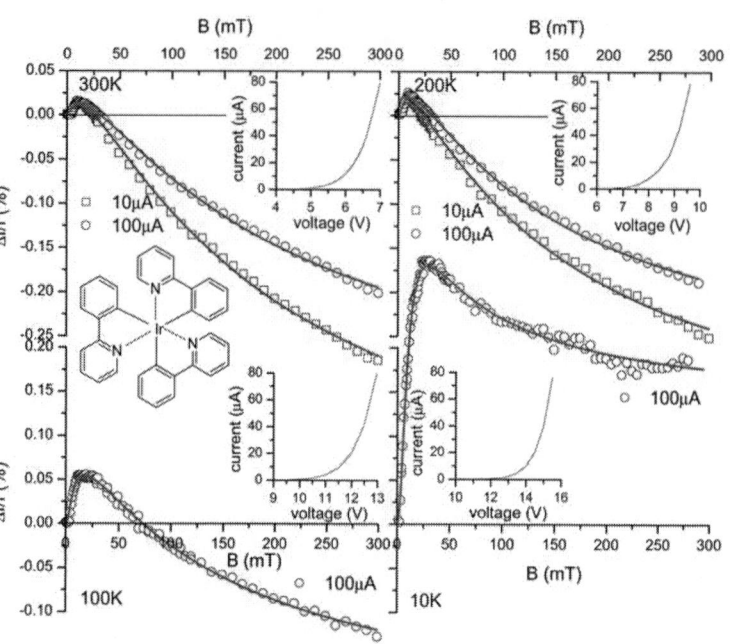

Magnetoconductance, $\Delta I / I$, curves measured in an $Ir(ppy)_3$ OLED at different voltages and different temperatures. $Ir(ppy)_3$ is special amongst organic semiconductors because of its large spin-orbit coupling strength. The insets show the current-voltage characteristics.

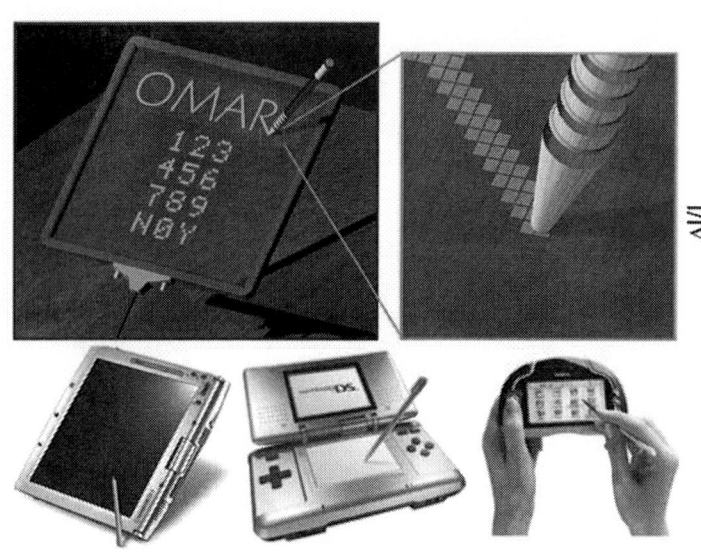

(Upper panel) An envisioned pen-input "touch"-screen based on organic magnetoresistive technology technology. (Lower panel) Typical applications of pen-input displays.

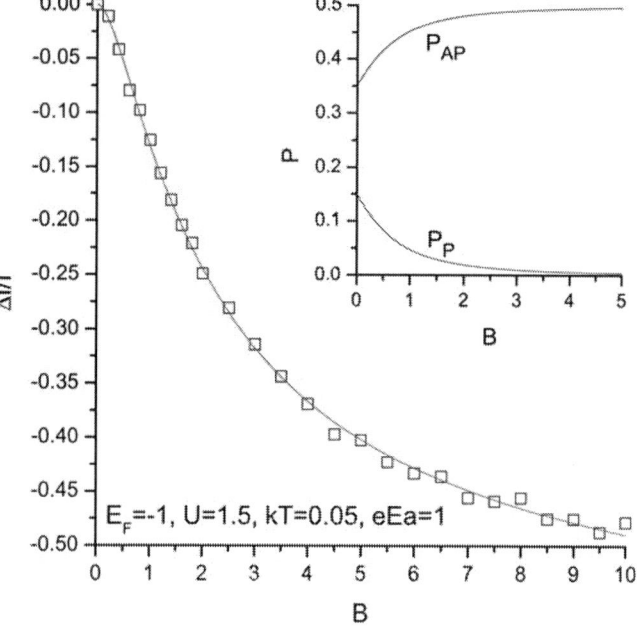

Theory of organic magnetoresistance: Monte-Carlo-simulated magnetoconductance traces as a function of the applied field, B [in units of the hyperfine magnetic field] (scatter plot). The model assumes that hopping sites may be doubly occupied only in an overall singlet state. The solid line shows a fit to the empirical law $B^2=(|B| + B0)^2$, which is known to accurately describe the experimental traces. The inset shows the average singlet probability, P, for a pair of spins as a function of B for "parallel" and "antiparallel" spins.

254

Metamaterials - Negative Indices with Positive Benefits?

M. C. K. Wiltshire

Imaging Sciences Department, Imperial College London, Hammersmith Hospital Campus, Du Cane Road,
London W12 0NN, UK

Email: michael.wiltshire@imperial.ac.uk   *Phone*: +44 (0)20 8383 1021, *Fax*: +44 (0)208743 5409

The electromagnetic behaviour of natural materials, described by their permittivity (ε) and permeability (μ), arises from the response of their individual atoms or molecules to an applied field. Thus, for example, the application of an electric field leads to a distortion of the atoms, which introduces an additional electric polarization and hence leads to the permittivity. Nature, though, does not provide us with materials that exhibit the whole gamut of possible responses. However, by building our own "atoms" and assembling them into metamaterials [1] we can now obtain an engineered response to electromagnetic radiation that is not available from the range of naturally occurring materials.

Metamaterials thus consist of arrays of structures in which both the individual elements and the unit cell are small compared to the wavelength of operation; homogenization of the structures then allows them to be described by the conventional electromagnetic constants ε and μ, but with values that could not previously be obtained. For example, arrays of very fine wires [2] can mimic a dilute plasma and produce a negative permittivity in the GHz frequency range (Fig. 1). A magnetic material with negative permeability in the same frequency range can be constructed from an array of split-ring resonators (SRRs) [3] (Fig. 2). Combining these two elements leads to a material whose permittivity and permeability are both negative.

In this talk, I will review metamaterials with particular emphasis on how they can deliver magnetism and hence a negative refractive index across the electromagnetic spectrum from the RF to the visible, and discuss the prospects for their applications.

For example, the combination of negative μ and negative ε was first discussed theoretically by Veselago in 1968 [4], who showed that the refractive index $n = \sqrt{\varepsilon\mu}$ must also be negative. Some 30 years later, the first practical demonstration of negative refraction was made by Smith and co-workers [5] (Fig. 3). The development of negative index metamaterials (NIM) opens up a new paradigm for conventional lens design: NIM lenses have less aberration for a given focal length than conventional lenses [6] and can weigh much less than their conventional equivalent [7]. Moreover, graded index lenses can be fabricated [8] with potentially even better lens performance.

Besides its ability to focus the propagating radiation (as is done by a conventional lens), Pendry [9] showed that a slab of NIM with $n = -1$ can also focus the evanescent radiation in the near field, thus generating an image with sub-wavelength resolution, the so-called "super" or "perfect" lens. Resolution enhancement factors of 3 for microwave [10], 6 for optical [11,12] and 64 for RF [13] (Fig. 4) have been demonstrated but are limited by the loss in the material. Improved performance may be achievable, but presents severe technological problems.

[1] D R Smith et al., *Science* **305**, 788 (2004)

[2] J B Pendry et al., *Phys. Rev. Lett.* **76**, 4773 (1996)

[3] J B Pendry et al., *IEEE Trans MTT*, **47**, 2075 (1999)

[4] V Veselago, *Sov. Phys. Usp.* **10**, 509 (1968)

[5] R Shelby et al., *Science* **292**, 77 (2001)

[6] D Schurig and D R Smith, *Phys. Rev. E* **70**, 065601 (2004)

[7] C G Parazzoli et al., *Appl. Phys. Lett.* **84**, 3232 (2004)

[8] R B Greegor et al., *Appl. Phys. Lett.* **87**, 091114 (2005)

[9] J B Pendry, *Phys. Rev. Lett.* **85**, 3966 (2000)

[10] A Grbic and G V Eleftheriades, *Phys. Rev. Lett.* **92**, 117403 (2004)

[11] N Fang et al. *Science* **308**, 534 (2005)

[12] D O S Melville and R J Blaikie, *Opt. Express* **13**, 2127 (2005)

[13] M C K Wiltshire et al., *J Phys: Conden. Mat.* **18**, L315 (2006)

Fig. 1 Schematic of a fine-wire grid metamaterial, an experimental example, and its effective permittivity

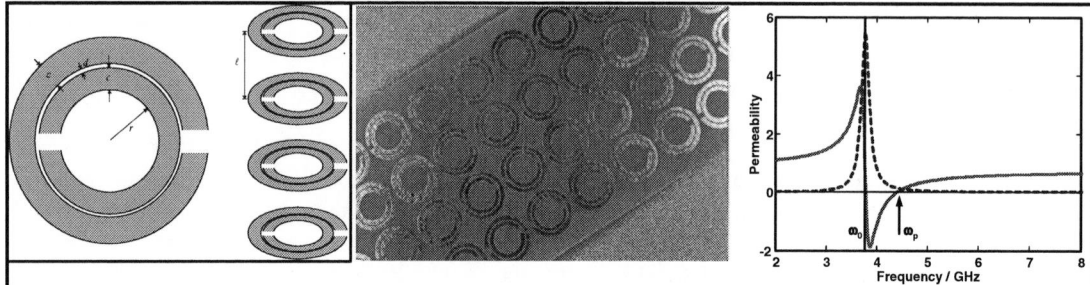

Fig. 2 Schematic of a split ring resonator, an experimental example, and its effective permeability

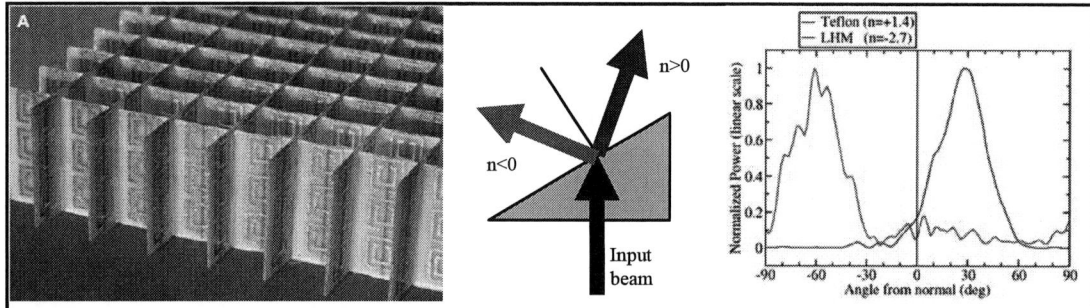

Fig. 3 A negative index metamaterial composed of fine wires and split ring resonators, a schematic diagram of the Snell's law experiment and the measured result showing that the NIM refracts in the opposite direction to a positive index material.

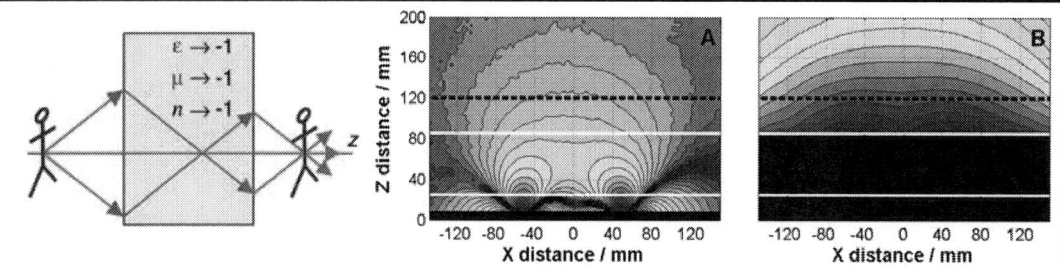

Fig 4 Schematic of the NIM superlens in operation: both the propagating and near fields are brought to a focus. (A) shows the magnetic field distribution in free space (at 24.5 MHz, $\lambda = 12$ m) arising from two dipole sources separated by 100 mm, and (B) the field distribution with a metamaterial slab placed between the two white lines. The intensity in the image plane denoted by the black dashed line shows an image resolution of $\lambda/64$.

HIGH PERFORMANCE POLYCRYSTALLINE DIAMOND MICRO RESONATORS

Nelson Sepúlveda, Jing Lu, Dean M. Aslam, and J. P. Sullivan[1]

Electrical and Computer Engineering, Michigan State University, E. Lansing, MI 48824, USA

[1] Sandia National Laboratories, Albuquerque, NM, 87185, USA

Phone: (517) 432-5648 Email Address: lujing1@msu.edu

Boron doped (5×10^{19} cm^{-3}) and undoped polycrystalline diamond (poly-C) films with a thickness of approximately 0.7 μm, grown at 600 °C or 780 °C, were used to fabricate resonators with dimensions in the range of 0.1 – 100 μm and highest quality factor (Q) of 116,000 measured in a vacuum chamber held at 10^{-8} torr, using piezoelectric actuation and laser detection, in the temperatures range of 23 - 400 °C. The measured values of temperature coefficient of resonance frequency are in the range of -1.59×10^{-5} to -2.56×10^{-5} °C^{-1} and are related to changes in the Young's modulus with temperature. While the highest Q values, reported for the first time for any polycrystalline material, are measured on undoped films, a thermally-activated relaxation process seems to limit the measured Q values for the highly doped samples.

As the performance of a microresonator is characterized by its resonant frequency and quality factor (Q), a number of recent studies have focused on increasing the resonance frequencies and Qs of poly-Si μresonators. For most applications, a GHz frequency range and a Q of at least 10,000 is required. While the miniaturization of poly-Si resonators has been used to achieve a resonance frequency of 1.169 GHz, the measured Q of 5,846 reported for this frequency [1] is substantially lower than the desired value. For most polycrystalline materials the Q is limited by intrinsic mechanisms that involve thermally activated relaxation processes [2]. These mechanisms are usually related to grain boundary sliding or internal friction, phonon-phonon interaction or defects within the resonator material [2]. It has been reported that temperature fluctuations can affect adversely the frequency and Q of microresonators made of nanocrystalline diamond [3]. Although a Q of 55,000 and a resonance frequency of 1.5 GHz have been reported for disc resonators made from nanocrystalline diamond [4], the thermal stability of poly-C resonators and the factors that cause Q degradation in such devices are not well understood.

In order to study the temperature dependence of frequency and Q in poly-C microresonators, we calculated their frequency shifts when tested at elevated temperatures. The poly-C μresonators used in this study have lengths, thicknesses and widths in the ranges of 100 – 500 μm, 0.6 - 0.7 μm and 10μm, respectively, and were fabricated using undoped and p-type poly-C films. Fig. 1 shows the fabrication process. Shown in Table 1 are the poly-C film growth parameters for all the three samples which have a broad characterization of poly-C microresonators. The fabrication technology developed in this work is capable of producing clean and smooth poly-C structures with a minimum feature size of 2μm as shown in Fig. 2. Fig. 3 shows the testing set up and SEM images of some of tested poly-C cantilevers with their dimensions. All measurements were made at a vacuum level below 1×10^{-5} mtorr in order to eliminate air damping losses. Table 2 shows a summary of the results obtained for poly-Si by other research groups and the results obtained for poly-C reported in this work. The values of temperature coefficient of resonant frequency (TC_f) were obtained by calculating the slopes of linear fits done to the measured frequency shift $\Delta f / f_o$ as a function of measurement temperature for a 300μm and 400μm long cantilever beams. Results on the TC_f values for poly-Si structures at temperatures above 247 °C have not been reported. For comparison purposes, the TC_f values for poly-C resonators reported in this work ranged from -1.37×10^{-6} to -2.74×10^{-6} in the temperature range of 23 °C – 100 °C and from -1.15×10^{-5} to 2.31×10^{-5} in the temperature range of 23 °C – 250 °C. This values show that poly-C μresonators have a TC_f values similar to those obtained for poly-Si bridges and free-free beams, and to those obtained from geometrically compensated poly-Si structures [3]. Fig. 4 shows a plot of the measured Q values as a function of frequency and the resonant peak with the highest Q value for each sample. The measured Q values for the cantilevers made from sample 1 are about 2 times larger than those for the other two poly-C films. The highest Q value of 116,000 that has been measured so far for any polycrystalline diamond structure or a cantilever made of any polycrystalline material was measured on the sample 1. This value is also much larger than nanocrystalline diamond resonators reported in the past [3][4] and is about half the value of 250,000 found for single crystal silicon resonators .

In conclusion, the highest Q factors for polycrystalline diamond resonators (also the highest Q factors for a cantilever beam made of any polycrystalline material) have been measured, whose undoped poly-C films grown at 780 °C. For boron doped poly-C films or poly-C films grown at a growth temperature of 600 °C, significantly lower Qs were observed. The doped poly-C film exhibited a peak in dissipation at 673 K, suggesting the existence of a dominant defect in these films that may be related to boron doping. In addition, the temperature dependence of the poly-C resonators was examined, and a temperature coefficient of resonance frequency in the range of -1.59×10^{-5} to -2.56×10^{-5} °C^{-1} was observed..

[1] S. S. Li, Y. W. Lin, Y. Xie, Z. Ren, and C. T-C. Nguyen "Micromechanic 'Hollow-Disk' Ring Resonators IEEE Int. Conference on MEMS pp.821-824, 2004.

[2] D. Czaplewski, J.P. Sullivan, T.A. Friedmann, D.W. Carr, B.E.N. Keeler and J.R. Wendt; *Journal of Applied Physics*, vol.97, 023517, 2005.

[3] L. Sekaric, J.M. Parpia, H.G. Craighead, T. Feygelson, B. H. Houston and J. E. Butler. "Nanomechanical resonant structures in nanocrystalline diamond"; *Applied Physics Letters*, Vol. 81, pp.4455-4457, December 2002.

[4] J. Wang, J. E. Butler, Tatyana Feygelsonand C. T.-C. Nguyen "1.51-GHz Nanocrystalline diamond micromechanical disk resonator with material-mismatched isolating support" IEEE Int. Conference on MEMS pp.731 – 734, 2004.

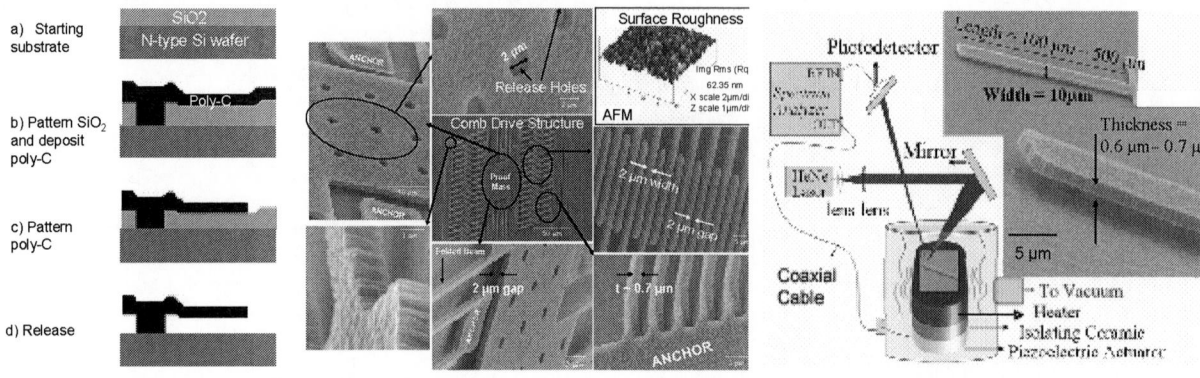

Fig. 1. Process Flow for the fabrication of poly-C microresonators

Fig. 2. Poly-C comb-drive structures produced using the described fabrication process flow. The AFM image is a representative sample of the poly-C film surface.

Fig. 3. Testing set-up for measurements using piezoelectric actuation and laser detection. Inset SEM image show poly-C cantilevers with its dimensions.

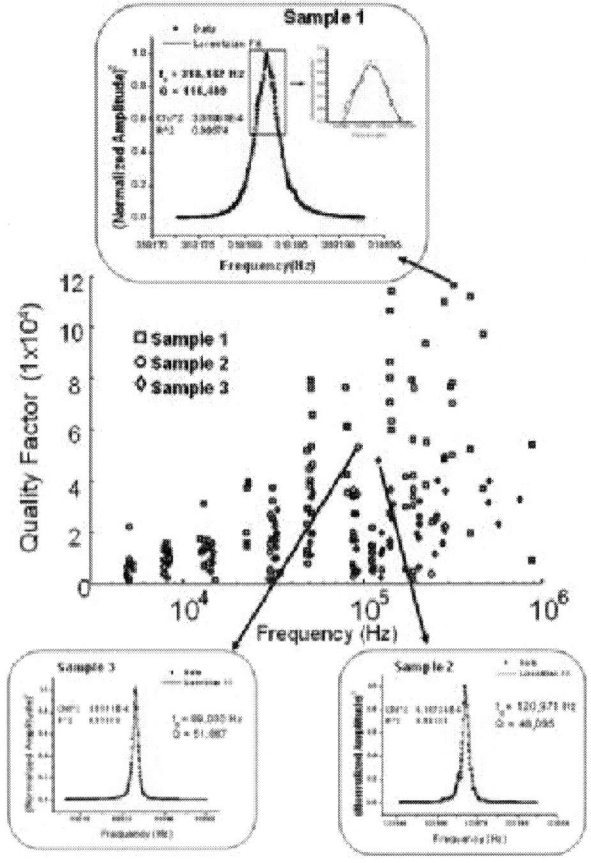

Fig. 4. Measured Q values as a function of frequency. Inset resonant peaks show the measurement with the highest Q value for each sample.

Poly-C Film Growth		
Gas Flow Rate (sccm)	**H$_2$**	**100[1,2,3]**
	CH$_4$	**1[1,2,3]**
	TMB	**4[2] ; 0[1,3]**
Temperature (°C)		**780[1,2] ; 600[3]**
Microwave Plasma Power (kW)		**2.0[1,2,3]**
MPCVD Gas Pressure (torr)		**35[1,2,3]**
Seeding Diamond Powder Size (nm)		**25[1,2,3]**
Dry Etching Parameters		
Gas Flow Rate (sccm)	**CF$_4$**	**1[1,2,3]**
	O$_2$	**30[1,2,3]**
Chamber pressure (mbar)		**4x10^{-2} [1,2,3]**
DC Power (W)		**400 [1,2,3]**

Table 1. Parameters used for poly-C growth and patterning. The superscript represents the sample on which the value was used.

Poly-Si				
Structure	$\Delta f/f_0$ (ppm)	Temperature Range (°C)	TC_f (ppm/°C^{-1})	Resonant Frequency (f_0) Range
Free-Free Beam [8]	≈ - 1000	30 – 247	-12.5	53.6 MHz
Clamped-Clamped [8]	≈ -3600	30 – 247	-16.7	4.2 MHz
Comb-Drive [33]	≈ -1200	30 – 107	-15.6	≈ 345 KHz
Comb-Drive (Compensated) [33]	≈ -600	30 – 107	-7.8	≈ 345 KHz
Clamped-Clamped (Compensated) [14];[15]	≈ - 240 ; ≈ - 18	30 – 107	-2.5 ; -0.24	≈ 10 MHz
Poly-C (cantilever beams reported in this work)				
Sample 1	≈ -8,450	30 – 400	≈-25.6	12 KHz – 50 KHz
Sample 2	≈ -5,900	30- 400	≈-15.9	8 KHz – 31 KHz
Sample 3	≈ -6,100	30 – 400	≈-18.6	8 KHz – 15 KHz

Table 2. Temperature dependent frequency performance for poly-Si and the poly-C µresonators reported in this work.

258

Session VII.B (DeBartolo Hall Room 136)

Carbon Nanotube and Graphene Devices

Wednesday PM, June 20th, 2007

Session Organizer: Prabhat Agarwal, NXP Semiconductors
Session Chair: Mark Stettler, Intel

1:30 PM VII.B-1 Invited Paper
Aligned Arrays Single Walled Carbon Nanotubes for Thin Film Electronics
J. A. Rogers, University of Illinois at Urbana/Champaign, Urbana, Illinois, USA

2:10 PM VII.B-2
Quantum Capacitance Measurement for SWNT FET with Thin ALD High-k Dielectric
Y. Lu[1], H. Dai[1] and Y. Nishi[2], [1]Department of Chemistry and Laboratory for Advanced Materials, Stanford University, Stanford, California, USA and [2]Electrical Engineering, Stanford University, Stanford, California, USA

2:30 PM VII.B-3 Student Paper
The study of low frequency noise of single-walled carbon nanotube transistors
S. Kim, D. Chang, Y. Xuan, P. Ye and S. Mohammadi, School of Electrical and Computer Engineering and Birck Nanotechnology Center, Purdue University, West Lafayette, Indiana, USA

2:50 PM VII.B-4
Semiconducting Graphene Ribbon Transistor
Z. Chen, P. Avouris, IBM T. J. Watson Research Center, Yorktown Heights, New York, USA

3:10 PM Break

3:30 PM VII.B-5 Invited Paper
Performance Limits of Nanocomposite Transistors & Nanobio Sensors: A Bottom-up Perspective
M. A. Alam, N. Pimparkar, P. Nair, S. Kumar, and J. Murthy, School of Electrical and Computer Engineering, Purdue University, West Lafayette, Indiana, USA

4:10 PM VII.B-6
Impact of Process Variation on Nanowire and Nanotube Device Performance
B. C. Paul[1,2], S. Fujita[2], M. Okajima[2], T. Lee[1], H.S.P. Wong[1], and Y. Nishi[1], [1]Center for Integrated Systems, Stanford University, Stanford, California, USA and [2]Toshiba America Research Inc., San Jose, California, USA

4:30 PM VII.B-7 Student Paper
Scaling Behaviors of Graphene Nanoribbon FETs
Y. Yoon, Y. Ouyang, and J. Guo, Electrical and Computer Engineering, University of Florida, Gainesville, Florida, USA

4:50 PM VII.B-8
Role of Electrical and Thermal Contact Resistance in the High-Bias Joule Breakdown of Single-Wall Carbon Nanotube Devices
E. Pop, Dept. of Electrical and Computer Engineering and Micro and Nanotechnology Lab, University of Illinois, Urbana-Champaign, Urbana, Illinois, USA

260

Aligned Arrays Single Walled Carbon Nanotubes for Thin Film Electronics

John A. Rogers

University of Illinois at Urbana/Champaign, Illinois 61822

Phone: 217-244-4979, Email: jrogers@uiuc.edu

The excellent electronic, thermal and mechanical properties of single-walled carbon nanotubes (SWNT) create interest in their possible use in various areas of electronics, ranging from heterogeneously integrated systems for applications in communications to large area distributed circuits for applications in flexible displays. In these cases, organized, horizontally aligned arrays of pristine SWNTs represent a scalable pathway to these and other device implementations of SWNTs. This talk describes (1) methods for guided growth of large scale, aligned arrays of long (>100 microns), linear SWNTs, where nearly perfect levels of alignment and linearity can be achieved, and (2) characteristics of thin film type field effect transistors that use these aligned arrays as the channel material, where excellent device level properties can be realized.

We generate the horizontally aligned arrays by use of optimized versions of guided growth procedures that use chemical vapor deposition on single crystal quartz substrates[1,2]. These new procedures employ patterned lines of evaporated sub-monolayer films of iron as the catalyst, ethanol as the feed gas with a slight background of hydrogen, and air annealed ST-cut single crystal quartz wafers as the growth substrate[3,4]. In this process, orientation dependent interaction energies between the growing tubes and the quartz create well aligned tubes in linear configurations. This surface interaction is critically important, as revealed by geometries of tubes grown on quartz substrates with 1-2 nm thick patterned lines of evaporated, amorphous SiO_2. Figure 1 shows results of such experiments[3]. Figure 2 presents representative images of SWNT arrays created by the optimized growth process[4].

Defining source, drain and gate electrode, and gate dielectrics on top of these arrays yields thin film type field effect transistors in which each SWNT provides an independent pathway for charge to transport between source and drain. Figure 3 provides a schematic illustration and a scanning electron micrograph (collected before deposition of the gate dielectric) of a typical device[4]. The as-grown tubes include both semiconducting and metallic tubes, at a ratio of approximately 2 to 1. To achieve high on/off current ratios in such devices, it is possible to electrically breakdown the metallic tubes by biasing the device into its off state and then slowly increasing the source/drain voltage. Figure 4 shows transfer curves collected before and after this breakdown process, and full current/voltage characteristics measured after breakdown. The devices exhibit well behaved, unipolar p channel response. The mobilities, computed using rigorous models for the capacitance, are greater than ~1000 cm^2/Vs[4]. These properties, together with the scalable route to circuit integration, might make such devices interesting for a variety of applications.

1. C. Kocabas et al., *Small* 1(11), 1110-1116 (2005).
2. C. Kocabas et al., *J. Am. Chem. Soc.* 128, 4540-4541 (2006).
3. C. Kocabas et al., submitted.
4. S.J. Kang et al, *Nature Nanotechn.*, in press.

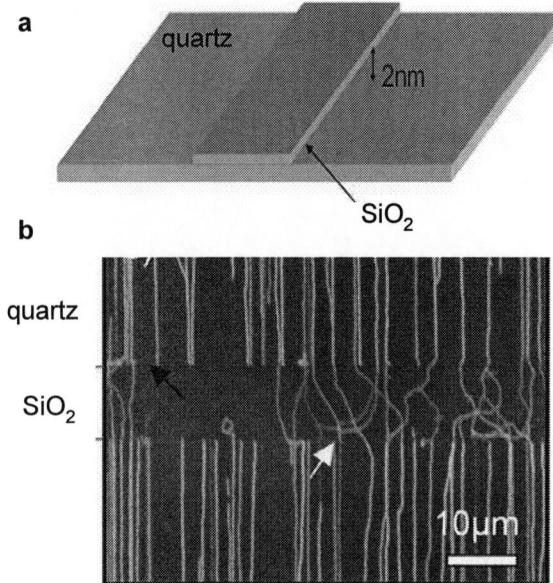

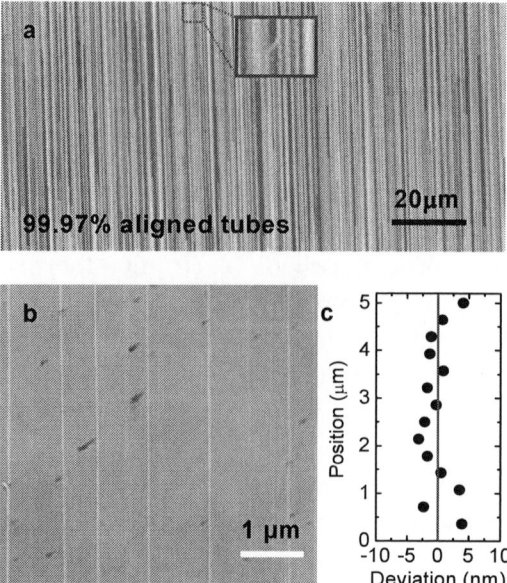

Fig. 1. Guided growth of SWNTs on a quartz substrate with a patterned line of amorphous SiO_2. The results illustrate the importance of surface interactions in the alignment mechanism. Parts (a) and (b) show schematic illustrations and scanning electron micrographs, respectively

Fig. 2. Scanning electron micrograph (a), atomic force micrograph (b) and line analysis (c) for arrays of SWNTs formed by optimized chemical vapor deposition growth on quartz. The levels of alignment and the degrees of linearity in the SWNTs are extremely high.

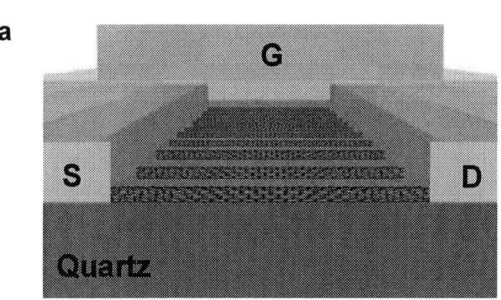

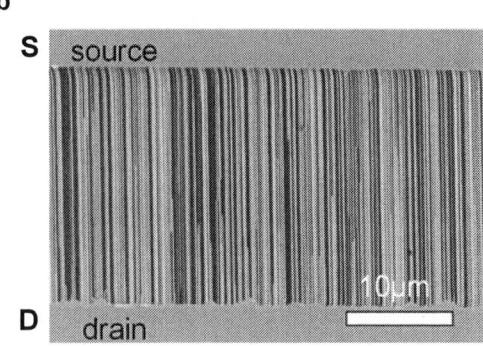

Fig. 3. Schematic illustration (a) and scanning electron micrograph (b), of a top gate transistor that uses aligned arrays of SWNTs as the semiconductor.

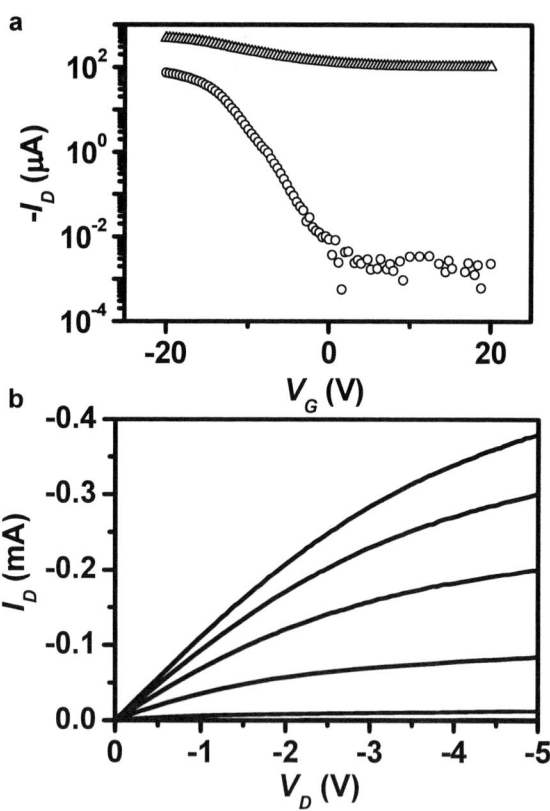

Fig. 4. Transfer characteristics before (triangles) and after (circles) electrical breakdown of a transistor that uses an array of SWNTs as the semiconductor. Part (b) shows current-voltage characteristics measured after breakdown.

Quantum Capacitance Measurement for SWNT FET with Thin ALD High-k Dielectric

Yuerui Lu, Hongjie Dai* and Yoshio Nishi**

*Department of Chemistry and Laboratory for Advanced Materials,
Stanford University, Stanford, CA 94305, USA.
**Electrical Engineering, Stanford University, Stanford, CA 94305, USA.
Phone: 650-796-4526, Fax: 650-725-0991, Email: yueruilu@stanford.edu

We used top-gated individual single walled carbon nanotube (SWNT) field effect transistors (FETs) (gate length $L\sim1\mu m$) with HfO_2 high-κ dielectrics (ε ~20) in under-lapping geometry (Fig.1a) [1]. The thin ($t_{ox}\sim$5nm), conformal HfO_2 dielectric coating on SWNTs afforded by DNA functionalization [2] was a key factor to study the quantum-capacitance of SWNT FETs, affording high geometric top-gate-capacitance C_{gg} (~5X of SiO_2 gate dielectrics), which is comparable to SWNT quantum capacitance C_q. This reached a regime that the experimentally measured total top-gate-capacitance (C_{TG}) (Fig.2b) $C_{TG}^{-1} = C_{gg}^{-1} + C_q^{-1}$ largely represented C_q. Consistent with the high gate-capacitance of the device, we got ideal 60mV/decade subthreshold switching of our FETs at room temperature with on/off $\sim10^5$ at a source-drain (S-D) bias of $V_{DS} = 0.6V$ (Fig.2a).

To measure small C_{TG} (<1fF) over large background parasitic capacitances, we used the capacitance bridge technique developed by McEuen et al. [3], reducing the background to ~10fF. By adding a ground metal plate between the S/D and TG probes (Fig.1b), we further reduced the background capacitance to $C_0 \sim$ 30aF. This was reflected in the measured C_{TG} (between TG and S/D leads) evolving from ~200aF to $C_0 \sim$30aF in a stepwise manner when the under-lapped regions of the SWNT was switched from electrical -on to -off by back-gate (Fig.3a). Reaching the ultra-low background C_0 is a key to quantum capacitance measurement for the SWNT FET.

We recorded C_{TG}-V_{TG} curves of SWNT FETs at $T = 65K$ by measuring the stepwise changes of C_{TG} to C_0 (negligible) under V_{BG} sweeps as in Fig.3a for various top-gate voltages (V_{TG}) (Fig.4b). Quantum capacitance C_q vs. V_{TG} was then obtained from $C_{TG}^{-1} = C_{gg}^{-1} + C_q^{-1}$ with geometrical capacitance $C_{gg} = 2\pi L/ln(4t_{ox}/d)$ (~0.4fF/μm) [3]. We observed pronounced oscillating peaks in the C_q vs. top-gate V_{TG} (Fig.4c), corresponding to the one-dimensional electron and hole VHSs of the 1st and 2nd sub-bands.

In summary, the thin conformal HfO_2 dielectric, which could provide very large geometric top-gate-capacitance C_{gg} comparable to SWNT quantum capacitance C_q, and the capacitance measurement technique developed by us, which could reduce the background capacitance down to $C_0 \sim$30aF, are two key promising factors for us to study the quantum capacitance of the SWNT FET. We successfully got the pronounced oscillating peaks in the C_q vs. top-gate V_{TG}, which will be very usefully for us to better characterize the performance of the SWNT FETs and to further study the low-dimensional electronic structure.

[1] Javey, A. et al. Nature Materials, vol.1, p.241 (2002).
[2] Lu, Y. R. et al. Journal of the American Chemical Society, vol. 128, p.3518 (2006).
[3] Ilani, S., & McEuen, P. L. et al. , Nature Physics, vol 2, p.687 (2006).

978-1-4244-1101-6/07/$25.00 ©2007 IEEE

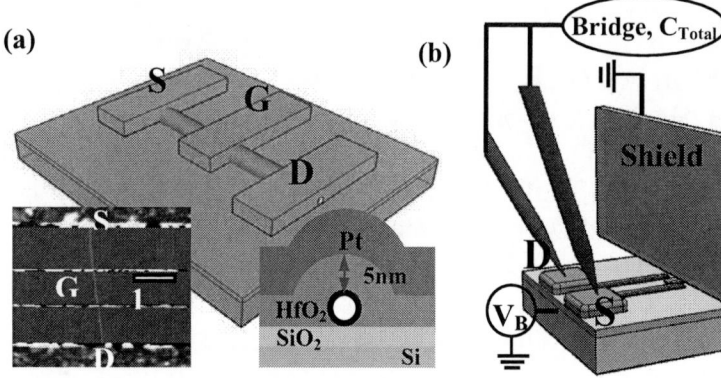

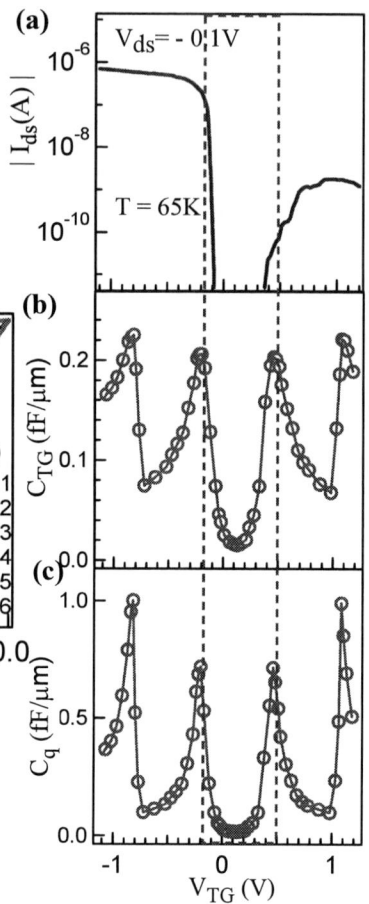

Fig.1 **(a)** Schematic and atomic force microscopy (AFM) image of a top-gated SWNT FET with 5nm thick HfO_2 as top-gate insulator and 1μm gate length (S: source; D: drain; G: gate). **(b)** A schematic illustration of the capacitance measurement setup. A grounded copper plate is positioned between the S/D and G probe tips, which reduces background capacitance from ~10fF to ~30aF, allowing for accurate measurement of small gate-capacitances of the SWNT FET.

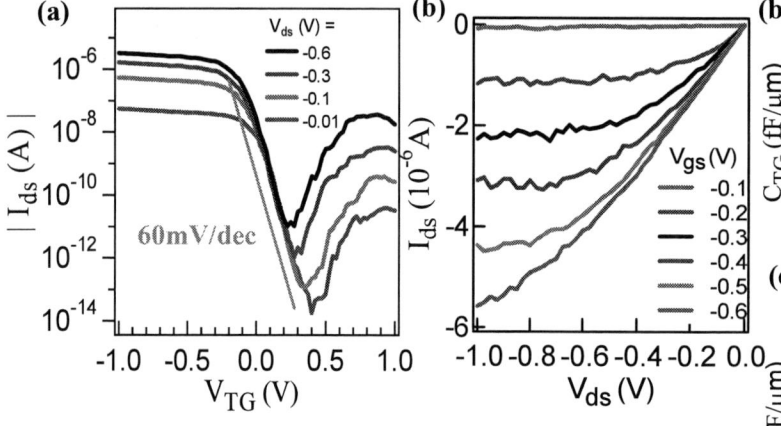

Fig.2 **(a)** Room temperature transfer characteristics (current vs. top gate-voltage) I_{ds}-V_{TG} curve. **(b)** I_{ds}-V_{ds} curves.

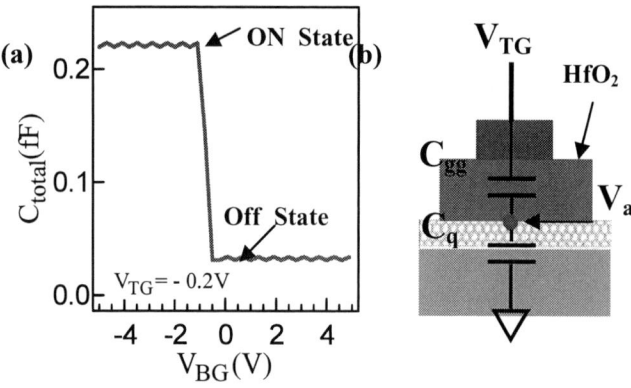

Fig.3 **(a)** Low temperature (65K) gate-capacitance of the SWNT device measured as a function of the back-gate voltage V_{BG} at a fixed top-gate voltage. The back ground capacitance, ~30aF, is measured when the under-lapped SWNT regions are turned off at high back-gate voltage. **(b)** An equivalent capacitance circuit model for the top-gate stack of the device, with top-gate geometric capacitance C_{gg} and quantum capacitance C_q in series. V_a is the local electrostatic potential of the SWNT.

Fig.4 Electrical transport and capacitance data of a SWNT FET **(a)** I_{ds}-V_{TG} curve of a SWNT FET device recorded at T=65K. **(b)** Top-gate capacitance vs. gate voltage C_{TG}-V_{TG} curve measured at 65K for the SWNT FET. C_{TG} was measured from such capacitance step as in Fig.3a at various top-gate voltages. **(c)** Quantum capacitance vs. top-gate voltage C_q-V_{TG} curve. C_q was obtained from C_{TG} using the equation $C_{TG}^{-1} = C_{gg}^{-1} + C_q^{-1}$ and a constant top gate geometric capacitance C_{gg}. The dashed lines highlight the band edges of the 1st sub-band of the nanotube.

The study of low frequency noise of single-walled carbon nanotube transistors

Sunkook Kim, David Chang, Yi Xuan, Peide Ye and Saeed Mohammadi

School of Electrical and Computer Engineering and Birck Nanotechnology Center, Purdue University,
West Lafayette, Indiana 47907
Email: saeedm@purdue.edu; Phone: (765)494-3557

Nano-scale single-walled carbon nanotube (SWNT) FETs with near ballistic electron and hole transport have attracted interest for potential application in ultra large scale integrated circuits. In addition to immature fabrication technology, hysteresis in the IV characteristics and reported large amplitudes of low frequency noise of SWNT-FETs have impeded their implementation into electronic and sensing systems [1-3]. In this paper, we report a top gate SWNT-FET with reduced hysteresis in the IV characteristics and extremely low 1/f noise. We have also investigated the source of 1/f noise in these devices and attributed the low noise property to low trap charges near the carbon nanotube substrate interface.

The SWNT-FET devices reported here and shown schematically in Fig.1(a), are fabricated on high resistivity Si substrate ($\rho \approx 10K\Omega$) with a 500nm thermal SiO_2. Catalyst patterns are defined by UV photo-lithography with a 10μm spacing and subsequent iron deposition and lift-off. Single-walled carbon nanotubes (SWNTs) are then synthesized by chemical vapor deposition (CVD) of methane on the substrate using iron catalyst. Source-drain contacts are formed by electron beam deposition of Pd metal. A 20nm high-k HfO_2 film is deposited at 300°C by using $HfCl_4$ and H_2O precursors in an ASM Microchemistry F-120 ALCVD^{TM} Reactor. Carbon nanotubes are bridged between source and drain as shown in Fig.1(b). Top Gate metal is defined by UV photolithography followed by the deposition of Cr/Au (10/50nm) with minimum gate length of 1.5μm. Cr/Au (20/450nm) metal interconnects are finally deposited on top of the source and drain Pd contacts.

Implementation of top gate structure allows the devices to be isolated from the environment and show stable DC and reduced low frequency noise characteristics. Back gating of SWNT-FETs while the nanotube is exposed to the environment [2-3] causes the nanotube-substrate interface to interact with various environmental factors, such as water molecule and mobile ions, resulting in large hysteresis. In our experiment, ALD HfO_2 gate oxide serves as a passivation layer for SWNT and helps stabilizing the electrical characteristics of SWNT-FET. Small hysteresis in I_d-V_{gs} characteristic observed in Fig.2(a) indicates that ALD HfO_2 suppresses traps at nanotube-substrate interface. Furthermore, a maximum on-current of -14 μA (Fig. 2(b)) and high transconductance of 0.8μS at a drain bias of 0.2V (Fig. 2(c)) indicate that SWNT-FETs based on ALD HfO_2 achieve high electrical performance.

Low frequency noise measurements of the fabricated SWNT-FETs are carried out to study the quality of nanotube-substrate interface and identify the source of the noise. Figure 3 shows measured low frequency noise as a function of drain (Fig. 3(a)) and gate bias (Fig. 3(b)). In Fig. 3(a), the amplitude of current noise (S_I) at f=100Hz is proportional to I_d^2 at a constant gate voltage, where the SWNT-FET channel is treated as a conventional resistor. The noise increases with the current in the linear regime and saturates as the device enters saturation regime. Figure 3(b) shows that the current noise amplitude (S_I) at f=100Hz is proportional to g_m^2 in off-state, subthreshold, and inversion regime. This is consistent with our low-frequency noise modeling of SWNT-FETs. Previous reports show a different 1/f noise trend, namely noise proportional to $1/I_d^2$ [2] and $1/I_d$ [3] in saturation regime. The difference in observed noise trends is that our top-gate device is self-passivated and show inherent low frequency noise characteristic of the nanotube. Devices reported in [2-3] are based on back-gate structure with exposed nanotubes with no effective passivation from the environment causing them to present much higher 1/f noise with different 1/f noise-drain current trends.

[1] W.Kim et al., *Nano Letters*, vol.3, p193 (2003) [2] E.S.Snow et al., *Applied Physics Letters*, vol. 85, p4172 (2004), [3] M. Ishigami et al., *Applied Physics Letters*, vol. 88, p203116 (2006)

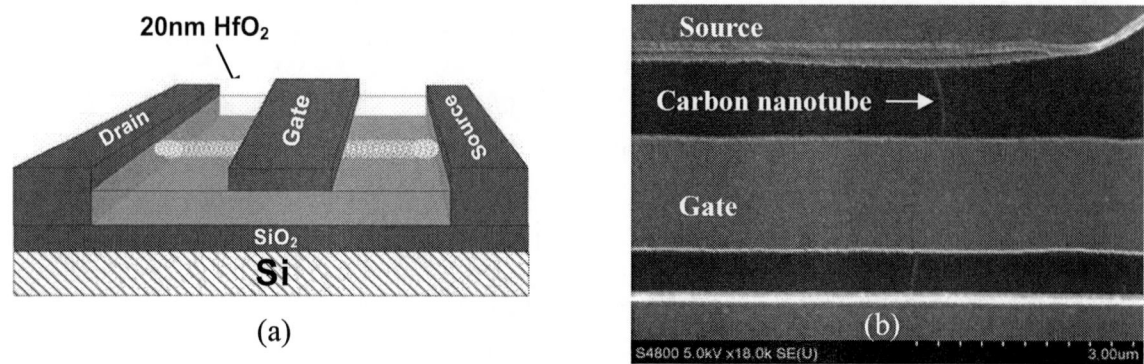

Figure 1. (a) The schematic view of the top gate SWNT-FETs with 20nm thick ALD-grown HfO_2 as gate dielectrics (b) The SEM image of a device with a single nanoutbe covered by ALD HfO_2.

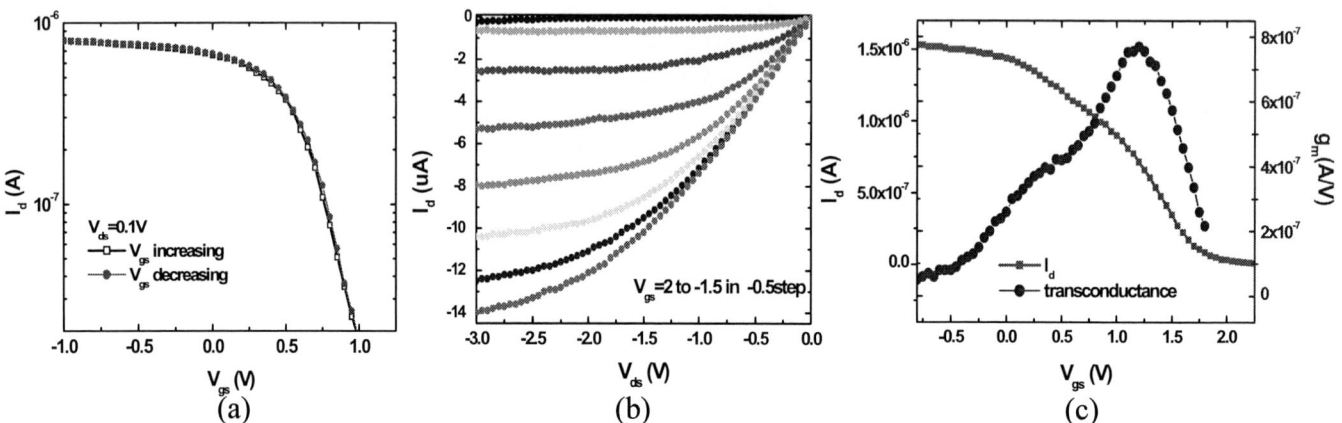

Figure 2. Characteristics of a p-type SWNT-FETs with high-k gate dielectric. (a) The transfer characteristics of nearly hysteresis-free ALD HfO2 SWNT-FETs. (b) Id-Vds curves of the FET at various top-gate voltages with 0.5V steps. (c) Id-Vgs curve at drain bias of Vds=0.2V(red dots) and extracted transconductance vs. Vgs (blue dots). Maximum transconductance is 8×10^{-7}S.

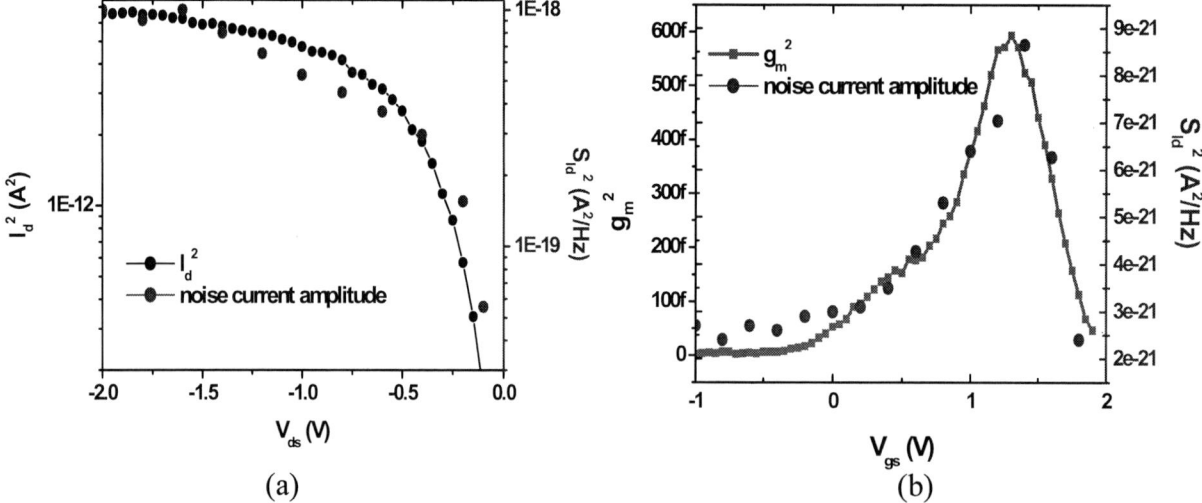

Figure 3. Measured low frequency noise current noise amplitude at 100Hz for a p-type SWNT-FETs with high-k gate insulator. (a) Measurement with constant gate voltage (gate bias = 1V) and varying drain voltage. The squared drain current (I_d^2) is also shown for comparison to demonstrate the noise trend. (b) Measurement with constant drain voltage (drain bias = 0.2V) and varying gate voltage. The transcouductance (gm) is also shown for comparison to demonstrate the noise trend.

266

Semiconducting Graphene Ribbon Transistor

Zhihong Chen, Phaedon Avouris

IBM T. J. Watson Research Center, Yorktown Heights, NY 10598, USA

Two-dimensional (2-D) graphene sheet was extensively studied theoretically and was used as the building block in the discussion of carbon nanotubes and other carbon-based materials. The recent success of extracting atomic layer thick graphene sheets enables detailed studies on the physical properties of this material[1,2] and also opens up a discussion whether graphene is useful for electronic applications. One of the proposals concerns the possibility of using graphene in device applications in a manner similar to carbon nanotubes. Two-dimensional graphene is a zero band gap semiconductor with linear dispersion at the Fermi level. By introducing quantization to one of the reciprocal vectors, the original 2-D energy dispersion will be split into a number of 1-D modes and a semiconductor gap can be formed depending on the quantization conditions. In this context, we present the fabrication of graphene ribbon transistors by conventional lithography and etching techniques and show measurements of the edge effects and the opening of semiconducting gaps.

We used the micromechanical cleavage method[1] to peel off single layer graphene from highly ordered pyrolytic graphite and deposited them onto heavily doped Si substrates covered with thermally grown SiO_2, as shown in Fig. 1. The inset is the cross-section measurement and shows the thickness of the graphene around 0.5nm. Fig. 2 presents the full fabrication process. Etching masks were defined on top of the graphene layer by e-beam lithography using HSQ resist. Oxygen plasma was employed to etch away the undesired areas and left graphene ribbons with different widths on the substrate. Source/drain contacts were patterned on top of the ribbons after the etching masks were removed.

On the same graphene flake, ribbons with different widths ranging from *20nm* to *200nm* were formed to provide a width dependence study. We measured the resistance of the ribbons as a function of Si back gate voltage at room temperature; an example is shown in Fig. 3 inset. The "V" shape modulation of the resistance is due to the shift of the Fermi level through the energy dispersion by the back gate. The maximum resistance is reached when the Fermi level goes through the point where the conduction and valence bands intersect and has the minimum density of states. We calculated the resistivity at the maximum resistance point for all devices and found an increasing trend as a function of the ribbon width for our small ribbons, as shown in Fig. 3. This trend indicates that the lithography and etching introduced boundaries play an important role in the transport properties of these narrow ribbons. Boundary roughness and edge states[3] induced scattering can be responsible for this resistivity increase trend.

Fig. 4 shows an example of temperature dependence current measurements for a *100nm* wide ribbon. It was found that for wide ribbons, the current modulation as a function of the back gate has very little temperature dependence. The limited difference observed is due to the temperature dependence of the carrier concentration. On the contrary, narrow ribbons show distinguishing temperature dependence in the current measurements. Fig. 5 (top) gives an example of a *20nm* ribbon device. At high temperatures, the current has small modulation as a function of the back gate. When the temperature drops, the curve starts to show more obvious modulation and it reaches more than two orders of magnitude on/off ratio at *4K*. These two obviously different behaviors between narrow and wide ribbons suggest that quantization is indeed successfully introduced to graphene by the intentional boundary confinement, and semiconducting gaps have been opened in these narrow graphene ribbons which result in the observed large on/off ratios at low temperatures. To quantitatively analyze how large the semiconducting gap is, we generated an Arrhenius plot of the minimum current, as shown in Fig. 5 (bottom) for the same *20nm* ribbon device measured in the top panel. In a semiconductor field-effect transistor, the off-state current is expressed in terms of thermal barriers: $I_{off} \propto exp(-E_g/2k_BT)$. At high temperatures, a clear linear relation between $logI_{off}$ and $1/T$ is observed. From the slope of the Arrhenius curve, we can extract an energy gap of *28meV* for this *20nm* graphene ribbon device. This value agrees well with the predictions from a density functional theory study[4]. The *4K* data point falls out of the linear dependence. It is understood that at such low temperatures, carrier transport is limited by tunneling rather than thermal injection.

In summary, we have demonstrated the fabrication of narrow graphene ribbons down to *20nm* width, and confirmed that a semiconducting gap can be introduced to graphene by introducing boundary confinement. Boundary roughness and edge state effects have been observed in the transport properties of narrow ribbons, which remind us to be more cautious when dealing with these narrow ribbons in the context of high performance device applications.

[1] K.S. Novoselov et. al., Nature (2005) 438, 197. [2] Y. Zhang et. al., Nature (2005) 438, 201. [3] K. Nakada et. al., Phys. Rev. B (1996) 54, 17954. [4] V. Barone et. al., Nano Lett. (2006) 6, 2748.

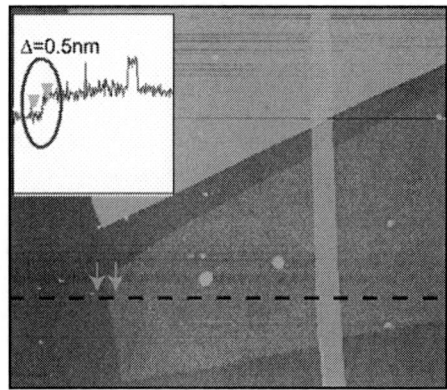

Fig.1 AFM image of a single layer graphene on SiO2 substrate. The cross section measurement shows the thickness to be 0.5nm.

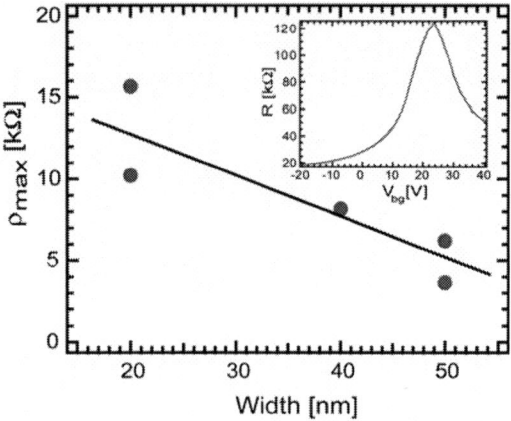

Fig.3 2-D resistivity vs. width for graphene ribbons smaller than 50nm. Inset shows an example of the resistance as a function of the back gate voltage for a 50nm ribbon.

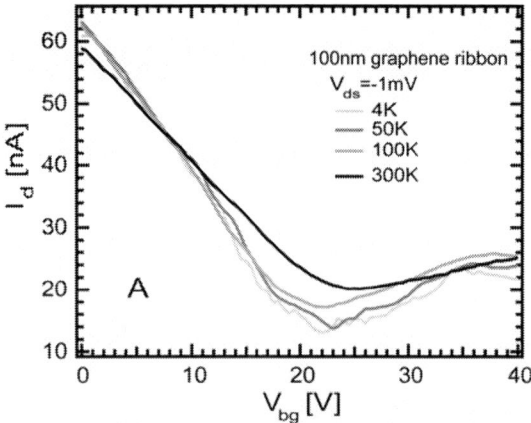

Fig. 4 Current vs. back gate as a function of temperature for a 100nm wide ribbon. The minimum current varies less than factor of 2 between 4K and 300K.

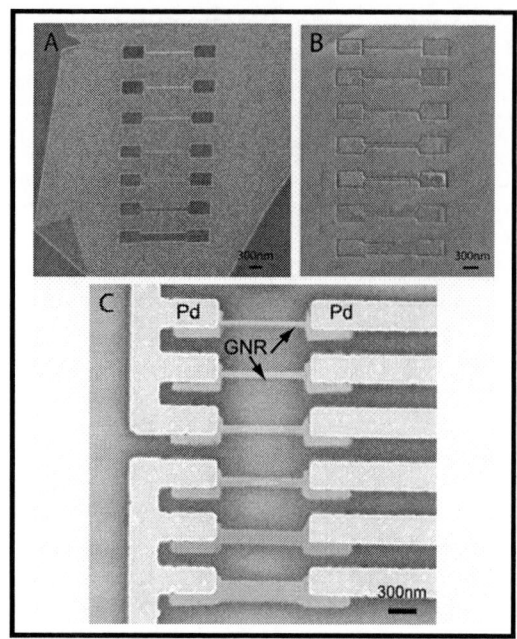

Fig.2 SEM images of the graphene ribbon device fabrication process. (A) Define etching masks on top of the graphene layer (B) Plasma etching to remove the undesired graphene area (C) Pd source/drain contacted ribbon devices with widths of 20nm, 30nm, 40nm, 50nm, 100nm and 200nm, from top to bottom.

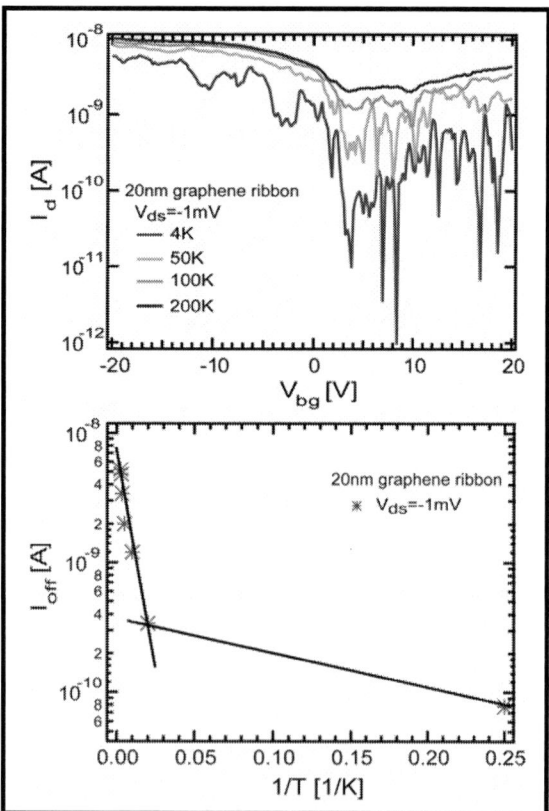

Fig. 5 (top) Current vs. back gate at different temperatures for a 20nm ribbon (bottom) Minimum current vs. 1/T for the same ribbon

Performance Limits of Nanocomposite Transistors & Nanobio Sensors: A Bottom-up Perspective

Muhammad Ashraful Alam, Ninad Pimparkar, Pradeep Nair, Satish Kumar, and Jayathi Murthy

School of Electrical and Computer Engineering, Purdue University, West Lafayette, IN 47907-1285, USA.

*Phone: (765) 494 5988   Fax: (765) 494 6441   Email: alam@purdue.edu

Background: In recent years, there have been many reports of remarkable performance of a new class of transistors and sensors based on percolating network of random/aligned Carbon nanotubes (CNT) and inorganic nanowires (NW) (Fig. 1) [1-6]. As type I transistors[7], they address low-drive current and placement issues associated with single CNT/NW transistors and may have applications in microelectronics. As type-II and type-III[8,9] transistors, they address grain-sized induced device-to-device fluctuation of poly-Si TFT, with potential applications in macroelectronic displays, solar cells, biosensors, *etc*. As type IV transistors, they address lithography-limitation and low-mobility (μ) of organic transistors and as such may have applications in flexible/wearable electronics. Despite these promise and almost monthly reports of extraordinary progress, the experimental results from various labs [2,5,6] often appear inconsistent and counter-intuitive, making it impossible to optimize the transistors/sensors for a disruptive technology. In this paper we trace the confusion in the field to inappropriate use of effective-media theories to characterize the transistors, and show that a 'bottom-up' view allows intuitive understanding of these devices as an elementary illustrations of stick-percolation theory (generalized to nonlinear regime) and diffusion-limited aggregation (generalized to transient regime). Our approach guides technology optimization, and in the process – as experimental testbeds – illuminates subtleties of stick-percolation theory and fractal geometry.

Nanocomposite TFTs and Stick Percolation Model. The Nanocomposite-TFT is characterized by the density (D), stick length (L_S), orientation of the tubes (θ), metal/semiconductor ratio, and the cladding matrix (insulating/conducting). Relative to percolation threshold (D_{perc}) and channel length (L_C), the transistors can be classified in four groups (Fig. 1). We simulate these transistors/sensors by constructing a first-principle numerical stick-percolation model[1]: The model populates a 2D grid by sticks of fixed length (L_S) and prescribed orientation (θ). I_{ON} through the network is computed by solving the percolating electron transport through individual sticks. Since L_C and L_S are much larger than the phonon mean free path, drift-diffusion description[1] of the transport problem is appropriate. A parameter of particular interest $C_{ij} = G_0/G_1$ is the dimensionless charge-transfer coefficient between tubes i and j at their intersection point where G_0 (~0.1 e^2/h) [10] and $G_1 (= qn\mu/\Delta x)$ [1] are mutual- and self-conductance of the tubes, respectively.

(a) Channel Length Scaling. Within this framework, it is easy to understand the puzzle of (geometry-specific!) μ reported in the literature. Classically one assumes $I_{ON} \sim k/L_C$, resulting in the popular definition of $\mu \sim (dI_{ON}/dV_G/V_D)L_C/(L_W C_{OX})$, yet both stick-percolation model and experimental data show (see Fig. 2a) that for NB-TFT, $I_{ON} \sim k\xi(L_C, L_S) = k/L_S(L_S/L_C)^{m(D)}$ – making the classical definition of μ meaningless[2,5,6]. The density-dependent nonlinear L_C-scaling reflects the increase of number of percolating paths with decreasing L_C (effectively widening the channel). Once this is accounted for, the puzzle of geometry-specific μ is resolved and the path to continued scaling is clearly defined, as explained above. For both type-II and type-III transistors, the metallic-CNT sub-network should be such that $D_{metal} < D_{perc}$ to avoid accidental shortening of the transistors and degradation of I_{OFF}. For this D (~ $3D_{perc}$), $m \sim 1.28$-1.5 such that superlinear scaling (i.e. $I_{ON} \sim 1/L_C^{1.28-1.5}$) is expected. For maximum I_{ON}, the obvious preference would be type-I transistors; however since local percolation density $D_{perc} \rightarrow 0$, there is no density at which I_{OFF} is reasonable. Only aggressive electrical/chemical filtering of the m-CNT makes such limiting performance achievable.

(b) Role of C_{ij}. An issue of particular concern for type II and III transistors is the role of C_{ij} because poor tube-tube contact can suppress I_{ON} dramatically (Fig. 2b). Standard fabrication techniques based on 'transfer printing' or 'solution processing' make it difficult for C_{ij} to reach its intrinsic value, providing another motivation for using type-I transistors.

(c) Role of Alignment: It has often been suggested that alignment of tubes along the transistor axis should increase performance dramatically. While this hypothesis appears intuitively true for type I transistors (×2 improvement over random network, Fig. 2c, dashed line), for type II transistors, alignment is actually counterproductive (Fig. 2c, solid line) because alignment increases D_{perc} and at a given D, reduces the probability of forming percolating paths. This counter-intuitive result is extensively verified by experiments (Fig. 2c, symbols). Finally, randomness of network in type IV transistors restricts the maximum size of the IC (Fig. 2d) and network alignment would improve IC size (i.e., higher D_{perc} reduces the chances of accidental shorts). In sum, the use of alignment should be application-specific and in some cases random structures would provide better performance than aligned ones.

Nanocomposite BioSensors and Reaction-Diffusion Model. Many groups have also explored the nanocomposite biosensors (Fig. 3a, actually a Type-II transistor in Fig. 1) for applications in Gemomics and Proteomics[11-14]. Just as D and L_S define the percolative transport in transistors, its 'response-time' for analyte capture (density ρ_0) is dictated by the fractal dimension ($1 < D_F(D) < 2$) of the composite (Fig. 3b). Specifically, one solves the reaction-diffusion equation (i.e. $d\rho/dt = \Delta^2\rho$ with $\rho(t,s(D_F) = 0)$ and $\rho(t=0,s<x<\infty)$) to obtain time-dependent response of the sensor[15]. Scaling arguments and numerical results confirm that $\rho_0^{1/DF} \sim$ constant such that (a) response of composite sensor is bracked by planar ($D_F = 2$) and cylindrical sensor ($D_F = 1$) (Fig. 3c) and (ii) the optimal D must balance of high sensitivity (requires low D) and noise-robust percolative response current (requires high D).

References: [1] S. Kumar, et al., Physical Review Letters 95 (2005). [2] E. S. Snow, et al., Applied Physics Letters 82, 2145 (2003). [3] L. Hu, et al., Nano Letters 4, 2513 (2004). [4] S. J. Kang, et al., Nature Nanotechnology (In Press). [5] C. Kocabas, et al., Nano Letters (In Press). [6] S. H. Hur, et al., Journal of the American Chemical Society 127, 13808 (2005). [7] N. Pimparkar, et al., IEEE Transactions of Electron Devices (In Press). [8] N. Pimparkar, et al., Electron Device Letters (In Review). [9] N. Pimparkar, et al., Electron Device Letters 28, 157 (2007). [10] M. S. Fuhrer, et al., Science 288, 494 (2000). [11] J. Hahm and C. M Leiber, *Nanoletters*, 4, 51 (2004). [12] W. U. Wang, C.Chen, K. Lin, Y Fang, and C. M. Lieber, *Nature Biotechnology*, 23, 1294 (2005). [13] Z. Li, Y. Chen, X. Li, T. I. Kamins, K. Nauka and R. S. Williams, *Nanoletters*, 4, 245 (2004). [14] Z. Li, B. Rajendran, T. I. Kamins, X. Li, Y. Chen and R. S. Williams, *Applied Physics A*, 80, 1257 (2005). [15] P. R. Nair and M. A. Alam, *Applied Physics Letters*, 88, 233021 (2006).

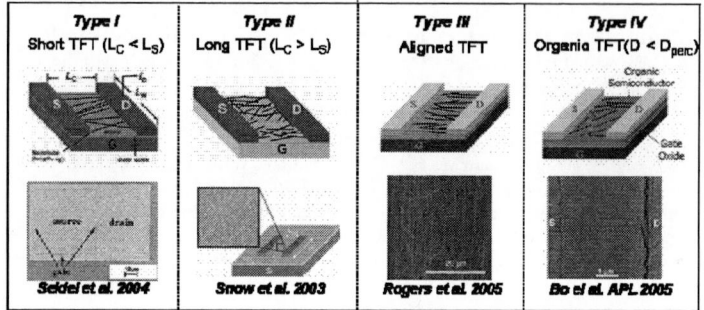

Fig. 1: Stick percolation model allows easy classification of composite transistors: (a) Type I short channel transistors with $L_C < L_S$ so that many tubes bridge S/D directly, (b) Type II long channel transistors ($L_C > L_S$) with random network of tubes such that electrons must percolate through the network to reach S/D, and (c) Type III long channel transistors with aligned network of tubes, and (d) Type IV long channel transistors in which sub-percolating sticks shorten the effective L_C, but the ON-OFF ratio is controlled by the organic host.

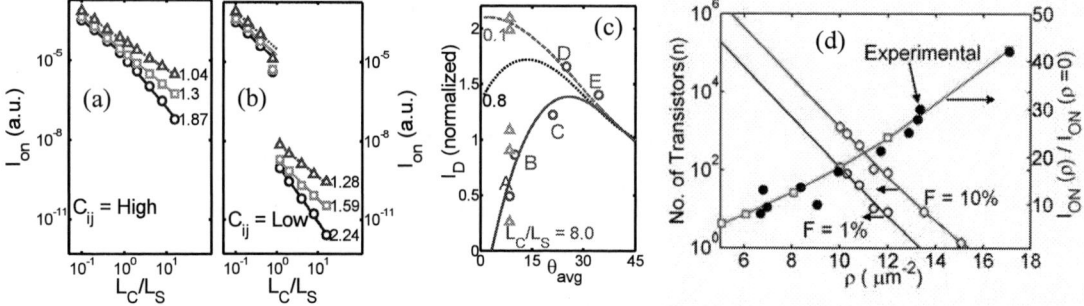

Fig. 2: (a) Scaling of I_{ON} with channel length provided the tube-tube contact is ideal. The power-law scaling of I_{ON} with L_C is evident. (b) If tube-tube is poor (small C_{ij}), there is an abrupt transition in I_{ON} form long-channel, G_0-limited conduction in type-II transistors vs. short-channel direct bridging transistors. (c) Increasing degree of alignment ($\theta = 45$ denotes random network, $\theta = 0$ is fully aligned network) is beneficial to short channel transistors, but is detrimental to long channel transistors. (d) The degree of randomness limit in Organic transistors that can be integrated within an IC with acceptable yield (F).

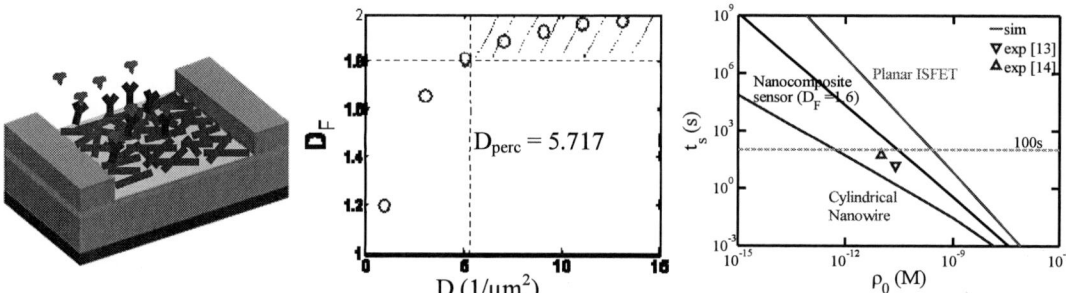

Fig. 3a. Type II transistor used as a nanobiosensor. The CNT-network is decorated with known receptors (blue). When analyte molecules (red) specifically bind to the receptor, the charges of the analyte molecules modulate the Source/Drain current.

Fig. 3b. Box counting method shows that the Fractal dimension (D_F) of the network is determined by the density of the sticks making up the nanocomposite film.

Fig. 3c. Trade-off between response time (t_s) and detectable analyte concentration (ρ_o). For a given response time, Cylidrical sensors (D$_F$=1, blue line) is far more sensitive than planar sensors (D$_F$=2, red line). Sensitivity of composite sensors (black line) fall somewhere in between.

Impact of Process Variation on Nanowire and Nanotube Device Performance

B. C. Paul[†‡], S. Fujita[‡], M. Okajima[‡], T. Lee[†], H.S.P. Wong[†], and Y. Nishi[†]

[†]Center for Integrated Systems, Stanford University
420 Via Palou, Stanford, CA 94305-4070

[‡]Toshiba America Research Inc.
2590 Orchard Parkway, San Jose, CA 95131

We present an in-depth analysis of the nanowire and nanotube device performance under process variability. While every process parameter variation drastically affects the conventional MOSFET performance, we found that nanowire/nanotube FETs are significantly (> 4X) less sensitive to many process parameter variations.

As conventional silicon technology approaches its limit, several emerging devices such as nanowire (NW) and carbon nanotube (CN) FETs are being extensively studied as possible alternatives [1, 2]. While inherent characteristics such as gate controllability, drive current etc, of these devices are shown to be superior to that of bulk MOSFETs, it is also widely believed that excessive variation in fabrication process, in particular the diameter and direction controlled growth, may significantly undermine those advantages. However, considering the recent progress [3], one can expect more sophisticated process and circuit solutions in near future to overcome this problem to a large extent. Further, due to cylindrical gate structure, NW/CN-FETs are largely unaffected due to variation in many lithography/geometry related parameters such as oxide thickness, gate width, etc., while variability in conventional Si MOSFET process is increasing drastically with scaling. It is hence, imperative to study the overall performance of these emerging devices under process variation. In this paper, we provide an in-depth analysis of the performance of CNFET and nanowire (Si and Ge) FETs under process parameter variations using novel compact device models and compare them with bulk and FinFET devices.

Fig. 1 shows the top view of a top gated structure of NW/CN-FET, which is used in our analysis. We assume ohmic source/drain contacts for both nanowire and nanotube FETs. Process parameters of such devices are divided into two categories, namely; lithography/geometry related parameters and NW/CN growth related parameters such as the diameter (D), nanotube chirality and wire/tube spacing. For low voltage operation (as in most digital applications), since chirality variation in nanotube does not significantly affect the device electrostatics [4], we therefore, neglect chirality variation in our analysis. The screening effect due to inter-tube/wire spacing variation is also neglected assuming sufficiently large spacing to avoid inter-wire coupling. Device parameters are chosen equivalent to 45nm and 32nm technologies with optimal diameters of nanotube (2nm) and NWs [2] and an 8nm thick HfO_2 as insulator. All analyses are done using ballistic transport model for nanotube and drift-diffusion model for nanowire [5]. Fringe capacitances are calculated using the analytical model in [6] with all parameters as independent random variables.

As mentioned earlier, most lithography/geometry related parameters variation does not significantly affect the drive current of NW/CN-FETs. For example, while variations in T_{ox} and W drastically affect I_{on} of bulk MOSFET, I_{on} of NWFET (Si or Ge) and CNFET is a weak function of T_{ox} variation due to cylindrical geometry and is independent of W (Fig. 2). *Note that W in NW/CN-FETs is not the effective channel width, and only contributes to the parasitic capacitances without affecting I_{on}* (see Fig. 1). On the other hand, though FinFET's response to T_{ox} variation is similar to NWFET, it however, has a strong dependency on W variation (Fig. 2b). Further, though I_{on} of NWFET is a function of L_{eff} variation, the impact is expected to be much less due to negligible short-channel effect (Fig. 3). Note that L_{eff} variation is attributed to both wire orientation variation and lithography variation. On the other hand, I_{on} of CNFET will be independent of L_{eff} variation considering ballistic transport. Variation in diameter (D) however, will considerably affect I_{on} of NW/CN-FET (Fig. 4). FinFET's negligible sensitivity (Fig. 4) to body thickness (T_{si}) variation is attributed to its large T_{si} (8.4nm). Consequently, its overall I_{on} sensitivity is comparable to NW/CN-FETs, despite its large sensitivity to W variation (Fig. 5). They are however, much less sensitive to variations than bulk.

Parameter variations also affect device capacitance. The effective capacitance in NW/CN-FETs, is dominated by fringe capacitances, C_{fr} [6], and will be less sensitive to most parameters variation such as L_{sd}, T_{ox}, for logarithmic dependency (Fig. 6). Fig. 7 shows the impact of variation on effective capacitance, C_g. C_g of NW/CN-FETs show less sensitivity due to the absence of overlap capacitance. We also, compare performance of a 2-input NAND gate under variation at 45nm and 32nm (predictive) technologies using HSPICE Monte-Carlo simulation (Fig. 8). As expected NW/CN-FETs show much lower sensitivity to variation than bulk (>4X) and FinFET (~2X) (Figs. 9, 10).

In conclusion, NW/CN-FETs are significantly less sensitive to process variations due to their geometrical structure. In other words, these devices will have larger margin to process parameters variation than bulk and FinFETs for an allowable performance variation limit.

----

[1] Javey, et al, Nano Lett, **4**, p. 1319, 2004
[2] J. Wang, et al, IEDM, p. 530, 2005.
[3] J. Deng, et al, ISSCC, p. 70, 2007.

[4] J. Mintmire, et al, PRL, **81**, p. 2506, 1998.
[5] B. Paul, et al, DAC, p. 717, 2006; Nanotech conf., 2007.
[6] B. Paul, et al, IEEE EDL, **27**, p. 380, 2006.

978-1-4244-1101-6/07/$25.00 ©2007 IEEE

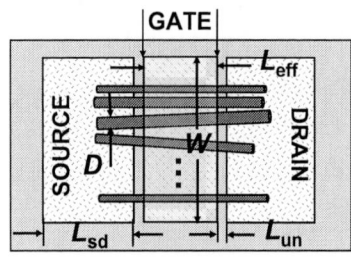
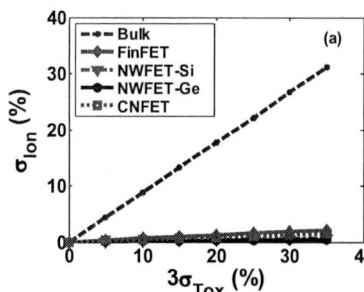
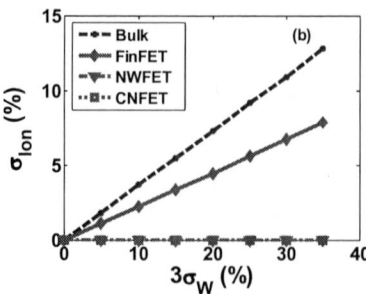

Fig. 1: Schematic top-view of nanotube/nanowire FET. Note that W here is the geometric gate width and ***NOT the effective channel width*** (*W_{eff}*). W_{eff} in these devices is obtained based on the diameter (D) of tube and no. of tubes (N) under the gate. W_{eff} variation is considered through D and N variations. Similarly, the variation in tube/wire orientation is considered through L_{eff} variation.

Fig. 2: Sensitivity of I_{on} of an NFET to T_{ox} (a) and W (b) variation. I_{on} of FinFET is less sensitive to W variation than bulk due to less narrow-width effect, while it is not sensitive in NWFET(<110>)/CNFET as explained above (Fig. 1). D_{CN}=2nm; $D_{NW,Si}$=3nm; $D_{NW,Ge}$=2nm [2].

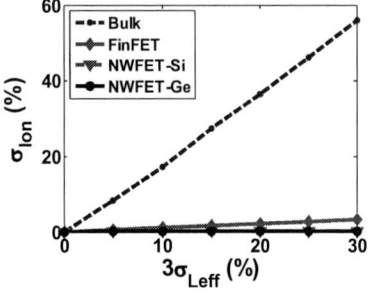

Fig. 3: I_{on} sensitivity to L_{eff} variation. Note that L_{eff} variation is attributed to variation in wire orientation and lithography.

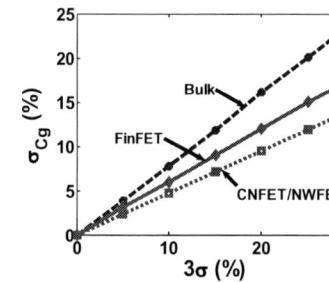

Fig. 4: I_{on} sensitivity to diameter variation of nanotube/nanowie and body thickness (T_{si}) variation of FinFET.

Fig. 5: Overall I_{on} sensitivity to all parameter variation. Random dopant effect is included in bulk and FinFET.

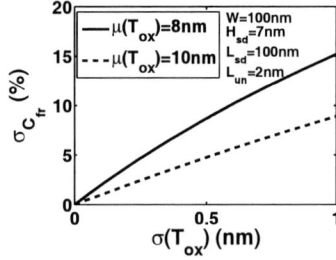

Fig. 6: Sensitivity of NW/CN-FET fringe capacitance to T_{ox} variation [6]. Due to logarithmic dependency $\sigma_{Cfr}/\sigma_{Tox} < 1$.

Fig. 7: Overall gate capacitance (C_g) sensitivity to all parameter variation. The approximate linear response demonstrates the dominance of W variation.

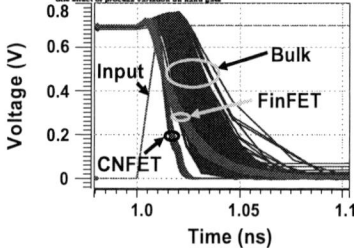

Fig. 8: Delay variation of a two input NAND with inverter load at 32nm technology node (V_{dd}=0.7V, 3σ=15%). NWFET result was similar to CNFET and is not shown due to clarity.

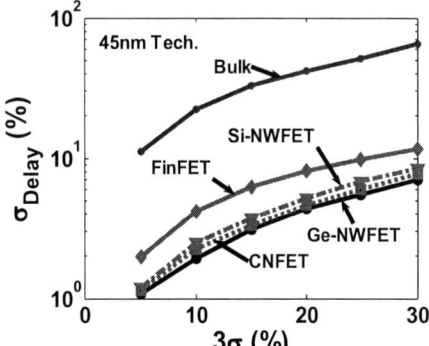
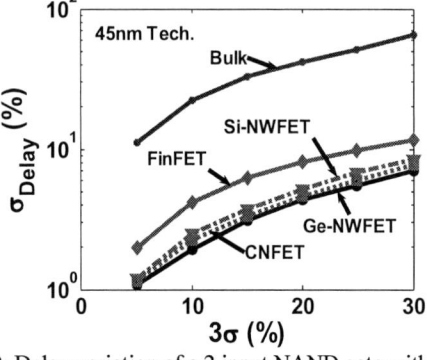

Fig. 9: Delay variation of a 2 input NAND gate with an inverter load. V_{dd}=0.9V.

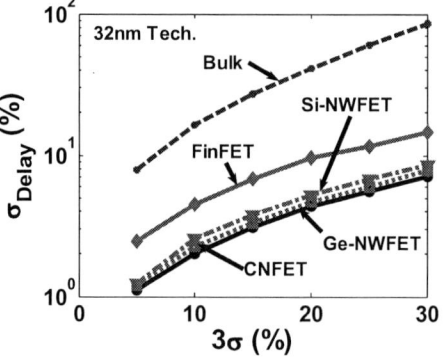

Fig. 10: Delay variation at 32nm technology with V_{dd}=0.7V.

Scaling Behaviors of Graphene Nanoribbon FETs
Youngki Yoon*, Yijian Ouyang, and Jing Guo
Electrical and Computer Engineering, University of Florida, Gainesville, FL 32611, USA
Phone: +1-352-392-0940 / e-mail: *ykyoon@ufl.edu

Introduction: With promising progress on fabricating and patterning a graphene layer recently [1-3], semiconducting graphene nanoribbon (GNR) is being intensively explored for potential transistor applications. The concept of all graphene circuits, in which GNRFETs are connected by metallic GNR interconnects, has been proposed [1, 4], and an exceptionally high mobility (~10,000 cm^2/V-s) of graphene and GNRs has been demonstrated experimentally [2] and theoretically [5]. In this study, we report the comprehensive 3-D quantum simulation of GNRFETs to explore scaling behaviors and ultimate scaling limits of GNRFETs. The dependence of the *I-V* characteristics, transconductance, subthreshold swing, and DIBL on the channel length is studied and compared for single gate, double gate, and wrapped around gate geometries, and the roles of gate insulator and contact size are examined.

Approach: Simulated device structures with different gate geometries are shown in Fig.1. The channel materials are GNRs, and source/drain contacts are metals. Schottky barrier height is a half band gap of the channel GNR. The DC characteristics of ballistic GNRFETs are simulated by solving the Schrödinger equation using the non-equilibrium Green's function (NEGF) formalism in the atomistic P$_Z$ orbital basis set self-consistently with Poisson equation [6]. Because the electric field varies in all dimensions for the simulated device structure, 3-D Poisson equation is numerically solved using the finite element method (FEM).

Device operation: I-V characteristics with gate and drain voltage variation are examined. Increasing the drain voltage ($V_D=V_{DD}$) leads to an exponential increase of the minimal leakage current, which indicates the importance of properly designed power supply voltage (Fig.2a). As V_G is increased, on-current is increased due to a larger energy range for carrier injection and thinner Schottky barrier for carrier injection (Fig.2b).

Switching characteristics: The switching-on and switching-off characteristics are described by transconductance g_m and subthreshold swing S, respectively, and the immunity of short channel effects is by DIBL. g_m and S remains approximately constant for long channel FETs, whereas g_m decreases and S increases for short channel FETs because of the worse gate control over the channel (Fig.3 and Fig.4a). The advantage of multiple gate geometry is obvious at a shorter channel length as shown in Fig. 4.

The scaling effect of channel length/width: For a channel length of 5nm, direct tunneling from source to drain leads to a large leakage current, and the gate voltage can hardly modulate the current. For channel length exceeding 15nm, however, GNRFETs operate at ballistic limit, and further increasing the channel length hardly changes the on-current and off-current (Fig.5). The scaling of channel width is achieved by changing the chiral index *N* of an armchair-edge GNR channel. As the channel width is increased, both off-current and on-current increase due to its smaller band gap and a smaller Schottky barrier height (Fig.6).

Gate oxide effect: Gate oxide dielectric constant does not affect much the subthreshold swing and the transconductance. However, the increase of the gate oxide thickness results in worse gate electrostatic control, and hence larger subthreshold swing. A thinner oxide, therefore, is desirable for a larger on-current (due to a larger transconductance) and a larger maximum on-off current ratio (due to a smaller subthreshold swing).

Contact size effect: As the size of contact is increased, both off-current and on-current decrease but on-off ratio keeps almost constant (Fig. 7) because of the increased fringe field between gate and contacts and the increased penetration length of source electrical field. In general, small size and low dimensional contacts not only improve the DC performance by increasing the on-current, but also improve the AC performance by decreasing the parasitic capacitance between the contacts, which leads to higher operation speed.

Summary: The results indicate that the transistor characteristics strongly depend on the GNR width because the band gap of the GNR is approximately inversely proportional to its width. Although multiple gate geometry improves immunity to short channel effects, it does not offer as much improvement as for Si FETs. The oxide thickness plays a more important role on the transistor characteristics than the gate insulator dielectric constant. Significant increase of the leakage current, subthreshold swing, DIBL, and decrease of output conductance are observed when the channel length is scaled down to 10nm, due to electrostatic short channel effects and direct source-drain tunneling.

References:
[1] R. V. Noorden, *Nature*, **442**, 228 (2006).
[2] Novoselov et al, *Science*, **306**, 666 (2004).
[3] Zhang et al, *Nature*, **438**, 201 (2005).
[4] Berger et al, *Science*, **312**, 1191 (2006).
[5] Obradovic et al, *Applied Physics Letters*, **88**, 142101 (2006).
[6] S. Datta, *Quantum transport: atom to transistor*. Cambridge, UK; Cambridge University Press (2005).

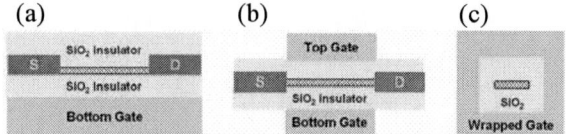

Fig. 1 Simulated device structures with different gate geometries. (a) single gate (SG), (b) double gate (DG), and (c) wrapped around gate (WG) graphene nanoribbon (GNR) FET. The channel materials are GNRs. Source and drain contacts are metals, and the Schottky barrier height is a half band gap of the channel material.

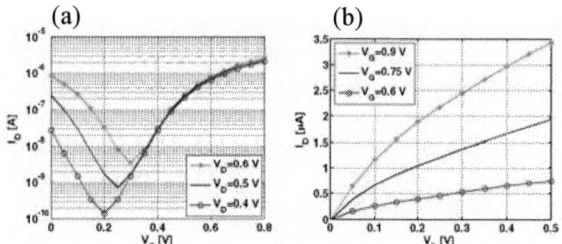

Fig. 2 (a) The I_D vs. V_G characteristics for nominal single gate device at different source-drain voltages. (b) The I_D vs. V_D characteristics for nominal single gate device at different gate voltages.

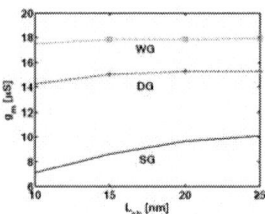

Fig. 3 The gate transconductance vs. channel length. Solid line is for SG, dashed line with crosses is for DG, and dash-dot line with squares is for WG GNRFET. A better gate control results in a larger transconductance.

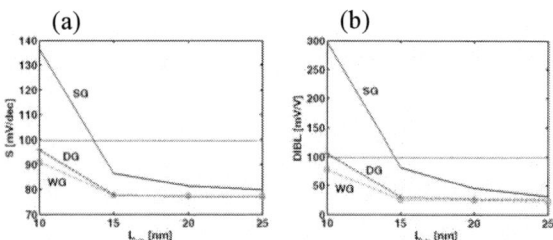

Fig. 4 Channel length dependence of (a) subthreshold swing and (b) DIBL. Solid line is for SG, dashed line with crosses is for DG, and dash-dot line with squares is for WG GNRFET. The dotted lines are guidelines of S=100 mV/dec or DIBL=100 mV/V, which are chosen as criteria. The multiple gate geometries can extend the scaling down of channel length below 10 nm.

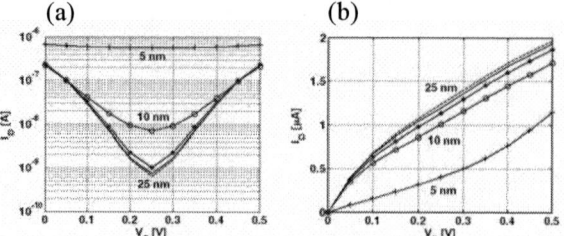

Fig. 5 (a) The I_D vs. V_G characteristics for single gate device with different channel lengths at V_D=0.5 V. (b) The I_D vs. V_D characteristics for single gate device with different channel lengths at V_G=0.75 V (The curve with crosses is for L_{ch} =5 nm; the curve with circles is for L_{ch} =10 nm; the curve with stars is for L_{ch} =15 nm; the curve without marks is for L_{ch} =20 nm; the dashed line is for L_{ch} =25 nm).

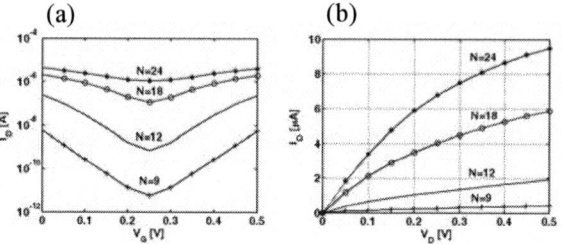

Fig. 6 (a) The I_D vs. V_G characteristics for single gate device with different channel widths at V_D=0.5 V. (b) The I_D vs. V_D characteristics for single gate device with different channel widths at V_G=0.75 V (The curves with crosses, circles, stars and without marks are for N=9, N=18, N=24 and N=12 armchair edge GNR, respectively. The channel length is 20 nm in all cases).

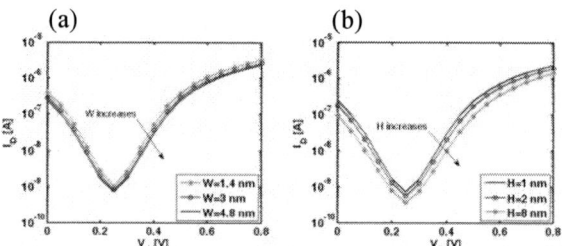

Fig. 7 (a) The I_D vs. V_G characteristics for single gate device with contact width equal to 1.4 nm, 3 nm, and 4.8 nm, respectively. (b) The I_D vs. V_G characteristics for single gate device with contact height equal to 1 nm, 2 nm and 8 nm, respectively.

Role of Electrical and Thermal Contact Resistance in the High-Bias Joule Breakdown of Single-Wall Carbon Nanotube Devices

Eric Pop

Dept. of Electrical and Computer Engineering, University of Illinois, Urbana-Champaign
2258 Micro and Nanotechnology Lab, 208 N Wright St, Urbana IL 61801. E-mail: epop@uiuc.edu

Several data sets of electrical breakdown *in air* of single-wall carbon nanotubes (SWNTs) on insulating substrates are collected and analyzed. These are data taken in different labs across the world on a wide range of SWNTs, spanning lengths 10 nm – 8 μm, diameters 0.8 – 3.2 nm and electrical contact resistance between 9 – 830 kΩ. A *universal* scaling of the Joule breakdown power with nanotube length is found, essentially independent of the insulating substrates used (here, SiO_2, Si_3N_4, Al_2O_3). The electrical *and* thermal resistance at the nanotube-electrode contacts regulate the breakdown behavior for short ($L < 0.6$ μm) SWNTs, whereas the breakdown power scales linearly with length for longer tubes.

Fig. 1 shows cross-sections of the typical two-terminal SWNT device considered here [1]. During *I-V* testing the voltage applied across the nanotube is raised until the power dissipated causes significant self-heating. The peak temperature occurs in the middle of the tube (Fig. 2b), and once this reaches the breakdown temperature the nanotube oxidizes (burns) irreversibly. This yields a sharp drop to zero in the *I-V* curve, and a physical "cut" in the nanotube itself (Figs. 2a and 2c). The breakdown temperature of SWNTs is approximately $T_{BD} \approx 600$ °C from thermogravimetric (TGA) analysis of bulk samples [2].

Published breakdown data from Refs. [1,3], [4] and [5] are collected and displayed in Figs. 3 and 4. These are labeled the "Stanford," "Caltech" and "Infineon" data sets, respectively. Only data for whom the complete *I-V* curve and the nanotube length are available are chosen. The electrical contact resistance (R_C) is estimated from the linear region of the *I-V* curve at low bias, and data sets for which this has significant bias dependence are eliminated. The aggregate data are shown both as breakdown voltage V_{BD} vs. length (Fig. 3), and breakdown power P_{BD} vs. length (Fig. 4). Figs. 3b and 4b present a "zoom-in" of the data for the shortest tubes. Note the effect of removing R_C from the breakdown data, i.e. subtracting IR_C and I^2R_C from V_{BD} and P_{BD} respectively. This accounts for the amount of voltage dropped and power dissipated at the contacts. The trends of V_{BD} and P_{BD} scaling appear more clearly once these are removed.

Solving the heat conduction along the nanotube, the breakdown power for lengths longer than about 0.6 μm can be approximated $P_{BD} \approx I(V_{BD}\text{-}IR_C) = g(T_{BD}\text{-}T_0)L$ [1]. This scales linearly with the length of SWNTs, as the dashed trend line in Figs. 3 and 4. The slope of this line gives a thermal conductance from nanotube to substrate $g \approx 0.16 \pm 0.03$ W/K/m across the aggregate data surveyed. This is significantly lower than the thermal conductance owed to any of the insulating substrates here (SiO_2, Si_3N_4 or Al_2O_3), and indicates that the heat flow is limited by the nanotube-substrate interface [1]. At the other extreme, the simple formula above does not work for very short nanotubes (Figs. 3b and 4b). In this range, the solution of the heat equation is better approximated by $P_{BD} \approx (T_{BD}\text{-}T_0)/(L/8kA)$ which predicts a $1/L$ dependence of the breakdown power (dash-dot line in Fig 4b). However, this implies an infinitely large breakdown power as the length approaches zero, which is *not* observed experimentally. The key is to realize there is a finite thermal resistance (R_T) associated with the two nanotube-electrode contacts. The breakdown power becomes $P_{BD} \approx (T_{BD}\text{-}T_0)/(L/8kA+R_T/2)$ which is shown with the solid line in Fig. 4b, where $R_T = 1.2 \times 10^7$ K/W (consistent with typical metal-dielectric interface thermal resistance for the small contact area here). This gives a finite $P_{BD} \approx 0.1$ mW for the shortest tubes. The competing effect of heat sinking through the contacts vs. the substrate also yields a minimum in P_{BD} for tubes with length around 0.6 μm $\approx 3L_H$, where $L_H = (kA/g)^{1/2} \approx 0.2$ μm is the thermal healing length along the SWNT [1].

In conclusion, this study analyzes *in-air* breakdown of single-wall nanotubes. The importance of the electrical and thermal nanotube-electrode resistance is shown, and simple scaling rules are given for breakdown power in the "short" and "long" length limits. The results are relevant for SWNT reliability, and the bottom-up approach to building SWNT circuits through controlled electrical breakdown [5].

[1] E. Pop *et al.*, *J. Appl. Phys.* (2007, in press), cond-mat/0609075

[2] I. W. Chiang *et al.*, *J. Phys. Chem. B* **105**, 8297 (2001)

[3] A. Javey *et al.*, *Phys. Rev. Lett.* **92**, 106804 (2004)

[4] H. Maune *et al.*, *Appl. Phys. Lett.* **89**, 013109 (2006)

[5] R. V. Seidel *et al.*, *J. Appl. Phys.* **96**, 6694 (2004)

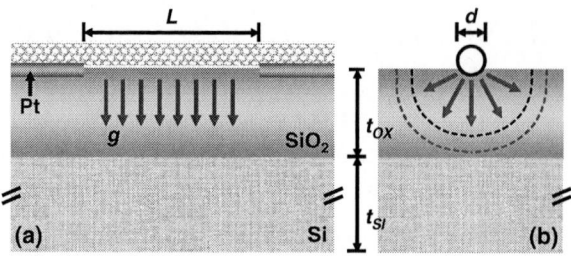

Fig. 1. (a) Longitudinal and (b) transverse cross-sections of the typical two-terminal single-wall nanotube (SWNT) device considered in this work. The arrows represent the direction of heat loss into the substrate (g term). Other symbols used in this work are T_0 = ambient temperature, k = nanotube thermal conductivity and A = transverse area.

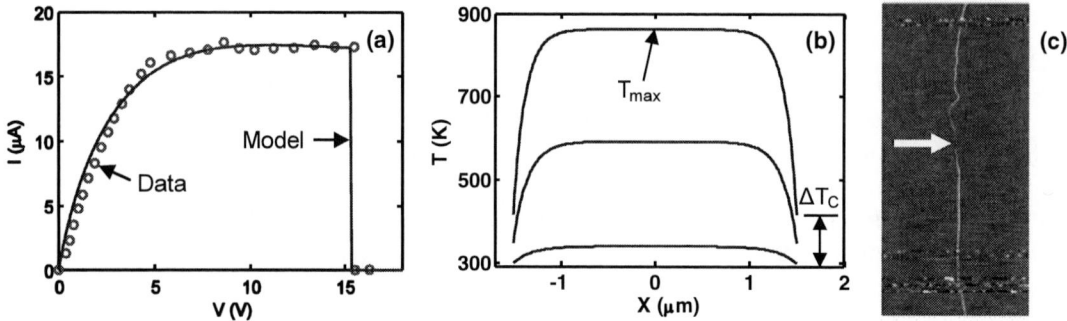

Fig. 2. (a) Typical I-V breakdown curve (here L=3 μm, d=2 nm). Data from [3] and model from [1]. (b) Calculated temperature profile at 3, 9 and 15 V bias (bottom to top). Note peak temperature at middle of tube and ΔT_C at the contacts. (c) AFM image of a similar tube after breakdown [1], showing the "cut" at the point of highest T.

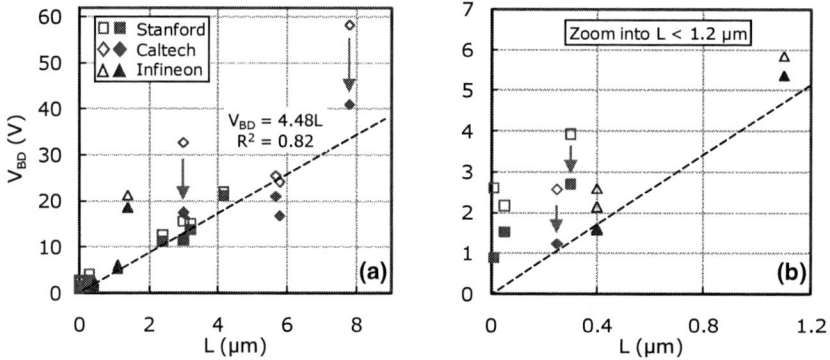

Fig. 3. (a) Breakdown voltage vs. SWNT length from the Stanford, Caltech and Infineon data sets. Empty symbols are before, and solid symbols are after removing the electrical contact resistance drop IR_C (arrows highlight some of the changes). (b) Same data, zoomed into the shorter nanotube range.

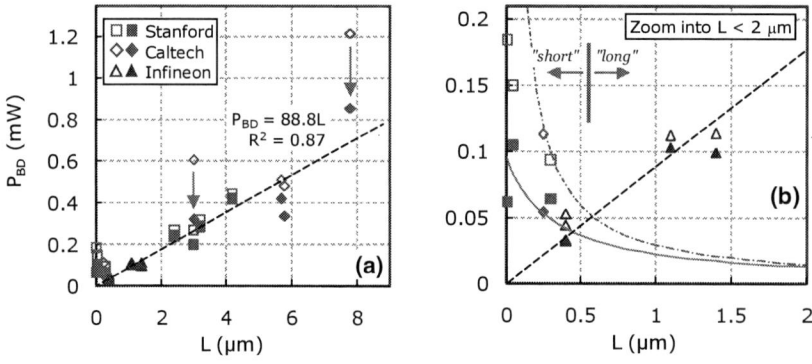

Fig. 4. (a) Breakdown power vs. SWNT length from the Stanford, Caltech and Infineon data sets. Empty symbols are before, and solid symbols are after removing the contact power dissipation I^2R_C (arrows highlight some of the changes). (b) Short nanotube range, and simple model without (dash-dot line) and with (solid line) the nanotube-electrode thermal resistance R_T. The latter correctly reproduces the finite breakdown power at near-zero length.

276

Author Index

A

Abe, K. 227
Abrokwah, J. 205
Adam, T. 51
Adesida, I. 39
Agarwal, A. 47
Ahn, Y. 215
Akarvardar, K. 103
Alam, M. 17, 267
Alam, M. T. 133
Allara, D. 141
Antcliffe, M. 33
Anthony, J. 23
Antoniadis, D. 219
Arthur, S. 121
Asada, M. 155
Aslam, D. 255
Avouris, P. 139, 265

B

Bach, H.-G. 187
Bai, X. 189
Bakhru, H. 203
Bakkers, E. 163
Balakrishnan, G. 73, 193
Balasubramanian, N. ..47
Balistreri, A. 35
Banerjee, K. 89
Banerjee, S. 49, 53, 93
Bank, S. 149
Bansal, A. 91
Barsky, M. 147
Basu, A. 39
Bayrakli, I. 197
Beling, A. 187
Berger, P. 153
Bernstein, G. 133
Bewley, W. 195
Biedenbender, M. 147
Bjoerk, M. 171
Boahen, K. 235
Boone, T. 247
Borgström, M. 163
Boutros, K. 143
Bowers, J. 185
Brar, B. 143
Buddharaju, K. 47
Burberry, M. 13
Butt, S. 51

C

Cai, L. 141
Campbell, J. 187, 189
Campion, A. 53
Canedy, C. 195
Carlin, J.-F. 109
Celler, G.. 15
Chabinyc, M. 21
Chang, D. 263
Chao, A. 79
Chao, P. 35
Chauhan, Y. 103
Chen, K. 77
Chen, P.-C. 169

Chen, S.-M. 83
Chen, T.-M. 81
Chen, Z. 265
Cheng, Z. 77
Cheng, Z. 117
Cherenack, K. 95
Cho, H. 173
Choe, J.-D. 215
Chow, D. 33
Chu, J. 175
Chu, K. 35
Chu, R. 127, 129
Chung, J. 107, 111
Chung, S.-Y. 153
Chuvilin, A. 109
Clarke, C. 29
Clemens, B. 59
Cohen, G. 175
Cohen, O. 185
Corrion, A. 37, 127, 129
Cowdery-Corvan, P. .13, 19
Crook, A. 149
Cywinski, L. 249

D

Dai, H. 261
Dalal, P. 249
Datta, S. 249
Davydov, A. 167
Dawson, L. 73, 193
De Souza, J. 209
Deal, M. 59
DeAngelis, J. 133
del Alamo, J. 35, 201
Deng, J. 43
Dentai, A. 183
Dery, H. 249
DeSalvo, G. 29
Dey, S. 49, 53
Domenicucci, A. 51
Dressler, S. 197
Droopad, R. 205
Dumka, D. 31, 35

E

Eggimann, C. 103
Eliashevich, I. 35

F

Facchetti, A. 169
Fang, A. 185
Fay, P. 151, 153, 211
Feltin, E. 109
Fichtenbaum, N. 127
Fincher, C. 69
Fogel, K. 209
Folks, L. 247
Freeman, D. 13, 19
Fromherz, P. 237
Fujita, S. 227, 269

G

Gao, X. 93
Gaquière, C. 109
Garrett, J. 121

Gaska, R. 43
Ghosh, S. 89
Gonschorek, M. 109
Gossard, A. 149
Grandjean, N. 109
Guo, J. 105, 271
Guo, S. 35
Guo-Qiang, P. 47
Gurney, B. 247

H

Haensch, W. 175
Hall, D. 115, 125, 211
Hanabe, M. 157
Harris, J. 191
Hashimoto, P. 33
Hashizume, T. 41
Hattori, J. 177
Hayden, O. 171
He, M. 167
Hearne, H. 29
Hekmatshoar, B. 95, 131
Helman, A. 163
Hill, R. 205
Hogg, R. 197
Hollister, M. 113
Holt, J. 51
Horio, K. 67
Hosako, I. 159
Hosono, Y. 157
Hou, T.-H. 221
Hu, M. 33
Hu, S. 133
Hu, X. 43
Huang, S. 73
Huffaker, D. 73, 193
Hurtt, S. 183

I

Indlekofer, M. 179
Ionescu, A. 103
Irving, L. 13
Ishibashi, K. 75
Isshiki, T. 123
Itagaki, K. 67

J

Jackson, T. 19, 23
Jallipalli, A. 193
Janes, D. 169
Jang, D. 215
Jeng, S.-J. 51
Jimenez, J. 35
Joh, J. 35
Johnson, J. 51
Jones, R. 185
Joshi, S. 49, 53
Joyner, C. 183
Ju, S. 169

K

Kaiser, U. 109
Kalavade, P. 173
Kambhampati, R. 203
Kan, E. 221
Kao, M.-Y. 35
Kapur, P. 79, 85, 173

Karam, N. 123
Katine, J. 247
Kattamis, A. 95
Kelkensberg, F. 163
Keller, S. 127
Kennedy, K. 197
Keys, D. 69
Khoshakhlagh, A. 73
Kiehl, R. 113
Kiewra, E. 209
Kim, C.-S. 193
Kim, D. 57
Kim, D. H. 201
Kim, D.-H. 39
Kim, H.-S. 99
Kim, M. 193
Kim, S. 101
Kim, S. 263
King, R. 123
Kirsch, P. 53
Kish, F. 183
Klauk, H. 69
Knez, M. 109
Knoch, J. 171, 179
Koester, S. 209
Kohn, E. 109
Kosel, T. 211
Kotani, J. 41
Kou, H.-Y. 83
Koudymov, A. 43
Kouwenhoven, L. 163
Koveshnikov, S. 203
Krishna, S. 89
Krishnamohan, T. 57
Krug, C. 53
Krut, D. 123
Kuan, H. 83
Kumar, S. 267
Kumar, V. 39
Kunkel, R. 187
kutty, M. 193
Kwong, D. 47

L
Lagally, M. 15
Lai, R. 147
Lam, C. 223
Larrabee, D. 195
Lau, K. 77
Laux, S. 175
Le, J. 113
Leburton, J. 87
Lee, C. 35
Lee, C. 93
Lee, C. 221
Lee, J. C. 99, 203
Lee, J. 215
Lee, S.-H. 49, 53
Lee, S.-H. 215
Lee, T. H...............227, 269
Lee, W.-C. 83
Lekshmanan, D. 91
Levy, D. 13, 19
Li, B. 217
Li, X. 205
Liang, D. 115
Likharev, K. 9

Lin, H. 207
Lin, Y.-M. 139
Lind, E. 149
Lindle, J. 195
Liu, H. 189
Liu, J. 51
Liu, J. 77
Liu, J. 217
Liu, P.-H. 147
Lochtefeld, A. 117
Long, K. 131
Löwgren, T. 165
Lu, C.-H. 59
Lu, G. 169
Lu, J. 255
Lu, X. 113
Lu, Y. 261

M
Ma, Z. 15
Madan, A. 51
Maitani, M. 141
Majhi, P. 49, 53, 85
Mallick, S. 89
Mallinger, M. 71
Mantey, K. 219
Marinero, E. 247
Marks, T. 169
Mårtensson, T. 135
Masselink, W. 197
Mathur, A. 183
Matsuda, T. 137
Mayer, T. 141
McCarthy, L. 37
McGuire, C. 33
Mcintosh, D. 189
McIntyre, D. 205
McNutt, T. 29
Medjdoub, F. 109
Mehrotra, V. 143
Mehta, M. 73, 193
Mei, X. 147
Mekonnen, G. 187
Meng, H. 69
Mermer, O. 251
Meyer, J. 195
Meziani, Y. 157
Micovic, M. 33
Minot, E. 163
Mishra, U. 37, 127, 129
Mohammad, S. 167
Mohammadi, S. 263
Monti, A. 97
Moon, J. 33
Moran, D. 205
Mori, N. 177
Mori, T. 75
Morifuji, E. 79
Motayed, A. 167
Mourey, D. 19
Moyer, H. 151
Murthy, J. 267
Muthiah, R. 183

N
Na, H.-J. 53

Nagarajan, R. 183
Nair, P. 267
Nakajima, A. 67
Nakazato, K. 177
Narasimha, S. 51
Nawaz, M. 87
Nayfeh, H. 51
Nayfeh, M. 219
Nayfeh, O. 219
Nechay, B. 29
Nelson, S. 13, 19
Nemeth, R. 179
Nguyen, T. 251
Niemier, M. 133
Nishi, Y. 59, 79, 227, 261, 269
Nolde, J. 195

O
Ohlson, J. 165
Ohno, M. 137
Ok, I. 99
Okajima, M. 269
Okayama, T. 121
Oktyabrsky, S. 203
Okumura, T. 81
Omura, K. 75
Otsuji, T. 157
Ouyang, Y. 105, 271

P
Pal, R. 51
Pala, N. 43
Palacios, T. 107, 111
Pang, H. 15
Parat, K. 85
Park, H. 185
Park, S. 99
Park, S. K. 19, 23
Park, S.-Y. 153
Passlack, M. 205
Patrashin, M. 159
Paul, B. 269
Pei, Y. 129
Pethe, A. 55
Pimparkar, N. 17, 267
Pleumeekers, J. 183
Poblenz, C. 37, 127, 129
Pop, E. 85, 123
Porod, W. 133
Putney, M. 133
Py, M. 109

R
Raday, O. 185
Rajagopalan, K. 205
Rajavel, R. 151
Rao, M. 121
Recht, F. 37, 129
Rehnstedt, C. 135
Riel, H. 171
Riess, W. 171
Rodriguez, J. 89
Rodwell, M. 149
Rogers, J. 259
Rooks, M. 175
Roy, K. 91

A-2

Rustagi, S. 47

S
Sadana, D. 209
Salahuddin, S. 249
Samuelson, L. 135165
Sands, T. 101
Sano, E. 157
Saraswat, K. 55, 57, 79, 85,173
Sato, S. 75
Saunier, P. 31, 35
Schmid, H. 171
Schmidt, D. 187
Schneider, R. 183
Schulman, J. 151
Scozzie, S. 29
Semtsiv, M. 197
Sepúlveda, N. 255
Serge Oktyabrsky, S. 125
Sham, L. 249
Shen, L. 37, 127, 129
Shen, T. 207
Sheng, Y. 251
Shi, T. 71
Shih, C.-H. 119
Shimizu, A. 81
Shur, M. 35, 43
Simin, G. 43, 97
Singh, N. 47
Singisetti, U. 149
Sivasubramani, P. 53
Slaughter, J. 245
Smith, L. 57
Smith, N. 247
Solomon, P. 175
Song, D. 77
Souzis, A. 35
Speck, J. 37, 127, 129
Stewart, E. 29
Street, R. 21
Sturm, J. 95, 131
Su, N. 151
Subramanian, S. 23
Sudharsanan, R. 123
Suemitsu, T. 157
Suh, C. 129
Suhara, M. 81
Sullivan, J. 255
Sun, F. 69
Sun, J. 19
Sun, Y. 209
Suzuki, M. 75
Sysak, M. 185

T
Tabakman, K. 51
Tajima, M. 41
Takagi, S. 5
Tang, H. 107
Tang, S. 93
Tang, W. 77
Tatebayashi, J. 73
Thayne, I. 205
Thelander, C. 135
Thompson, P. 153
Thoms, S. 205

Tipirneni, N. 97
Tokranov, V. 125, 203
Troadec, D. 109
Tsai, C.-C. 83
Tsai, W. ....
Tsamados, D. 103
Tserng, H. 35
Tutt, L. 13, 19

U
Uang, K.-M. 83
Uchida, K. 75
Ueda, D. 27
Uno, S. 177

V
van Dam, J. 163
van den Einden, W. ..163
van Kouwen, M. 163
van Weert, M. 163
Van Zeghbroeck, B. ..71
Veeraraghavan, G. ...251
Veliadis, V. 29
Verheijen, M. 163
Verma, S. 85
Voros, J. 241
Vurgaftman, I. 195

W
Wagner, S. 95, 131
Waite, A. 51
Wan, G. 103
Wang, B. 97
Wang, J. 125
Wang, S.-J.. 83
Wang, W. 53
Wang, Z. 227
Weitz, R. 69
Welch, D. 183
Wernersson, L.-E. 135, 165
Wienold, M. 197
Willadsen, P. 33
Wiltshire, M. 253
Wistey, M. 149
Wohlgenannt, M. 251
Wong, D. 33
Wong, G. 59
Wong, H.-S. 57, 103, 269
Wu, Y. 117, 207
Wunnicke, O. 163

X
Xiao, G. 113
Xuan, Y. 117, 207, 263

Y
Yablonovitch, E. 3
Yajima, H. 75
Yakimov, M. 203
Yang, Z. 43
Ye, P. 117, 207, 263
Yeh, S.-P. 119
Yoh, K. 137
Yoon, H. 141

Yoon, Y. 271
Yoon, Y.-B. 215
Yu, R. 153
Yuan, H.-C. 15
Yum, J. 99

Z
Zhang, J. 211
Zhang, M. 99
Zhang, Z. 151
Zhao, D. 19
Zhao, F. 71
Zhao, H. 99
Zhao, X. 107, 111
Zhiqiang, G. 47
Zhou, C. 169
Zhou, H. 205
Zhu, F. 99, 203
Zhu, Y. 217
Ziegler, M. 197
Zimmerman, J. 149
Zschieschang, U. 69
Zurcher, P. 205
Zwiller, V. 163